Algebra

A First Course

▲ Third Edition ▲

John D. Baley
Martin Holstege
Cerritos College

The McGraw-Hill Companies, Inc.
College Custom Series

New York St. Louis San Francisco Auckland Bogotá
Caracas Lisbon London Madrid Mexico Milan Montreal
New Delhi Paris San Juan Singapore Sydney Tokyo Toronto

McGraw-Hill's **College Custom Series** consists of products that are produced from camera-ready copy.
Peer review, class testing, and accuracy are primarily the responsibility of the author(s).

McGraw-Hill
A Division of The McGraw-Hill Companies

ALGEBRA: *A FIRST COURSE*

This book was previously published by Wadsworth, Inc.

2 3 4 5 6 7 8 9 0 GDP GDP 9 0 9 8 7

ISBN 0-07-006090-8

Editor: Julie Kehrwald
Cover Design: Maggie Lytle
Printer/Binder: Greyden Press

ALGEBRA

A First Course

THIRD EDITION

About the authors: John Baley has an Ed.D. from the University of Southern California; Martin Holstege has an M.S. from State University of Iowa. Both teach at Cerritos College in Norwalk, California, and are also the authors of *Basic Mathematics: A Program for Semi-independent Study* (D.C. Heath), *Understanding Algebra* (Random House/McGraw-Hill), and *Trigonometry* (Random House/McGraw-Hill).

PREFACE

This text was written to make mastery of a standard elementary algebra course as easy as possible, without reducing the level of mathematics. Explanations of basic concepts are carefully sequenced and presented in simple prose. Bubbles are used to call attention to important points or common difficulties. And of special note, students are not asked to solve problems that have not been fully explained.

FEATURES OF THIS BOOK

This book is divided into fifteen units to allow instructors greater flexibility in arranging course outlines to conform to academic calendars and student needs. Supplements that review arithmetic and provide an overview of plane geometry are available through your sales representative.

Each unit includes the following:

Unit Objectives: These objectives give students an overview of the skills they are expected to master after completing each unit.

Sections: Each unit is divided into sections of material that approximate one hour of classroom lecture time.

Numerous Examples: Over 600 examples illustrated with graphics demonstrate the concepts presented in the text. Students are given one or more examples of each task they are expected to perform.

Ample Exercises: Over 4400 problems give the student the opportunity to apply the concepts and practice the skills taught in the text. Problem sets are a critical part of the learning process because they not only give the student needed practice, but they also show the student the results of subtle variations. By doing problems, a student gets an opportunity to internalize mathematics and see the effects of changes in a parameter.

Unit Reviews: Unit reviews give students a summary of key ideas and definitions to aid them in organizing their knowledge and in preparing for exams.

Review Tests: To further ensure students' mastery of the concepts of each unit, every unit has a review test that closely approximates the questions that are likely to be asked on an exam.

SPECIAL FEATURES OF THIS TEXT

Cumulative Reviews: There are cumulative reviews after Units 4, 8, and 12 and at the end of the book. These reviews are an excellent opportunity for students to get an overview of the course as they study for midterm and final exams.

Conversational Bubbles: These bubbles are spread throughout the book but generally appear in the context of a worked example. Teachers will quickly recognize that these bubbles anticipate and verbalize the questions that students are likely to ask as they are learning the material. Answering bubbles provide the information that experienced teachers are likely to give.

Highlighted Definitions, Properties, Theorems, and Rules: These features, which are essential to remember and understand, are highlighted in boxes throughout the text, both to draw students' attention to important concepts and to make it easy for students to reference key ideas.

Pointers for Better Understanding: Twelve special boxes are spread throughout the text to clarify ideas or give students helpful hints about how to write or visualize certain algebra problems. These pointers deal with many ideas that are needed to work algebra but are frequently never explicitly taught to students. The pointers show how to deal with fractions, how to solve equations, and how to write the solution in a way that can be understood and followed by others. The pointers also help to develop critical thinking and provide hints for easier factoring, faster solutions, and better understanding.

Help with Word Problems: This book makes a special effort to help students develop the skills and understanding needed to apply algebra to practical situations. Use of formulas and percents is introduced in Unit 4.

Unit 5 is devoted to word problems to help students (a) develop skills to express in precise mathematical terms ideas that are imprecisely expressed in English; (b) critically analyze a problem and identify both relevant and irrelevant pieces of information in the problem; (c) build a knowledge-base of formulas that can be applied in this book and throughout life; and (d) visualize or draw mental pictures based on information given in a problem. Skills learned in Unit 5 are then reinforced throughout the remainder of the text, particularly in Units 7, 8, 12, 13, and 14.

Class Testing: This book has been class tested through two previous editions at over 30 colleges in both lecture and semi-independent mathematics classes.

NEW FEATURES IN THE THIRD EDITION

The comments of students and instructors who have used the first and second editions have been incorporated into this edition. As a result, many improvements have been made in order to make this edition even more useful to students.

- Factoring of trinomials is now taught using the AC or key number method in Sections 10.2 and 10.3. Many teachers find this systematic approach speeds student learning. Those who prefer the traditional trial-and-error method may use Alternate Sections 10.2 and 10.3 in the Appendix. Both methods develop the skills needed to manipulate algebraic fractions and to solve quadratic equations.
- Twenty percent more problems have been added to give students extra practice.
- Cumulative reviews have been added after Units 4 and 12, so there are now four cumulative reviews throughout the text.
- Spiral reviews have been added at the end of each unit review starting with Unit 3.

- "Review Your Skills" sections have been added at the end of many problem sets. These sections reinforce ideas from earlier units that will be used in the next section.
- Examples have been added and minor refinements have been made to clarify explanations and ease understanding.
- A supplement that reviews arithmetic is now available for students who need to strengthen their basic skills before beginning the traditional topics of algebra.
- A supplement that provides an overview of plane geometry is also available to help students who are preparing for statewide tests similar to the California Entry Level Mathematics Test.
- A review of percent remains in Unit 4, where the powerful ideas of equation solving can be used to clarify the solution of percent problems.
- A section on parallel and perpendicular lines has been added as an option.

NOTES TO STUDENTS ON HOW TO USE THE BOOK

Problems that help you check your understanding usually follow the explanations in this book. Answers to these problems appear in the left-hand margins of the page. Cover them with a piece of paper while you write your own answers in the blanks. A written answer requires more commitment than a mental answer, and it fills the blank, making the next line easier to understand. The following procedures are important:

1. Do not look at an answer before you have arrived at one of your own. This forces you to think through the problem rather than just agree with the solution.
2. Check your answer immediately after you have written it. Do not complete an entire page before checking the answers. This ensures that you are on the right path in solving a problem.
3. Ask for help when you fail to understand a basic concept. Don't go through an entire unit if there is some part you don't understand. Continuing to work ahead without understanding every section causes confusion and makes it difficult for someone else to help you.

There are practice problems throughout the textbook. Answers to the odd-numbered problems are given in the back of the book. Be sure to check that you have worked all problems correctly. Do not just work for an answer; become completely involved in each problem. It will take less time to complete the course if you work all of the problems than if you work only some of them. By solving all of the problems, you will develop skills that will help you understand the text and work the harder problems that come later.

For easy reference, the left-hand margin contains not only answers to various questions, but also titles (highlighted in color) for rules, definitions, properties, and topics. The definitions, rules, and properties have also been boxed in color for easy reference.

At the end of each unit, there is a review and a review test. The test, which covers each section, is designed to help determine if you understand the entire unit. If you can complete the review test without referring to material in the unit, you should then have a good grasp of the material just completed. The cumulative reviews at the end of Units 4, 8, 12, and 15 are designed to assist you in unifying your knowledge and in studying for midterm and final examinations.

SOLUTIONS MANUAL

A separate Solutions Manual contains complete solutions and additional explanations for the even-numbered problems. It also gives the solutions for all problems in the review tests. We would like to thank Sharon Bird, Richmond College, for checking the accuracy of the Solutions Manual.

TEST ITEMS

A test items booklet containing several forms of tests for each unit, midterms, and finals is available to adopters of the text.

TESTING SYSTEM

EXPTEST©, a computerized test-generating program and test bank, is available to adopters of the text.

ACKNOWLEDGMENTS

We would like to thank the following reviewers for their many helpful suggestions: John Alberghini, Manchester Community College; Sharon Bird, Richland College; Barbara Branum, Richland College; Susan H. Brown, Unity College; Al Calkin, Richland College; Carmy Carranza, Indiana University of Pennsylvania; John Chatfield, Southwest Texas State University; Doris Edwards, Motlow State Community College; Tyrone D. Gormley, Austin Community College; Vern Greenwood, Southeastern Illinois College; Garry Hart, California State University—Dominguez Hills; Jackson N. Henry, California State University, Dominquez Hills; L. Grant Hinchcliff, Dixie College; Norma Agras Innes, Miami-Dade Community College; Bruce Jacobs, Laney College; Rosemary Karr, Eastern Kentucky University; Cathy Mania, Alice Lloyd College; Jerry Matlock, Richland College; Anthony Monteith, College of Marin; Jack D. Murphy, Lycoming College; John C. Murphy, Cayuga Community College; Mohammed Rajah, Miracosta College; Gary Sarell, Cerritos College; Marsha Schoonover, Chattanooga State Technical Community College; Betsy Darken Smith, University of Tennessee, Chattanooga; Gerald M. Smith, Cayuga Community College; Richard Spangler, Tacoma Community College; Margaret B. Swavely, Glendale Community College; Froylan Tiscareno, Mt. San Antonio College; Shelda Warren, Moorhead State University; and Carolyn Volpe, Mesa College. We, of course, are responsible for any possible errors or omissions.

John D. Baley
Martin Holstege

CONTENTS

UNIT 1

INTEGERS

OBJECTIVES

After you have successfully completed this unit, you will be able to:

1. Add and subtract integers (1.2)
2. Multiply and divide integers (1.3)
3. Evaluate a numerical expression using exponents and order of operations for whole numbers (1.1) and for integers (1.4)
4. Evaluate a formula when given the values of the variables (1.4)

1.1 INTRODUCTION: DEFINITIONS AND PROPERTIES

Many people think that mathematics is something that man "discovered" out in the real world. Actually mathematics, including algebra, can be viewed as a very practical game invented by man involving numbers, operations, and relations. Numbers, or symbols that stand for numbers, are the equipment used to play the game like balls are used to play tennis. Operations are the plays we can make in the game. The basic operations or plays are addition, subtraction, multiplication, and division. Relations are the way we identify the players and keep score. The basic relations in the game of algebra are less than, greater than, and equal.

Set
Element

As sports equipment often comes in a set, so do numbers. A *set* is simply a collection of objects. The objects in the set are called the *elements* of the set. One way to represent a set is to list its elements and enclose the set in braces. An example of a set containing the elements 1, 2, and 3 is

$$\{1, 2, 3\}$$

The basic set of numbers we deal with is the set of natural numbers, sometimes called the counting numbers. They are the numbers that are learned first as a child develops the skill of counting.

Natural numbers

1.1A DEFINITION

The set of *natural numbers* is the set of counting numbers. They are

$$\{1, 2, 3, 4, 5, \ldots\}$$

The three dots (. . .) mean to continue in the same manner.

Variable

> **1.1B DEFINITION**
>
> A *variable* is a symbol that can represent any number in a set of numbers.

We usually use letters as variables—for instance a, b, c, m, n, x, y, z—but you can use any symbol like □, △, *0*, *a*. However, in algebra you must use only a single symbol to represent a variable.

a, b, □, and Q are legal variable names

A2, CAT, BOX are not legal variable names in algebra.

Constant

> **1.1C DEFINITION**
>
> A *constant* is a symbol that represents a single number.

Ways to Express Multiplication in Algebra

In arithmetic, the product of three and four is indicated as 3×4. Since x is a popular variable in algebra, we prefer to avoid using $\times$ as a multiplication sign. One simple alternative is to replace the multiplication sign with a raised dot ($\cdot$).

$3 \cdot 4$ indicates 3 times 4

Another way to indicate multiplication is to enclose each factor in parentheses:

$(3)(4)$ also indicates 3 times 4

These methods can also be used to indicate multiplication of a times b:

$a \cdot b$ stands for a times b

$(a)(b)$ stands for a times b

Sometimes multiplication of variables is shown without using any operation symbol:

ab is understood to mean a times b

$5x$ or $5 \cdot x$ or $(5)(x)$ all indicate 5 times x

Can I use 33 to express 3 times 3?

Only if you invent a new expression for thirty-three. Use $3 \cdot 3$ or $(3)(3)$ to mean 3 times 3.

Factors
Product

Numbers or variables that are multiplied are called *factors*. The result of a multiplication is called a *product*.

$$3 \cdot 4 = 12$$

Factors — Product

Terms
Sum

Numbers or variables that are added are called *terms*. The result of an addition is called a *sum*.

$$3 + 4 = 7$$

Terms Sum

Exponents

An easy way to show repeated multiplication of a factor like $2 \cdot 2 \cdot 2 \cdot 2$ is to use an exponent.

Base/Exponent

1.1D DEFINITION

The *base* is the factor that is to be multiplied repeatedly. The *exponent* is the number written to the right and a little above the base. Natural number exponents indicate how many times the base is to be used as a factor.

Exponent

$$3^2 = 3 \cdot 3 = 9$$

Base 2 Factors

The expression 3^2 is called the square of 3 or the second power of 3.

EXAMPLE 1 $4^2 = 4 \cdot 4$

$= 16$

Use 4 as a factor twice because the exponent is 2.

Whenever you see a blank ______, you should fill it in, then check your answer in the left margin.

EXAMPLE 2 $2^3 = 2 \cdot 2 \cdot 2$

8

$=$ ______

Use 2 as a factor three times because the exponent is 3.

EXAMPLE 3 $x^4 = x \cdot x \cdot x \cdot x$

Use x as a factor four times because the exponent is 4.

x^2 is the second power of x, which is sometimes called "x squared." Are there other special words for exponents?

Only for x^3. x to the third power can be called "x cubed."

EXAMPLE 4 Use exponential notation to show that x is the product of six times six.

Because 6 is multiplied by itself, we use an exponent of 2.

$$x = 6^2$$

PROBLEM SET 1.1A

Find the value of each of the following.

1. 5^2
2. 3^3
3. 4^3
4. 2^2
5. 3^4
6. 5^3
7. 3^2
8. 2^4
9. 10^2
10. 10^3
11. 10^4
12. 7^2
13. 2^5
14. 4^4
15. 5^1
16. 2^5

Use exponents to express each of the following.

17. x is the product of four times four.
18. x is the number that results when five is used as a factor seven times.

Order of Operations

What does $4 + 2 \cdot 3$ equal? If you first add $4 + 2$ and then multiply the result by 3, you get 18. If you first multiply $2 \cdot 3$, and then add the product, 6, to 4, the result is 10. To avoid this sort of confusion, mathematicians have agreed on the order in which operations are to be performed.

Order of operations

1.1E DEFINITION: ORDER OF OPERATIONS

When finding the value of an expression, perform the operations in the following order.

1. Simplify all expressions within symbols of grouping. If one set of grouping symbols is inside another, perform the operations in the innermost grouping symbols first. Sometimes, this step is called removing parentheses.

$$\underbrace{(4 + 2)}_{6} \cdot 3 = 6 \cdot 3 = 18 \quad \text{but} \quad 4 + \underbrace{(2 \cdot 3)}_{6} = 4 + 6 = 10$$

2. Evaluate any expressions with exponents.

$$2 + 3^2 = 2 + 9 = 11$$

3. Perform any multiplications or divisions, working from left to right, as they occur.

$$\underbrace{8 \div 2}_{4} \cdot 4 = 4 \cdot 4 = 16$$

4. Perform any additions or subtractions, working from left to right, as they occur.

$$\underbrace{8 - 2}_{6} + 4 = 6 + 4 = 10$$

Some grouping symbols are

Grouping symbols

( ) parentheses

[ ] brackets

{ } braces

Usually we use parentheses first, then brackets, then braces.

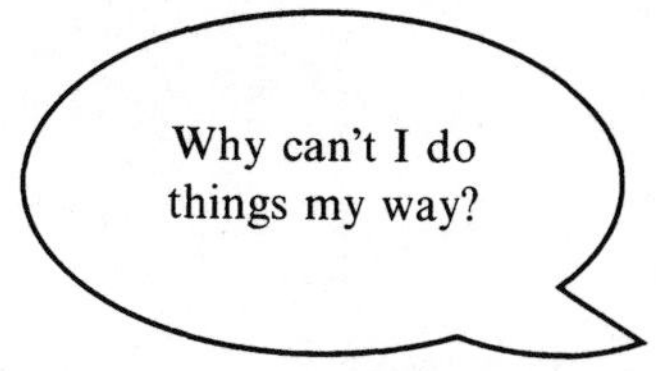

If you want to talk to others, you need to speak their language.

EXAMPLE 5 Evaluate $3^2 + 4(1 + 2)$.

$3^2 + 4\underbrace{(1 + 2)} =$	Simplify expressions within parentheses
$3^2 + 4 \cdot (3) =$	Evaluate expressions with exponents
$9 + \underbrace{4 \cdot (3)} =$	Multiply
$9 + 12 = 21$	Add

EXAMPLE 6 Evaluate $12 \div 3 \cdot 2^2$.

Remember: First fill in the blank, then check your answer in this column.

$12 \div 3 \cdot 2^2 =$	There are no parentheses to simplify
$12 \div 3 \cdot 2^2 =$	Evaluate expressions with exponents
$\underbrace{12 \div 3} \cdot 4 =$	Multiply and divide from left to right as they occur
$4 \cdot 4 =$ ___16___	

16

EXAMPLE 7 Evaluate $[9 - (2 \cdot 3)]^2$.

$[9 - \underbrace{(2 \cdot 3)}]^2 =$	Simplify inside grouping symbols
$\underbrace{[9 - 6]}^2 =$	Simplify outside grouping symbols
______ = ___9___	Evaluate expressions with exponents

3^2, 9

EXAMPLE 8 Write an expression to indicate that x is found by first finding the difference of ten and seven and then multiplying the result by the sum of four and five.

The difference of ten and seven is $10 - 7$.

The sum of four and five is ___4+5___.

$4 + 5$

Use parentheses to indicate that these operations are to be performed before the product is found.

$$x = (10 - 7)(4 + 5)$$

PROBLEM SET 1.1B

Evaluate the following expressions.

1. $2 \cdot 3 + 4$
2. $3 \cdot 5 + 2$
3. $3 + 2 \cdot 5$
4. $2 \cdot 5 + 3$
5. $2 + 3 \cdot 4$
6. $2 + 3 \cdot 5$
7. $2 \cdot (3 + 4)$
8. $5 \cdot (2 + 3)$
9. $(2 + 3) \cdot 4$
10. $(2 + 3) \cdot 5$
11. $(3 + 4)(6 + 4)$
12. $(5 + 2)(1 + 4)$
13. $(5 + 3)(12 - 4)$
14. $(15 - 4)(18 - 15)$
15. $2 + 3^2 \cdot 4$
16. $5 + 2^2 \cdot 3$
17. $(2 + 3)^2 \cdot 4$
18. $(5 + 2)^2 \cdot 3$
19. $10 - 5 + 2$
20. $5^2 + 2^2$
21. $15 - (3 + 5)$
22. $15 - 3 + 5$
23. $10 - (5 + 2)$
24. $(5 + 2)^2$
25. $8 \div 2 \cdot 2$
26. $9 \div 3 \cdot 3$
27. $8 \div (2 \cdot 2)$
28. $9 \div (3 \cdot 3)$
29. $12 - 6 \div 2 \cdot 3$
30. $12 - (6 \div 2) \cdot 3$
31. $12 - 6 \div (2 \cdot 3)$
32. $(12 - 6) \div (2 \cdot 3)$
33. $4^2 \div 4 - 2 + 1$
34. $2 \cdot 9 - 6 \div 3$
35. $4^2 \div (4 - 2) + 1$
36. $(2 \cdot 9) - (6 \div 3)$
37. $4^2 \div 4 - (2 + 1)$
38. $2 \cdot (9 - 6) \div 3$
39. $2^3 \cdot (9 - 6) \div 3$
40. $2 \cdot [(9 - 6) \div 3]$
41. $[2^3 \cdot 9 - 6] \div 3$
42. $2 \cdot [9 - (6 \div 3)]$
43. $2[(9^3 - 6) \div 3]$
44. $[(2 \cdot 9) - 6] \div 3$

Write an equation to indicate the following.

45. x is two plus the product of three and four.
46. y is the value obtained if first you find the product of two and five and then increase the result by four.
47. z is the value obtained if first you find the sum of eight and eighteen and then multiply the result by five.
48. Use an exponent to indicate that the sum of fifteen and two is to be multiplied by itself. The result of these operations is to be named y.
49. Subtract four from sixteen; then use exponents to indicate that the difference is to be used as a factor five times. Call the final product z.
50. a is the result if you divide ten by two and then multiply the quotient by three.
51. x is the quotient of eighteen divided by the product of two and three.
52. z is the difference of twenty and the quotient of six divided by two.

Some Basic Properties of a Mathematical System

Sometimes the order in which you perform an operation is important. Other times it is not. When the order in which an operation is performed is unimportant, the operation is said to be *commutative*.

EXAMPLE 9 Answer the following before you state which operations are commutative.

Yes, No

Is $7 + 6 = 6 + 7$? YES Is $7 - 6 = 6 - 7$? NO

Yes, No

Is $4 \cdot 2 = 2 \cdot 4$? YES Is $4 \div 2 = 2 \div 4$? NO

Because it does not matter what order we add or multiply, we say addition and multiplication are commutative. But it does matter in what order we subtract or divide, therefore we say subtraction and division are not commutative.

Commutative property

> **1.1F COMMUTATIVE PROPERTY FOR ADDITION AND MULTIPLICATION**
>
> $a + b = b + a$ and $a \cdot b = b \cdot a$
>
> where a and b are variables that can stand for any number or algebraic expression.

Binary operations

Most mathematical operations are *binary operations*, that is, they deal with only two elements at a time and yield a single result. If you wish to perform a binary operation with three elements, you have a problem.

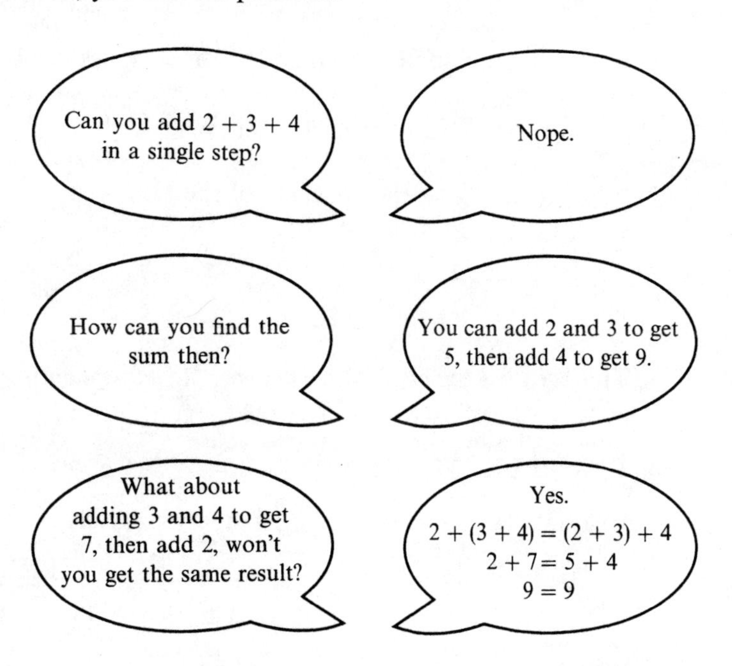

If it doesn't matter which way the elements are grouped, the operation is said to be *associative.*

EXAMPLE 10 Answer the following before you state which algebraic operations are associative.

Yes, No

Is $(8 + 4) + 2 = 8 + (4 + 2)$? YES Is $(8 - 4) - 2 = 8 - (4 - 2)$? NO

Yes, No

Is $(8 \cdot 4) \cdot 2 = 8 \cdot (4 \cdot 2)$? YES Is $(8 \div 4) \div 2 = 8 \div (4 \div 2)$? NO

Because how the elements are grouped does not matter for addition or multiplication, we say that addition and multiplication are associative. How the elements are grouped does matter for subtraction and division, therefore we say subtraction and division are not associative.

Associative property

1.1G ASSOCIATIVE PROPERTY FOR ADDITION AND MULTIPLICATION

$$(a + b) + c = a + (b + c) \qquad \text{and} \qquad (a \cdot b) \cdot c = a \cdot (b \cdot c)$$

where a, b, and c are variables that can stand for any number or algebraic expression.

EXAMPLE 11 For the following operations, write a C if the operation is commutative, an A if it is associative, or an N if it is neither.

Addition for natural numbers is C, A

N — Subtraction for natural numbers is N

C, A — Multiplication for natural numbers is C,A

N — Division for natural numbers is N

There is also a property that combines multiplication and addition. Suppose we had three boxes, each containing a circle and a triangle. We might describe our holdings like this:

$$3 \cdot \boxed{\bigcirc + \triangle} = \boxed{\bigcirc + \triangle} + \boxed{\bigcirc + \triangle} + \boxed{\bigcirc + \triangle}$$

If we take the contents out of the three boxes, we have

$$3 \cdot \boxed{\bigcirc + \triangle} = \bigcirc + \triangle + \bigcirc + \triangle + \bigcirc + \triangle$$

Now, rearranging the contents of the box,

$$3 \cdot \boxed{\bigcirc + \triangle} = \bigcirc + \bigcirc + \bigcirc + \triangle + \triangle + \triangle$$

$$= 3\bigcirc + 3\triangle$$

This is called the *distributive property of multiplication over addition.* It works for numbers as well as for circles and triangles.

EXAMPLE 12 Does $3(2 + 4) = 3 \cdot 2 + 3 \cdot 4$?

Evaluate each side of the equal sign separately.

Left side	Right side
$3(2 + 4) =$	$3 \cdot 2 + 3 \cdot 4 =$
$3 \cdot 6 = 18$	$6 + 12 = 18$

Both sides equal 18, therefore $3(2 + 4) = 3 \cdot 2 + 3 \cdot 4$.

Distributive property

1.1H THE DISTRIBUTIVE PROPERTY

$$a(b + c) = a \cdot b + a \cdot c$$

It also works when the multiplier is on the right.

$$(b + c)a = b \cdot a + c \cdot a$$

where a, b, and c are variables that can stand for any number or algebraic expression.

EXAMPLE 13 Use the distributive property to rewrite $4(10 + 2)$.

$$4(10 + 2) = 4 \cdot 10 + 4 \cdot 2$$

EXAMPLE 14 Use the distributive property to rewrite $6(x + y)$.

6, 6

$6(x + y) =$ 6 $\cdot x +$ 6 $\cdot y$

PROBLEM SET 1.1C

Use the distributive property to rewrite the following expressions.

1. $3(4 + 5)$
2. $5(10 + 2)$
3. $7(2 + 8)$
4. $9(6 + 3)$
5. $4(x + y)$
6. $10(a + b)$
7. $3(x + 2)$
8. $8(z + 3)$

Whole numbers

1.1I DEFINITION

The set of *whole numbers* is the set of natural numbers combined with the number zero. They are

$\{0, 1, 2, 3, 4, \ldots\}$

Isn't zero a natural number?

No. In fact, zero wasn't used until the Middle Ages.

Identity element for addition

1.1J DEFINITION

Zero is the *identity element for addition.* That is, if you add zero to any number, the result is identical to the original number.

$4 + 0 = 4$ $\quad 0 + 9 = 9$

$2783 + 0 = 2783$ $\quad 0 + 0 = 0$

EXAMPLE 15 Replace each variable below with a whole number that will make each sentence true.

Answer	Sentence	Variable
0	$3 + a = 3$	$a =$ 0
6	$3 + b = 9$	$b =$ 6
5	$0 + x = 5$	$x =$ 5
8	$y + 0 = 8$	$y =$ 8
0	$a + 0 = 0$	$a =$ 0
4	$a + 0 = 4$	$a =$ 4

We can find numbers to make sentences like $a + 0 = 4$ true. But, there is no whole number that will make $a + 4 = 0$ true. What we do in this case is invent a new kind of number that makes the sentence true. Since this is a new number, we need a new symbol to represent it. We will call it -4 or "negative four." -4 is the number with the property

$$-4 + 4 = 0$$

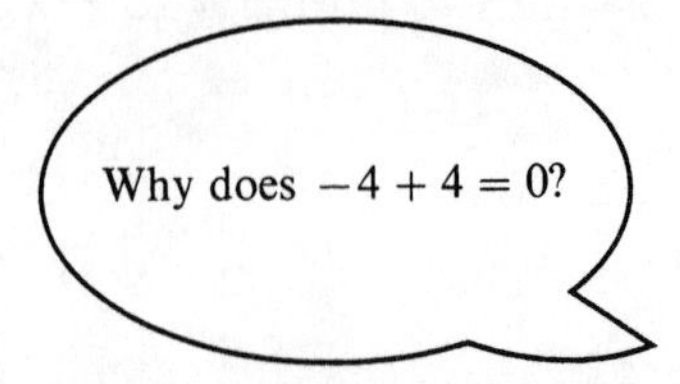

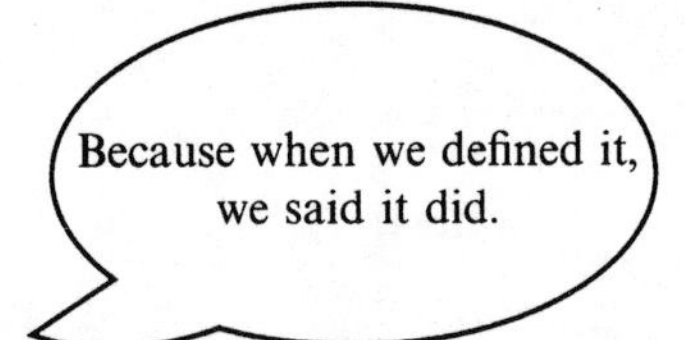

Negative number

1.1K DEFINITION

Every natural number n has a *negative*, $-n$, such that

$$n + (-n) = 0$$

For example,

$$3 + (-3) = 0 \qquad -2 + 2 = 0$$

$$-4125 + 4125 = 0 \qquad 78 + (-78) = 0$$

The negative of a number can be viewed as the opposite of the number. Taking two steps forward can be represented as 2. Taking two steps backward can be represented as -2.

EXAMPLE 16 Give the number you would use to represent each of the following.

Answer	Item
$+4$	Going up four floors in an elevator. $+4$
-3	Going down three floors in an elevator. -3
-5	A loss of 5 yards in football. -5
$+15$	A gain of 15 yards in football. $+15$
-20	A debt of twenty dollars. -20
-50	The depth of a submarine 50 feet below the surface. -50
0	The distance covered by a car that doesn't move. 0
0	The position of a football team that gained 6 yards then lost 6 yards. 0
0	The location of a car that backed up 50 feet then drove forward 50 feet. 0

The opposite of any number is called its *additive inverse.*

Additive inverse

1.1L DEFINITION

If two numbers have a sum of zero, they are called *additive inverses* of each other.

$$3 + (-3) = 0 \qquad -3 + 3 = 0$$

Additive inverses   Additive inverses

EXAMPLE 17 Write the additive inverse of each number below.

Answer	
-8	The additive inverse of 8 is $\underline{-8}$
8	The additive inverse of -8 is $\underline{+8}$
-5	The additive inverse of 5 is $\underline{-5}$
$-a$	The additive inverse of a is $\underline{-a}$
y	The additive inverse of $-y$ is $\underline{y}$
-6	The additive inverse of 6 is $\underline{-6}$
6	The additive inverse of -6 is $\underline{+6}$
0	The additive inverse of 0 is $\underline{0}$

Integers

> **1.1M DEFINITION**
>
> *Integers* are the set of numbers made up of the natural numbers, zero, and the negatives of the natural numbers. They are
>
> $\{\ldots, -3, -2, -1, 0, 1, 2, 3, \ldots\}$

Number line

Numbers are sometimes represented as points on a *number line.* To make a number line, draw a horizontal line. Mark off equal spaces on the line and number them in order.

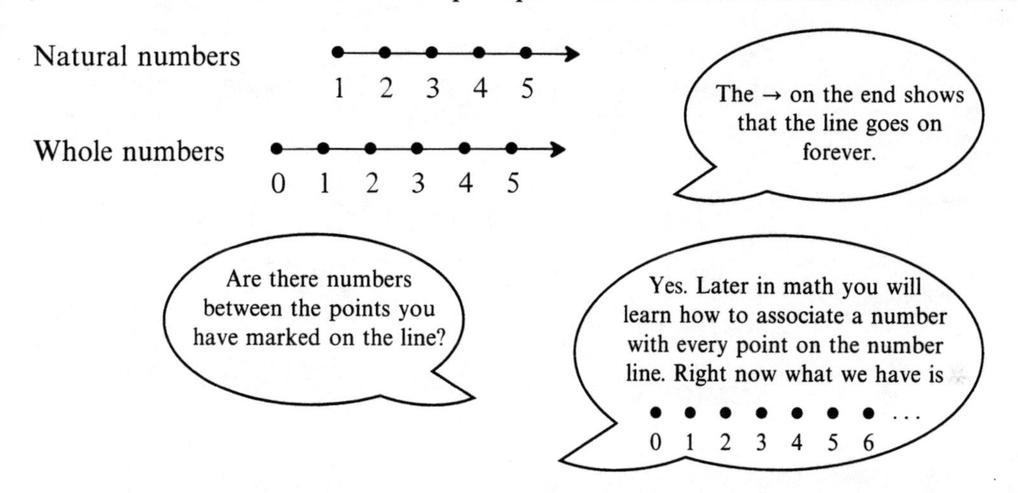

To make a number line for the integers, select one point on the line and call it zero. From the zero point, number the points to the right with natural numbers and to the left with the negatives of the natural numbers.

Integers

$-3 \quad -2 \quad -1 \quad 0 \quad 1 \quad 2 \quad 3$

Negative integers Positive integers

Positive integers
Negative integers

The integers to the right of zero are called *positive integers.* The integers to the left of zero are called *negative integers.* Zero is neither positive nor negative.

PROBLEM SET 1.1D

Fill in the blank to make the statement true.

1. $2 + (-2) =$ ____ 2. $2 +$ ____ $= 0$ 3. $-2 +$ ____ $= 0$

4. ____ $+ (-2) = 0$ 5. ____ $+ (-3) = 0$ 6. $3 +$ ____ $= 0$

7. $-9 + 9 =$ ____ 8. $x + (-x) =$ ____ 9. $x +$ ____ $= 0$

10. ____ $+ (-x) = 0$

Give the additive inverse of each of the following.

11. 3 12. -3 13. -9 14. 9 15. -7

16. $+6$ 17. -1 18. 1 19. $+1$ 20. -15

21. -10 22. 5 23. x 24. $-x$

25. The additive inverse of 13 is ____.
26. The negative of 13 is ____.
27. The number you would add to 13 to get zero for an answer is ____.

1.2 ADDITION AND SUBTRACTION OF INTEGERS

Addition of Integers

We add positive integers just as we do in arithmetic.

We don't need to write the positive signs but sometimes we do it for emphasis.

EXAMPLE 18

$$(+4) + (+6) = (+10)$$
$$(+3) + (+31) = (+34)$$

We add two negative numbers similarly.

EXAMPLE 19 $-4 + (-6) = -10$

This is like owing your brother 4 dollars and your friend 6 dollars. Therefore, you owe a total of 10 dollars.

Addition of numbers with like signs

1.2A RULE: ADDITION OF NUMBERS WITH LIKE SIGNS

To add two positive numbers, add as in arithmetic. To add two negative numbers, add as in arithmetic keeping the negative sign with the answer.

How come some negative numbers are enclosed in parentheses (-6) and others aren't?

That's so you don't confuse the operation symbols for addition $(+)$ or subtraction $(-)$ with the positive or negative signs $+$ or $-$.

I think I just confused $-$ with $-$.

When $+$ or $-$ is between two numbers, it is an operation such as $8 - 6 = 2$. When the sign is in front of a number, it tells us whether the number is positive or negative. Since the combination $+\ -$ doesn't mean anything, we use parentheses to show we have an operation on a signed number.

Oh! Then $-4 + (-6)$ means add negative six to negative four.

EXAMPLE 20 Add the following.

10, −10, −8

$6 + 4 =$ 10 $\qquad (-6) + (-4) =$ −10 $\qquad -3 + (-5) =$ −8

−4, 7, −7

$-2 + (-2) =$ −4 $\qquad 4 + 3 =$ 7 $\qquad (-4) + (-3) =$ −7

To add a positive and a negative number, we rely on the definition of additive inverses.

$$n + (-n) = 0$$

$$6 + (-6) = 0$$

Because these are additive inverses of each other, their sum is zero.

EXAMPLE 21 To add $8 + (-6)$ think of 8 as $2 + 6$.

$$\overbrace{8} + (-6) =$$

$$2 + \underbrace{6 + (-6)} =$$

$$2 + \quad 0 \quad = 2$$

Therefore, $\qquad 8 + (-6) = +2$

Notice the use of the associative property of addition here.

EXAMPLE 22 Add the following.

$$\overbrace{-8} \quad + (+6) =$$

$$-2 + \underbrace{(-6) + (+6)} =$$

$$-2 + \quad 0 \quad = -2$$

Think of -8 as $(-2) + (-6)$.

EXAMPLE 23 Add the following.

$$+2 + \quad \overbrace{(-5)} \quad =$$

$$\underbrace{+2 + (-2)} + (-3) =$$

$$0 \quad + (-3) = -3$$

EXAMPLE 24 Add the following.

$$+96 \quad + (-6) =$$
$$\overbrace{+90 + (+6)} + (-6) =$$
$$+90 + \underbrace{(+6) + (-6)} =$$
$$+90 + \quad 0 \quad = +90$$

Magnitude
Absolute value

Magnitude is the value of a number if you ignore its sign or how far the number is from zero on the number line. Sometimes the magnitude of a number is called its *absolute value*.

Adding signed numbers by using additive inverses is quite easy to understand, but it is a slow process. Here is a faster method to add numbers with unlike signs. The previous examples can be used to show why this rule is true.

Addition of numbers with unlike signs

> **1.2B RULE: ADDITION OF NUMBERS WITH UNLIKE SIGNS**
>
> To find the sum of two numbers with *unlike* signs, find the difference between their magnitudes and use the sign of the number with the larger magnitude.

EXAMPLE 25 Add $2 + (-5)$.

-5	The number with the larger magnitude is -5.
Negative	The sign of the sum will be Negative.
3	The difference between 5 and 2 is 3.
-3	Therefore, the sum of $2 + (-5)$ is -3.

EXAMPLE 26 Add $96 + (-6)$.

96	The number with the larger magnitude is 96.
Positive	The sign of the sum will be Positive.
90	The difference between 96 and 6 is 90.
90	Therefore, the sum of $96 + (-6)$ is 90.

Summary of rules for addition of signed numbers

> **1.2C SUMMARY OF RULES FOR ADDITION OF SIGNED NUMBERS**
>
> To add two positive numbers , add as in arithmetic.
>
> $6 + 8 = 14$
>
> To add two negative numbers, add as in arithmetic keeping the negative sign with the answer.
>
> $-6 + (-8) = -14$
>
> To add two numbers with unlike signs, find the difference between their magnitudes and use the sign of the number with the larger magnitude.
>
> $6 + (-8) = -2$
>
> $-6 + 8 = 2$

PROBLEM SET 1.2A

Add the following.

1. $6 + (-1)$	2. $3 + (-7)$	3. $3 + (-2)$	4. $1 + (-5)$
5. $-9 + 5$	6. $-4 + 8$	7. $-2 + (-4)$	8. $-3 + (-6)$
9. $-4 + (-7)$	10. $-5 + (-9)$	11. $2 + 5$	12. $-2 + (-5)$
13. $-2 + 5$	14. $2 + (-5)$	15. $-7 + (-9)$	16. $-7 + 9$
17. $7 + (-9)$	18. $7 + 9$	19. $-9 + 7$	20. $9 + (-7)$
21. $-6 + (-6)$	22. $8 + (-8)$	23. $-5 + 5$	24. $-4 + 0$
25. $-12 + 8$	26. $-5 + 19$	27. $-25 + (-35)$	28. $25 + (-43)$
29. $25 + (0.4)$	30. $32 + (0.05)$	31. $-5 + (-0.07)$	32. $14 + (-0.04)$
33. $0.24 + (1.24)$	34. $3.45 + (2.35)$	35. $-4.17 + (2.17)$	36. $-5.19 + (2.21)$
37. $-64.58 + (-4.68)$		38. $-75.99 + (-24.01)$	

Subtraction of Integers

The easy way to subtract is to avoid subtraction. To subtract a number, we add its negative.

Subtraction

1.2D DEFINITION: SUBTRACTION

To subtract a number, add its additive inverse (opposite).

$$a - b = a + (-b)$$

EXAMPLE 27 Subtract $6 - 4$.

$6 - 4 =$

$6 - (+4) =$

$6 + (-4) = 2$ Rewriting as an addition problem

The $-$ sign between the 6 and the 4 means subtract. Think of the 4 as $+4$.

EXAMPLE 28 Subtract $12 - 5$.

$12 - 5 =$

$12 - (+5) =$

$12 + (-5) = 7$

The operation changed from subtraction to addition and the sign on the 5 changed from + to −.

EXAMPLE 29 Subtract $15 - 7$.

$15 \quad - \quad 7 \quad =$

Rewriting as an addition _____ + (_____) = _____.

15, −7, 8

Now let's subtract a negative number.

EXAMPLE 30 Subtract $6-(-4)$.

$6-(-4)=$

$6+(+4)=+10$

The sign on the 4 changed from $-$ to $+$. $+4$ is the additive inverse of -4.

Yes, and the subtraction sign changed to addition.

EXAMPLE 31 Subtract $18-(-5)$.

$+5$

18, (+5), 23

The additive inverse of -5 is ______.

Rewrite as an additive problem and add: ______ + ______ = ______

This method also works when you subtract a number from a negative number.

EXAMPLE 32 Subtract $-4-(-6)$.

$-4-(-6)=$

$\downarrow\ \downarrow$

$-4+(+6)=+2$

The operation changed from subtraction to addition and the sign on the 6 changed from $-$ to $+$.

EXAMPLE 33 Subtract $-4-(+6)$.

$-4-(+6)=$

$\downarrow\ \downarrow$

$-4+(-6)=-10$

EXAMPLE 34 Rewrite each of the following as an addition problem.

$5+(-4)$, $-5+(-4)$

$5+4$, $-5+4$

$5-4=$ __________ $-5-4=$ __________

$5-(-4)=$ __________ $-5-(-4)=$ __________

PROBLEM SET 1.2B

Subtract the following by first rewriting as an addition problem.

1. $2-3$	2. $4-2$	3. $1-(-6)$	4. $7-(-2)$	5. $8-(-3)$
6. $3-(-8)$	7. $-5-3$	8. $-1-8$	9. $-4-3$	10. $-5-9$

11. $2-5$	12. $-2-(-5)$	13. $-2-5$	14. $2-(-5)$
15. $-8-(-4)$	16. $-8-4$	17. $8-(-4)$	18. $8-4$
19. $-5-5$	20. $-6-(-6)$	21. $0-(-7)$	22. $0-15$
23. $20-6$	24. $-17-8$	25. $-9-(-14)$	26. $7-(-14)$
27. $56-(-36)$	28. $12-48$	29. $36-48$	30. $48-(36)$

31. $5 - 0$ 32. $-37 - 22$ 33. $0 - 5$ 34. $0 - (-15)$

35. $5.06 - 2$ 36. $8.59 - 6$ 37. $7.35 - (-4)$ 38. $9.12 - (-3)$

39. $6.75 - 8.42$ 40. $10.15 - 25.48$ 41. $8.69 - (-3.84)$

42. $19.21 - (-3.86)$ 43. $3.94 - (-12.16)$ 44. $7.06 - (-9.80)$

1.3 MULTIPLICATION AND DIVISION OF INTEGERS

Multiplication of Integers

Zero is a special integer. An important property involves multiplying by zero.

Multiplication by zero

> **1.3A MULTIPLICATION BY ZERO**
>
> $a \cdot 0 = 0 \cdot a = 0$
>
> where a is a variable that can stand for any number or algebraic expression.

This says that anything multiplied by zero is equal to zero.

$0 \cdot 4 = 0$ and $-8 \cdot 0 = 0$

When you multiply by a positive number, the proper sign for the product is fairly clear because you can view multiplication as repeated addition.

$$3 \cdot (+2) = (+2) + (+2) + (+2) = +6$$

and

$$3 \cdot (-2) = (-2) + (-2) + (-2) = -6$$

You did $3 \cdot (-2)$. What is $(-2) \cdot 3$?

This time the negative number is first. Read on.

$(-2)(+3)$ is a case where the first factor is negative. Because multiplication is commutative, we can change the order of the factors.

$$(-2)(+3) = (+3)(-2)$$

Since we know $(+3)(-2) = -6$

Therefore, $(-2)(+3) = -6$

So far we have seen that multiplying two positive numbers gives a positive answer and multiplying two numbers with unlike signs (a positive times a negative or a negative times a positive) gives a negative answer.

This leaves the case of negative number multiplied by another negative number. To cover this case we need to maneuver a bit.

We start with a statement that we know is true.

$(-2) \cdot 0 = 0$	This is true because multiplication by zero always gives a zero result
$(-2) \cdot [3 + (-3)] = 0$	We've used the definition of additive inverses to rewrite zero
$(-2)(3) + (-2)(-3) = 0$	Use the distributive property to multiply by (-2)
$-6 + (-2)(-3) = 0$	Next replace $(-2)(3)$ with -6
Now $-6 + \quad ? \quad = 0$	? stands for whatever $(-2)(-3)$ is

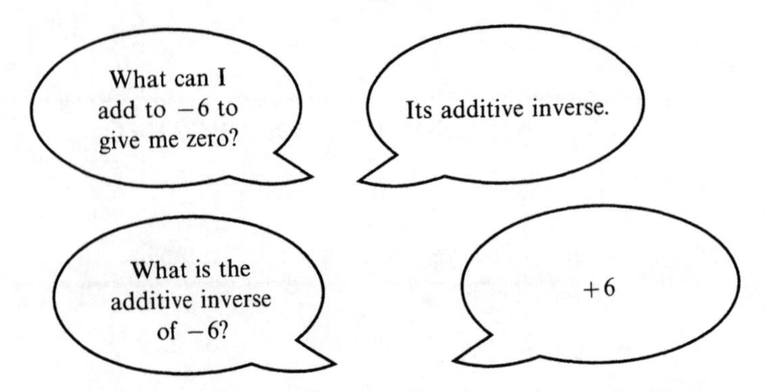

Therefore, $(+6)$ must be the right thing to replace the ?. The line above the bubbles reads: $-6 + ? = 0$. Replacing the ? with $(+6)$ we have $-6 + (+6) = 0$.

Therefore, $(-2)(-3) = +6$.

Multiplication of signed numbers

1.3B RULE: MULTIPLICATION OF SIGNED NUMBERS

When two numbers with like signs are multiplied, the product is positive. When two numbers with unlike signs are multiplied, the product is negative.

$(+) \cdot (+)$, $(-) \cdot (-)$ ⇨ $+$

$(+) \cdot (-)$, $(-) \cdot (+)$ ⇨ $-$

EXAMPLE 35 To make the rules for multiplying signed numbers seem reasonable, imagine a tank is being filled with water at the rate of $+3$ gallons per minute.

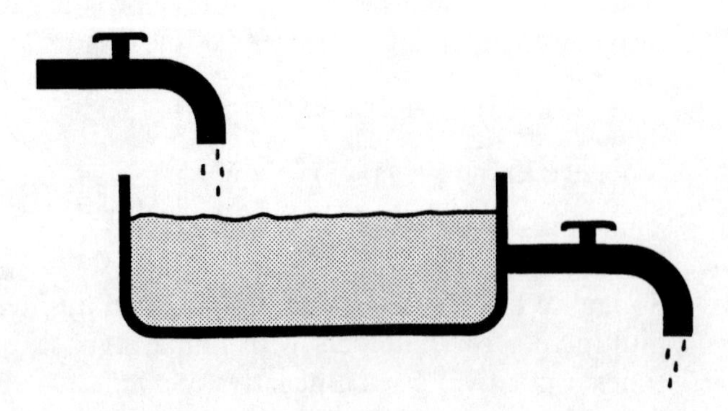

Two minutes from now $(+2)$, there will be 6 gallons more water in the tank.

$(+2) \cdot (+3) = +6$ More water

Two minutes earlier (-2), there were 6 gallons less (-6) in the tank.

$(-2) \cdot (+3) = -6$ Less water

Now, suppose we shut off the filling valve and turn on the drain valve so that water flows out at 3 gallons per minute. The flow is -3 gallons per minute.

Two minutes from now $(+2)$, the tank will hold 6 gallons less (-6).

$(+2) \cdot (-3) = -6$ Less water

Two minutes earlier (-2), the tank contained 6 gallons more $(+6)$.

$(-2) \cdot (-3) = +6$ More water

Another way to help visualize the rules for multiplication is to use a movie projector.

EXAMPLE 36

Suppose you had a motion picture showing a car traveling forward (+). If you ran the projector backward (−), the car would go ____________.

Backward (−)

Now, suppose the film was of a car going backward (−). If the projector were running forward (+), the car would be going ____________.

Backward (−)

If the projector were running backward (−), the car would be going ____________.

Forward (+)

Car	Projector	Picture
+	+	+
+	−	−
−	+	−
−	−	+

PROBLEM SET 1.3A

Multiply the following.

1. (2)(−4)
2. (−3)(7)
3. (−6)(0)
4. (8)(−2)
5. (−3)(−6)
6. (−3)(−8)
7. (7)(−3)
8. (−5)(−2)
9. (−9)(−4)
10. (−7)(−8)
11. (3)(0)
12. (−3)(−4)
13. (−3)(4)
14. (3)(−4)
15. (6)(−8)
16. (6)(8)
17. (−6)(−8)
18. (−6)(8)
19. (1)(−7)
20. (−1)(−5)
21. (9)(−1)
22. (−12)(0)
23. (15)(−10)
24. (−30)(−10)
25. (−5)(12)
26. (−10)(−27)
27. (−23)(−36)
28. (23)(−36)
29. (−1.2)(5)
30. (3.4)(−4)
31. (−3.6)(−8)
32. (−4.7)(−6)
33. (3.5)(8.6)
34. (6.2)(3.6)
35. (−2.2)(4.6)
36. (−5.1)(4.2)
37. (−5.9)(−3.2)
38. (−7.2)(−3.5)

Division of Integers

We say $\frac{+6}{+2} = +3$, because $(+2)(+3) = +6$.

There is an intimate connection between multiplication and division. Division can be thought of as the inverse—or opposite—operation of multiplication. Therefore, the same rules for signs apply to division as for multiplication.

$$\frac{-6}{-2} = +3 \quad \text{because} \quad (-2)(+3) = -6$$

$$\frac{-6}{+2} = -3 \quad \text{because} \quad (+2)(-3) = -6$$

$$\frac{+6}{-2} = -3 \quad \text{because} \quad (-2)(-3) = +6$$

Division of signed numbers

1.3C RULE: DIVISION OF SIGNED NUMBERS

When two numbers with like signs are divided, the quotient is positive. When two numbers with unlike signs are divided, the quotient is negative.

$(+) \div (+)$ and $(-) \div (-)$ ⇨ $+$

$(+) \div (-)$ and $(-) \div (+)$ ⇨ $-$

EXAMPLE 37

+2, −2

$\frac{+8}{+4} =$ ______ $\quad \frac{+8}{-4} =$ ______

+2, −2

$\frac{-8}{-4} =$ ______ $\quad \frac{-8}{+4} =$ ______

What happens when zero is involved?

$\frac{0}{6} = 0$ because $(6)(0) = 0$, but $\frac{6}{0}$ is impossible because no number times 0 equals 6.

Zero divided by any integer is zero, but **division by zero is not allowed!**

PROBLEM SET 1.3B

Divide the following.

1. $\frac{-6}{2}$
2. $\frac{36}{-9}$
3. $30 \div (-5)$
4. $-24 \div 6$
5. $\frac{-16}{-4}$
6. $\frac{-56}{-8}$
7. $-63 \div (-7)$
8. $72 \div (-6)$
9. $-24 \div (-3)$
10. $-54 \div (-9)$
11. $\frac{12}{3}$
12. $\frac{-12}{-3}$
13. $-12 \div 3$
14. $12 \div (-3)$
15. $\frac{-48}{-6}$
16. $\frac{-48}{6}$

17. $48 \div (-6)$

18. $48 \div 6$

19. $\frac{-16}{-2}$

20. $\frac{13}{-1}$

21. $\frac{-14}{1}$

22. $\frac{-26}{-1}$

23. $17 \div (-1)$

24. $0 \div (-1)$

25. $\frac{-72}{-6}$

26. $\frac{75}{-15}$

27. $-120 \div 24$

28. $-104 \div (-8)$

29. $\frac{-5.4}{6}$

30. $\frac{-8.4}{7}$

31. $\frac{6.8}{-4}$

32. $\frac{9.6}{-6}$

33. $\frac{-8.1}{-0.9}$

34. $\frac{-7.2}{-0.8}$

35. $-1.15 \div 0.5$

36. $-1.04 \div 0.4$

37. $-1.68 \div (-1.2)$

38. $-1.28 \div (-1.6)$

PROBLEM SET 1.3C

Mixed Practice: Perform the indicated operations.

1. $\frac{-8}{4}$

2. $4 - 8$

3. $\frac{8}{-4}$

4. $8 - 4$

5. $8 \cdot (-4)$

6. $8 + (-4)$

7. $(-8) \cdot 4$

8. $-8 + 4$

9. $-8 - (-4)$

10. $-4 - 8$

11. $10 - 5$

12. $\frac{-10}{5}$

13. $5 - 10$

14. $\frac{10}{-5}$

15. $10 - (-5)$

16. $\frac{-10}{-5}$

17. $-10 - (-5)$

18. $(-10) \cdot (-5)$

19. $-10 + (-5)$

20. $(10) \cdot (-5)$

21. $-10 + 5$

22. $(-10) \cdot (5)$

23. $5 + (-10)$

24. $-10 \div 5$

25. $-5 + (-10)$

26. $10 \div (-5)$

27. $18 - (-18)$

28. $-18 + (-10)$

29. $-12 + (-8)$

30. $14 - (-8)$

31. $\frac{-24}{-6}$

32. $\frac{36}{-4}$

33. $6 \cdot (-5)$

34. $7 \cdot (-6)$

35. $30 - (-2)$

36. $22 + (-4)$

37. $18 \div (-1)$

38. $0 \div (-2)$

39. $14 + (-16)$

40. $-16 - (-4)$

41. $-8 \cdot 0$

42. $(-7) \cdot (-4)$

43. $-16 - 4$

44. $-10 + (-6)$

Write an expression to indicate the following.

45. x is the sum of four and negative five.

46. x is the amount by which eight exceeds ten.

47. y is the sum of twelve and the additive inverse of five.

48. z is the difference when the negative of ten is subtracted from six.

49. x is the product of fifteen and the identity element for addition.

50. y is the quotient of negative seven and the additive inverse of four.

51. x is thirty divided by six.

52. x is six divided by thirty.

53. x is the quotient of thirty and six.

54. x is the same as six divided into thirty.

1.4 EVALUATING EXPRESSIONS

In Section 1.1, we gave the rules for the order of operations for natural numbers. The same rules hold true when dealing with signed numbers. However, special care must be given to the use of negative numbers, especially when raising negative numbers to a power. For example,

$$(-3)^2 = 9 \text{ because } (-3)^2 \text{ means } (-3)(-3) = 9$$

but

$$-3^2 = -9 \text{ because } -3^2 \text{ means } -(3)(3) = -9$$

EXAMPLE 38

16, +16

$(4)^2 =$ _____ $\quad (-4)^2 =$ _____

−16, −16

$-4^2 =$ _____ $\quad -(-4)^2 =$ _____

Review the rules for the order of operations.

Order of operations

1.4 WHEN EVALUATING AN EXPRESSION, PERFORM THE OPERATIONS IN THE FOLLOWING ORDER

1. Simplify all expressions within symbols of grouping. If one set of grouping symbols is inside another, perform the operations in the innermost grouping symbols first. Sometimes this step is called removing parentheses.

$$\underbrace{(4+2)}\cdot 3 = \qquad \text{but} \qquad 4 + \underbrace{(2\cdot 3)} =$$
$$6 \cdot 3 = 18 \qquad\qquad 4 + 6 = 10$$

2. Evaluate any expressions with exponents.

$$2 + 3^2 =$$
$$\downarrow$$
$$2 + 9 = 11$$

3. Perform any multiplications or divisions, working from left to right, as they occur.

$$\underbrace{8 \div 2}\cdot 4 =$$
$$4 \cdot 4 = 16$$

4. Perform any additions or subtractions, working from left to right, as they occur.

$$\underbrace{8 - 2} + 4 =$$
$$6 + 4 = 10$$

EXAMPLE 39 Evaluate $3 \cdot (-1) + 8$.

$\underbrace{3 \cdot (-1)} + 8 =$ Multiply first

$\_\_\_\_ + 8 = 5$ Add signed numbers

-3

EXAMPLE 40 Evaluate $32 \div (-8) \cdot 4$.

$\underbrace{32 \div (-8)} \cdot 4 =$ Work multiplication and division from left to right as they occur

$-4 \quad \cdot 4 = \_\_\_\_$

-16

EXAMPLE 41 Evaluate $(7 - 3) \cdot 4$.

$\underbrace{(7 - 3)} \cdot 4 =$ Parentheses first

$4 \quad \cdot 4 = \_\_\_\_$

16

EXAMPLE 42 Evaluate $(-18) \div 6 - (-3)$.

$\underbrace{(-18) \div 6} - (-3) =$ Division before subtraction

$-3 \quad - (-3) =$

$-3 + 3 = \_\_\_\_$

0

EXAMPLE 43 Evaluate $4 - 6 \cdot 2 + 3$.

$4 - \underbrace{6 \cdot 2} + 3 =$

$4 - 12 + 3 =$ Multiplication first

$\underbrace{4 + (-12)} + 3 =$ Rewrite subtraction as addition

$-8 \quad + 3 = -5$ Perform additions from left to right

EXAMPLE 44 Evaluate $3 \cdot 2^3$.

$3 \cdot 2^3 =$ Evaluate expressions with exponents before multiplying

$3 \cdot \_\_\_\_ = 24$

8

EXAMPLE 45 Evaluate $3 \cdot (-2)^3$.

Be careful with the negative sign.

$3 \cdot (-2)^3 =$

$3 \cdot \_\_\_\_ = -24$

-8

EXAMPLE 46 Evaluate $2^2 + 3^2$.

$2^2 + 3^2 =$ Do operations with exponents before adding

$\_\_\_\_ + \_\_\_\_ = 13$

4, 9

EXAMPLE 47 Evaluate $(2 + 3)^2$.

$(2 + 3)^2 =$ Evaluate parentheses first

$(5)^2 = 25$

PROBLEM SET 1.4A

(3) Hint: $-(-3) = +3$

Evaluate the following.

1. $-2 + (-3) + 4$
2. $-2 + (-3) - 4$
3. $-2 - (-3) - 4$
4. $(-8) \cdot (4) \cdot (-2)$
5. $(-8) \div (4) \cdot (-2)$
6. $(-8) \div 4 \div (-2)$
7. $(3) \cdot (-4) + 6$
8. $2 + (-5) \cdot (3)$
9. $-2 \cdot 3 + (-5)$
10. $4 \cdot (-3) - 4$
11. $-2 + 3 \cdot (-2)$
12. $(-12) \div (-2) + 4$
13. $-12 + 8 \div (-2)$
14. $15 - (-10) \div (-5)$
15. $2 - 4 \cdot 3 + 6$
16. $3(-2) + 4 \cdot 4$
17. $12 \div (-2) \cdot (-3) - 2$
18. $9 - (-1) + (-2) \cdot 4$
19. $5 - 7 - 3 \cdot 4$
20. $-6 - 6 \div 3 - (-3)$
21. $-2 \cdot (4 - 7)$
22. $3 \cdot (-3 - 4)$
23. $8 \div 4 \cdot 2$
24. $18 \div 3 \cdot 2$
25. $-5 \cdot [3 - (-2)]$
26. $[5 - (-3)] \div 4 - 2$
27. $-18 \div 3 - 2 \cdot (-4)$
28. $[7 + (-1)] \cdot (-5)$
29. $-1.2 + (-3.4) - 5$
30. $-5.6 - (-2.4) - 3$
31. $(1.2)(-3) + 6$
32. $5 + (-3.4)(-4)$
33. $6 - (-3.4) \div (-2)$
34. $(-6.8) \div (-4) + 4$
35. $4.6 - (-2.8) \div (-0.7) - (-4)$
36. $5.8 + (-4.8) \div (-0.4) - (-3)$

EXAMPLE 48 Evaluate $36 \div 3^2 \cdot 4 - (-6)$.

	$36 \div 3^2 \cdot 4 - (-6) =$	Evaluate expressions with exponents first
9	$\underbrace{36 \div \_\_\_\_}_{4} \cdot 4 - (-6) =$	Multiply and divide from left to right as they occur
	$4 \cdot 4 - (-6) =$	Multiplication before subtraction
16	$\_\_\_\_ - (-6) =$	
6	$16 + \_\_\_\_ = 22$	

EXAMPLE 49 Evaluate $(-28 + 8) \div 5 \cdot (-3)^2 - 4$.

	$(-28 + 8) \div 5 \cdot (-3)^2 - 4 =$	Parentheses first
	$-20 \div 5 \cdot (-3)^2 - 4 =$	Evaluate expressions with exponents
9	$-20 \div 5 \cdot \_\_\_\_ - 4 =$	Multiply and divide from left to right as they occur
	$-4 \cdot 9 - 4 =$	Multiplication before subtraction
	$-36 - 4 =$	
(-4), -40	$-36 + \_\_\_\_ = \_\_\_\_$	Adding signed numbers

PROBLEM SET 1.4B

Simplify the following expressions.

1. $2^2 + 9$
2. $(-2)^2 + 9$
3. $(-2)^3 + 9$
4. $10 - 3^2$
5. $10 - (-3)^2$
6. $10 - (-3)^3$

7. $(2-3)^2$

8. $(2-3)^3$

9. $(3-2)^3$

10. $(-3-2)^2$

11. $(-2-3)^3$

12. $6-(-4)^2+8$

13. $-8-3^2\cdot 2$

14. $3^2\cdot 4-2$

15. $(-3)^2\cdot 2+(-3)$

16. $(-4)^2-(-5)$

17. $(-3)^2-(2)^2$

18. $(-3+2)^2$

19. 2^3-3^3

20. $(-2-1)^2+5$

21. $(2-3)^3$

22. $(4-6)^3-(-6)\cdot(-2)$

23. $(-6+3)^3-(-3)^3$

24. $[-9(-1)+(-3)]+(-4)^2$

25. $-5+(-3)^2-(-2)^3$

26. $(-3)^2-3^2+2^3$

27. $(-4+1)^3-(6-3)^3$

28. $(-3)^2\cdot(-2)^3+(-10)\cdot(-1)$

Formulas

Formulas frequently involve the use of parentheses, exponents, and the four basic operations of arithmetic. One or more variables occur in formulas. To use a formula, you must be able to replace the variable(s) with a numerical value(s) and then evaluate it.

Perimeter

The *perimeter* of a figure is the distance around its border.

The perimeter of a rectangle can be expressed as

$$P = l + w + l + w$$

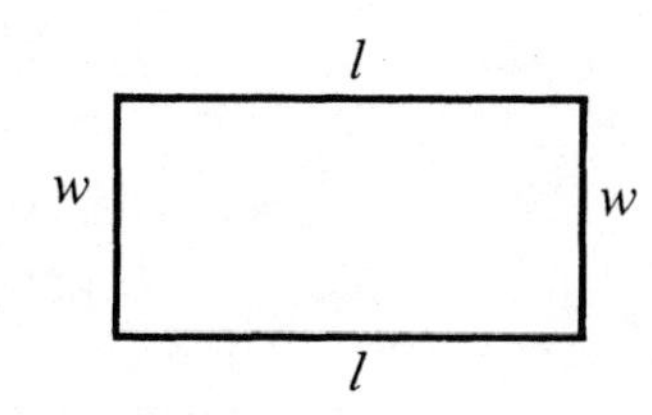

or

$$P = 2l + 2w$$

where l is the length of the rectangle, and w is the width of the rectangle.

EXAMPLE 50 Find the perimeter of a 3-foot by 4-foot rectangle.

$$P = 2\cdot l + 2\cdot w$$
$$P = 2\cdot(4\text{ ft}) + 2\cdot(3\text{ ft})$$
$$= 8\text{ ft} + 6\text{ ft}$$
$$= 14\text{ ft}$$

Perimeter

Perimeter is a measure of length. Think of perimeter as the length of a fence needed to go around the outside of a figure.

The perimeter of a triangle can be expressed as

$$P = a + b + c$$

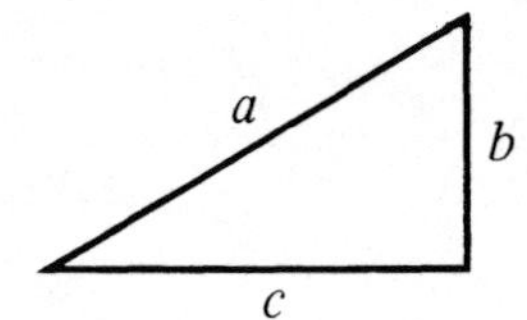

where a, b, and c are the lengths of the sides of a triangle.

EXAMPLE 51 Find the perimeter of a triangle with sides of 3 meters, 5 meters, and 6 meters.

$P = a + b + c$

$P = (3\text{ m}) + (5\text{ m}) + (6\text{ m})$

$= \_\_\_\_\_$

14 m

To make a substitution, replace the variable with a set of parentheses and then write the value of the letter inside the parentheses.

Area

Think of area as the number of square tiles that would be required to cover the figure. In the case of a 5-foot by 8-foot bathroom, you would have to buy forty 1-square-foot tiles to cover the floor.

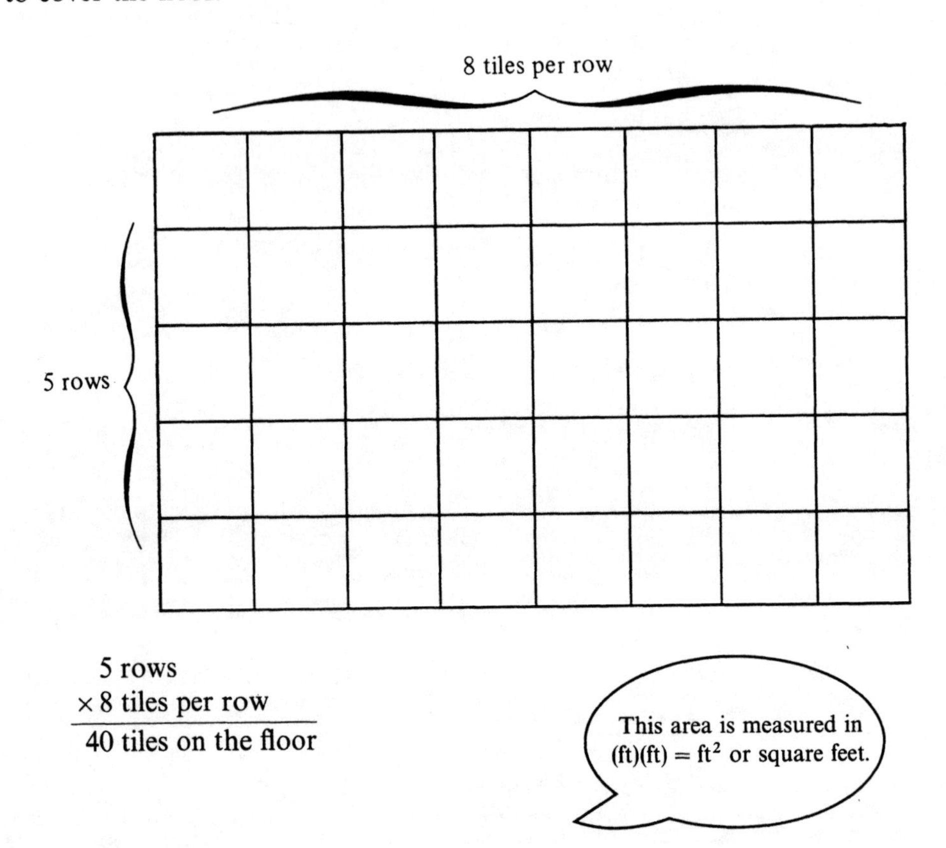

5 rows
× 8 tiles per row
40 tiles on the floor

This area is measured in (ft)(ft) = ft^2 or square feet.

Area

Area is the measure of the surface of a geometric figure. Area is measured in square units.

The area of a rectangle can be expressed as

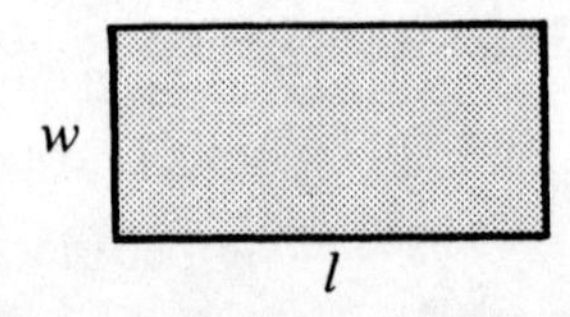

$A = l \cdot w$

where l is the length and w is the width.

EXAMPLE 52 Find the area of a kitchen floor that is 9 feet by 12 feet.

$A = l \cdot w$

$= (\_\_\_\_\_)(\_\_\_\_\_)$

12 ft, 9 ft

$= \_\_\_\_\_\ \text{ft}^2$

108

Trapezoid

The area of a trapezoid can be expressed as

$$A = \frac{1}{2} h \cdot (B + b)$$

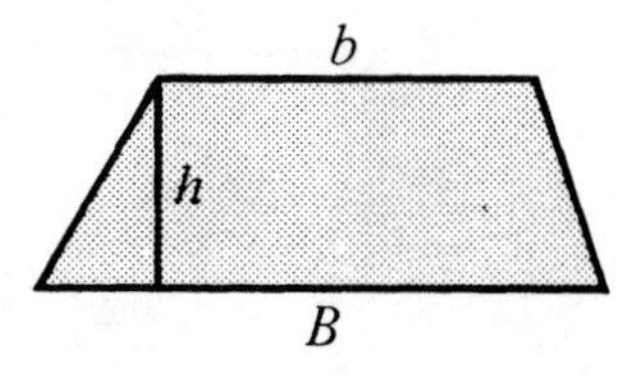

A trapezoid is a four-sided figure with at least two sides parallel. The two parallel sides are called bases.

where

h is the height

B is the length of the lower base

b is the length of the upper base

EXAMPLE 53 Find the area of the trapezoid with the following dimensions: height 8 meters, lower base 12 meters, and upper base 10 meters. Use the formula

$$A = \frac{1}{2} h \cdot (B + b)$$

$$A = \frac{1}{2} (8 \text{ m}) \cdot (12 \text{ m} + 10 \text{ m})$$ Substituting numerical values

22 m | $$A = \frac{1}{2} (8 \text{ m}) \cdot (\_\_\_\_\_)$$ Perform operations in parentheses first

4 m, 22 m | $$= \_\_\_\_\_ \cdot \_\_\_\_\_$$

m^2 | $$A = 88 \; \_\_\_\_\_$$ Area is given in square measure

Parts of a Circle

The perimeter of a circle is called the *circumference*. It is the distance around the outside.

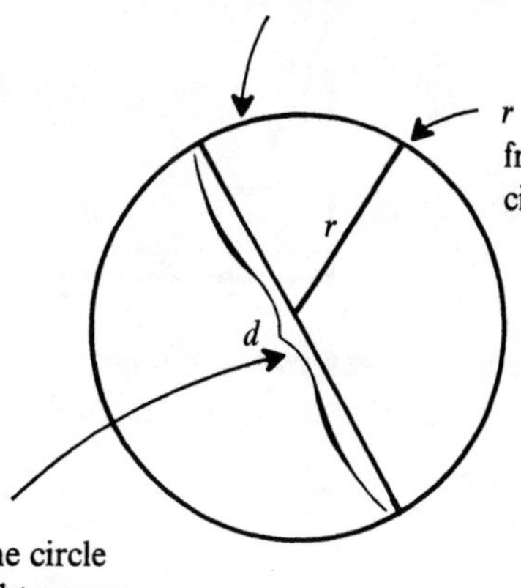

r is the *radius* of the circle or the distance from the center of the circle to its circumference.

d is the *diameter* of the circle or the distance straight across the circle through the center.

The length of the diameter is always twice the length of the radius.

$d = 2r$

π

π (pronounced "pie") is a Greek letter used to represent the ratio of the circumference of a circle to its diameter. π is approximately equal to 3.14.

EXAMPLE 54 Find the circumference of a circle with radius 4 yards.

$$C = 2 \cdot \pi \cdot r$$

$$C \approx 2(3.14)(4 \text{ yd})$$ Substitution

25.12 yd | $$C \approx \_\_\_\_\_\_\_$$

Area of a Circle

To find the area of a circle, think of it as four squares. The length of each square is one radius.

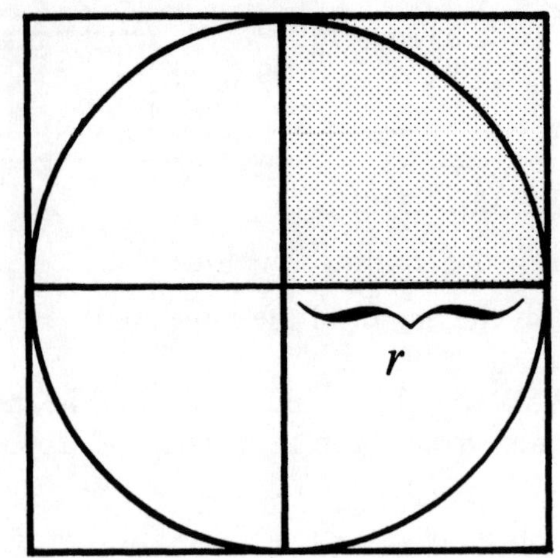

The area of the shaded part is $r \cdot r = r^2$. Since we have four of these squares, $4r^2$ could be used as an approximation of the area of the circle.

If we use $4r^2$ as the area of the circle, we will have the extra area shown by the shaded area below.

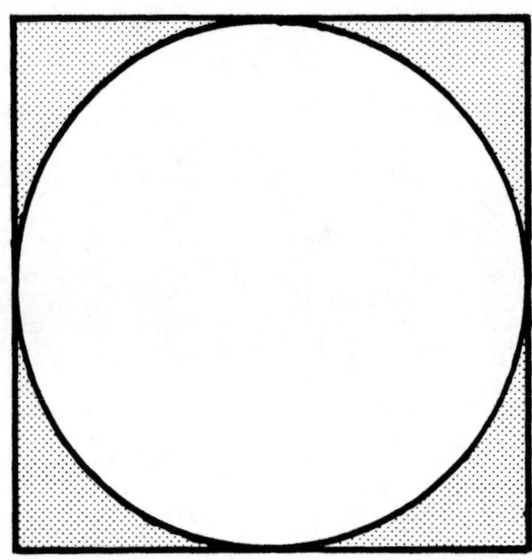

To discount the extra area in the corners, we multiply r^2 by π instead of by 4. π is really an unending decimal but we usually round it to 3.14 or approximate π with $\frac{22}{7}$.

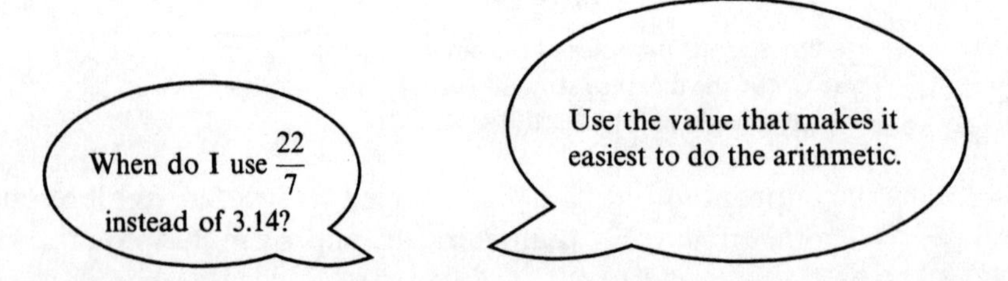

EXAMPLE 55 Find the area of a circle with a radius of 5 centimeters.

$A = \pi \cdot r^2$

$A = (3.14)(5 \text{ cm})^2$

$= 3.14 \cdot \_\_\_\_\_ \cdot \_\_\_\_\_$

$= 78.5 \text{ cm}^2$

5 cm

5 cm, 5 cm

Other Formulas

Formulas occur in nearly every field of endeavor.

In the field of physics, the formula for the distance an object falls during a period of time is

$$d = \frac{1}{2}gt^2$$

where d is the distance, g is the acceleration of gravity, which is 32 feet per second squared, and t is the time in seconds.

EXAMPLE 56 Find the distance a ball will fall in 3 seconds.

$$d = \frac{1}{2}gt^2$$

$$d = \frac{1}{2}(32) \cdot (3)^2 \quad \text{Substitute numerical values}$$

$$= \frac{1}{2}(32) \cdot 9 \quad \text{Evaluate expressions with exponents first}$$

16

$$= (\_\_\_\_) \cdot 9$$

144

$$d = \_\_\_\_ \text{ ft}$$

PROBLEM SET 1.4C

1. Use the formula $A = l \cdot w$ to find the area of a rectangle with a length of 8 inches and a width of 6 inches.
2. Use the formula $P = 2l + 2w$ to find the perimeter of a rectangle with a length of 8 inches and a width of 6 inches.
3. Use the formula $C = \pi d$, with $\pi = 3.14$, to find the circumference of a circle with a diameter of 12 feet.
4. Use the formula $F = \frac{9}{5}C + 32$ to find the boiling point of water on the Fahrenheit scale if the boiling point on the Celsius scale is 100°.
5. Use the formula $A = \pi r^2$, with π equal to 3.14, to find the area of a circle with a radius of 10 feet.
6. Use the formula $A = \frac{1}{2}bh$ to find the area of a triangle with a base of 12 inches and a height of 8 inches.
7. Find the equivalent body temperature on the Celsius scale (metric system) for 98.6° Fahrenheit using the formula $C = \frac{5}{9}(F - 32)$.
8. Use the formula $V = \pi r^2 h$, with π equal to 3.14, to find the volume of a can with a radius of 3 inches and a height of 10 inches.
9. A stone is dropped from the top of the Empire State Building. How far will it fall in 4 seconds? Use the formula $d = \frac{1}{2}gt^2$, with $g = 32$ ft/sec^2.
10. Use the formula $d = vt$, where d is the distance, v is the velocity, and t is the time, to find the distance traveled by a car moving at a velocity of 30 miles per hour for 6 hours.
11. Use the formula $C = 2\pi r$, with $\pi = 3.14$, to find the circumference of a circular pool with a radius of 11 feet.
12. Use the formula $P = a + b + c$ to find the amount of barbed wire needed to enclose a triangular-shaped field with sides 15 meters by 20 meters by 25 meters.
13. Use the formula $A = \pi r^2$ to find the area of a circle with a radius of 4 feet.
14. Use the formula $A = \frac{1}{2}h(B + b)$ to find the area of a trapezoid with a height of 6 feet, a lower base of 8 feet, and an upper base of 4 feet.

1 Review

NOW THAT YOU HAVE COMPLETED UNIT 1, YOU SHOULD BE ABLE TO:

Use the following definitions in your vocabulary.

Term—A number or expression that is added.

Factor—A number or expression that is multiplied.

Set—A collection of elements.

Elements—Objects in a set.

Natural numbers—The counting numbers $\{1, 2, 3, \ldots\}$.

Whole numbers—Counting numbers and zero $\{0, 1, 2, 3, \ldots\}$.

Integers—Counting numbers, zero, and the negative of the counting numbers $\{\ldots, -3, -2, -1, 0, 1, 2, 3, \ldots\}$.

Variable—Symbol that can represent any number in a set of numbers.

Constant—Symbol that represents a single number.

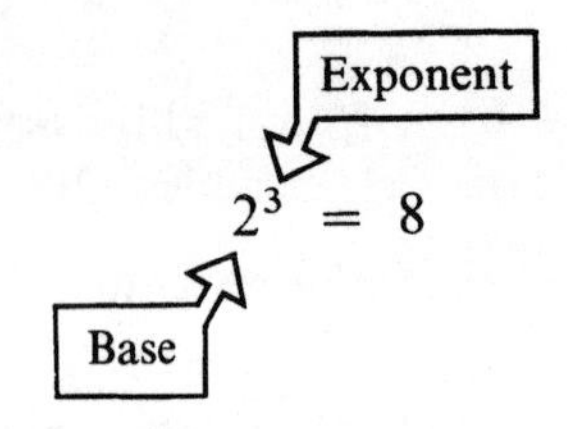

Base—Factor that is to be multiplied repeatedly.

Exponent—Raised number that indicates how many times the base is to be used as a factor.

Identity element for addition—Zero, because for any number a, $a + 0 = a$.

Identity element for multiplication—One, because for any number a, $a \cdot 1 = a$.

Additive inverses—A pair of numbers whose sum is zero: $a + (-a) = 0$.

Absolute value or magnitude—The value of a number if you ignore its sign.

Use the following properties to simplify algebraic expressions.

Commutative Property of Addition	$a + b = b + a$
Commutative Property of Multiplication	$a \cdot b = b \cdot a$
Associative Property of Addition	$(a + b) + c = a + (b + c)$
Associative Property of Multiplication	$(a \cdot b) \cdot c = a \cdot (b \cdot c)$
Distributive Property	$a(b + c) = ab + ac$
Zero Factor Law	$a \cdot 0 = 0 \cdot a = 0$

Use the following formulas.

Rectangle

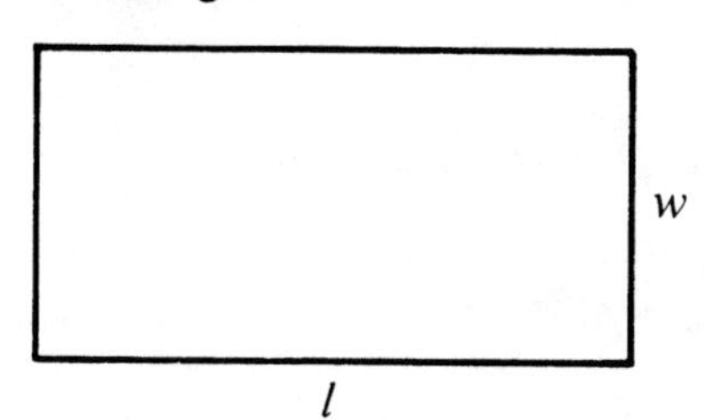

Perimeter: $P = 2l + 2w$

Area: $A = l \cdot w$

Circle

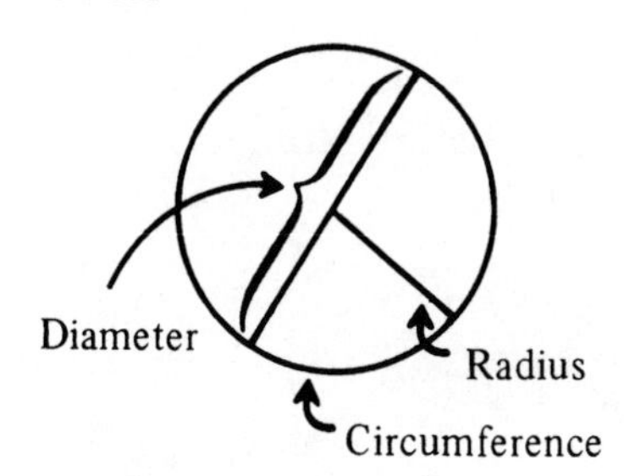

Diameter: $d = 2r$

Circumference: $C = \pi d$

$C = 2\pi r$

Area: $A = \pi r^2$

Triangle

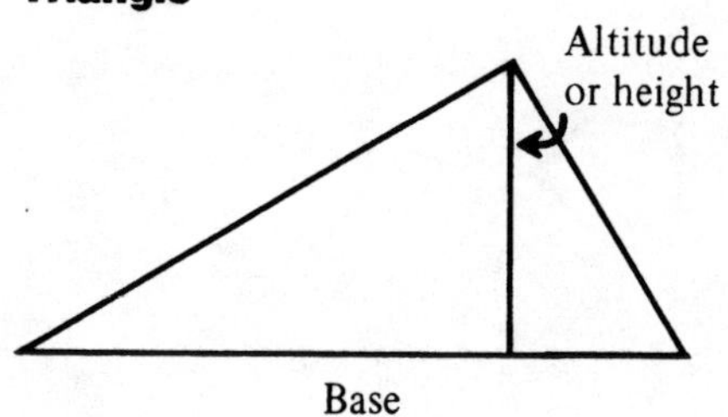

Perimeter: $P = a + b + c$

Area: $A = \frac{1}{2}bh$

Trapezoid

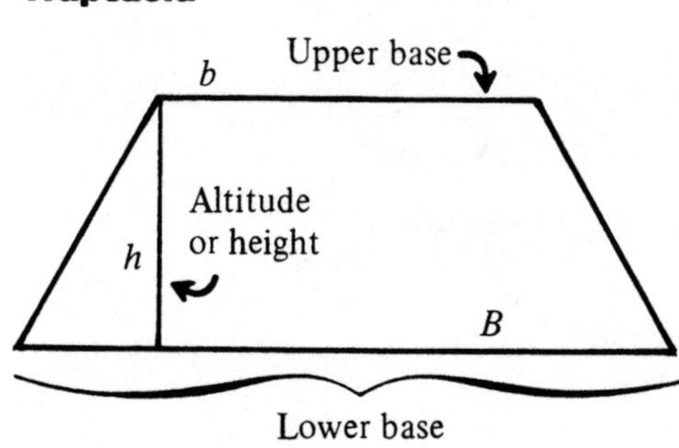

Area: $A = \frac{1}{2}h(B + b)$

Apply the following rules and definitions.

Addition of Signed Numbers

To add two positive numbers, add as in arithmetic.

To add two negative numbers, add as in arithmetic keeping the negative sign with the answer.

To add two numbers with unlike signs, find the difference between their magnitudes and use the sign of the number with the larger magnitude.

Subtraction of Signed Numbers

To subtract a number, add its additive inverse (negative).

$$a - b = a + (-b)$$

Multiplication of Signed Numbers

When two signed numbers are multiplied, the product is positive if the signs of the factors are alike and the product is negative if the signs of the factors are different.

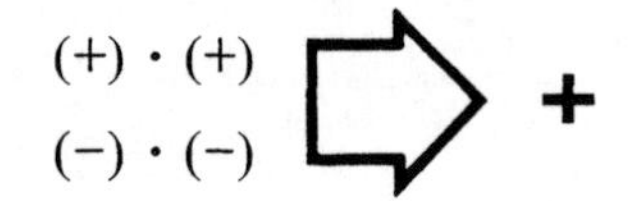

(+) · (−)
(−) · (+) ⇨ −

Division of Signed Numbers

When two numbers with like signs are divided, the quotient is positive.

When two numbers with unlike signs are divided, the quotient is negative.

(+) ÷ (+)
(−) ÷ (−) ⇨ +

(+) ÷ (−)
(−) ÷ (+) ⇨ −

Order of Operations

When evaluating an expression, do the operations in the following order.

1. Simplify all expressions within symbols of grouping. If one set of grouping symbols is inside another, perform the operations in the innermost grouping symbols first. Sometimes this step is called removing parentheses.
2. Evaluate any expressions with exponents.
3. Perform any multiplications or divisions, working from left to right, as they occur.
4. Perform any additions or subtractions, working from left to right, as they occur.

Review Test 1

Add or subtract the following. *(1.2)*

1. $12 + 17 =$ ________
2. $22 + (-43) =$ ________
3. $8 - 17 =$ ________
4. $-9 - 15 =$ ________
5. $-13 + (-21) =$ ________
6. $-8 + 14 =$ ________
7. $-12 - (-17) =$ ________
8. $0 - (-12) =$ ________

Multiply or divide the following. *(1.3)*

9. $(-5)(12) =$ ________
10. $(9)(11) =$ ________
11. $-48 \div (-4) =$ ________
12. $66 \div (-3) =$ ________
13. $(8)(-13) =$ ________
14. $(-17) \cdot 0 =$ ________
15. $(-105) \div (7) =$ ________
16. $(-18)(-12) =$ ________

Evaluate the following. *(1.1 and 1.4)*

17. $8 \div (-2) \cdot 2 =$ ________
18. $6 - 15 \div 3 \cdot 2 =$ ________
19. $-3^2 + 8 =$ ________
20. $8 - 3^2 =$ ________
21. $6^2 \div 3 - 1 + 2 =$ ________
22. $[2^3 \cdot 8 - 4] \div 2 =$ ________
23. $(3 - 5)^3 - (-6) \cdot (6 - 8) =$ ________
24. Use the formula $C = \frac{5}{9}(F - 32)$ to find the room temperature on the Celsius scale for 68° Fahrenheit.
25. Use the formula $A = \pi r^2$ to find the area of a circle with radius 6 centimeters ($\pi = 3.14$).

UNIT 2

Algebraic Expressions—Polynomials

OBJECTIVES

After you have successfully completed this unit, you will be able to:

1. Write algebraic expressions in exponential form (2.1)
2. Evaluate a polynomial for the given replacement set (2.1)
3. Multiply a monomial by a monomial using the multiplication law of powers (2.2)
4. Multiply a polynomial by a monomial (2.3)
5. Add and subtract polynomials (2.4)

2.1 POLYNOMIALS

Exponents

In Unit 1 we wrote the product of a times b as

$a \cdot b$ or ab

To show repeated multiplication of the same number, we used exponents, that is, a times a is

$a \cdot a$ or a^2

Exponential form

2.1A DEFINITION

a^n is written in *exponential form.*

a^n means that a is to be used as a factor n times.

$$a^n = \underbrace{a \cdot a \cdot a \cdot \ldots \cdot a}_{n \text{ factors}}$$

a is the base and n is the exponent.

a^1 means that there is only one factor of a. We write it as a.

$a^1 = a$ or $a = a^1$

EXAMPLE 1 Write the following in exponential form.

$$3 \cdot 3 \cdot 3 \cdot 3 = 3^4$$

3 $$(-2) \cdot (-2) \cdot (-2) = (-2)^{\_}$$

$$y \cdot y \cdot y = y^3$$

5 $$x \cdot x \cdot x \cdot x \cdot x = x^{\_}$$

$$2 \cdot 2 \cdot 2 \cdot x \cdot x \cdot x \cdot x = 2^3 \cdot x^4$$

2, 3 $$4 \cdot 4 \cdot a \cdot a \cdot a = 4^{\_}a^{\_}$$

2, 3, 1 $$x \cdot x \cdot y \cdot y \cdot y \cdot z = x^{\_}y^{\_}z^{\_}$$

Remember z appears as a factor only once. $z = z^1$

PROBLEM SET 2.1A

Write the following in exponential form.

1. $3 \cdot 3$
2. $6 \cdot 6 \cdot 6$
3. $(-2)(-2)$
4. $(-3)(-3)(-3)$
5. $(-4)(-4)(-4)(-4)$
6. $(-5)(-5)(-5)$
7. $2 \cdot 2 \cdot 2 \cdot 3 \cdot 3$
8. $2 \cdot 2 \cdot 7 \cdot 7 \cdot 7$
9. $x \cdot x \cdot x$
10. $y \cdot y \cdot y \cdot y \cdot y$
11. $x \cdot x \cdot y \cdot y$
12. $a \cdot a \cdot a \cdot b \cdot b$
13. $x \cdot y \cdot y \cdot y$
14. $r \cdot r \cdot r \cdot r \cdot s$
15. $2 \cdot 2 \cdot x \cdot x \cdot x$
16. $3 \cdot 3 \cdot a \cdot b \cdot b$
17. $(-2) \cdot a \cdot a \cdot b$
18. $(-3)(-3) \cdot a \cdot b \cdot b \cdot b$
19. $(-5)(-5) \cdot a \cdot a \cdot b$
20. $(-2)(-2)(-2) \cdot x \cdot y \cdot y \cdot y$
21. $2 \cdot 2 \cdot a \cdot a \cdot b \cdot c \cdot c \cdot c$
22. $7 \cdot r \cdot r \cdot r \cdot s \cdot s \cdot t$

Polynomials

Recall from Unit 1 that a variable is a symbol that can represent any element in a set of numbers.

Factors
Product

Numbers or variables that are multiplied are called *factors.* The result of a multiplication is called a *product.*

$$3 \cdot 4 \cdot b = \quad 12b$$

Factors   Product

Terms
Sum

Numbers or variables that are added are called *terms.* The result of an addition is called a *sum.*

$$3 + 4 = \quad 7$$

Terms   Sum

Expressions like

$$7, \quad a, \quad 3b, \quad x^3, \quad x^2 + y^2, \quad \text{and} \quad 2x^2 + x + 1$$

are called polynomials.

Polynomial

Polynomial in one variable

2.1B DEFINITIONS

A *polynomial* is a sum of one or more terms, each term consisting of a constant or a product of a constant and one or more variables. The variables may be raised to any whole number power.

A *polynomial in one variable* is a polynomial whose terms contain only one variable. The variable may be raised to different whole number powers in the separate terms.

Some typical polynomials are

$3b, \quad y + 4, \quad 2x^2 + x + 1$

Monomial

2.1C DEFINITION

A polynomial with only one term is called a *monomial.*

Some typical monomials are

$3, \quad -17, \quad \frac{3}{4}, \quad a, \quad -a, \quad a^2, \quad 3a^2, \quad -7a^3$

Coefficient

In the monomial $3a$ there are two factors, 3 and a. A numerical factor is referred to as the numerical *coefficient* of the term. 3 is the numerical coefficient of $3a$.

If there is no number in front of the letter, what is its coefficient?

The coefficient is understood to be 1. a means $1 \cdot a$. You can also consider the coefficient of $-a$ to be -1. $-a$ means $-1 \cdot a$.

EXAMPLE 2

The coefficient of $6y$ is ___6___

-7 — The coefficient of $-7a$ is ______

-7 — The coefficient of $-7a^2$ is ______

1, since $1 \cdot x = x$ — The coefficient of x is ______

1 — The coefficient of ab^2 is ______

-1 — The coefficient of $-ab^2$ is ______

What's the difference between terms and factors?

Terms are added. Factors are multiplied.

Binomial

2.1D DEFINITION

A *binomial* is a polynomial with exactly two terms. Some typical binomials are

$a + 2, \quad a + b, \quad x^2 + 6, \quad a^2 - b^2$

Trinomial

> **2.1E DEFINITION**
>
> A *trinomial* is a polynomial with exactly three terms. Some typical trinomials are
>
> $$x^2 + 2x + 1, \qquad 2y^2 - 3y + 7, \qquad ab^2 - ab - a^2$$

EXAMPLE 3 Identify each of the following as a monomial, binomial, or trinomial.

	$2x$	monomial
monomial	x^2	______
binomial	$x^2 + 2x$	______
monomial	ab	______
binomial	$a + b$	______
trinomial	$x^2 - 2x + 1$	______

Evaluating Polynomials

What does the polynomial $3x, + 1$ equal?

That depends on the value used to replace x. To evaluate an expression containing a variable, first replace *all* occurrences of the variable in the expression with its current value. Use parentheses when substituting. Then follow the rules of order of operations to evaluate.

EXAMPLE 4 Evaluate $3x + 1$ if $x = 4$.

$$3x + 1 =$$

Replace x with its value (4) $\qquad 3(4) + 1 =$

13 $\qquad$ Follow the rules of order of operations $\qquad 12 + 1 =$ ______

EXAMPLE 5 Evaluate $3x + 1$ if $x = -3$.

$$3x + 1 =$$

Replace x with its current value (-3) $\qquad 3(-3) + 1 =$

$-9, -8$ $\qquad$ Follow the rules of order of operations $\qquad$ ______ $+ 1 =$ ______

EXAMPLE 6 Evaluate $2a + 3b$ if $a = 3$ and $b = -4$.

$$2a + 3b =$$

Replace a with (3) and b with (-4) $\qquad 2(3) + 3(-4) =$

$6, (-12), -6$ $\qquad$ Evaluate $\qquad$ ______ + ______ = ______

EXAMPLE 7 Evaluate $y^2 + y + 2$ if $y = -6$.

$$y^2 + y + 2 =$$

Replace *each* y with (-6) $\qquad (-6)^2 + (-6) + 2 =$

32 $\qquad$ Evaluate $\qquad 36 + (-6) + 2 =$ ______

EXAMPLE 8 Evaluate $b^2(a - c^3)$ if $a = 2$, $b = -2$, and $c = -3$.

	$b^2(a - c^3) =$
Replace variables	$(-2)^2(2 - (-3)^3) =$
Evaluate exponents	$4(2 - (-27)) =$
Write subtraction as addition	$4(2 + 27) =$
Simplify parentheses	$4(29) = 116$

POINTERS FOR BETTER STYLE

Evaluate: $y^2 + y + 2$ if $y = -6$

How it's written	What to notice
$y^2 + y + 2 =$	First recopy the problem without any changes.
$(\ \ )^2 + (\ \ ) + 2 =$	Recopy again, this time leave empty parentheses where the variables were. Notice the exponent is on the first parentheses just like the exponent on y^2.
$(-6)^2 + (-6) + 2 =$	Now fill in the empty parentheses with the value of the variable. You can combine this step with the previous step. There's no need to write it again.
$36 + (-6) + 2 =$	$(-6)^2$ was replaced with its value directly below. Try to keep operation signs aligned when possible.
$30 + 2 = 32$	The 30 came from $36 + (-6)$ directly above. Write it next to the $+2$.
Try to keep the = signs aligned. Only one = sign to a line.	Nothing is done with the $+2$ through most of the problem. Since $+2$ came down in a straight line, it's easy to see it's not used until the last step.

PROBLEM SET 2.1B

Replace the variable in the following polynomials with the given value, then evaluate.

1. $x + 3$ if $x = 7$
2. $b + 4$ if $b = -4$
3. $x - 3$ if $x = -1$
4. $4x - 2$ if $x = 3$
5. $5y + 4$ if $y = -2$
6. $2x - 3$ if $x = 3$
7. $3x - 5$ if $x = -3$
8. $4y - 2$ if $y = -6$
9. $a - (a - 6)$ if $a = 2$
10. $x - (x + 9)$ if $x = -5$
11. $(z - 8) + (8 - z)$ if $z = -3$
12. $x(2 - x)$ if $x = 4$
13. $y(y + 6)$ if $y = -1$
14. $(8 - m)m$ if $m = -4$
15. $x^2 + 1$ if $x = 4$
16. $x^3 - 6$ if $x = -2$
17. $x^2 + x$ if $x = -5$
18. $2a^2$ if $a = 3$
19. $3x^2 - 4$ if $x = 2$
20. $5x^2 - x - 6$ if $x = -2$
21. $5 - 4x - 2x^2$ if $x = -3$
22. $8 + 2x - 3x^2$ if $x = -4$

Evaluate the following polynomials for $a = -2$, $b = 3$, and $c = -4$.

23. $a + b$
24. $a + b + c$
25. $a + b - c$
26. $2b - a^2$
27. $3c^2 + b$
28. $4b^2 - 2c$
29. ab^2c
30. $a^3b - c^2$
31. $a^3 + b^2c$
32. $a^2b + a^3c$
33. $a(b^3 + c)$
34. $b(a^3 - c)$
35. $a^3(a - b)$
36. $c^2(b - a)$

EXAMPLE 9 Evaluate $3x^2 - 4$ for the following values of x: 2, 1, 0, -1, -2, -3.

First set up a table to make it easier.

Replacement values for x go here

x	$3x^2 - 4$
2	$3(2)^2 - 4 = 8$
1	$3(1)^2 - 4 = -1$
0	$3(0)^2 - 4 = -4$
-1	$3(-1)^2 - 4 = -1$
-2	$3(-2)^2 - 4 = 8$
-3	$3(-3)^2 - 4 = 23$

The expression with the variable replaced goes in this column.

PROBLEM SET 2.1C

Complete the tables.

1.

x	$2x + 6$
-4	
-2	
1	
2	
3	

2.

x	$x^2 - 1$
-2	
-1	
0	
1	
2	

3.

x	$x^3 + 1$
-2	
-1	
0	
1	
2	

4.

x	$x^2 + 1$
-2	
-1	
0	
1	
2	

5.

x	$x^3 - 1$
-2	
-1	
0	
1	
2	

6.

x	$x^3 + 2$
-3	
-1	
0	
1	
3	

7.

x	$x^2 + 2x + 3$
-2	
0	
1	
2	
3	

8.

x	$x^2 - 3x + 4$
-4	
-2	
0	
2	
4	

9.

x	$2x^3 + x - 3$
-3	
-1	
0	
1	
2	

10.

x	$3x^2 - x + 4$
-3	
-1	
0	
1	
2	

11.

x	$2x^3 + 4x^2 + 4$
-3	
-1	
0	
1	
2	

12.

x	$3x^3 - 2x^2 - 4$
-2	
-1	
0	
1	
2	

2.2 MULTIPLICATION OF MONOMIALS

To multiply the monomial $3x$ by the monomial $4y$, we use the commutative and associative properties, which allow us to change the order and grouping of factors.

$3x \cdot 4y = 3 \cdot x \cdot 4 \cdot y$	Meaning of $3x \cdot 4y$
$= 3 \cdot 4 \cdot x \cdot y$	Commutative property of multiplication
$= (3 \cdot 4)(x \cdot y)$	Associative property of multiplication
$= 12 \cdot xy$	

A quicker method, with the same result, would be to multiply the coefficients and then multiply the variables.

EXAMPLE 10

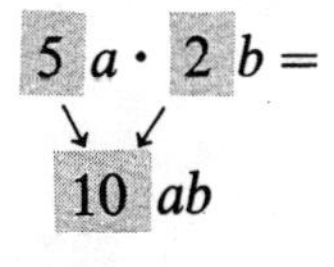

It takes fewer steps this way.

That is why we showed you the shorter method. But remember, this is only a short cut.

EXAMPLE 11

$-4x \cdot 4y^2 =$

$-16xy^2$

EXAMPLE 12

$(-2a^3) \cdot (-6) =$

$12a^3$

PROBLEM SET 2.2A

Use multiplication to simplify the following monomials.

1. $x \cdot 2y$	2. $3x \cdot 6y$	3. $2x \cdot 3y$	4. $4a \cdot 5b$
5. $-2x \cdot 4y$	6. $(6s)(-3t)$	7. $(-7a)(-2x)$	8. $(-100)(-5d)$
9. $4 \cdot 6z$	10. $(8)(-4x)$	11. $(-9y)(-1)$	12. $2 \cdot 3a^2$
13. $4t^2 \cdot 5$	14. $(-3x^4)(6)$	15. $(-x^2)(2^2)$	16. $(-3^3)(2a^3)$
17. $2^2 \cdot 3y^2$	18. $7a^2 \cdot 4b^3$	19. $s^2 \cdot 7t^4$	20. $x^2 \cdot 5y$
21. $(-2m)(-8n^5)$	22. $(-5a^2)(-5b)$	23. $a^2 \cdot m^2$	24. $-a^3 \cdot (3s^4)$
25. $-x^4 \cdot (-6y^3)$	26. $-x^3 \cdot y^2$	27. $-a^2 \cdot (-b^3)$	28. $x^2 \cdot (-y^4)$

Perform the indicated operations.

29. $-8 - (-4)$ 30. $(-8) \div (-4)$ 31. $(-6 + 3)^3 - (-3)^2$

Multiplication of Monomials with Like Bases

We can multiply two monomials with the same variable such as a^2 times a^3 if we use the definition of exponents.

$$a^2 \cdot a^3$$

Two a's · Three a's — Five a's

$$a \cdot a \cdot a \cdot a \cdot a = \overbrace{a \cdot a \cdot a \cdot a \cdot a}^{\text{Five } a\text{'s}} = a^5$$

Rather than write out all the factors and then count the number of times the base is used, just add the exponents.

$$a^{2} \cdot a^{3} = a^{2+3} = a^{5}$$

Multiplication law of powers

2.2A MULTIPLICATION OF POWERS WITH LIKE BASES

When multiplying monomials with the same base, add the exponents.

$$a^x \cdot a^y = a^{x+y}$$

EXAMPLE 13

$2^3 \cdot 2^2 = 2^5$ $\quad 3^2 \cdot 3^4 = 3^6$ $\quad x^3 \cdot x^4 = x^7$

$b^5 \cdot b^4 = b^9$ $\quad y^3 \cdot y^2 = y^5$ $\quad p^2 \cdot p^{100} = p^{102}$

EXAMPLE 14

$$a \cdot a^3 = ?$$

Since $\quad a = a^1$

then $\quad a \cdot a^3 =$

$$a^1 \cdot a^3 = a^{1+3} = \_\_\_\_$$

a^4

EXAMPLE 15

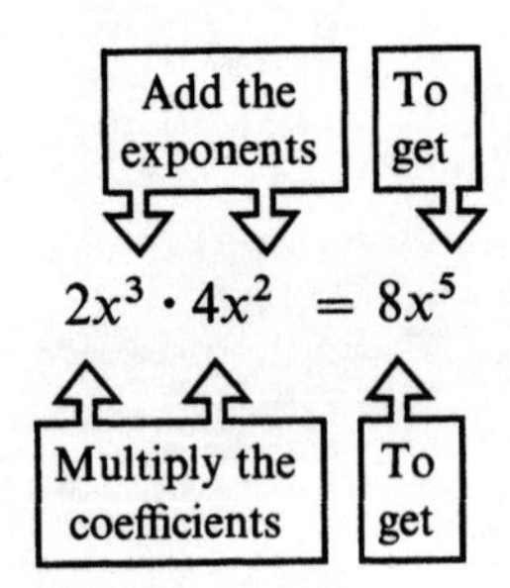

EXAMPLE 16

$$-3x \cdot 5x^2 = \_\_\_\_ x^{\_}$$

$-15, ^3$

EXAMPLE 17

$a^2b^2 \cdot a^2b^3 = a^2 \cdot b^2 \cdot a^2 \cdot b^3$

$= a^2 \cdot a^2 \cdot b^2 \cdot b^3$ Commutative property of multiplication

$= a^4 \cdot b^5$ Multiplication law of powers

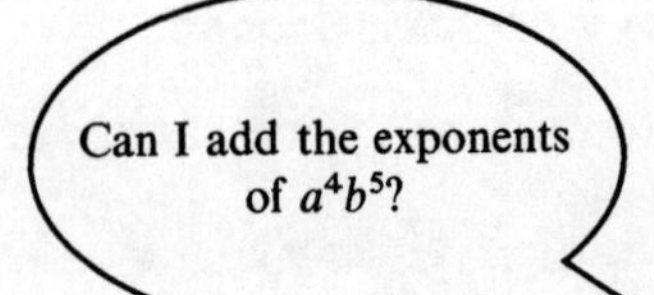

No, to use the multiplication law of powers, the bases must be the same.

EXAMPLE 18

16, 9, 3 — $4x^4y^2 \cdot 4x^5y = \_\_\_\_\_ x^{—} y^{—}$

$15a^5b$ — $3a^2 \cdot 5a^3b = \_\_\_\_\_\_$

$-8p^3q^2$ — $-2pq \cdot 4p^2q = \_\_\_\_\_\_$

Zero as an Exponent

The multiplication law of powers tells us that to multiply two powers of the same base, add the exponents. Using this law, even though we don't know what a^0 means yet,

$$a^3 \cdot a^0 = a^{3+0}$$
$$= a^3$$

Because 1 is the identity element for multiplication,

$$a^3 \cdot 1 = a^3$$

Now, because both expressions $a^3 \cdot 1$ and $a^3 \cdot a^0$ are equal to a^3, they are equal to each other.

Same

$$a^3 \cdot a^0 = a^3 \cdot 1$$

Must be equal

Therefore, the following definition seems reasonable.

Zero exponent

2.2B DEFINITION: ZERO EXPONENT

Any number except zero raised to the zero power equals one.

$a^0 = 1 \qquad x^0 = 1 \qquad (-3)^0 = 1 \qquad (2x)^0 = 1$

PROBLEM SET 2.2B

Use the multiplication law of powers to find the following products.

1. $x \cdot x$
2. $x \cdot x^4$
3. $x^2 \cdot x$
4. $a \cdot a^3$
5. $b^2 \cdot b^3$
6. $3a \cdot a^4$
7. $-2x^2 \cdot x^4$
8. $y^2(-3y^5)$
9. $2a^3 \cdot 3a^2$
10. $-4a \cdot 5a^3$
11. $(3x^2)(-4x^3)$
12. $(4a^3)(-5a^4)$
13. $(-5x^4)(-6x^3)$
14. $6x \cdot 7x^2$
15. $-9x^2 \cdot 8y^3$
16. $(-5x^4)(-6a^5)$
17. $2ab^2 \cdot 3a^3b^2$
18. $-x^2y \cdot 5x^3y^4$
19. $(-3x^4y)(-4xy)$
20. $4a^0b \cdot 5ab^2$
21. $-a^2 \cdot b^0 \cdot 3a^4 \cdot b^2$
22. $(-2a^0b^3)(-3a^0b^2)$
23. $6a \cdot 7x^2y$
24. $-8a \cdot 9a^2x^3$

25. $(-6a^0b)(-9a^3)$ 26. $a^0 \cdot 5ab^2$ 27. $-x^0 \cdot y^0 \cdot 7xa^2$

28. $(-3x^0)(-4x^0y^2)$ 29. $(x)(-y^0)(-6x^2a)$ 30. $(-a^0)(b^0)(-6x^3y^0)$

Write in symbols.

31. x is the product of negative three times the third power of b multiplied by the second power of c.

32. x is the fourth power of p times the negative of the second power of q.

2.3 MULTIPLICATION OF A POLYNOMIAL BY A MONOMIAL

Now, suppose we wish to multiply the monomial 2 by the binomial $x + 3$.

$$2 \cdot (x + 3) = ?$$

Because a binomial has two terms, we will need to use the distributive property to simplify it.

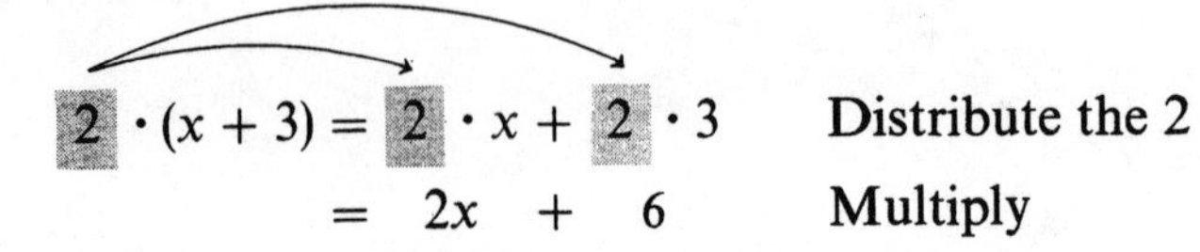

$$2 \cdot (x + 3) = 2 \cdot x + 2 \cdot 3 \quad \text{Distribute the 2}$$
$$= 2x + 6 \quad \text{Multiply}$$

Multiply each term inside the parentheses by 2.

EXAMPLE 19 Simplify $2(3a - 4b)$.

$$2(3a - 4b) = 2 \cdot 3a - 2 \cdot 4b$$
$$= 6a - 8b$$

What does simplify mean?

It means to perform all indicated operations including removal of parentheses.

EXAMPLE 20 Simplify $-3(x + 4)$.

$$-3(x + 4) = (-3)x + (-3)4 \quad \text{Distribute the } -3$$
$$= -3x + (-12) \quad \text{Multiply}$$
$$= -3x - 12 \quad \text{Remove parentheses}$$

Why rewrite $+(-12)$ as -12?

Because to simplify means to perform all operations and to remove all parentheses.

EXAMPLE 21A Simplify $-1(5x - 3y)$.

$$
\begin{aligned}
-1(5x - 3y) &= (-1)(5x) - (-1)(3y) \\
&= -5x - (-3y) \\
&= -5x + 3y
\end{aligned}
$$

EXAMPLE 21B This is an alternate way to handle negative signs.

$-1(5x - 3y) = -1[5x + (-3y)]$	First, rewrite as an addition
$(-1)(5x) + (-1)(-3y) =$	Now multiply each term by (-1)
$-5x + 3y =$	Simplify each term

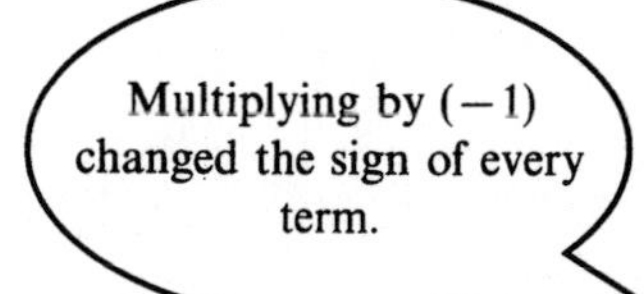

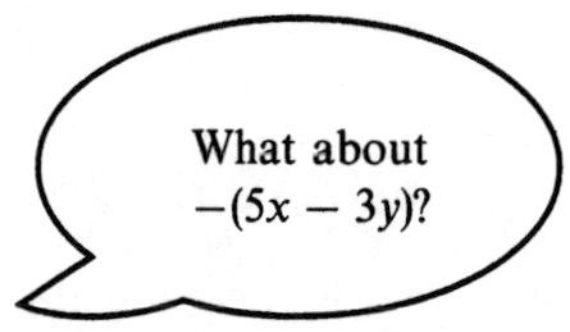

Negative of an expression

2.3 NEGATIVE OF AN EXPRESSION

The negative of any expression can be found by multiplying the expression by negative one.

$$-a = -1 \cdot a$$

for any nonzero number a.

EXAMPLE 22 Simplify $-(5x - 3y)$.

$$
\begin{aligned}
-(5x - 3y) &= (-1)(5x - 3y) \\
&= (-1)(5x) - (-1)(3y) \\
&= -5x - (-3y) \\
&= -5x + 3y
\end{aligned}
$$

The negative sign in front of the polynomial changed the sign of every term.

EXAMPLE 23

$$-(a + 2) = -a - 2$$

$-(2x - 3) =$ _______ ($-2x + 3$)

$-(-3p + 4q - 6) =$ _______ ($3p - 4q + 6$)

We can multiply a binomial by any monomial if we use the distributive property and the multiplication law of powers.

EXAMPLE 24 Simplify $a^3(a^2 + a)$.

$a^3 \cdot (a^2 + a) = a^3 \cdot a^2 + a^3 \cdot a$	Distribute
$= a^5 + a^4$	Add exponents

EXAMPLE 25 Simplify $2ab(3a^2b - ab)$.

$$
\begin{aligned}
2ab(3a^2b - ab) &= (2ab)(3a^2b) - (2ab)(ab) \\
&= 6a^3b^2 - 2a^2b^2
\end{aligned}
$$

EXAMPLE 26 Simplify $-3x(6x - 4y)$.

$$-3x(6x - 4y) = -3x[6x + (-4y)]$$ Rewrite the subtraction as an addition

$$= (-3x)(6x) + (-3x)(-4y)$$ Distribute the $-3x$

$$= -18x^2 + 12xy$$ Simplify each term

The distributive property can be extended to a monomial times any polynomial.

EXAMPLE 27 Simplify $3x^2(x^2 - 2x + 6)$.

$$3x^2(x^2 - 2x + 6) = 3x^2 \cdot x^2 - 3x^2 \cdot 2x + 3x^2 \cdot 6$$
$$= 3x^4 - 6x^3 + 18x^2$$

EXAMPLE 28 Simplify $-2a(a^2 - 3a + 5)$.

$$-2a(a^2 - 3a + 5) =$$ Copy the problem

$$= -2a[a^2 + (-3a) + 5]$$ Rewrite the subtraction of $3a$ as an addition

$$= -2a(a^2) + (-2a)(-3a) + (-2a)(5)$$ Multiply

$$= -2a^3 + 6a^2 - 10a$$ Simplify

PROBLEM SET 2.3

Simplify the following.

1. $2(a + 4)$
2. $3(x - 5)$
3. $6(x - 8)$
4. $-3(x + 6)$
5. $-1(a - 9)$
6. $-4(-y - 7)$
7. $4(3a + 2b)$
8. $7(5x - 6y)$
9. $-8(3a - 2y)$
10. $-5(4a - 5b)$
11. $-(-3a + 4x)$
12. $-(3a - 4x)$
13. $-(-2a - 3b)$
14. $-(5a - 6b)$
15. $-(-4x + y)$
16. $-(-6x - 8y)$
17. $x^2(x^4 + x)$
18. $x^3(x^2 - x + 3)$
19. $2xy(7xy^2 + 3x^2y)$
20. $3a^2b(2a^2b - 5ab^3)$
21. $-4xy^2(3xy - 6x^2y)$
22. $-6x^2y^2(-2xy^4 + 3x^2y^5)$
23. $-3ab^2(5abx - 6ay^3)$
24. $-2axy^2(4ax^2y^3 - 6a^4x)$
25. $-3ab(-4a^2bx - 6bx^2)$
26. $-x^2y^3(4axy^2 - 8a^2y^3)$
27. $2x(3x^2 - 4x + 1)$
28. $3x^2(3x^4 - 5x^2 + 7x - 2)$
29. $-4x^3(7x^3 - 8x + 1)$
30. $-6a^3(-7a^3 + 8a^2 - a)$
31. $9ax^2(9ax^4 - 8a^2x + 6ax)$
32. $6a^2b(3ab^3 - 2a^4b^2 + a)$
33. $-3a^3b^4c(3abc - 5ab^2c + 4a^4b)$
34. $-2abc^4(-6a + 7b - 9c)$
35. $-6x^2ya(-2xa^2 + 3x^3y^2 - 5x^4y^2a)$
36. $-4xy^3a^3(3x^2a - 6x^4y^2 - 5xy^4a^3)$

2.4 ADDITION AND SUBTRACTION OF POLYNOMIALS

Combining Like Terms

If two monomials such as $3x$ and $5x$ are the same except for their numerical coefficients, we say that they are like terms.

Like terms

> **2.4A DEFINITION**
>
> *Like terms* are terms that have identical letters and exponents. Their numerical coefficients may differ.

$3ab^2$ and $-7ab^2$ are like terms.

$3ab$ and $3ab^2$ are not like terms.

Notice that the variables and the exponents on the variables must be exactly the same to be called like terms.

EXAMPLE 29 On each row of terms, circle the term that is like the underlined term.

Answer	Underlined			
$5x^2$	$\underline{3x^2}$	$3x$	$5x^2$	x^3
$3y$	$\underline{-2y}$	y^2	2	$3y$
$-z^4$	$\underline{z^4}$	$4z$	$-z^4$	$-4z^3$
$12ab^2$	$\underline{-2ab^2}$	$-4a^2b^2$	$4ab$	$12ab^2$

If we use the distributive property on like terms, we can simplify some algebraic expressions that involve addition and subtraction.

EXAMPLE 30 Simplify $2x + 3x$.

$$2x + 3x = (2 + 3)x \quad \text{Distributive property in reverse}$$
$$= 5x$$

EXAMPLE 31 Simplify $-8a + 5a$.

$$-8a + 5a = (-8 + 5)a$$
$$= -3a$$

EXAMPLE 32 Simplify $b^2 - 12b^2$.

b^2

$$b^2 - 12b^2 = (1 - 12)\_\_\_\_$$
$$= -11b^2$$

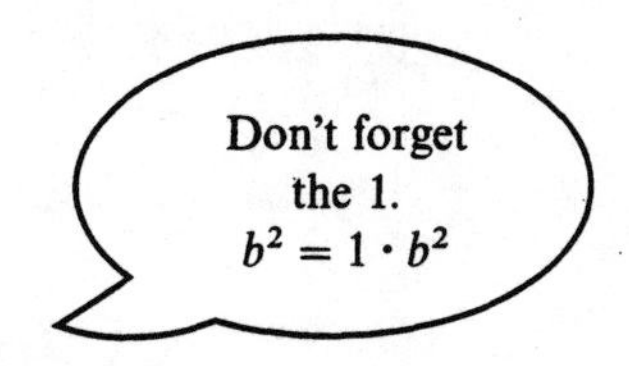

PROBLEM SET 2.4A

Use the distributive property to combine like terms.

1. $3a + 5a$
2. $7x - 2x$
3. $8y - 10y$
4. $-2ab - 3ab$
5. $x - 2x$
6. $-x - 3x$
7. $8a^2 - 9a^2$
8. $7a^2b - 12a^2b$
9. $14abc - 7abc$

Combining like terms by using the distributive property can become a long process. Sometimes it is easier to view the process as adding or subtracting like terms.

2 [apples] + 3 [apples] = 5 [apples]

12 [apples] + 3 [oranges] − 4 [apples] = 8 [apples] + 3 [oranges]

Combining like terms

2.4B COMBINING LIKE TERMS

To combine like terms, add their numerical coefficients and keep the common variables.

EXAMPLE 33 Simplify $-4a + 7a$.

Think $-4 + 7$ | Variable common to both like terms

$$-4a + 7a = 3 \cdot a$$

EXAMPLE 34 Simplify $4y + 3 + 2y$.

Rearrange like terms to be combined

$$\begin{aligned} 4y + 3 + 2y &= (4y + 2y) + 3 \\ &= 6y + 3 \end{aligned}$$

EXAMPLE 35 Simplify $4a + 8b + 2a - 3b$.

Rearrange like terms to be combined | Rearrange like terms to be combined

$$\begin{aligned} 4a + 8b + 2a - 3b &= (4a + 2a) + (8b - 3b) \\ &= 6a + 5b \end{aligned}$$

EXAMPLE 36 Simplify $8x^2 + x - x^2 + 3x$.

Like terms | Like terms

$$\begin{aligned} 8x^2 + x - x^2 + 3x &= (8x^2 - x^2) + (x + 3x) \\ &= 7x^2 + 4x \end{aligned}$$

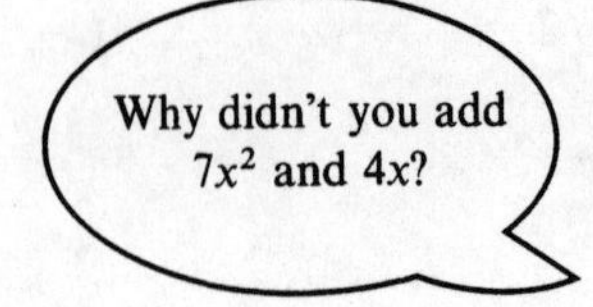

You can only add like terms. Like terms must have the same exponents as well as the same letters.

When simplifying it is possible to avoid rearranging the terms; just pick out the like terms and combine their coefficients.

EXAMPLE 37 Simplify $10x + 3 - 6x - 7$.

Think $10 + (-6)$ | Think $3 + (-7)$

Remember: To subtract, add the opposite: $3 - 7 = 3 + (-7)$.

$$10x + 3 - 6x - 7 = 4x - 4$$

EXAMPLE 38 Simplify $a^2 - 3a - 7 + 4a^2 - 6a + 2$.

Think $1 + 4$ | Think $-3 + (-6)$ | Think $-7 + 2$

$$a^2 - 3a - 7 + 4a^2 - 6a + 2 = 5a^2 - 9a - 5$$

PROBLEM SET 2.4B

Simplify the following.

1. $x + 2x$
2. $3x - x$
3. $4x + 5x$
4. $7a - 2a$
5. $-9b + 3b$
6. $-3x^2 - 4x^2$
7. $3a^2 - 12a^2$
8. $-4a^3b + 7a^3b$
9. $2x - 6y + 3x + y$
10. $3a + 7b - 2b + 7a$
11. $-x + 8a - 9x - a$
12. $-3x - 8y + 10x - 2y$
13. $a - 2b + 3a - 4b + 1$
14. $-x + 4y + x - 5y - 8$
15. $x^2 - 2x + 1 - 3x^2 + 4x - 1$
16. $3x^3 - 2x + 8 - 6x^3 - 4x + 5$
17. $7x^3 + 2x^2 - 4 + 3x^3 - 5x + 2$
18. $-10x^4 + 8x^2 - 5x^3 - 6x^2 + 2x^4$
19. $6a^2b - 7ab + 4ab^2 - a^2b + 5ab^2$
20. $-x^2y + 3x - 4xy^2 + 4x^2y - y - 3xy^2$
21. $-3a^2b + 5ab^3 - 7ab + 2a^2b + ab$
22. $12a + 2b^2 - 3ab + 9ab^2 - 3b^2 - 11a$
23. $-2x^2 - 4x + 1 - x^2 + x - 1$
24. $xy^2 + 3xy - 6 + 3xy - 4xy^2 - 1$
25. $9x^2y^2 - 4xy + 3x - 3xy - 6x^2y$
26. $7ab^2 - 6ab + 3ab^2 - ab + 3ab^2$
27. $-x^2y - 3xy^2 + 2xy - x + 6xy^2 - 2xy + y$
28. $-6x + 7y - 8xy + 2x^2y - 4y + 6xy + 7x - 3x^2y$
29. $-5a + 4b - ab + a^2 - 5b + 8a - 7ab - 3a^2$
30. $5a^2b - 3ab^2 + 7a - 8b + 9ab^2 - 9b + 5a - 5a^2b$

Addition of Polynomials

Adding polynomials is just an exercise in combining like terms.

EXAMPLE 39A Add and simplify $(2a^2 + a + 7) + (a^2 - 3a - 4)$.

$$(2a^2 + a + 7) + (a^2 - 3a - 4) = 2a^2 + a + 7 + a^2 - 3a - 4 \quad \text{Remove parentheses}$$
$$= 3a^2 - 2a + 3 \quad \text{Combine like terms}$$

EXAMPLE 39B Add and simplify $(2a^2 + a + 7) + (a^2 - 3a - 4)$.

Sometimes it is easier to add polynomials vertically than horizontally. To do this, write the polynomials so that the like terms are in columns.

$$(2a^2 + a + 7) + (a^2 - 3a - 4) \longrightarrow \begin{array}{r} 2a^2 + a + 7 \\ +a^2 - 3a - 4 \\ \hline 3a^2 - 2a + 3 \end{array}$$

Subtraction of Polynomials

Recall the definition of subtraction. To subtract a number, add its additive inverse (opposite).

$$a - b = a + (-b)$$

Subtracting polynomials is done in a similar manner; that is, to subtract one polynomial from another, add its additive inverse (opposite).

Remember: To write the negative of a polynomial, we change the sign of each term of the polynomial.

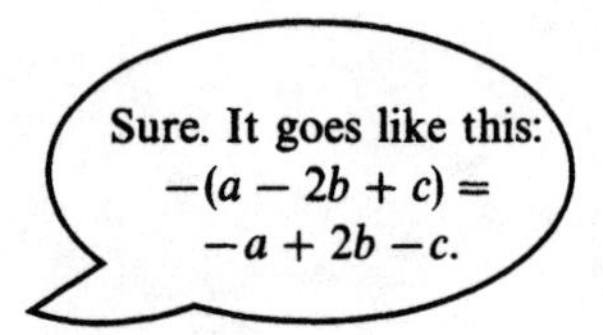

EXAMPLE 40 Subtract and simplify $(4x + 3) - (3x + 6)$.

$$(4x + 3) - (3x + 6) = 4x + 3 - 3x - 6 \quad \text{Remove parentheses}$$
$$= \_\_\_\_\_\_\_\_ \quad \text{Combine like terms}$$

$x - 3$

EXAMPLE 41 Subtract and simplify $(3x - 18) - (4x + 8)$.

$$(3x - 18) - (4x + 8) = 3x - 18 - 4x - 8 \quad \text{Remove parentheses}$$
$$= \_\_\_\_\_\_\_\_ \quad \text{Combine like terms}$$

$-x - 26$

EXAMPLE 42 Subtract and simplify $(2x^2 - 3x + 1) - (-x^2 + 4x - 1)$.

$$(2x^2 - 3x + 1) - (-x^2 + 4x - 1) = 2x^2 - 3x + 1 + x^2 - 4x + 1 \quad \text{Remove parentheses}$$
$$= \_\_\_\_\_\_\_\_ \quad \text{Combine like terms}$$

$3x^2 - 7x + 2$

EXAMPLE 43 Perform the indicated operations.

$$(5x - 6) + (4x + 2) - (6x - 3) = 5x - 6 + 4x + 2 - 6x + 3 \quad \text{Remove parentheses}$$
$$= \_\_\_\_\_\_\_\_ \quad \text{Combine like terms}$$

$3x - 1$

EXAMPLE 44 Perform the indicated operations.

$$(2x^2 - 3x + 5) + (5x^2 - 6x + 4) - (3x^2 - 8x + 4)$$

$= 2x^2 - 3x + 5 + 5x^2 - 6x + 4 - 3x^2 + 8x - 4$ Remove parentheses

$4x^2 - x + 5$

$=$ ____________ Combine like terms

EXAMPLE 45 Subtract $(12x^2 - 3x + 4) - (-x^2 + 4x - 1)$.
Subtraction can also be done vertically.

$$\begin{array}{r} 12x^2 - 3x + 4 \\ -(-x^2 + 4x - 1) \\ \hline \end{array} \Longrightarrow \begin{array}{r} 12x^2 - 3x + 4 \\ +x^2 - 4x + 1 \\ \hline 13x^2 - 7x + 5 \end{array}$$

Always write as addition using the additive inverse.

EXAMPLE 46 Subtract $(-5x^2 + 6x - 7) - (2x^2 - 3x + 4)$.

$$\begin{array}{r} -5x^2 + 6x - 7 \\ -(2x^2 - 3x + 4) \\ \hline \end{array} \Longrightarrow \begin{array}{r} -5x^2 + 6x - 7 \\ -2x^2 + 3x - 4 \\ \hline -7x^2 + 9x - 11 \end{array}$$

PROBLEM SET 2.4C

Add the following polynomials.

1. $(3x + 4) + (-6x + 5)$
2. $(8y^2 - 1) + (-6y^2 + 2)$
3. $(-4x^2 + 2) + (-6x^2 - 5)$
4. $(-3x^2 - 5) + (x^2 + 9)$
5. $(x^2 + 3x - 1) + (3x^2 - 5x - 6)$
6. $(a^2 - 3a + 5) + (-6a^2 + 7a - 9)$
7. $(7x^2 + 8x - 3) + (3x^2 - 10x - 9)$
8. $(-12 + 4y - 8x) + (3y - 3 + 4x)$
9. $(-y^2 + xy - 8x^2) + (y^2 - 3xy + x^2)$
10. $(a - b + 2ab) + (-3ab + 5b - 7a)$
11. $(2x - 1) + (-4x + 5) + (6x - 8)$
12. $(x^2 - 1) + (3x^2 - 4) + (-8x^2 + 9)$
13. $(x^2 - 2x + 1) + (x - 5) + (3x^2 + 4x - 8)$
14. $(-6y^2 + 7y - 2) + (3y^2 - 5y - 4) + (2y^2 - 3y)$

Find the difference of the following polynomials.

15. $(2x - 8) - (7x + 9)$
16. $(-3a + 4) - (5a + 7)$
17. $(-10xy + 9) - (-6xy + 8)$
18. $(12y - 4) - (-13y - 6)$
19. $(2x^2 - 3x + 1) - (x^2 - 4x - 6)$
20. $(-3x^2 - 3x + 2) - (-4x^2 + 7x - 2)$
21. $(x^2 - 3xy + y^2) - (x^2 + 5xy - y^2)$
22. $(-2x^2 + 5x - 6) - (10x^2 - 7x + 4)$
23. $(a^2 - 8a - 6) - (2a - 5a^2 + 1)$
24. $(3x - x^2 + y^2) - (y^2 - 9x^2 - 6x)$
25. $(2y + y^2 - x^2) - (3x^2 - 4y + 5y^2)$
26. $(b^2 + a^2 - 4ab) - (a^2 - b^2 + 2ab)$

Write in simplest form.

27. $(6x - 1) + (20x - 16) - (6x + 14)$

28. $(-21x - 6) + (12x + 25) - (8x - 5)$

29. $(3x - 3y) + (3x - 9y) - (14x + 8y)$

30. $(x^2 + 2x - 1) + (12x^2 - 16x + 5) - (-8x^2 + 9x + 5)$

31. $(3x^2 - 5xy + y^2) - (18x^2 - 12xy + 7y^2) - (-10x^2 + 7xy - 8y^2)$

32. $(8x^2 + 3xy + 4y^2) + (18x^2 - 24xy + 9y^2) - (-10x^2 - 7xy - 20y^2)$

Write an algebraic expression for

33. x is the sum of three times a number, represented by b, plus four added to a quantity that consists of the negative of six times the same number added to five.

34. b is equal to the quantity x minus four subtracted from the quantity twice x squared plus x.

REVIEW 2

NOW THAT YOU HAVE COMPLETED UNIT 2, YOU SHOULD BE ABLE TO:

Use the following definitions in your working vocabulary.

Polynomial—A sum of one or more terms, each consisting of a constant or product of a constant and one or more variables raised to a whole number power.

Polynomial in one variable—A polynomial whose terms contain only one variable.

Monomial—A polynomial with exactly one term.

Binomial—A polynomial with exactly two terms.

Trinomial—A polynomial with exactly three terms.

Numerical coefficient—The numerical factor in a monomial.

Like terms—Terms that have identical letters and exponents.

Zero exponent—Any quantity except zero raised to the zero power equals one.

Apply the following rules and definitions.

To combine like terms, add the numerical coefficients of the like terms, then multiply by the common variables.

To multiply powers with like bases, add the exponents.

$$a^x \cdot a^y = a^{x+y}$$

To evaluate polynomials, substitute the given value for the variable and follow the rules for order of operations.

To multiply monomials, first multiply the coefficients and then multiply the variables.

To multiply a polynomial by a monomial, multiply every term in the polynomial by the monomial.

To remove a symbol of grouping preceded by a negative sign, multiply the quantity within the grouping symbols by -1.

To add polynomials, remove the parentheses and combine like terms, either horizontally or vertically.

To subtract polynomials, add the additive inverse or negative of the quantity to be subtracted, either horizontally or vertically.

Review Test 2

Write the following in exponential form. (2.1)

1. $7 \cdot 7 \cdot 7 \cdot 7 =$ ________

2. $(-6)(-6)(-6) =$ ________

3. $3 \cdot 3 \cdot a \cdot b \cdot b =$ ________

4. $(-2)(-2) \cdot a \cdot a \cdot a \cdot b =$ ________

Evaluate the following if $a = -2$, $b = 3$, $c = -4$. (2.1)

5. $4a - 5 =$ ________

6. $a^3 - 3c =$ ________

7. $2a^2 + a - 6 =$ ________

8. $a^3(a^3b^2 - c) =$ ________

Evaluate the following polynomial for the given replacement values of x by completing the table. (2.1)

9.

x	$x^2 - 2$
-4	
-2	
0	
1	
3	

10.

x	$2x^3 - x + 4$
-2	
-1	
0	
1	
2	

Multiply the following. (2.2)

11. $(-4x^3)(3y^2) =$ ________

12. $-x^2 \cdot (5a^4) =$ ________

13. $(-9x^4) \cdot (-3y) =$ ________

14. $2xy \cdot 3x^2y =$ ________

15. $(-3a^0b^2c)(-5a^3b^0c^5) =$ ________

16. $-a^0 \cdot b^0 \cdot 6a^2b^0c^3 =$ ________

***Multiply using the distributive property.* (2.3)**

17. $-3(4x - 6) =$ ________

18. $ab^2(a^2b + 2ab) =$ ________

19. $x^3(x^2 - x + 3) =$ ________

20. $-2x(x^2 + 4x + 6) =$ ________

21. $-8x^3y^3(3xy^4 - 4x^2y^5) =$ ________

22. $-2a^2bc^4(-ba^2 + 5b^3 - 4a^4b^0c^2) =$ ________

***Add or subtract the following polynomials.* (2.4)**

23. $(6x - 1) + (-2x + 4) =$ ________

24. $(3x + 1) - (4x - 5) =$ ________

25. $(2a^2 + a + 3) + (a^2 - 4a + 4) =$ ________

26. $(6y^2 - y - 1) - (y^2 - y + 1) =$ ________

27. $(x - 2x^2 + 2) + (4x^2 + x - 5) =$ ________

28. $(10a + 6b + 5) - (3a - 10 + 6b) =$ ________

29. $(3x + 4) + (6 - 2x) - (3x - 1) =$ ________

30. $(-2x^2 + 3y - 5) - (-10x^2 + 6x + 5 - 9y) - (7x^2 - 6x + 9y) =$ ________

***Simplify the following.* (2.4)**

31. $-8x^4 + 5x^2 - 3 - 7x^2 - x^4 + 2x =$ ________

32. $-5x + 4y - 7xy + 2x^2y - 3y + 6xy + 4x - 3x^2y =$ ________

UNIT 3

RATIONAL NUMBERS

OBJECTIVES

After you have successfully completed this unit, you will be able to:

1. Write a fraction in standard form (3.1)
2. Write a fraction with a specific denominator that is equivalent to a given fraction (3.1)
3. Reduce a fraction to lowest terms (3.1)
4. Multiply two rational numbers (3.2)
5. Use the distributive property with rational numbers (3.2)
6. Find the multiplicative inverse or reciprocal of a nonzero rational number (3.2)
7. Divide two rational numbers (3.2)
8. Add or subtract rational numbers with like denominators (3.3)
9. Add or subtract rational numbers with unlike denominators (3.3)
10. Evaluate algebraic expressions involving rational numbers (3.4)

3.1 EQUIVALENT FRACTIONS

So far we have worked only with natural numbers, whole numbers, and integers. The answer to any addition, subtraction, or multiplication problem using integers is an integer. However, integers do not provide an answer for all division problems.

$$3 + 7 = 10 \qquad 3 - 7 = -4 \qquad 3 \cdot 7 = 21 \qquad 3 \div 7 = ?$$

To provide an answer for $3 \div 7$, we define a way to write the answer to any division. The result of $3 \div 7$ will be written as $\frac{3}{7}$. Similarly, $-2 \div 3$ will be written $\frac{-2}{3}$. This leads to the definition of rational numbers.

3.1A DEFINITION

Rational number

A *rational number* is any number that can be written as the quotient of two integers.

A rational number can be written in the form $\frac{a}{b}$, where a and b are integers and b does not equal zero, written $b \neq 0$.

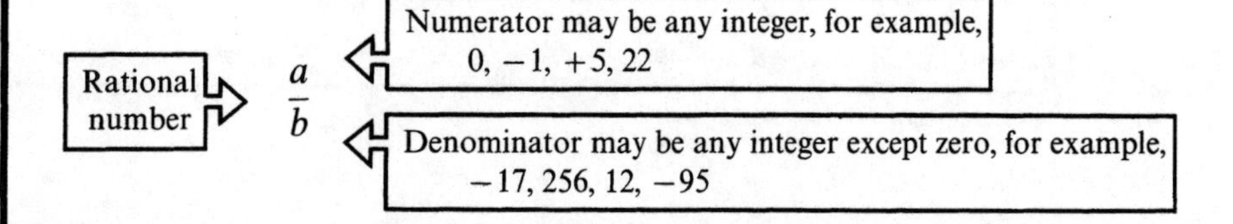

Rational numbers contain all the fractions that can be formed by using integers for the numerator and the denominator.

$\frac{2}{5}, \frac{-7}{8}, \frac{2375}{-16}, \frac{17}{3}$ are all rational numbers

Because the rational number $\frac{3}{1} = 3 \div 1 = 3$, the integer 3 is a rational number.

The set of rational numbers contains the set of integers.

Although 0 is a rational number $\left(0 = \frac{0}{1}\right)$, $\frac{1}{0}$ is not a number because division by zero is not allowed.

Before defining equivalent fractions, we need some preliminary definitions.

Just as there is an identity element for addition, there is an identity element for multiplication.

Identity element for multiplication

3.1B DEFINITION

The *identity element for multiplication* is 1.

$a \cdot 1 = a$ or $1 \cdot a = a$

That is, any number multiplied by 1 yields the original number.

Why do we need so many definitions?

So we know exactly what we are talking about.

EXAMPLE 1

$3 \cdot 1 = 3$ $\qquad (x + y) \cdot 1 =$ ____

$-5 \cdot 1 =$ ____ $\qquad 1 \cdot (3 + a) =$ ____

$x \cdot 1 =$ ____ $\qquad \frac{3}{2} \cdot 1 =$ ____

$x + y$

-5 $\quad 3 + a$

x $\quad \frac{3}{2}$

Product of rational numbers

3.1C DEFINITION

To find the product of two rational numbers, first multiply the numerators and then multiply the denominators.

The *product of two rational numbers* $\frac{a}{b}$ and $\frac{c}{d}$ is $\frac{a \cdot c}{b \cdot d}$.

$$\frac{a}{b} \cdot \frac{c}{d} = \frac{a \cdot c}{b \cdot d}$$

EXAMPLE 2

$$\frac{3}{8} \cdot \frac{5}{2} = \frac{3 \cdot 5}{8 \cdot 2}$$
$$= \frac{15}{16}$$

Notice the left side of each of these equations is the product of two separate fractions. The right side is a single fraction with the two factors in the numerator and denominator.

$$\frac{2}{3} \cdot \frac{x}{5} = \frac{2 \cdot x}{3 \cdot 5}$$
$$= \frac{2x}{15}$$

To change the form of a fraction, that is, to reduce the fraction to lower terms or build it to higher terms, use the following rule.

Equivalent fractions

3.1D DEFINITION: EQUIVALENT FRACTIONS

The rational number $\frac{a}{b}$ is equivalent to the rational number $\frac{c}{d}$ if

$$\frac{a}{b} = 1 \cdot \frac{c}{d}$$

But multiplying by 1 doesn't change anything.

That's the point! However, if we write 1 as $\frac{2}{2}, \frac{3}{3}, \frac{-4}{-4}$, things will look different.

When two rational numbers are equal but are written in different forms, they are said to be equivalent.

The multiplication property of 1 is used to write fractions that are equivalent to a given fraction.

EXAMPLE 3 Write a fraction equivalent to $\frac{3}{4}$ with a denominator of 8.

First ask yourself "What would I multiply 4 by to get 8?"

2

$$4 \cdot \_\_\_\_ = 8$$

Therefore, write 1 as $\frac{2}{2}$, and multiply $\frac{3}{4}$ by $\frac{2}{2}$ to produce a fraction with a denominator of 8.

$$\frac{3}{4} = 1 \cdot \frac{3}{4}$$
$$\downarrow$$
$$\frac{3}{4} = \frac{2}{2} \cdot \frac{3}{4}$$
$$\frac{3}{4} = \frac{6}{8}$$

Can I use any numbers I want to write 1?

Yes, except for zero, as long as the numerator is the same as the denominator. $\frac{-4}{-4}, \frac{23}{23}, \frac{5}{5}, \frac{-7}{-7}$ are all equal to 1.

EXAMPLE 4 Write a fraction equivalent to $\frac{-3a}{5}$ with a denominator of 15.

$3, \frac{3}{3}$

$$\frac{-3a}{5} = \frac{?}{15} \qquad 5 \cdot \_\_\_\_ = 15 \qquad \text{Therefore, write 1 as } \_\_\_\_.$$

$$\frac{-3a}{5} = \frac{-3a}{5} \cdot \boxed{1}$$

$$\frac{-3a}{5} = \frac{-3a}{5} \cdot \boxed{\frac{3}{3}}$$

That's because
$\frac{-3a}{5} \cdot 1 = \frac{-9a}{15}$

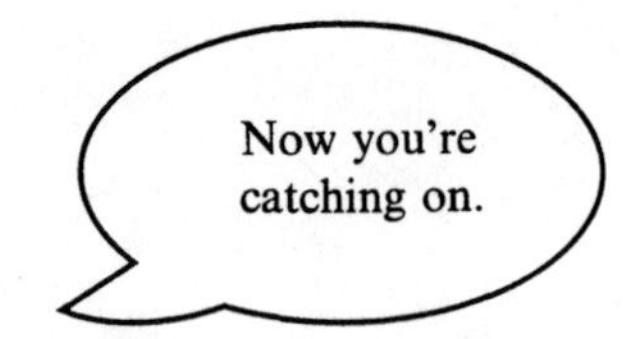

$\frac{-9a}{15}$

$$\frac{-3a}{5} = \_\_\_\_$$

PROBLEM SET 3.1A

Write a fraction equivalent to the given fraction with the denominator indicated.

1. $\frac{1}{2}$ with denominator 6
2. $\frac{1}{2}$ with denominator 22
3. $\frac{2}{3}$ with denominator 21
4. $\frac{2}{3}$ with denominator 24
5. $\frac{5}{6}$ with denominator 24
6. $\frac{5}{6}$ with denominator 48
7. $\frac{-6}{7}$ with denominator 56
8. $\frac{-6}{7}$ with denominator 28
9. $\frac{-3a}{4}$ with denominator 32
10. $\frac{-3a}{4}$ with denominator 24
11. $\frac{-4x}{-7}$ with denominator 28
12. $\frac{-4x}{7}$ with denominator 28
13. $\frac{4x}{-7}$ with denominator 28
14. $\frac{3a}{-4}$ with denominator 24

Find the missing numerator or denominator.

15. $\frac{3}{4} = \frac{}{24}$
16. $\frac{5}{6} = \frac{}{30}$
17. $\frac{5}{8} = \frac{}{40}$
18. $\frac{-7}{8} = \frac{}{72}$
19. $\frac{6}{11} = \frac{}{99}$
20. $\frac{4}{13} = \frac{12}{}$
21. $\frac{-3}{5} = \frac{-12}{}$
22. $\frac{-2}{7} = \frac{}{42}$
23. $\frac{-5x}{9} = \frac{}{54}$
24. $\frac{4x}{5} = \frac{}{40}$
25. $\frac{-8a}{9} = \frac{56a}{}$
26. $\frac{-5y}{12} = \frac{-30y}{}$
27. $\frac{-6xy}{7} = \frac{}{56}$
28. $\frac{-2ax}{3} = \frac{}{45}$
29. $\frac{8xa}{9} = \frac{}{81}$
30. $\frac{5ay}{14} = \frac{}{70}$
31. $\frac{-10ay}{11} = \frac{}{88}$
32. $\frac{-13bx}{18} = \frac{}{72}$

Standard Form of Fractions

We write rational numbers with as few negative signs as possible. $\frac{-a}{-b}$ can be simplified by multiplying it by 1 written in the form $\left(\frac{-1}{-1}\right)$.

$$\frac{-a}{-b} = \frac{-a}{-b}\left(\frac{-1}{-1}\right) = \frac{a}{b}$$

We also like to avoid negative signs in the denominator. $\frac{a}{-b}$ can be simplified using multiplication by 1.

$$\frac{a}{-b} = \frac{a}{-b}\left(\frac{-1}{-1}\right) = \frac{-a}{b}$$

Standard form of a fraction

In general, if a and b represent positive numbers, the *standard form of a fraction* is

$\frac{a}{b}$ for positive fractions

$\frac{-a}{b}$ for negative fractions

What does $-\frac{1}{2}$ mean? The negative sign isn't in either the numerator or the denominator.

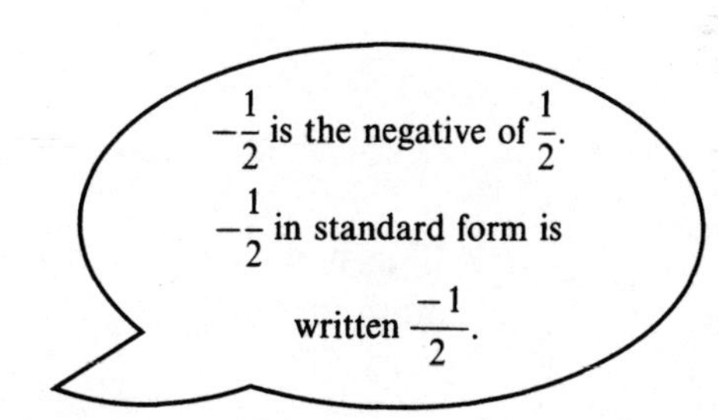

EXAMPLE 5 Indicate whether the following equalities are true (T) or false (F).

T, F, **F**

1. $\frac{-5}{-6} = \frac{5}{6}$ ____
2. $\frac{-5}{6} = \frac{5}{6}$ ____
3. $\frac{5}{-6} = \frac{5}{6}$ ____

T, T, F

4. $-\frac{6}{7} = \frac{-6}{7}$ ____
5. $\frac{6}{-7} = \frac{-6}{7}$ ____
6. $\frac{-6}{7} = \frac{-6}{-7}$ ____

T, F, T

7. $\frac{x}{-2} = \frac{-x}{2}$ ____
8. $\frac{a}{3} = \frac{-a}{3}$ ____
9. $\frac{-4x}{-5} = \frac{4x}{5}$ ____

EXAMPLE 6 Write the following fractions in standard form.

$\frac{-1}{2}, \frac{1}{2}, \frac{-1}{2}$

1. $-\frac{1}{2} =$ ____
2. $\frac{-1}{-2} =$ ____
3. $\frac{1}{-2} =$ ____

$\frac{3}{4}, \frac{-x}{3}, \frac{-2x}{3}$

4. $\frac{-3}{-4} =$ ____
5. $\frac{-x}{3} =$ ____
6. $\frac{2x}{-3} =$ ____

Reducing Fractions

A fraction with no common factor except 1 in the numerator and denominator is called a fraction in lowest terms. Such a fraction is said to be reduced.

Reducing fractions

3.1E RULE: TO REDUCE FRACTIONS TO LOWEST TERMS

1. Write the numerator and denominator in factored form.
2. Remove the common factors from the numerator and denominator or divide the numerator and denominator by the same number. Continue the process until there is no common factor except 1 in the numerator and denominator.

EXAMPLE 7 Reduce $\frac{9}{12}$ to lowest terms.

$$\begin{aligned}\frac{9}{12} &= \frac{3 \cdot 3}{4 \cdot 3}\\ &= \frac{3}{4} \cdot \frac{3}{3}\\ &= \frac{3}{4} \cdot 1\\ &= \frac{3}{4}\end{aligned}$$

The work can be shortened by dividing the numerator and denominator by the same number.

$$\frac{\overset{3}{\cancel{9}}}{\underset{4}{\cancel{12}}} = \frac{3}{4}$$ Divide numerator and denominator by 3

EXAMPLE 8 Reduce $\frac{-18a}{30}$ to lowest terms.

$$\frac{-18a}{30} = \frac{-3a \cdot 6}{5 \cdot 6}$$ Rewrite showing largest common factor

$$= \frac{-3a}{5} \cdot \frac{6}{6}$$ Write as product of two fractions

$$= \frac{-3a}{5} \cdot 1$$ Identify multiplication by one

$$= \frac{-3a}{5}$$

Using the shorter method, we have

6

$$\frac{\overset{-3a}{\cancel{-18a}}}{\underset{5}{\cancel{30}}} = \frac{-3a}{5}$$ Divide numerator and denominator by ______

EXAMPLE 9 Reduce $\frac{-4x}{-12}$ to lowest terms. Write the answer in standard form.

$\frac{-4x}{-12} = \frac{-4 \cdot x}{-4 \cdot 3}$ Identify largest common factor

$= \frac{-4}{-4} \cdot \frac{x}{3}$ Write as product of two fractions

$= 1 \cdot \frac{x}{3}$ Identify multiplication by one

$= \frac{x}{3}$

Using the shorter method, we have

$\frac{x}{3}$

$$\frac{\overset{-x}{\cancel{-4x}}}{\underset{-3}{\cancel{-12}}} = \frac{-x}{-3} = \text{\_\_\_\_\_\_ in standard form}$$

PROBLEM SET 3.1B

Reduce the following fractions to lowest terms by first finding a number equal to 1. Write the answers in standard form.

1. $\frac{8}{12}$ 2. $\frac{9}{15}$ 3. $\frac{12}{16}$ 4. $\frac{15}{18}$ 5. $\frac{-18}{21}$

6. $\frac{24}{-28}$ 7. $\frac{-33}{-77}$ 8. $\frac{-30}{-48}$ 9. $-\frac{36}{48}$ 10. $-\frac{56}{72}$

Reduce the following fractions to lowest terms using the shorter method. Write the answers in standard form.

11. $\frac{12}{16}$ 12. $\frac{12}{18}$ 13. $\frac{18}{24}$ 14. $\frac{24}{33}$ 15. $\frac{-34}{51}$

16. $\frac{42}{-48}$ 17. $\frac{-84x}{72}$ 18. $-\frac{96a}{64}$ 19. $\frac{-72}{-64}$ 20. $\frac{-56b}{-96}$

21. $-\frac{39}{65}$ 22. $\frac{-12ab}{60}$ 23. $-\frac{88xy}{33}$ 24. $\frac{60}{-96}$

3.2 MULTIPLICATION AND DIVISION OF RATIONAL NUMBERS

Multiplication of Rational Numbers

In the previous section, we were reminded that to multiply fractions we multiply their numerators first and then multiply their denominators. That is,

$$\frac{a}{b} \cdot \frac{c}{d} = \frac{a \cdot c}{b \cdot d}$$

EXAMPLE 10 Multiply $\frac{3}{4} \cdot \frac{7}{8} = \frac{3 \cdot 7}{4 \cdot 8}$.

21

$$= \frac{\quad}{32}$$

EXAMPLE 11 Multiply $\frac{8}{9} \cdot \frac{5}{4} = \frac{40}{36}$.

Because this result is not in lowest terms, we reduce it.

$$\frac{40}{36} = \frac{10 \cdot 4}{9 \cdot 4}$$

$$= \frac{10}{9} \cdot \frac{4}{4}$$

$$= \frac{10}{9} \cdot 1$$

$$= \frac{10}{9}$$

In this example, we first multiplied two fractions and then reduced the result. A shorter method is to divide out common factors before multiplying.

$$\frac{\overset{2}{\cancel{8}}}{9} \cdot \frac{5}{\underset{1}{\cancel{4}}} = \frac{10}{9}$$ Divide out the common factor 4

EXAMPLE 12 Multiply $\frac{12a}{15} \cdot \frac{10y}{7}$.

Start by looking for factors of any denominator that will divide into any numerator. 3 and 5 are factors of 15. 3 will divide into 12 and 5 will divide into 10.

8ay

$$\frac{\overset{4a}{\cancel{12}a}}{\underset{\underset{1}{\cancel{5}}}{\cancel{15}}} \cdot \frac{\overset{2y}{\cancel{10}y}}{7} = \frac{\quad}{7}$$ Divide out the common factor 3, then divide out the common factor 5

-$6x^3$

EXAMPLE 13 Multiply $-8x \cdot \frac{3x^2}{4}$.

Before we multiply, we write $-8x$ as the fraction $\frac{-8x}{1}$

$$-8x \cdot \frac{3x^2}{4} = \frac{\overset{-2x}{\cancel{-8x}}}{1} \cdot \frac{3x^2}{\underset{1}{\cancel{4}}}$$

$$= \frac{-6x^3}{1}$$

$$= \_\_\_\_\_$$

Many people like to determine the sign of the product first when multiplying expressions involving a combination of positive and negative signs. The sign of any product will be positive if there is an even number of negative signs and negative if there is an odd number of negative signs in the factors of the product.

EXAMPLE 14 Multiply $\frac{-15}{8} \cdot \frac{4}{-3}$.

$$\frac{-15}{8} \cdot \frac{4}{-3} = +\left(\frac{15}{8} \cdot \frac{4}{3}\right)$$ The problem has two negative signs, therefore the answer is positive

$$= \frac{\overset{5}{\cancel{15}}}{\underset{2}{\cancel{8}}} \cdot \frac{\overset{1}{\cancel{4}}}{\underset{1}{\cancel{3}}}$$ Recopy the problem before dividing out like factors

$$= \frac{5}{2}$$ Find the product

EXAMPLE 15 Multiply $\frac{-4a}{3} \cdot \frac{5}{6}$.

$$\frac{-4a}{3} \cdot \frac{5}{6} = -\left(\frac{4a}{3} \cdot \frac{5}{6}\right)$$ First determine the sign of the product. It is negative

$$= -\left(\frac{\overset{2a}{\cancel{4a}}}{3} \cdot \frac{5}{\underset{3}{\cancel{6}}}\right)$$ Divide out like factors

$$= \frac{-10a}{9}$$ Find the product

PROBLEM SET 3.2A

Determine the sign *only* of the following products. Do not find the products.

1. $\frac{2}{3} \cdot \frac{3}{4}$
2. $\frac{-3a}{5} \cdot \frac{6}{7}$
3. $\frac{4x}{-9} \cdot \frac{5}{7}$
4. $\frac{5y}{6} \cdot \frac{-8}{12}$
5. $\frac{8x}{5} \cdot \frac{-7b}{-9}$
6. $\frac{-5a}{4} \cdot \frac{6}{-7}$
7. $\frac{6a}{-5} \cdot \frac{-11}{-14}$
8. $\frac{3x}{-4} \cdot \frac{5}{6}$
9. $\frac{-10y}{-3} \cdot \frac{-4}{5}$
10. $\frac{-7x}{-4} \cdot \frac{-3}{-5}$

PROBLEM SET 3.2B

Multiply the following fractions. Write all answers in lowest terms. Use the standard form for fractions.

1. $\frac{2}{5} \cdot \frac{3}{4}$
2. $\frac{3}{5} \cdot \frac{10}{7}$
3. $\frac{3}{7} \cdot \frac{5}{6}$
4. $\frac{4}{9} \cdot \frac{3}{8}$
5. $\frac{-4}{5} \cdot \frac{5}{12}$
6. $\frac{7}{-8} \cdot \frac{3}{14}$
7. $16 \cdot \frac{7}{8}$
8. $-6 \cdot \frac{5}{12}$
9. $\frac{-15}{8} \cdot \frac{8a}{-5}$
10. $\frac{6x}{-7} \cdot \frac{-14}{15}$
11. $\frac{-12x}{5} \cdot \frac{10}{7}$
12. $\frac{8a}{15} \cdot \frac{5}{-12}$
13. $\frac{-4x}{25} \cdot 5$
14. $\frac{6a}{11} \cdot \frac{22a}{36}$
15. $\frac{-8y}{-15} \cdot \frac{-21}{14}$
16. $\frac{12x}{-18} \cdot \frac{-9}{-10}$
17. $\frac{12a}{-16} \cdot \frac{-8b}{-15}$
18. $\frac{-22x^2}{15} \cdot \frac{25y}{-33}$
19. $\frac{16x}{-7} \cdot \frac{49x}{-4}$
20. $-7m \cdot \frac{-18m^2}{28}$
21. $\frac{-9a}{12} \cdot 4ab$
22. $\frac{27ax}{-16} \cdot \frac{-24x^2}{9}$
23. $-3ax \cdot \frac{-44ax^2}{-18}$
24. $\frac{14ab}{-16} \cdot \frac{-a}{21}$

Review Your Skills

Perform the indicated operations.

25. $(7)(6)$
26. $(-8)(-4)$
27. $(-9)\left(-\frac{2}{3}\right)$
28. $\left(-\frac{5}{6}\right)(12)$

Distributive Property Used with Fractions

In later work with equations, we will need to multiply the sum of two fractions by another number. To do this, we use the distributive property first introduced in Unit 1.

Distributive property

3.2A DISTRIBUTIVE PROPERTY OF MULTIPLICATION OVER ADDITION

For all algebraic expressions a, b, and c,

$$a(b + c) = a \cdot b + a \cdot c \quad \text{and} \quad (b + c)a = b \cdot a + c \cdot a$$

Here are some examples of how to use the distributive property with fractions.

EXAMPLE 16 Use the distributive property to simplify $6\left(\frac{1}{2}+\frac{1}{3}\right)$.

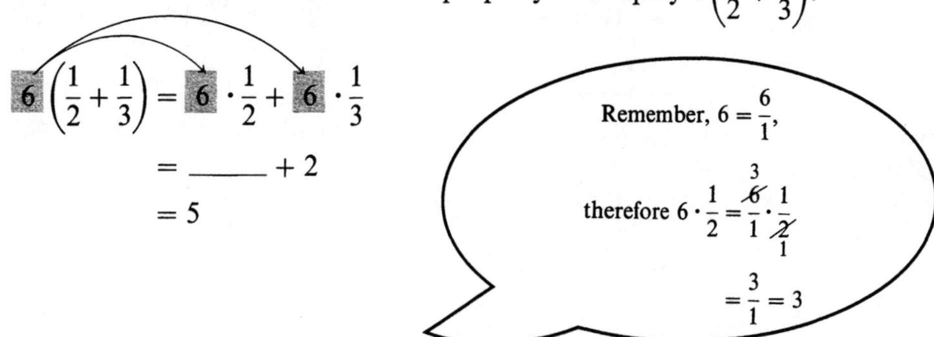

3 $\quad = \underline{\qquad} + 2$

$= 5$

EXAMPLE 17 Use the distributive property to write the product $8\left(\frac{a}{2}+\frac{1}{4}\right)$ as a sum.

$$8\left(\frac{a}{2}+\frac{1}{4}\right) = 8\cdot\frac{a}{2} + 8\cdot\frac{1}{4}$$

$$= \frac{8}{1}\cdot\frac{a}{2} + \frac{8}{1}\cdot\frac{1}{4}$$

4a $\quad = \underline{\qquad} + 2$

EXAMPLE 18 Use the distributive property to write the product $\frac{1}{6}(12a - 36b)$ as a sum or difference.

$$\frac{1}{6}(12a - 36b) = \frac{1}{6}(12a) - \frac{1}{6}(36b)$$

$$= \frac{1}{6}\left(\frac{12a}{1}\right) - \frac{1}{6}\left(\frac{36b}{1}\right)$$

2a − 6b $\quad = \underline{\qquad\qquad}$

EXAMPLE 19 Use the distributive property to write the product $-12\left(\frac{x}{6}-\frac{1}{4}\right)$ as a sum.

$$-12\left(\frac{x}{6}-\frac{1}{4}\right) = (-12)\left(\frac{x}{6}+\frac{-1}{4}\right)$$ Rewrite the subtraction as addition of the negative

$$= \left(\frac{-12}{1}\right)\left(\frac{x}{6}\right) + \left(\frac{-12}{1}\right)\left(\frac{-1}{4}\right)$$

$$= -2x + 3$$

In Example 19 you rewrote the subtraction as an addition. In Example 18 you left it as a subtraction. Why?

Notice in Example 19 we're multiplying by a negative number. In this case I find it easier to rewrite the quantity inside the parentheses as a sum.

EXAMPLE 20 Use the distributive property to rewrite $-\frac{2}{5}(10x - 15)$.

$$-\frac{2}{5}(10x - 15) = \left(\frac{-2}{5}\right)[(10x) + (-15)]$$

$$= \frac{-2}{5}(10x) + \frac{-2}{5}(-15)$$

$$= -4x + 6$$

Remember $-\frac{2}{5} = \frac{-2}{5}$

EXAMPLE 21 Use the distributive property to rewrite $3a\left(4a - \frac{1}{12}\right)$.

$$3a\left(4a - \frac{1}{12}\right) = 3a \cdot 4a - 3a \cdot \frac{1}{12}$$

$$= 12\_\_\_\_\_ - \_\_\_\_\_$$

$a^2, \frac{a}{4}$

EXAMPLE 22 Use the distributive property to rewrite $-\frac{a}{6}(12a - 9x)$.

$$-\frac{a}{6}(12a - 9x) = -\frac{a}{6}[12a + (-9x)]$$ Rewrite the subtraction as an addition

$$= \left(-\frac{a}{6}\right)12a + \left(-\frac{a}{6}\right)(-9x)$$

$$= \frac{-a}{\cancel{6}_1} \cdot \frac{\cancel{12a}^{2a}}{1} + \frac{a}{\cancel{6}_2} \cdot \frac{\cancel{9x}^{3x}}{1}$$

$$= \_\_\_\_\_ + \frac{3ax}{2}$$

$-2a^2$

EXAMPLE 23 Use the distributive property to rewrite $-\left(\frac{a}{3} - \frac{b}{4}\right)$ without parentheses. Since $-x$ is equivalent to $-1 \cdot x$,

$$-\left(\frac{a}{3} - \frac{b}{4}\right) = -1\left(\frac{a}{3} + \frac{-b}{4}\right)$$

$$= \_\_\_\_\_ \cdot \frac{a}{3} + \_\_\_\_\_\left(\frac{-b}{4}\right)$$

$$= \frac{-a}{3} + \frac{b}{4}$$

$$= \_\_\_\_\_\_\_$$

$-1, -1$

$\frac{-a}{3} + \frac{b}{4}$

We can also treat this as removing a grouping symbol preceded by a minus sign: $-\left(\frac{a}{3} - \frac{b}{4}\right) = \frac{-a}{3} + \frac{b}{4}$

POINTERS FOR BETTER STYLE

We get valuable hints about what was done (or not done) to an equation by the way we write each step. There are two basic ways to modify an equation—across and down.

Consider this example:

How it's written	**What to notice**
$-a\left(12a - \frac{1}{5}\right) = -a\left[12a + \left(\frac{-1}{5}\right)\right]$	Changes were made ACROSS the equal sign. Compare each side, symbol by symbol. $-a(12a$ on the left is equivalent to $-a[12a$ on the right. Therefore, the rest of the line on each side must also be equivalent. This makes it easier to see that $\frac{-1}{5}$ is equivalent to $+\left(\frac{-1}{5}\right)$.
$= (-a)(12a) + (-a)\left(\frac{-1}{5}\right)$	Changes were made DOWN the column. In this case the distributive property was used. When we write the result directly below each term, it is easier to see $(-a)$ was distributed.
$= \quad -12a^2 + \frac{a}{5}$	This is the line above, simplified. $(-a)(12a)$ was replaced with $-12a^2$. $(-a)\left(\frac{-1}{5}\right)$ was replaced with $\left(\frac{a}{5}\right)$ directly below.

PROBLEM SET 3.2C

Use the distributive property to write the following without parentheses.

1. $4\left(\frac{1}{2} + \frac{1}{4}\right)$
2. $8\left(\frac{1}{4} + \frac{3}{8}\right)$
3. $8\left(\frac{1}{2} + \frac{3}{4}\right)$
4. $6\left(\frac{1}{3} + \frac{1}{2}\right)$
5. $10\left(\frac{3}{5} - \frac{1}{2}\right)$
6. $12\left(\frac{5}{6} - \frac{2}{3}\right)$
7. $12\left(\frac{a}{4} + \frac{5}{6}\right)$
8. $15\left(\frac{3}{5} - \frac{2}{3}\right)$
9. $-3\left(\frac{x}{3} - 1\right)$
10. $-4\left(x + \frac{1}{2}\right)$
11. $-\frac{1}{2}(12 + 6a)$
12. $-\frac{1}{3}(15a - 12)$
13. $-\left(\frac{2a}{3} + \frac{b}{2}\right)$
14. $-\left(\frac{6x}{7} - \frac{7y}{9}\right)$
15. $-12\left(\frac{1}{2} - \frac{a}{3}\right)$
16. $-8\left(\frac{5a}{2} - b\right)$
17. $-\frac{3}{4}(-12 + 4a)$
18. $-\frac{5}{6}(-12b - 18)$
19. $-\frac{3}{4}[16 - (-8a)]$

20. $-6\left(-\frac{1}{2}-\frac{1}{3}\right)$

21. $-\frac{7}{8}(-4a+12)$

22. $-\frac{8}{9}(-6a-18)$

23. $2a\left(\frac{a}{8}+\frac{1}{4}\right)$

24. $-3a\left(\frac{2}{3}a-\frac{5}{6}\right)$

25. $\frac{2a}{3}(6a-9x)$

26. $-\frac{x}{8}\left(-24x+\frac{12a}{25}\right)$

27. $-\frac{3x}{7}\left(-21x-\frac{14a}{3}\right)$

28. $-\frac{5a}{12}\left(6a-\frac{28x}{15}\right)$

Division of Rational Numbers

To carry out the division of rational numbers, we need to define multiplicative inverse. Before we do, recall that the additive inverse of a number is what you would add to the number to get zero (the additive identity) for an answer. The symbol for the additive inverse of a is a with a negative sign in front of it. And the additive inverse of $-a$ is a; hence,

$$a+(-a)=0$$

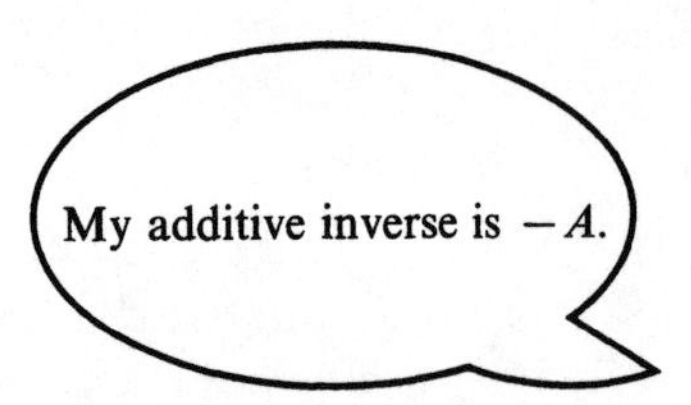

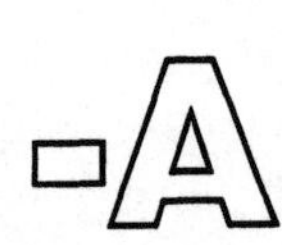

Similarly, there is a multiplicative inverse for every number except zero. The symbol for the multiplicative inverse of a number a is $\frac{1}{a}$.

Multiplicative inverse

3.2B DEFINITION

Two rational numbers are *multiplicative inverses* of each other if their product is 1.

$$a \cdot \frac{1}{a} = 1 \qquad \frac{a}{b} \cdot \frac{b}{a} = 1$$

Multiplicative inverses | Multiplicative inverses

Reciprocal

Multiplicative inverses are sometimes called *reciprocals.*

The sum of a number and its additive inverse gives the additive identity element, zero. The product of a number and its multiplicative inverse gives the multiplicative identity element, 1.

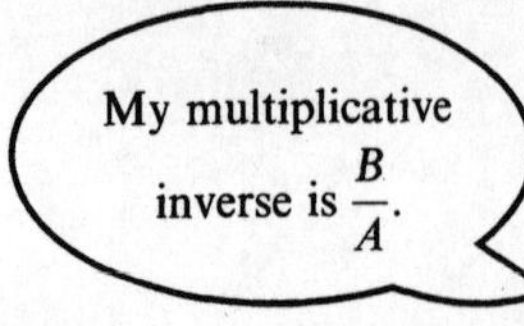

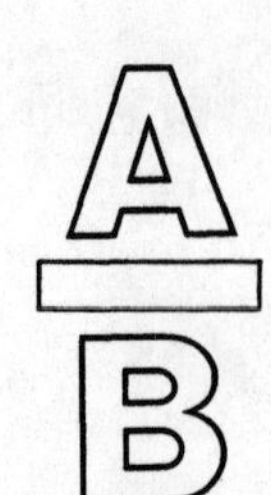

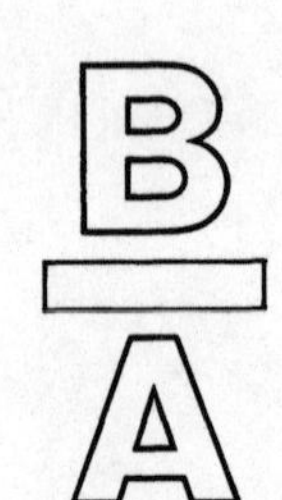

My multiplicative inverse is $\frac{A}{B}$.

EXAMPLE 24 The multiplicative inverse of 3 is $\frac{1}{3}$.

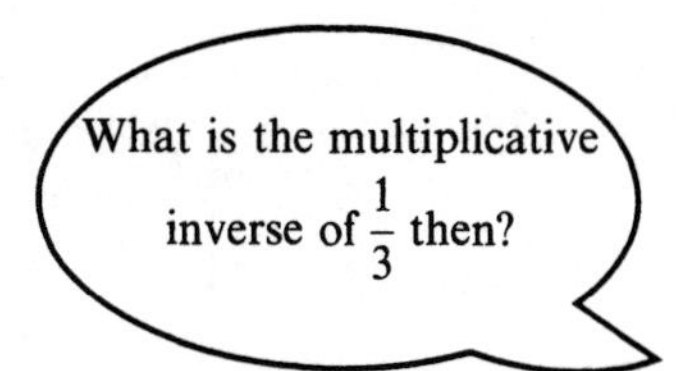

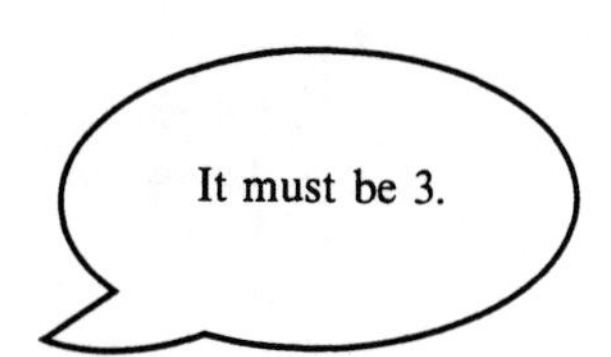

The multiplicative inverse of $x + y$ is $\frac{1}{x + y}$.

EXAMPLE 25

The reciprocal of 3 is $\frac{1}{3}$.

The reciprocal of $\frac{1}{4}$ is 4.

The reciprocal of a is $\frac{1}{a}$.

The reciprocal of $\frac{3}{5}$ is $\frac{5}{3}$.

The reciprocal of $\frac{-2}{3}$ is $\frac{-3}{2}$.

By definition, the reciprocal of $\frac{-2}{3}$ is $\frac{3}{-2}$. However, we prefer to write fractions in standard form. The standard form of $\frac{3}{-2}$ is $\frac{-3}{2}$.

We now use the multiplicative inverse to define the division of rational numbers.

Division using the multiplicative inverse

3.2C RULE

To divide by a number, multiply by its multiplicative inverse.

$$a \div b = a \cdot \frac{1}{b} \qquad \text{also} \qquad \frac{a}{b} \div \frac{c}{d} = \frac{a}{b} \cdot \frac{d}{c}$$

Here is a brief proof of the rule for division. (It is optional.)

Another way to write $a \div b$ is $\frac{a}{b}$, because the fraction bar "—" means division.

$a \div b = \frac{a}{b}$

$= \frac{a}{b} \cdot 1$ Multiplication by 1 doesn't change anything

$= \frac{a \cdot 1}{b \cdot 1}$ Definition of multiplication of rational numbers

$= \frac{a \cdot 1}{1 \cdot b}$ Commutative property of multiplication

$= \frac{a}{1} \cdot \frac{1}{b}$ Definition of multiplication of rational numbers

$a \div b = a \cdot \frac{1}{b}$ Anything divided by 1 equals itself

An example from arithmetic illustrates that to divide by a number we multiply by its reciprocal. Six divided by two is the same as six times one-half.

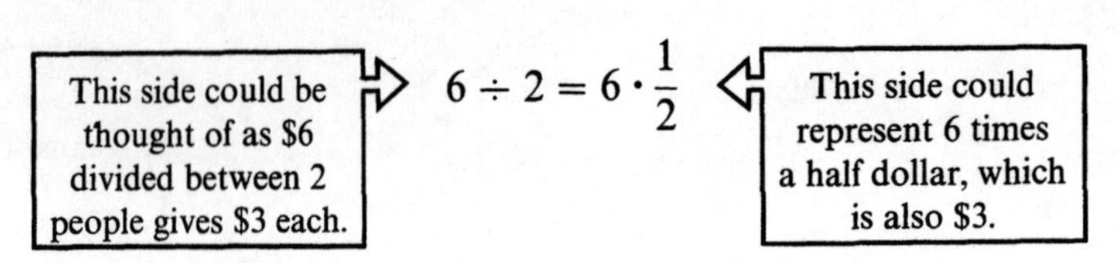

Notice we have a way to convert division problems to multiplication problems.

EXAMPLE 26 Divide $\frac{1}{2} \div 6$.

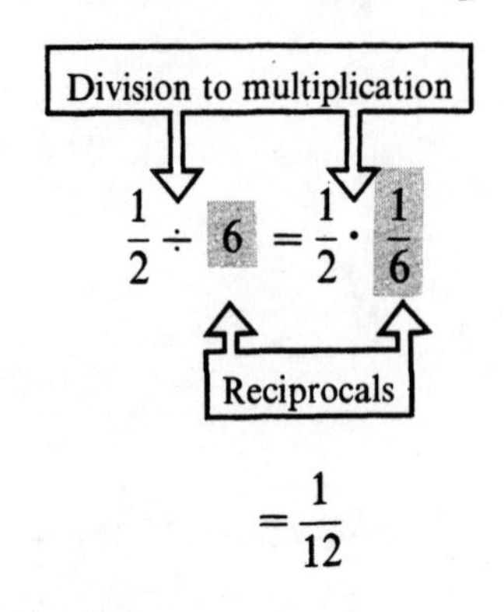

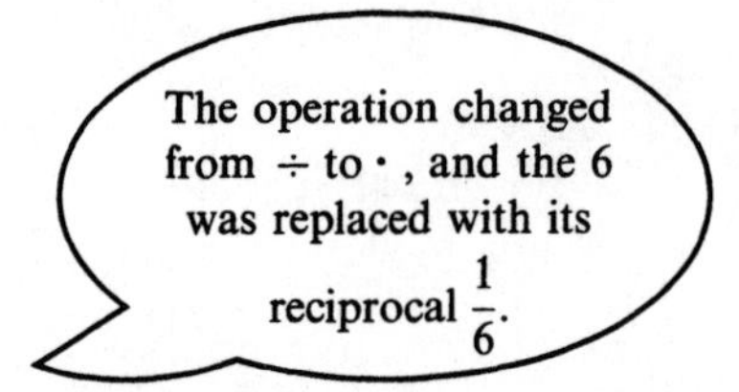

$$= \frac{1}{12}$$

EXAMPLE 27 Divide $12 \div \frac{1}{3}$.

Division to multiplication

$$12 \div \frac{1}{3} = 12 \cdot \_\_\_\_$$

Reciprocals

$\frac{3}{1}$

$$= \frac{36}{1}$$

$$= \_\_\_\_$$

36

EXAMPLE 28 Divide $\frac{3}{4} \div \frac{5}{6}$.

$$\frac{3}{4} \div \frac{5}{6} = \frac{3}{4} \cdot \_\_\_\_$$ First, convert from division to multiplication

$\frac{6}{5}$

$$= \frac{3}{\not{4}_2} \cdot \frac{\not{6}^3}{5}$$ Divide out common factors

$$= \frac{\_\_\_\_}{10}$$

9

EXAMPLE 29 Divide $\frac{3a}{4} \div \frac{7}{8}$.

$$\frac{3a}{4} \div \frac{7}{8} = \frac{3a}{\underset{1}{\cancel{4}}} \cdot \frac{\overset{2}{\cancel{8}}}{7}$$

$$= \_\_\_\_\_$$

$\frac{6a}{7}$

EXAMPLE 30 Divide $\frac{-12}{15} \div \frac{3}{4}$.

Remember to fill in the blanks. The more often you participate, the longer you will remember.

$$\frac{-12}{15} \div \frac{3}{4} = \frac{-12}{15} \cdot \_\_\_\_\_$$

$\frac{4}{3}$

$$= \_\_\_\_\_$$

$\frac{-16}{15}$

EXAMPLE 31 Divide $\frac{-15}{18} \div \frac{3}{-4}$.

$$\frac{-15}{18} \div \frac{3}{-4} = \frac{\overset{-5}{\cancel{-15}}}{\underset{9}{\cancel{18}}} \cdot \frac{\overset{-2}{\cancel{-4}}}{\underset{1}{\cancel{3}}}$$

$$= \frac{(\_\_\_\_)(\_\_\_\_)}{9}$$

$-5, -2$

$$= \frac{10}{9}$$

An alternate way is to determine the sign of the product first.

$$\frac{-15}{18} \div \frac{3}{-4} = \frac{-15}{18} \cdot \frac{-4}{3}$$

$$= +\frac{\overset{5}{\cancel{15}}}{\underset{9}{\cancel{18}}} \cdot \frac{\overset{2}{\cancel{4}}}{\underset{1}{\cancel{3}}}$$

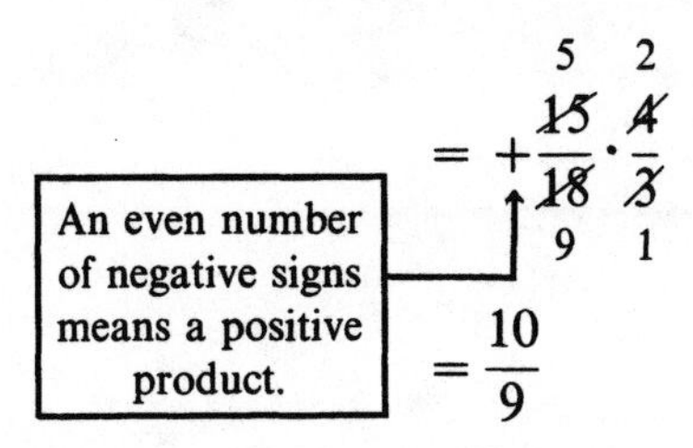

$$= \frac{10}{9}$$

PROBLEM SET 3.2D

Determine the multiplicative inverse of the following.

1. 4
2. 8
3. $\frac{1}{3}$
4. $\frac{1}{4}$
5. -3
6. -10
7. $\frac{3}{4}$
8. $\frac{5}{6}$
9. $\frac{-8}{9}$
10. $-\frac{12}{7}$

Determine the reciprocal of the following.

11. 6 12. 8 13. a 14. $-b$ 15. 0

16. $\frac{1}{2}$ 17. $-\frac{1}{3}$ 18. $\frac{-1}{4}$ 19. $\frac{-3}{4}$ 20. $-\frac{19}{18}$

Divide the following fractions. Reduce all answers to lowest terms and write all fractions in standard form.

21. $\frac{1}{4} \div 8$ 22. $\frac{1}{8} \div 4$ 23. $4 \div \frac{1}{2}$ 24. $9 \div \frac{1}{3}$

25. $\frac{1}{3} \div 9$ 26. $\frac{1}{2} \div 7$ 27. $8 \div \frac{1}{2}$ 28. $27 \div \frac{1}{3}$

29. $\frac{2}{3} \div \frac{3}{4}$ 30. $\frac{5}{6} \div \frac{7}{8}$ 31. $\frac{-4}{9} \div \frac{7}{15}$ 32. $\frac{-6}{7} \div \frac{12}{14}$

33. $\frac{7a}{-12} \div \frac{3}{4}$ 34. $\frac{-12x}{11} \div \frac{15}{22}$ 35. $\frac{-24x}{15} \div \frac{-12}{25}$ 36. $\frac{-15x}{26} \div \frac{-10}{39}$

37. $\frac{18a}{-15} \div \frac{27}{-25}$ 38. $\frac{-30}{21} \div \frac{-15}{14}$ 39. $\frac{-24a}{-35} \div \frac{12}{-49}$ 40. $\frac{-48x}{42} \div \frac{-72}{-56}$

3.3 ADDITION AND SUBTRACTION OF RATIONAL NUMBERS

Addition of Rational Numbers with Like Denominators

As you learned in arithmetic, we can only add fractions if they have the same denominators.

Addition of rational numbers

3.3A DEFINITION: ADDITION OF RATIONAL NUMBERS

If a, b, and c are integers, where c does not equal zero,

$$\frac{a}{c} + \frac{b}{c} = \frac{a+b}{c}$$

EXAMPLE 32 Add $\frac{2}{7} + \frac{3}{7}$.

$$\frac{2}{7} + \frac{3}{7} = \frac{2+3}{7}$$

5

$$= \frac{\quad}{7}$$

If the denominators of a fraction are the same, we add only the numerators. Why not add the denominators?

Remember that the denominator tells us what kind of thing the fraction is. The numerator tells us how many we have. Two sevenths plus three sevenths are five sevenths. Like 2 apples plus 3 apples are 5 apples.

EXAMPLE 33 Add $\frac{1}{15}+\frac{4}{15}+\frac{7}{15}$.

$$\frac{1}{15}+\frac{4}{15}+\frac{7}{15}=\frac{1+4+7}{15}$$

$$=\frac{12}{15}$$

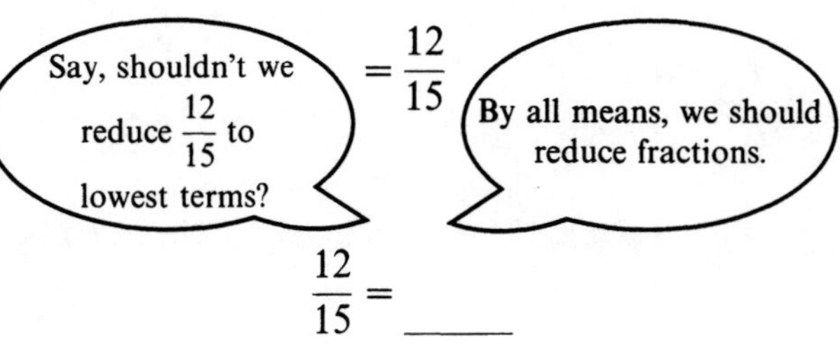

$\frac{4}{5}$

$$\frac{12}{15}= \_\_\_\_\_$$

EXAMPLE 34 Add $\frac{6}{7}+\frac{-2}{7}$.

$$\frac{6}{7}+\frac{-2}{7}=\frac{6+(-2)}{7}$$

4

$$=\frac{\quad}{7}$$

EXAMPLE 35 Add $\frac{2}{9}+\frac{8}{9}+\frac{-7}{9}$.

$$\frac{2}{9}+\frac{8}{9}+\frac{-7}{9}=\frac{2+8+(-7)}{9}$$

$$=\frac{10+(-7)}{9}$$

$$=\frac{3}{9} \qquad \text{Reduces to}$$

$\frac{1}{3}$

$$= \_\_\_\_\_$$

EXAMPLE 36 Add $\frac{1}{6}+\frac{-5}{6}$.

$$\frac{1}{6}+\frac{-5}{6}=\frac{1+(-5)}{6}$$

-4

$$=\frac{\quad}{6} \qquad \text{Reduces to}$$

$\frac{-2}{3}$

$$= \_\_\_\_\_$$

EXAMPLE 37 Add $\frac{a}{6}+\frac{-5}{6}$.

$$\frac{a}{6}+\frac{-5}{6}=\frac{a+(-5)}{6}$$

$$=\frac{a-5}{6}$$

EXAMPLE 38 Add $\frac{a}{2}+\frac{b}{2}$.

Can't we combine $a+b$ in Example 38?

No. They are not like terms.

$$\frac{a}{2}+\frac{b}{2}=\frac{a+b}{2}$$

EXAMPLE 39 Add $\frac{a}{7}+\frac{5a}{7}$.

$$\frac{a}{7}+\frac{5a}{7}=\frac{a+5a}{7}$$ Because a and $5a$ are like terms, they can be added

$\frac{6a}{7}$

$= \_\_\_\_\_$

EXAMPLE 40 Add $\frac{2x}{9}+\frac{7x}{9}$.

$$\frac{2x}{9}+\frac{7x}{9}=\frac{2x+7x}{9}$$

$\frac{9x}{9}$

$= \_\_\_\_\_$ $2x$ and $7x$ are like terms

x

$= \_\_\_\_\_$

EXAMPLE 41 Add $\frac{3a}{5}+\frac{-a}{5}$.

$$\frac{3a}{5}+\frac{-a}{5}=\frac{3a+(-a)}{5}$$

$\frac{2a}{5}$

$= \_\_\_\_\_$

EXAMPLE 42 Add $-\frac{3a}{12}+\frac{-5a}{12}$.

$$-\frac{3a}{12}+\frac{-5a}{12}=\frac{-3a}{12}+\frac{-5a}{12}$$ Write the negative sign of the first term in the numerator

$$=\frac{-3a+(-5a)}{12}$$

$\frac{-8a}{12}$

$= \_\_\_\_\_$

$\frac{-2a}{3}$

$= \_\_\_\_\_$

PROBLEM SET 3.3A

Add the following fractions. Reduce all answers to lowest terms.

1. $\frac{5}{7}+\frac{1}{7}$
2. $\frac{3}{11}+\frac{4}{11}$
3. $\frac{3}{8}+\frac{-1}{8}$
4. $\frac{-1}{6}+\frac{5}{6}$
5. $\frac{-1}{4}+\frac{3}{4}$
6. $\frac{5}{8}-\frac{-1}{8}$
7. $\frac{7}{9}+\frac{-4}{9}$
8. $\frac{-2}{9}+\frac{5}{9}$
9. $\frac{11}{16}+\frac{-9}{16}$
10. $\frac{3}{10}+\frac{-7}{10}$
11. $\frac{13}{24}+\frac{-7}{24}$
12. $\frac{-5}{18}+\frac{11}{18}$

13. $\frac{-23}{30} + \frac{-17}{30}$ 14. $\frac{-13y}{36} + \frac{-19}{36}$ 15. $\frac{-29a}{21} + \frac{5}{21}$ 16. $\frac{-37x}{28} + \frac{y}{28}$

17. $\frac{x}{8} + \frac{3x}{8}$ 18. $\frac{7a}{12} + \frac{-5a}{12}$ 19. $\frac{-4x}{15} + \frac{11x}{15}$ 20. $\frac{-7x}{18} + \frac{-11x}{18}$

21. $-\frac{7ab}{9} + \frac{5ab}{9}$ 22. $-\frac{11rs}{12} + \frac{2rs}{12}$

23. $\frac{-19ax}{24} + \frac{-5ax}{24}$ 24. $\frac{-7xy}{12} + \frac{5xy}{12}$

Additive Inverses

Just as integers have additive inverses, so do rational numbers.

Additive inverse

3.3B ADDITIVE INVERSE OF A RATIONAL NUMBER

The *additive inverse* of a rational number is the number you add to it to get zero for an answer.

The additive inverse of $\frac{a}{b}$ is $\frac{-a}{b}$, because $\frac{a}{b} + \frac{-a}{b} = \frac{0}{b}$

The additive inverse of $\frac{-a}{b}$ is $\frac{a}{b}$, because $\frac{-a}{b} + \frac{a}{b} = \frac{0}{b}$

EXAMPLE 43 Give the additive inverse of each of the following.

Number	*Additive inverse*
$\frac{1}{3}$	$\frac{-1}{3}$
$\frac{-2}{5}$	$\frac{2}{5}$
$\frac{4}{7}$	______
$\frac{-x}{y}$	______
$\frac{b}{a}$	______

$\frac{-4}{7}$

$\frac{x}{y}$

$\frac{-b}{a}$

Subtraction of Rational Numbers with Like Denominators

Subtraction of rational numbers is defined just like subtraction of integers.

Subtraction of rational numbers

3.3C SUBTRACTION OF RATIONAL NUMBERS

To subtract, add the additive inverse.

$\frac{a}{c} - \frac{b}{c} = \frac{a}{c} + \frac{-b}{c}$, where a, b, and c are integers and $c \neq 0$

EXAMPLE 44 Subtract $\frac{4}{5} - \frac{1}{5}$.

$$\frac{4}{5} - \frac{1}{5} = \frac{4}{5} + \frac{-1}{5}$$

$$= \frac{4 + (-1)}{5}$$

3

$$= \frac{\quad}{5}$$

EXAMPLE 45 Subtract $\frac{a}{3} - \frac{2}{3}$.

$$\frac{a}{3} - \frac{2}{3} = \frac{a}{3} + \frac{-2}{3}$$

$$= \frac{a + (-2)}{3}$$

$$= \frac{a - 2}{3}$$ Write the numerator as a subtraction so there are fewer signs

EXAMPLE 46 Subtract $\frac{4}{7} - \frac{-5}{7}$.

$$\frac{4}{7} - \frac{-5}{7} = \frac{4}{7} + \frac{+5}{7}$$ Change subtraction to an addition

$$= \frac{4 + 5}{7}$$

9

$$= \frac{\quad}{7}$$

After an addition or subtraction, reduce the answer to lowest terms if possible.

EXAMPLE 47 Subtract $\frac{5x}{12} - \frac{7x}{12}$.

$$\frac{5x}{12} - \frac{7x}{12} = \frac{5x}{12} + \frac{-7x}{12}$$ Change to an addition

$$= \frac{5x + (-7x)}{12}$$

$-2x$

$$= \frac{\quad}{12}$$ Add like terms after reducing

$$= \frac{-x}{6}$$

EXAMPLE 48 Subtract $\frac{8a}{21} - \frac{-8a}{21}$.

$$\frac{8a}{21} - \frac{-8a}{21} = \frac{8a}{21} + \frac{8a}{21}$$

$$= \frac{8a + 8a}{21}$$

$\frac{16a}{21}$

$$= \underline{\qquad}$$

EXAMPLE 49 Subtract $\frac{-19ax}{30} - \frac{-11ax}{30}$.

$\frac{11ax}{30}$

$$\frac{-19ax}{30} - \frac{-11ax}{30} = \frac{-19ax}{30} + \underline{\qquad}$$

$$= \frac{-19ax + 11ax}{30}$$

$$= \frac{-8ax}{30} \quad \text{Reduces to}$$

$\frac{-4ax}{15}$

$$= \underline{\qquad}$$

PROBLEM SET 3.3B

Perform the following subtractions by rewriting each as an addition problem. Reduce all answers to lowest terms.

1. $\frac{3}{5} - \frac{2}{5}$
2. $\frac{5}{7} - \frac{3}{7}$
3. $\frac{7}{9} - \frac{4}{9}$
4. $\frac{5}{6} - \frac{1}{6}$
5. $\frac{5}{9} - \frac{8}{9}$
6. $\frac{5}{12} - \frac{7}{12}$
7. $\frac{6a}{7} - \frac{2a}{7}$
8. $\frac{11b}{12} - \frac{7b}{12}$
9. $\frac{7x}{16} - \frac{13x}{16}$
10. $\frac{9y}{14} - \frac{13y}{14}$
11. $\frac{8}{11} - \frac{-6}{11}$
12. $\frac{7}{13} - \frac{-8}{13}$
13. $\frac{-7}{15} - \frac{3}{15}$
14. $\frac{-8}{15} - \frac{2}{15}$
15. $\frac{-6a}{12} - \frac{7}{12}$
16. $\frac{8x}{14} - \frac{-7}{14}$
17. $\frac{17ax}{18} - \frac{-11}{18}$
18. $\frac{13xy}{16} - \frac{-9}{16}$
19. $\frac{-19xy}{24} - \frac{-23}{24}$
20. $\frac{-23ax}{30} - \frac{-29}{30}$
21. $\frac{7x}{10} - \frac{3x}{10}$
22. $\frac{8xy}{15} - \frac{-4xy}{15}$
23. $\frac{-23ab}{27} - \frac{19ab}{27}$
24. $\frac{19xy}{30} - \frac{11xy}{30}$
25. $\frac{25ax}{29} - \frac{-4ax}{29}$
26. $\frac{13ay}{21} - \frac{-8ay}{21}$
27. $\frac{7x^2}{18} - \frac{-11x^2}{18}$
28. $\frac{-4y^2}{27} - \frac{-14y^2}{27}$

Addition and Subtraction with Unlike Denominators

Least common denominator

Before adding fractions whose denominators are not alike, express both fractions as equivalent fractions with a common denominator. The easiest common denominator to use is the smallest number that contains all the factors of each denominator. It is called the *least common denominator* (LCD). Frequently, the least common denominator can be found by inspection. When the LCD is not obvious, you can find it using this method.

EXAMPLE 50 Find the least common denominator for $\frac{5}{6}$ and $\frac{3}{8}$.

Completely factor both denominators

$6 = 2 \cdot 3$

$8 = 2 \cdot 2 \cdot 2$

$$\overbrace{2 \cdot 2 \cdot 2}^{\text{Factors of 8}} \cdot 3 = 24 \qquad \text{Common denominator}$$

(with $\underbrace{2 \cdot 3}_{\text{Factors of 6}}$ marked under the last two factors)

24 is the least common denominator because it is the smallest number that contains all the factors of 6 and all the factors of 8.

Least common denominator

3.3D TO FIND THE LEAST COMMON DENOMINATOR

1. Completely factor each denominator.
2. Write each factor that appears in *any* denominator.
3. Build the LCD by raising each factor to the highest power it has in any *single* denominator.

EXAMPLE 51 Find the least common denominator for $\frac{5}{24}$ and $\frac{7}{30}$.

Step 1. Completely factor both denominators

$24 = 2 \cdot 2 \cdot 2 \cdot 3 = 2^3 \cdot 3^1$

$30 = \quad 2 \cdot 3 \cdot 5 \quad = 2^1 \cdot 3^1 \cdot 5^1$

Step 2. Write each factor that appears in *any* denominator

$2 \cdot 3 \cdot 5$

Step 3. Raise each factor to the highest power it has in any single denominator

$$\begin{aligned} \text{LCD} &= 2^3 \cdot 3^1 \cdot 5^1 \\ &= 8 \cdot 3 \cdot 5 \\ &= 120 \end{aligned}$$

Notice that the least common denominator is the smallest number that can be divided by each of the denominators.

EXAMPLE 52 Add $\frac{3}{4} + \frac{2}{3}$.

12 The least common denominator is ____.

Convert each fraction to an equivalent fraction with a denominator of 12

$$\frac{3}{4} + \frac{2}{3} = \frac{3}{4} \cdot \frac{3}{3} + \frac{2}{3} \cdot \frac{4}{4}$$

$$= \frac{9}{12} + \frac{8}{12}$$

$$= \frac{9+8}{12}$$

$$= \frac{17}{12}$$

Remember, we multiply each of the original fractions by some form of 1 to convert the original denominator to the LCD.

EXAMPLE 53 Subtract $\frac{5}{6} - \frac{2}{3}$.

Rewrite as an addition problem.

$$\frac{5}{6} - \frac{2}{3} = \frac{5}{6} + \frac{-2}{3}$$

6 The least common denominator is ____.

$$= \frac{5}{6} + \frac{-2}{3} \cdot \frac{2}{2}$$ Multiply by 1 to make the denominators common

$$= \frac{5}{6} + \frac{-4}{6}$$

$$= \frac{5+(-4)}{6}$$

$\frac{1}{6}$ $= \_\_\_\_$

EXAMPLE 54 Add $\frac{a}{3} + \frac{b}{2}$.

6 The least common denominator is ____.

$$\frac{a}{3} + \frac{b}{2} = \frac{2}{2} \cdot \frac{a}{3} + \frac{3}{3} \cdot \frac{b}{2}$$ Multiply by 1 to make the denominators common

$$= \frac{2a}{6} + \frac{3b}{6}$$

$2a + 3b$ $= \frac{\_\_\_\_}{6}$

EXAMPLE 55 Subtract $\frac{x}{8} - \frac{-5}{12}$ by rewriting as an addition problem.

$$\frac{x}{8} - \frac{-5}{12} = \frac{x}{8} + \frac{5}{12}$$

24

The least common denomina·tor is ____.

$$= \frac{3}{3} \cdot \frac{x}{8} + \frac{2}{2} \cdot \frac{5}{12}$$

$$= \frac{3x}{24} + \frac{10}{24}$$

$3x + 10$

$$= \frac{\qquad}{24}$$

EXAMPLE 56 Add $\frac{3x}{4} + \frac{5x}{6}$.

12

The least common denominator is ____.

$$\frac{3x}{4} + \frac{5x}{6} = \frac{3}{3} \cdot \frac{3x}{4} + \frac{2}{2} \cdot \frac{5x}{6}$$

$$= \frac{9x}{12} + \frac{10x}{12}$$

$$= \frac{9x + 10x}{12}$$

$19x$

$$= \frac{\qquad}{12} \quad \text{Combine like terms}$$

EXAMPLE 57 Subtract $\frac{7xy}{12} - \frac{3xy}{8}$ by rewriting as an addition problem.

$$\frac{7xy}{12} - \frac{3xy}{8} = \frac{7xy}{12} + \frac{-3xy}{8}$$

24

The least common denominator is ____.

$\frac{3}{3}$

$$= \frac{2}{2} \cdot \frac{7xy}{12} + \underline{\qquad} \cdot \frac{-3xy}{8}$$

$$= \frac{14xy}{24} + \frac{-9xy}{24}$$

$$= \frac{14xy + (-9xy)}{24}$$

$5xy$

$$= \frac{\qquad}{24}$$

EXAMPLE 58 Add $\frac{7ab}{15} + \frac{-5ab}{6}$.

30

The least common denominator is ______.

$\frac{2}{2}, \frac{5}{5}$

$$\frac{7ab}{15} + \frac{-5ab}{6} = \underline{\quad\quad} \cdot \frac{7ab}{15} + \underline{\quad\quad} \cdot \frac{-5ab}{6}$$

$$= \frac{14ab}{30} + \frac{-25ab}{30}$$

$$= \frac{14ab + (-25ab)}{30}$$

$\frac{-11ab}{30}$

$$= \underline{\quad\quad\quad}$$

EXAMPLE 59 Subtract $\frac{-5x}{6} - \frac{-11x}{15}$.

$$\frac{-5x}{6} - \frac{-11x}{15} = \frac{-5x}{6} + \frac{11x}{15}$$

30

The least common denominator is ______.

$\frac{2}{2}$

$$= \frac{-5x}{6} \cdot \frac{5}{5} + \frac{11x}{15} \cdot \underline{\quad\quad}$$

$$= \frac{-25x}{30} + \frac{22x}{30}$$

$$= \frac{-25x + 22x}{30}$$

$$= \frac{-3x}{30} \quad \text{Reduces to}$$

$\frac{-x}{10}$

$$= \underline{\quad\quad\quad}$$

PROBLEM SET 3.3C

Perform the indicated operations. Reduce all answers to lowest terms and write fractions in standard form.

1. $\frac{1}{8} + \frac{3}{4}$
2. $\frac{2}{3} + \frac{1}{6}$
3. $\frac{3}{5} + \frac{1}{4}$
4. $\frac{3}{4} + \frac{1}{3}$
5. $\frac{8}{9} - \frac{2}{3}$
6. $\frac{7}{8} - \frac{1}{2}$
7. $\frac{3}{8} - \frac{11}{12}$
8. $\frac{1}{8} - \frac{5}{6}$
9. $\frac{7}{10} + \frac{-13}{15}$
10. $\frac{5}{6} + \frac{-11}{15}$
11. $\frac{5}{8} - \frac{-5}{12}$
12. $\frac{5}{6} - \frac{7}{8}$

13. $\frac{4a}{6} - \frac{7}{10}$ 14. $\frac{8b}{15} - \frac{5}{6}$ 15. $\frac{7}{9} - \frac{a}{6}$ 16. $\frac{9}{10} - \frac{4a}{15}$

17. $\frac{-7}{8} - \frac{-5a}{6}$ 18. $\frac{-11}{12} - \frac{-5x}{8}$ 19. $\frac{11a}{15} - \frac{7}{9}$ 20. $\frac{13x}{14} - \frac{10}{21}$

21. $\frac{-19a}{12} + \frac{-7}{15}$ 22. $\frac{-11x}{15} + \frac{-7}{10}$ 23. $\frac{-7xy}{18} - \frac{-11}{12}$ 24. $\frac{-11ax}{12} - \frac{7}{9}$

25. $\frac{4a}{9} + \frac{5a}{6}$ 26. $\frac{7x}{8} - \frac{5x}{6}$ 27. $\frac{3x}{8} + \frac{-7x}{12}$ 28. $\frac{5a}{9} - \frac{-7a}{15}$

29. $\frac{-5bx}{14} + \frac{-3bx}{4}$ 30. $\frac{-7by}{8} - \frac{11by}{20}$ 31. $\frac{-ax^2}{6} + \frac{9ax^2}{14}$

32. $\frac{-11by^2}{18} - \frac{-3by^2}{4}$ 33. $\frac{a}{6} + \frac{-5a}{8}$ 34. $\frac{-11ab}{12} - \frac{-3ab}{8}$

35. $\frac{x}{12} + \frac{-x}{8}$ 36. $\frac{ax^2}{15} - \frac{-5ax^2}{6}$ 37. $\frac{-3by}{8} - \frac{-5by}{6}$

38. $\frac{11ab}{12} - \frac{13ab}{15}$ 39. $\frac{13a^2b}{18} - \frac{7a^2b}{8}$ 40. $\frac{-19x^2y}{20} - \frac{-5x^2y}{8}$

Review Your Skills

Simplify the following.

41. $\left(\frac{-1}{2}\right)(-4)$ 42. $\left(\frac{-1}{2}\right)\left(\frac{2}{3}\right)\left(\frac{-3}{5}\right)$

43. $(-2)^2$ 44. -2^2

3.4 EVALUATING RATIONAL EXPRESSIONS

Evaluating expressions

Here is a brief reminder of the procedure you should use when evaluating expressions and the rules of order you should follow.

1. Replace all occurrences of the variable with its current value.
2. Simplify all expressions within symbols of grouping.
3. Evaluate expressions with exponents.
4. Perform any multiplications and divisions, working from left to right, as they occur.
5. Perform any additions and subtractions, working from left to right, as they occur.

For a more detailed set of rules, refer to Unit 1, page 4.

EXAMPLE 60 Evaluate $2 - \frac{1}{2}(a - b)$ if $a = 2$ and $b = -4$.

First, replace the variables with the values they represent.

$$2 - \frac{1}{2}(a - b) =$$

$$2 - \frac{1}{2}[2 - (-4)] = 2 - \frac{1}{2}[2 + 4]$$

$$= 2 - \frac{1}{2}[6]$$

$$= 2 - 3$$

(−3)

$$= 2 + \_\_\_\_$$

$$= -1$$

Sometimes we use brackets [ ] or braces { } so that it's easier to detect groupings inside of groupings.

EXAMPLE 61 Evaluate $6x - (2y + 3)$ if $x = \frac{2}{3}$ and $y = \frac{-1}{2}$.

$$6x - (2y + 3) =$$

−1

$$6\left(\frac{2}{3}\right) - \left[2\left(\frac{-1}{2}\right) + 3\right] = 4 - [\_\_\_\_ + 3]$$ Replace variables with values

$$= 4 - [2]$$ Simplify parentheses

(−2)

$$= 4 + \_\_\_\_$$ Rewrite subtraction as addition

$$= 2$$

EXAMPLE 62 Evaluate $\frac{2}{3}a^2 - \frac{1}{2}b^3$ if $a = 6$ and $b = -2$.

$$\frac{2}{3}a^2 - \frac{1}{2}b^3 =$$

6, −2

$$\frac{2}{3}(\_\_\_\_)^2 - \frac{1}{2}(\_\_\_\_)^3 = \frac{2}{3} \cdot 36 - \frac{1}{2} \cdot (-8)$$

24

$$= \_\_\_\_ - (-4)$$

$$= 24 + 4$$

28

$$= \_\_\_\_$$

Evaluate expressions with exponents first.

EXAMPLE 63 Evaluate $6ab + 3ax - \frac{3}{4}bx$ if $a = \frac{1}{2}$, $b = \frac{1}{3}$, and $x = -8$.

$$6 \cdot a \cdot b + 3 \cdot a \cdot x - \frac{3}{4} \cdot b \cdot x =$$

−12, −2

$$6\left(\frac{1}{2}\right)\left(\frac{1}{3}\right) + 3\left(\frac{1}{2}\right)(-8) - \frac{3}{4}\left(\frac{1}{3}\right)(-8) = 1 + (\_\_\_\_) - (\_\_\_\_)$$

$$= 1 + (-12) + 2$$

−9

$$= \_\_\_\_$$

PROBLEM SET 3.4A

Evaluate the following if $a = -2$, $b = -3$, $c = 0$, $x = \frac{2}{3}$, and $y = \frac{-1}{2}$.

1. $3 + (a - b)$
2. $4 + 3(a + b)$
3. $6 - (3x - b)$
4. $-7 - 2(a - 2y)$
5. $5 - \frac{1}{2}(6x - 4y)$
6. $8 - \left(-\frac{1}{3}\right)(2y - 3x)$
7. $a - (3 - b + a)$
8. $2b - 2(4 - c + b)$
9. $b - \frac{3}{4}(7 - c + b)$
10. $9c + \frac{5}{6}(2b + 12 - 3a)$
11. $\frac{1}{2}a^2 - b^2 + \frac{1}{4}c^2$
12. $\frac{1}{4}a^3 + \frac{2}{3}b^2$
13. $ay + bx$
14. $9x + 2ay - 4y$
15. $3ax - 4by + \frac{1}{3}ab$
16. $6xy + \frac{2}{3}by + 2c$
17. $24x^2y - \frac{1}{3}ab^2 + cxy$
18. $\frac{3}{4}a^3b - 5ac^2x + 7abxy$

Evaluate the following, using the indicated values.

19. $8b - \frac{1}{3}a^2 + ac$ if $a = -6$, $b = \frac{1}{2}$, $c = \frac{1}{3}$
20. $\frac{1}{3}(9b - 6bx - 12a^2)$ if $a = \frac{-1}{2}$, $b = -\frac{1}{3}$, $x = 8$
21. $5(a + b + c) - \frac{8}{9}abc - 8bc$ if $a = 0$, $b = \frac{1}{2}$, $c = \frac{-1}{2}$
22. $\frac{1}{2}x^2 - 12ax - xy^2$ if $x = -4$, $y = -\frac{1}{2}$, $a = \frac{1}{3}$
23. $\frac{1}{3}a^2x - 15bx - a^2b^2x$ if $a = -3$, $b = -5$, $x = \frac{2}{3}$
24. $\frac{5}{6}b^2c + 6b^2c - \frac{4}{3}abc$ if $a = -\frac{1}{2}$, $b = 4$, $c = -\frac{3}{4}$

POINTERS ABOUT REDUCING

To reduce or divide out a factor, you must be able to identify a multiplication by 1.

$\frac{12x}{4}$ can be reduced because it can be rewritten with 1 as a factor.

$$\frac{12x}{4} = \frac{\overset{1}{\cancel{4}} \cdot 3x}{\underset{1}{\cancel{4}}} \qquad \frac{4}{4} \text{ is } 1$$

$$= 3x$$

$\frac{12 + x}{4}$ cannot be reduced because there is not a common factor in the numerator and denominator.

Generally, fractions are reduced to lowest terms by dividing numerator and denominator by the same quantity. The common factors must be a part of a product. <u>We cannot reduce a fraction if it is part of a sum or difference.</u>

Problem	Possible to reduce	How
$\frac{6x + 5}{6x}$	No	Never reduce part of a sum.
$\frac{12x}{5} \cdot \frac{15y}{2x}$	Yes	$\frac{12x}{5} \cdot \frac{15y}{2x} = \frac{\overset{1}{\cancel{2x}} \cdot 6}{\underset{1}{\cancel{5}}} \cdot \frac{\overset{1}{\cancel{5}} \cdot 3y}{\underset{1}{\cancel{2x}}}$ $= 18y$
$\frac{12x}{5} + \frac{15y}{2x}$	No	Never reduce part of a sum.
$\frac{8(2x + 3)}{16x}$	Yes	8 is a common factor. $\frac{8(2x + 3)}{16x} = \frac{\overset{1}{\cancel{8}}(2x + 3)}{\underset{1}{\cancel{8}} \cdot 2x}$ $= \frac{2x + 3}{2x}$ Don't reduce further. $2x$ is part of a sum.
$\frac{12x}{5} \div \frac{25}{4x}$	No	Never reduce in a division problem. You may be able to reduce after you change to multiplication.

PROBLEM SET 3.4B

When possible, simplify the following expressions.

1. $\dfrac{48x}{16y}$
2. $\dfrac{12x}{5} \cdot \dfrac{10}{3x}$
3. $\dfrac{48 + x}{16y}$
4. $\dfrac{12x + 5}{5 + 3x}$
5. $\dfrac{48x}{16 + y}$
6. $\dfrac{12x}{5} \div \dfrac{10}{3x}$
7. $\dfrac{48x}{16(x + y)}$
8. $\dfrac{12x + 10}{3x}$
9. $\dfrac{4(12x + 1)}{4x}$
10. $\dfrac{3(4x + 3)}{3x}$
11. $\dfrac{16x^2}{3y} \div \dfrac{3y}{4x}$
12. $\dfrac{6(2x + 1)}{3xy}$

Review Your Skills Simplify the following.

13. $7x - 8x$
14. $9x + 4 - 8x - 6$
15. $5x - 8 - 12 - 4x$
16. $3x + 7 - 12 - 4x$

REVIEW 3

NOW THAT YOU HAVE COMPLETED UNIT 3, YOU SHOULD BE ABLE TO:

Use the following definitions in your working vocabulary.

Rational number—Any number that can be written as the quotient of two integers $\frac{a}{b}$, where a and b are integers and $b \neq 0$.

Standard form of a fraction:

$\frac{a}{b}$ for positive fractions

$\frac{-a}{b}$ for negative fractions

Equivalent fractions—Two fractions are equivalent if one can be obtained from the other by multiplying by 1.

If $\frac{a}{b} = 1 \cdot \frac{c}{d}$, then $\frac{a}{b} = \frac{c}{d}$

Identity element for multiplication is 1.

$a \cdot 1 = a$ or $1 \cdot a = a$

Multiplicative inverse—Two rational numbers are multiplicative inverses of each other if their product is 1.

$a \cdot \frac{1}{a} = 1$ (Multiplicative inverses) $\frac{a}{b} \cdot \frac{b}{a} = 1$ (Multiplicative inverses)

Reciprocal—Another name for multiplicative inverse.

Additive inverse—The additive inverse of a rational number is the number you add to it to get zero for an answer.

The additive inverse of $\frac{a}{b}$ is $\frac{-a}{b}$, because $\frac{a}{b} + \frac{-a}{b} = \frac{0}{b}$

The additive inverse of $\frac{-a}{b}$ is $\frac{a}{b}$, because $\frac{-a}{b} + \frac{a}{b} = \frac{0}{b}$.

Apply the following rules and definitions.

To reduce fractions to lowest terms

Write the numerator and denominator in factored form, then remove the common factors from the numerator and denominator.

or

Divide the numerator and denominator by the same number. Continue the process until there is no common factor except 1 in the numerator and denominator.

To find the product of two rational numbers

1. Multiply their numerators.
2. Multiply their denominators.
3. Reduce your answers to lowest terms.

or

Use the shorter method of dividing out common factors before multiplying.

To find the quotient of two rational numbers, first change from division to multiplication. That is, to divide by a number, multiply by its multiplicative inverse or reciprocal.

$$\frac{a}{b} \div \frac{c}{d} = \frac{a}{b} \cdot \frac{d}{c}$$

then proceed as in multiplication of rational numbers.

To find the least common denominator

1. Completely factor each denominator.
2. Write each factor that appears in *any* denominator.
3. Build the LCD by raising each factor to the highest power it has in any *single* denominator.

To find the sum of two rational numbers

1. Find a common denominator for the fractions.
2. Convert each fraction into an equivalent fraction with the common denominator.
3. Add the numerators of the fraction and put the result over the common denominator.
4. Reduce the answer to lowest terms.

To find the difference between two rational numbers, first write the subtraction problem as an addition problem

$$\frac{a}{b} - \frac{c}{d} = \frac{a}{b} + \frac{-c}{d}$$

then proceed as in the addition of rational numbers.

To evaluate an expression, replace all occurrences of the variable with its current value and follow the rules for order of operations.

Review Test 3

1. The multiplicative inverse of $-\frac{2}{3}$ is ______. *(3.2)*

2. The reciprocal of 5 is ______. *(3.2)*

3. Write a fraction equivalent to $\frac{-4}{5}$ with a denominator of 45. ______ *(3.1)*

4. Write a fraction equivalent to $\frac{7a}{8}$ with a denominator of 48. ______ *(3.1)*

Reduce the following fractions to lowest terms. Write the answer in standard form. ***(3.1)***

5. $-\frac{16}{24} =$ ______

6. $\frac{-48}{-60} =$ ______

7. $\frac{88a}{-33} =$ ______

8. $\frac{-15xy}{55} =$ ______

Perform the indicated operations. Reduce all answers to lowest terms, and write the fractions in standard form. ***(3.2)***

9. $\frac{20}{7} \cdot \frac{-14}{15} =$ ______

10. $\frac{-9a}{16} \cdot \frac{24x^2}{-3} =$ ______

11. $\frac{36ab}{-25} \cdot \frac{-75a^2}{27} =$ ______

12. $-7x^2 \cdot \frac{-15x}{21} =$ ______

13. $\frac{-2}{3} \div 12 =$ ______

14. $\frac{-15x}{22} \div \frac{-12}{33} =$ ______

15. $\frac{-52x}{42} \div \frac{26}{-28} =$ ______

16. $\frac{-24x^2}{18} \div \frac{-20}{-27} =$ ______

17. $18\left(\frac{5a}{6} - \frac{4}{9}\right) =$ ______

18. $\frac{-3}{4}(-24a + 12x) =$ ______

19. $-5a\left(\frac{3}{10}a - \frac{7}{15}\right) =$ ______

20. $-\frac{a^2}{6}\left(-24a + \frac{14x}{9}\right) =$ ______

Add or subtract the following. Reduce all answers to lowest terms and write the fractions in standard form. (3.3)

21. $\dfrac{-9ab}{16} + \dfrac{5ab}{16} =$ ______

22. $\dfrac{12x}{15} - \dfrac{-7}{15} =$ ______

23. $\dfrac{7}{8} + \dfrac{-5}{6} =$ ______

24. $\dfrac{-11b}{12} - \dfrac{13}{8} =$ ______

25. $\dfrac{-17ax}{12} + \dfrac{-11}{15} =$ ______

26. $\dfrac{-4x^2}{9} - \dfrac{7x^2}{15} =$ ______

27. $\dfrac{-7ax}{15} + \dfrac{-11ax}{6} =$ ______

28. $\dfrac{-5x}{6} - \dfrac{-2x}{15} =$ ______

29. $\dfrac{-9x^2}{8} - \dfrac{-5x^2}{12} =$ ______

30. $\dfrac{-ax^2}{8} - \dfrac{10ax^2}{14} =$ ______

Evaluate the following, using the indicated values. (3.4)

31. $4x - (6y + 3) =$ ______ if $x = \dfrac{3}{4}$ and $y = \dfrac{-1}{2}$

32. $\dfrac{1}{2}a - \dfrac{2}{3}ab =$ ______ if $a = -2$ and $b = 3$

33. $4(a + b + c) - \dfrac{4}{3}abc =$ ______ if $a = -2$, $b = \dfrac{3}{4}$, $c = -1$

34. $\dfrac{1}{3}x^2 - 10ax - 6xy^2 =$ ______ if $a = -\dfrac{1}{2}$, $x = -6$, $y = \dfrac{2}{3}$

REVIEW YOUR SKILLS

Simplifying the following.

35. $18 \div (-3) \cdot (3)$

36. $15 - 9 \div 3 \cdot 2$

37. $12^2 \div 4 - 3 + 2$

38. $[3^2 \cdot 4 - 2] \div 2$

UNIT 4

First–Degree Equations

OBJECTIVES

After you have successfully completed this unit, you will be able to:

1. Solve a linear equation by using the addition and subtraction property of equivalent equations (4.1)
2. Solve a linear equation by using the multiplication and division property of equivalent equations (4.2)
3. Solve a linear equation by using a combination of the addition, subtraction, multiplication, and division properties of equivalent equations (4.2)
4. Solve a literal equation for the specified variable (4.3)
5. Solve for the value of the specified variable in a formula when given the numerical values of the other variables (4.4)

4.1 SOLVING EQUATIONS CONTAINING ADDITION AND SUBTRACTION

Conditional equation

An equation is made up of a left member and a right member separated by an equal sign. $3x + 3 = 15$ is an equation. It is a *conditional equation* because it is true only on the condition that you replace x with 4.

Solution set

4.1A DEFINITION

The *solution set* of an equation is the set of values that makes the equation true.

In this unit, solution sets will have one member. In later units, we will deal with equations that have more than one solution.

The solution set of the equation $x - 7 = 10$ is 17, because replacing x with 17 makes the equation true.

Equivalent equations

4.1B DEFINITION

Equivalent equations are equations that have the same solution set.

$2x + 7 = 15$ and $2x = 8$ are equivalent equations because $x = 4$ is the solution to both equations.

Additive property of equality

4.1C ADDITIVE PROPERTY OF EQUALITY

If the same quantity is added to both sides of a true equation, the resulting equation is also true.

In symbols, if $a = b$

then $a + c = b + c$

An illustration of this property is

because $2 = 2$

then $2 + 3 = 2 + 3$

Another illustration is

because $3 + 2 = 5$

then $3 + 2 + 4 = 5 + 4$

The additive property of equality and the definition of equivalent equations allow us to write a very useful rule.

Addition and subtraction property of equivalent equations

4.1D ADDITION AND SUBTRACTION PROPERTY OF EQUIVALENT EQUATIONS

If the same quantity is added to or subtracted from both sides of an equation, the new equation is equivalent to the original equation.

This property allows us to find the solution to many equations.

To solve the equation $x - 8 = 11$, we want to find the value of x that will make the equation true. Therefore, we want the left side, or member, to be x, not $x - 8$. Notice that 8 is subtracted from x. Therefore, we will eliminate the 8 by doing the inverse operation, addition, to both sides. Add an 8 to both sides.

EXAMPLE 1 Solve $x - 8 = 11$ for x.

$x - 8 = 11$ Notice 8 is subtracted from x

$x - 8 + 8 = 11 + 8$ Add 8 to both sides

$x = 19$ Simplify

EXAMPLE 2 Solve $x + 8 = 11$ for x.

$x + 8 = 11$ Notice 8 is added to x

$x + 8 - 8 = 11 - 8$ Do the inverse (subtract) on both sides

3

$x =$ _____ Simplify

EXAMPLE 3 Solve $16 = a - 4$ for a.

$16 = a - 4$ 4 is subtracted from a

$16 + 4 = a - 4 + 4$ Add 4 to both sides

20, a

_____ = _____ Simplify

In this example, we have a good illustration of the ways in which an answer can be written.

$20 = a$ can be written $a = 20$

A formal statement that $a = 20$ also means $20 = a$ is called the symmetric property of equality.

Symmetric property of equality

4.1E SYMMETRIC PROPERTY OF EQUALITY

If $a = b$, then $b = a$

where a and b represent any algebraic expression.

The symmetric property tells us that the members of an equation can be interchanged. The equation $16 = a - 4$ could have been solved as follows.

$$16 = a - 4$$

$$a - 4 = 16 \quad \text{Symmetric property of equality}$$

$$a - 4 + 4 = 16 + 4 \quad \text{Add 4 to both sides}$$

$$a = 20$$

To this point the variable has appeared only once in the equation. When the variable appears more than once in an equation, combine like terms to simplify as in Example 4.

EXAMPLE 4 Solve $5b - 4b + 3 = 12$ for b.

Notice, we always write each step in the solution of an equation below the previous step, with the equal signs lined up in a column.

$$\underbrace{5b - 4b} + 3 = 12$$

$$b \quad + 3 = 12 \quad \text{Combine like terms}$$

$$b + 3 \_\_\_\_ = 12 \_\_\_\_ \quad \text{Subtract 3 from both sides}$$

$$b = \_\_\_\_ \quad \text{Simplify}$$

$-3, -3$

9

EXAMPLE 5 Solve $8y + 5 = 7y - 9$ for y.

What do I do if the equation has the variable on both sides and constants on both sides?

Work to get all the variable terms on one side of the equation and all the constants on the other side.

$$8y + 5 = 7y - 9$$

$$8y - 7y + 5 = 7y - 7y - 9 \quad \text{Subtract } 7y \text{ from both sides}$$

$$y + 5 = -9 \quad \text{Simplify}$$

$$y + 5 - 5 = -9 - 5 \quad \text{Subtract 5 from both sides}$$

$$y = \_\_\_\_$$

-14

EXAMPLE 6 Solve $5x - 7 = 6x - 5$ for x.

$$5x - 7 = 6x - 5$$

$$5x - 5x - 7 = 6x - 5x - 5 \quad \text{Subtract } 5x \text{ from both sides}$$

$$-7 = x - 5$$

5, 5

$$-7 + \_\_\_\_ = x - 5 + \_\_\_\_ \quad \text{Add 5 to both sides}$$

−2

$$\_\_\_\_ = x$$

Why subtract $5x$ instead of $6x$ from both sides?

That way you'll get $+x$ in the last step instead of $-x$.

EXAMPLE 7 Solve $2(5z + 2) = 4(2z + 5) + z$ for z.

$$2(5z + 2) = 4(2z + 5) + z \quad \text{Use the distributive property to eliminate the parentheses}$$

$$10z + 4 = 8z + 20 + z$$

$$10z + 4 = 9z + 20 \quad \text{Combine like terms}$$

$$10z - 9z + 4 = 9z - 9z + 20 \quad \text{Subtract } 9z \text{ from both sides}$$

$$z + 4 = 20 \quad \text{Combine like terms}$$

−4, −4

$$z + 4 \_\_\_\_ = 20 \_\_\_\_ \quad \text{Subtract 4 from both sides}$$

16

$$z = \_\_\_\_$$

How can I be sure that 16 is the correct solution to the equation?

Check your solution by using it to replace the variable in the original equation.

Check

$$2(5 \cdot z + 2) = 4(2 \cdot z + 5) + z \quad \text{Original equation}$$

$$2[5 \cdot (\mathbf{16}) + 2] \stackrel{?}{=} 4[2 \cdot (\mathbf{16}) + 5] + (\mathbf{16}) \quad \text{Replace all } z\text{'s with 16}$$

$$2[80 + 2] \stackrel{?}{=} 4[32 + 5] + 16 \quad \text{Simplify each side separately}$$

$$2[82] \stackrel{?}{=} 4[37] + 16$$

$$164 \stackrel{?}{=} 148 + 16$$

$$164 = 164$$

Since replacing the variable in the equation resulted in an identity, we say that the solution set checks, and 16 is the correct solution.

PROBLEM SET 4.1

Solve for the variable. Check your solutions.

1. $x - 3 = 9$
2. $x - 5 = -2$
3. $x + 5 = 17$
4. $x + 6 = -12$
5. $-2 = y - 3$
6. $9 = a - 7$
7. $9 - 5 + x = 0$
8. $10 + z - 12 = 0$
9. $2t - t - 3 = 6$
10. $4 = 2x + 1 - x$
11. $5x - 5 - 4x = 6 - 10$
12. $6 - 8m + 9m = 1 + 5$
13. $3x = 2x - 3$
14. $8x = 4 + 7x$
15. $4b + 2 = 3b - 9$
16. $17x - 4 = 8 + 16x$
17. $15x - 4 = 14x + 5$
18. $5 - 8x = 12 - 7x$
19. $2(c + 1) = c$
20. $4(x - 2) = 3x$
21. $5(x - 3) = 4x - 1$
22. $10(2 + x) = 9x - 15$
23. $3(a + 2) = 2(a - 3)$
24. $7(4 + x) = 6(x + 2)$
25. $6 + 4(x - 4) = 3(x + 1)$
26. $8(2 + y) = 1 + 7(y - 3)$
27. $6(x + 3) = 2(x - 3) + 3x$
28. $2x + 3(x - 3) = 4(x - 5)$
29. $3x + 5(x - 4) + 6 = 2 + 8(x + 4)$
30. $5(x - 4) + 3x = 2x + 5(x + 2) + 4$

Review Your Skills

Simplify the following.

31. $\left(-\frac{3}{2}\right) \cdot 4$
32. $(3)(-18)$
33. $\frac{-8}{-1}$
34. $\frac{12}{-2}$

4.2 SOLVING EQUATIONS CONTAINING MULTIPLICATION AND DIVISION

Multiplication property of equality

4.2A MULTIPLICATION PROPERTY OF EQUALITY

If both sides of a true equation are multiplied by the same quantity, the resulting equation is also true.

In symbols, if $a = b$

then $a \cdot c = b \cdot c$

An illustration of this property is

since $8 = 8$

then $2 \cdot 8 = 2 \cdot 8$

or

since $3 + 2 = 5$

then $4 \cdot (3 + 2) = 4 \cdot (5)$

The multiplication property of equality allows us to write a very useful rule.

Multiplication and division property of equivalent equations

> **4.2B MULTIPLICATION AND DIVISION PROPERTY OF EQUIVALENT EQUATIONS**
>
> If both sides of an equation are multiplied or divided by a nonzero quantity, the new equation is equivalent to the original equation.

EXAMPLE 8 Solve $3x = 5$ for x.

$3x = 5$ — x is multiplied by 3

$\frac{3x}{3} = \frac{5}{3}$ — Dividing both sides by 3

$x = \frac{5}{3}$

Dividing by 3 is the same as multiplying by the multiplicative inverse of 3, which is $\frac{1}{3}$.

EXAMPLE 9 Solve $\frac{x}{3} = -18$ for x.

$\frac{x}{3} = -18$ — x is divided by 3

3, 3

$\_\_\_\_ \frac{x}{3} = \_\_\_\_(-18)$ — Multiplying by 3 is the inverse of dividing by 3

−54

$x = \_\_\_\_$

EXAMPLE 10 Solve $\frac{-2}{3}a = 4$ for a.

Notice that a is multiplied by $\frac{-2}{3}$. Since multiplication by the multiplicative inverse is the same as division, we can eliminate the $\frac{-2}{3}$ by multiplying both sides by $\frac{-3}{2}$.

$$\frac{-2}{3}a = 4$$

$$\frac{-3}{2} \cdot \frac{-2}{3}a = \frac{-3}{2} \cdot 4$$

$$a = \frac{-3}{\cancel{2}_1} \cdot \overset{2}{\cancel{4}}$$

$$a = -6$$

An alternate solution to this equation views $\frac{-2}{3}a$ as $-2a$ divided by 3.

$$\frac{-2a}{3} = 4 \qquad a \text{ is divided by 3}$$

$$3 \cdot \frac{-2a}{3} = 4 \cdot 3 \qquad \text{Therefore, multiply by 3}$$

$$-2a = 12 \qquad a \text{ is multiplied by } -2$$

$$\frac{-2a}{-2} = \frac{12}{-2} \qquad \text{Divide by } -2$$

-6 $\quad a =$ ______

$\frac{-2}{-2} = 1$

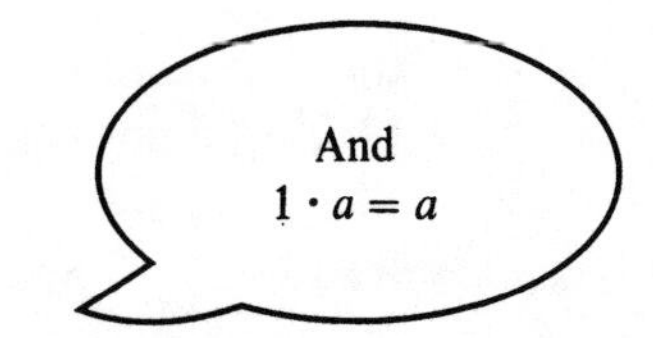

EXAMPLE 11 Solve $-x = -8$.

$$-x = -8 \qquad -x \text{ can be thought of as } -1 \cdot x$$

$$\frac{-1 \cdot x}{-1} = \frac{-8}{-1} \qquad \text{Divide both sides by } -1$$

$$x = 8$$

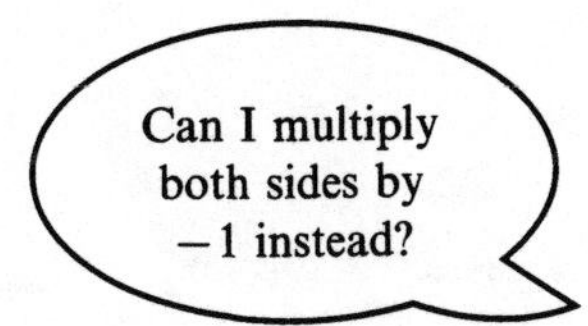

Since
$(-1) \cdot (-x) = +x$,
that will work.
Try it.

EXAMPLE 12 Solve $\frac{4}{x} = 5$ for x.

This time the variable appears in the denominator. As long as we insist that the denominators can never equal zero, there is no problem.

$$\frac{4}{x} = 5$$

$$\frac{4}{x} \cdot x = 5 \cdot x \qquad \text{Multiply both sides by } x$$

$$4 = 5x \qquad \frac{x}{x} = 1$$

$$\frac{4}{5} = \frac{5x}{5} \qquad \text{Divide both sides by 5}$$

$$\frac{4}{5} = x$$

PROBLEM SET 4.2A

Solve for the variable. Check your solutions.

1. $3x = 12$
2. $4x = -5$
3. $-6 = -4x$
4. $-12 = -8x$
5. $-1 = -2x$
6. $-5x = 0$
7. $-x = 6$
8. $-12 = -x$
9. $\frac{x}{4} = -2$
10. $\frac{x}{-3} = -2$
11. $0 = \frac{x}{-6}$
12. $\frac{-x}{4} = -2$
13. $\frac{3x}{4} = 6$
14. $-4 = \frac{4x}{5}$
15. $\frac{-2x}{7} = -1$
16. $\frac{-3x}{4} = 6$
17. $-2 = \frac{-4x}{9}$
18. $-4 = \frac{-12}{x}$
19. $-12 = \frac{-36}{x}$
20. $-8 = \frac{-48}{x}$

Now that we have the basic rules for equation solving, we can combine them to solve the following equations.

EXAMPLE 13 Solve $7x = 2x + 20$ for x.

This equation cannot be solved using either addition/subtraction or multiplication/division alone. First get all the terms involving variables on one side and all other terms on the other side.

$$7x = 2x + 20$$

$$7x - 2x = 2x - 2x + 20 \quad \text{Subtract } 2x \text{ from both sides}$$

$$5x = 20 \quad x \text{ is being multiplied by 5}$$

$$\frac{5x}{5} = \frac{20}{5} \quad \text{Divide both sides by 5}$$

4

$$x = \_\_\_\_\_$$

EXAMPLE 14 Solve $3x - 4 = 6 - 2x$.

$$3x - 4 = 6 - 2x$$

$$3x + 2x - 4 = 6 - 2x + 2x \quad \text{Add } 2x \text{ to both sides}$$

$$5x - 4 = 6$$

$$5x - 4 + 4 = 6 + 4 \quad \text{Add 4 to both sides}$$

$$5x = 10$$

$$\frac{5x}{5} = \frac{10}{5} \quad \text{Divide both sides by 5}$$

$$x = 2$$

PROBLEM SET 4.2B

Solve for the variable. Check your solutions.

1. $2x + 3 = 9$
2. $3x + 2 = 17$
3. $6x - 7 = 5$
4. $8x - 5 = 3$
5. $7 - y = 4$
6. $-12 = 3 - x$
7. $14 = 8 - 3x$
8. $2 - 5a = -1$
9. $5z = 3z + 6$
10. $4x = x - 12$
11. $3x + 2 = x$
12. $6m - 1 = 2m$
13. $5b = 12 - b$
14. $14 - 3x = 4x$
15. $2x + 3 = x - 4$
16. $5x + 4 = 2x - 11$
17. $3x + 6 = x + 16$
18. $6t - 7 = 2t + 9$
19. $2x + 4 = 6 - x$
20. $8 - 2x = 3x - 7$

EXAMPLE 15 Solve $\frac{5x - 3}{7} = 1$ for x.

$$\frac{5x-3}{7} = 1 \qquad \frac{5x-3}{7} \text{ represents a single fraction}$$

$$7 \cdot \frac{5x-3}{7} = 1 \cdot 7 \qquad \text{Multiply both sides by 7}$$

$$5x - 3 = 7 \qquad \text{Simplify}$$

$$5x - 3 + 3 = 7 + 3 \qquad \text{Add 3 to both sides}$$

$$5x = 10 \qquad \text{Simplify}$$

$$\frac{5x}{5} = \frac{10}{5} \qquad \text{Divide both sides by 5}$$

$$x = 2$$

EXAMPLE 16 Solve $\frac{5y}{6} - 3 = 7$ for y.

$$\frac{5y}{6} - 3 = 7 \qquad \text{First, clear the denominator}$$

$$6\left[\frac{5y}{6} - 3\right] = 7 \cdot 6 \qquad \text{Multiply both sides by 6}$$

$$6 \cdot \frac{5y}{6} - 6 \cdot 3 = 42 \qquad \text{Distributive property}$$

$$5y - 18 = 42 \qquad \text{Simplify}$$

$$5y - 18 + 18 = 42 + 18 \qquad \text{Add 18 to both sides}$$

$$5y = 60 \qquad \text{Simplify and notice } y \text{ is multiplied by 5}$$

$$\frac{5y}{5} = \frac{60}{5} \qquad \text{Divide both sides by 5}$$

$$y = 12$$

EXAMPLE 17 Solve $2 - \frac{3x}{4} = -1$ for x.

$$2 - \frac{3x}{4} = -1$$

$$4\left[2 - \frac{3x}{4}\right] = -1 \cdot 4$$ Multiply both sides by 4 to clear the denominator

$$4 \cdot 2 - 4 \cdot \frac{3x}{4} = -4$$ Distributive property

$$8 - 3x = -4$$ Simplify

$$(-8) + 8 - 3x = -4 + (-8)$$ Add -8 to both sides

$$-3x = -12$$ Simplify and notice x is multiplied by -3

$$\frac{-3x}{-3} = \frac{-12}{-3}$$ Divide both sides by -3

$$x = 4$$

EXAMPLE 18 Solve $\frac{7y}{5} - 2 = \frac{4}{5}$ for y.

$$\frac{7y}{5} - 2 = \frac{4}{5}$$ We'd like to avoid the denominators

$$5\left(\frac{7y}{5} - 2\right) = \frac{4}{5} \cdot 5$$ Multiply both sides by 5

$$5\left(\frac{7y}{5}\right) + 5(-2) = \frac{4}{5} \cdot 5$$ Use the distributive property to remove the parentheses on the left

$$7y - 10 = 4$$ Divide out fractions

$$7y - 10 + 10 = 4 + 10$$ Add 10 to both sides

$$7y = 14$$ 7 is multiplying the variable

$$\frac{7y}{7} = \frac{14}{7}$$ Divide both sides by 7

$$y = 2$$

PROBLEM SET 4.2C

Solve for the variable. Check your solutions.

1. $\frac{3r - 5}{4} = 1$
2. $\frac{2r - 5}{3} = 1$
3. $\frac{4r}{5} + \frac{1}{5} = 1$
4. $\frac{3r}{4} - \frac{5}{4} = 1$
5. $\frac{3r}{4} - 5 = 1$
6. $\frac{2x}{3} + 1 = -2$
7. $1 - \frac{2x}{3} = -2$
8. $-7 = 3 - \frac{5x}{2}$

9. $5s - \frac{3s}{4} = 5$ 10. $-6 = 4 - \frac{5v}{2}$ 11. $6 - \frac{3}{2}x = 9$ 12. $7 + \frac{5}{4}x = 2$

13. $2x = \frac{3x}{5} + 21$ 14. $-1 = 7 - \frac{4x}{3}$ 15. $6 - \frac{4x}{5} = 2$ 16. $4 - \frac{7x}{3} = -3$

Review Your Skills

Remove the parentheses in the following.

17. $4(x + 3)$ 18. $-2(x + 1)$ 19. $-(2a - 3)$ 20. $-(-3x + 4)$

Solving First-Degree Equations

First-degree equations

4.2C FIRST-DEGREE EQUATIONS

Equations with only a single variable raised to the first power are called *first-degree* or *linear equations.*

POINTERS FOR SOLVING FIRST-DEGREE EQUATIONS

There is no hard and fast rule concerning the order in which the steps must be taken. Keep things as simple as possible. Some ways to make things simpler are:

1. As you look at each line of the equation check to see if there are like terms that can be combined. You may find additional like terms resulting from some of the steps. If so, combine them.
2. Use multiplication to make the denominators of any fraction 1.
3. Use the distributive property to remove parentheses.
4. Use addition or subtraction to get all loose constants on one side of the equation (generally on the right side).

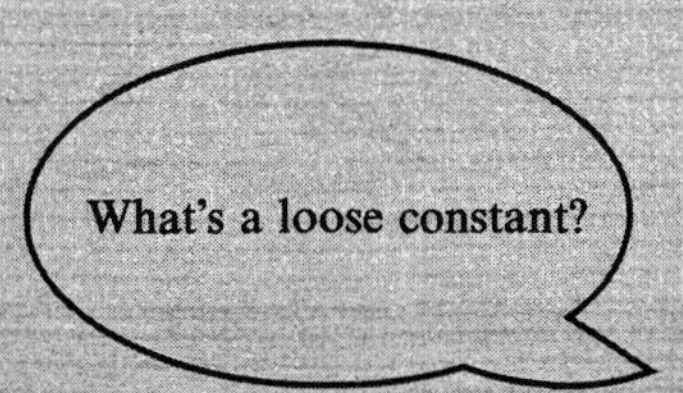

In $3x + 4$, 3 is part of a term, 4 is loose.
In $9x + 5 - \frac{x}{7}$, 9 and 7 are parts of terms, 5 is loose.

5. Use addition or subtraction to get all terms with the specified variable on one side of the equation (generally on the left side).
6. Use multiplication or division to make the coefficient of the variable 1.

Do you treat constants as like terms?

Yes, they are as much alike as a group of terms containing the same letter are alike.

EXAMPLE 19 Solve $4(a + 3) + 5 = -2(a + 1) - 5$ for a.

$4(a + 3) + 5 = -2(a + 1) - 5$

$4a + \underbrace{12 + 5} = -2a \underbrace{- 2 - 5}$ Use the distributive property to remove parentheses

$4a + 17 = -2a - 7$ Combine like terms

$4a + 2a + 17 = -2a + 2a - 7$ Eliminate variable terms on the right

−7 $6a + 17 =$ ____

$6a + 17 - 17 = -7 - 17$ Eliminate constant terms on the left

−24 $6a =$ ____

$\frac{6a}{6} = \frac{-24}{6}$ Divide both sides by 6

−4 $a =$ ____

It is a good policy to test the solution of any equation in the original equation to be sure that you have not made a mistake. Simply substitute the proposed solution for the variable.

Check.

$4(a + 3) + 5 = -2(a + 1) - 5$

−4, −4 $4[\_\_\_\_ + 3] + 5 \stackrel{?}{=} -2[\_\_\_\_ + 1] - 5$ Substituting −4 for a

−3 $4[-1] + 5 \stackrel{?}{=} -2[\_\_\_\_] - 5$ Simplify each side separately

$-4 + 5 \stackrel{?}{=} 6 - 5$

1, 1 ____ = ____ The solution is correct

Must I check
or test the solution
for all equations?

It's good practice to
make sure that no mistake
was made. However, some kinds
of equations should always be checked.
Those equations will be covered
in a later unit.

A POINTER FOR SOLVING EQUATIONS

Students often ask "What do I do first to solve an equation?" One approach is to think of solving an equation as a problem similar to the problem of a parent undressing a child in the winter. First, the outer layers of clothing like the overcoat must be removed before the inner layers can be reached. With equations, we are trying to undress the equation down to the variable.

EXAMPLE 20 Solve $\frac{3x - 8}{5} = 2$ for x.

Notice that 3 is directly next to the x. 8 is separated from the x by a subtraction sign, and 5 is on the bottom of a fraction bar. The 2 on the right-hand side of the equation is not operating on the variable at all, therefore we do not need to remove it.

$$\frac{3x-8}{5} = 2$$ 5 is the outer layer

$$\frac{3x-8}{5} \cdot \frac{5}{1} = \frac{2}{1} \cdot \frac{5}{1}$$ To remove the 5, multiply both sides by $\frac{5}{1}$

$$3x - 8 = 10$$

On the side with the variable, 8 is now farthest from the variable.

$$3x - 8 + 8 = 10 + 8$$ To remove -8, add 8 to both sides

$$3x = 18$$

The final layer to remove is 3.

$$\frac{3x}{3} = \frac{18}{3}$$ Divide both sides by 3

$$x = 6$$

EXAMPLE 21 Solve $\frac{3x}{5} - 8 = -2$.

This time the outer layer is 8, which is subtracted from the left side.

$$\frac{3x}{5} - 8 + 8 = -2 + 8$$ To remove the outer layer, add 8 to both sides

$$\frac{3x}{5} = 6$$

5 is separated from $3x$ by the fraction bar.

$$\frac{3x}{5} \cdot 5 = 6 \cdot 5$$ To remove the 5, multiply both sides by 5

$$3x = 30$$

$$x = 10$$ Divide both sides by 3

PROBLEM SET 4.2D

Solve for the variable. Check your solutions.

1. $2(x + 1) = 12$
2. $3(x + 1) = 6$
3. $2(x - 5) = -4$
4. $4(x - 3) = -2$
5. $6 - a = 3(a - 3)$
6. $8x - 5 = 5(2 + x)$
7. $2x = 7 - (x + 1)$
8. $7 - (2k + 3) = 3k$
9. $2(h + 3) = 3(2 + h)$
10. $4(2x - 5) = -(x + 2)$
11. $3(2x + 4) = 2(x - 1) + 6$
12. $4(3x - 1) = -3(x - 2) + 20$
13. $5(x - 6) + 20 = 3(x - 1) - 3$
14. $2(2x + 1) + x = 2(x - 4) + 1$
15. $3y - 2(y - 1) = 6 - (4 + y)$
16. $20 - 6(3 - n) = n - (3n + 1)$
17. $4(x + 3) - 2(x + 3) = 0$
18. $5(2c - 3) - 3(3 - c) = 0$

Review Your Skills

Simplify the following.

19. $3a - 6 + 6$
20. $-8 + 8 - 4x$
21. $3 + 5a - 5a$
22. $7a - 7a + 3$

4.3 SOLVING EQUATIONS FOR A SPECIFIED VARIABLE

When working with formulas, we don't try to remember all possible forms of the formula. Instead, we memorize one form and solve the equation for the variable that is needed. Consider the formula for simple interest.

$$I = PRT$$

where I = Interest in dollars

P = Principal

R = Annual percentage rate

T = Time in years

If you want to know the rate of interest on a loan you could memorize another formula, or you could just solve the formula above for R.

EXAMPLE 22 Solve $I = PRT$ for R.

$$I = PRT$$

P, T

What factors are multiplying R? ________

To isolate R, we will divide both sides by each factor multiplying R.

$$I = P \cdot R \cdot T$$

$$\frac{I}{PT} = \frac{P \cdot R \cdot T}{P \cdot T} \qquad \text{Divide by } P \cdot T$$

$$\frac{I}{PT} = R$$

EXAMPLE 23 Solve $I = PRT$ for T.

To solve the original formula for T, simply divide both sides by $P \cdot R$.

$$I = PRT \qquad T \text{ is multiplied by } P \cdot R$$

$$\frac{I}{PR} = \frac{P \cdot R \cdot T}{P \cdot R} \qquad \text{Divide by } P \cdot R$$

$$\frac{I}{PR} = T$$

Is it all right if I think of the letters that aren't T as constants?

It won't hurt, but try thinking of them as fixed, but not specified, numbers.

To solve any formula for a specific variable, use the same procedures you used in solving first-degree equations. The only difference is more than one letter will remain in the solved equation.

Equations with all letters seem harder to solve.

They will until you get used to it. Remember, the same rules apply. Just treat the letters as though they were numbers.

EXAMPLE 24 Solve $p = 4q(s + t)$ for q.

$$p = 4q(s + t)$$

$4(s + t)$ What factors are multiplying q? ________

$4(s + t)$ Therefore, we will divide both sides of the equation by ________.

$$\frac{p}{4(s+t)} = \frac{4q(s+t)}{4(s+t)}$$

$$\frac{p}{4(s+t)} = \frac{\cancel{4}q\cancel{(s+t)}}{\cancel{4}\cancel{(s+t)}}$$

$$\frac{p}{4(s+t)} = q$$

EXAMPLE 25 Solve $y = mx + b$ for x.

$y = mx + b$ — b is added to x

$y - b = mx + b - b$ — Remove the b by subtraction

mx $y - b =$ _____ — Observe that x is multiplied by m

$\dfrac{y-b}{m} = \dfrac{mx}{m}$ — Divide both sides by m

$\dfrac{y-b}{m} = x$ — Be careful to show that the entire side is divided by m

or

$$x = \frac{y-b}{m}$$

EXAMPLE 26 Solve $a = 8b - 4c + 6$ for b.

$$a = 8b - 4c + 6$$

6 Since 6 is added to $8b$ on the side with $8b$, we will subtract ______ from both sides of the equation.

$$a - 6 = 8b - 4c + 6 - 6$$

$$a - 6 = 8b - 4c$$

$4c$ Since $4c$ is being subtracted from $8b$, we will add ______ to both sides of the equation.

$$a - 6 + 4c = 8b - 4c + 4c$$

$$a - 6 + 4c = 8b$$

8 The factor multiplying b is ______.

8 Therefore, we will divide both sides of the equation by ______.

$$\frac{a - 6 + 4c}{8} = \frac{8b}{8}$$

$$\frac{a - 6 + 4c}{8} = b$$

EXAMPLE 27 The formula for conversion from the *Fahrenheit* temperature scale to the *Celsius* scale is

$$C = \frac{5}{9}(F - 32)$$

Find the formula for converting from the Celsius scale to the Fahrenheit scale. In other words, solve the equation for F.

Because $\frac{5}{9}$ is multiplying the quantity in parentheses with F, we will remove it by first multiplying both sides by 9 and then dividing both sides by 5.

$$C = \frac{5}{9}(F - 32)$$

$$9 \cdot C = 9 \cdot \frac{5}{9}(F - 32)$$ Multiply by 9

$$9C = 5(F - 32)$$ Observe that 5 is multiplying the quantity containing F

$$\frac{9C}{5} = \frac{5}{5}(F - 32)$$ Divide both sides by 5

$$\frac{9}{5}C = F - 32$$ Parentheses are no longer useful

$$\frac{9}{5}C + 32 = F - 32 + 32$$ Add 32 to both sides

$$\frac{9}{5}C + 32 = F$$

Water boils at 100°C. If we put this into our formula and evaluate it, we will find the Fahrenheit temperature of boiling water.

$$\frac{9}{5} \cdot C + 32 = F$$

$$\frac{9}{5}(100) + 32 = F$$

180 $$\_\_\_\_\_ + 32 = F$$

$$212 = F$$

212° Water boils at _____ Fahrenheit.

100° Celsius is an easier number to remember than 212° Fahrenheit.

Most numbers in the Metric System are easier to remember than those in the U.S. System.

EXAMPLE 28 Solve $S = \frac{n}{2}(a + l)$ for l.

$S = \frac{n}{2}(a + l)$	We want to remove the denominator
$2 \cdot S = 2 \cdot \frac{n}{2}(a + l)$	Remove the denominator by multiplying both sides by 2
$2S = n(a + l)$	l is still inside the parentheses. This time we will remove the multiplier of the parentheses
$\frac{2S}{n} = \frac{n}{n}(a + l)$	Divide both sides by n
$\frac{2S}{n} = a + l$	l would be alone if a wasn't on the right side
$\frac{2S}{n} - a = a + l - a$	Subtract a from both sides
$\frac{2S}{n} - a = l$	

In this example, we could have multiplied both sides by $\frac{2}{n}$ as the second step.

PROBLEM SET 4.3

Solve for the indicated variable.

1. $rt = d$ for t
2. $rt = d$ for r
3. $f = ma$ for m
4. $f = ma$ for a
5. $y = kx$ for k
6. $y = kx$ for x
7. $A = bh$ for b
8. $A = \frac{bh}{2}$ for b
9. $V = lwh$ for l
10. $V = lwh$ for w
11. $C = 2\pi r$ for r
12. $V = \pi r^2 h$ for h
13. $a + b = c$ for a
14. $a - b = c$ for a
15. $ax + b = 0$ for x
16. $ax + b = c$ for x
17. $P = 2l + 2w$ for l
18. $v = a + gt$ for t
19. $V = a + gt$ for g
20. $s = a + (n - 1)d$ for a
21. $s = a + (n - 1) \cdot d$ for d
22. $F = \frac{9}{5}C + 32$ for C
23. $A = \frac{h}{2}(B + b)$ for h
24. $A = \frac{h}{2}(B + b)$ for b
25. $S = \frac{h}{2}(a + n)$ for a

4.4 REVIEW OF PERCENT (OPTIONAL)

Many uses of algebra in business and science require an understanding of percent. Percents are simply fractions with a denominator of 100. Therefore, 25% means $\frac{25}{100}$.

Percent is a handy way to compare the relative size of parts by imagining everything is divided into 100 pieces.

To shade a percentage of an area, divide it into 100 equal parts and then shade the number of parts equal to the percentage desired.

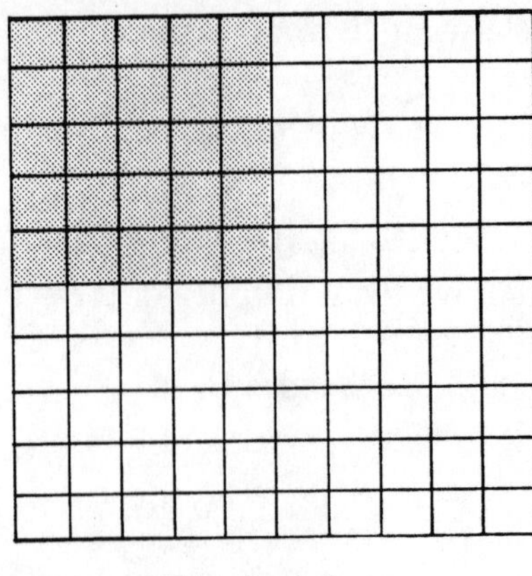

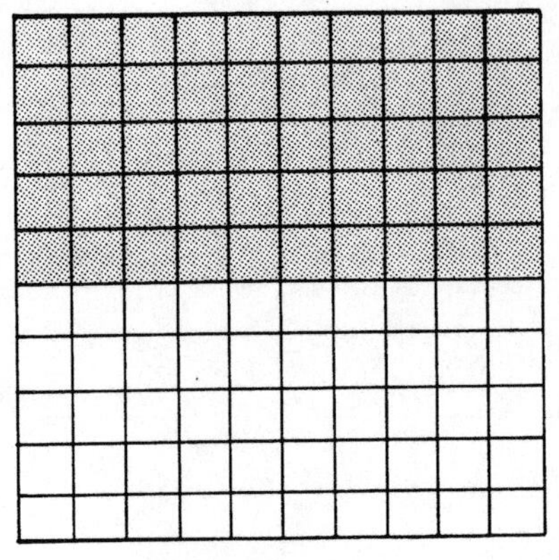

25%	Shaded
+75%	Unshaded
100%	The whole thing

50%	Shaded
+50%	Unshaded
100%	The whole thing

PROBLEM SET 4.4A

Shade the specified percent of each block below.

1. 7%

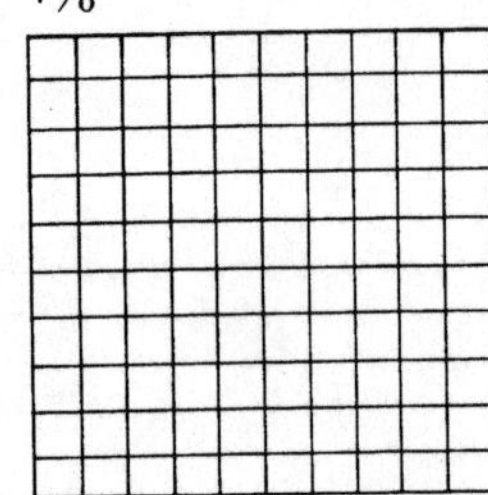

2. 70%

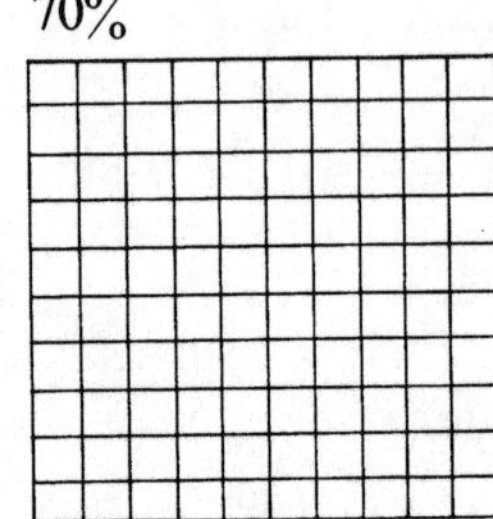

3. 30%

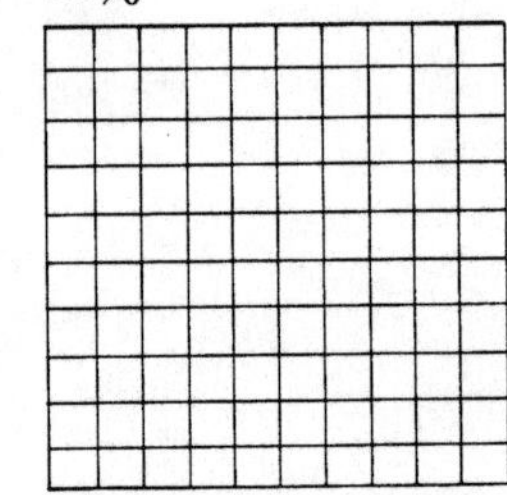

4. 1%

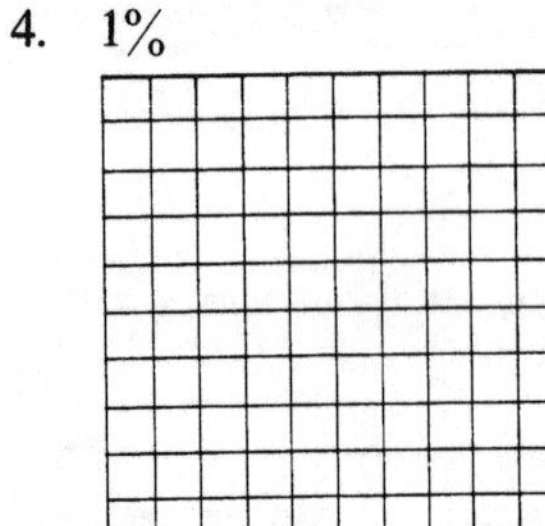

5. 10%

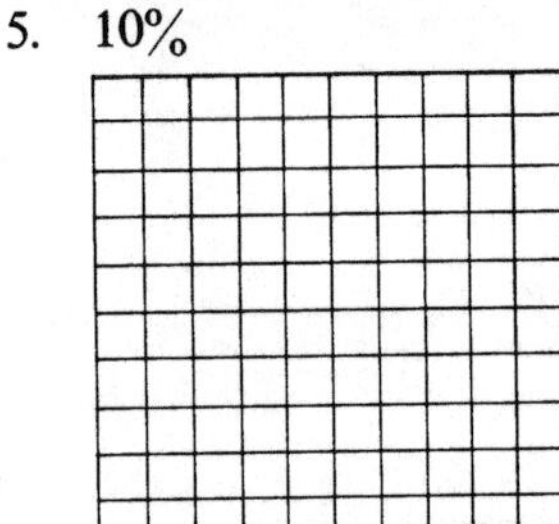

6. 90%

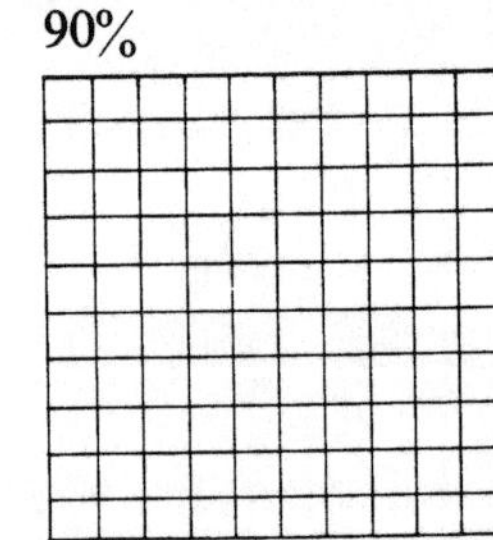

It's nice to be able to use mental images of percent. That is, 95% of something is almost all of it; 2% of something is a small part. Here are a few images for you.

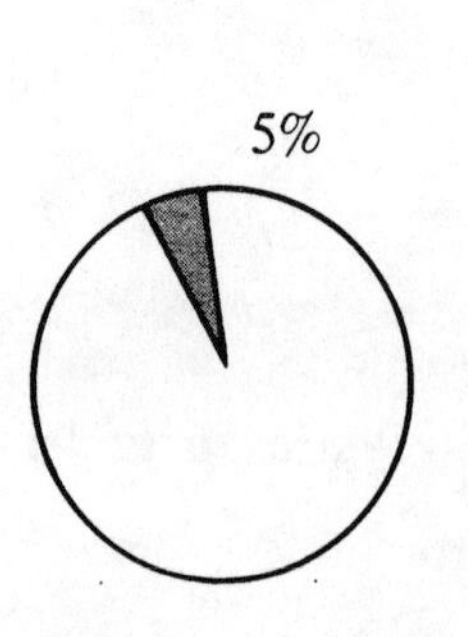

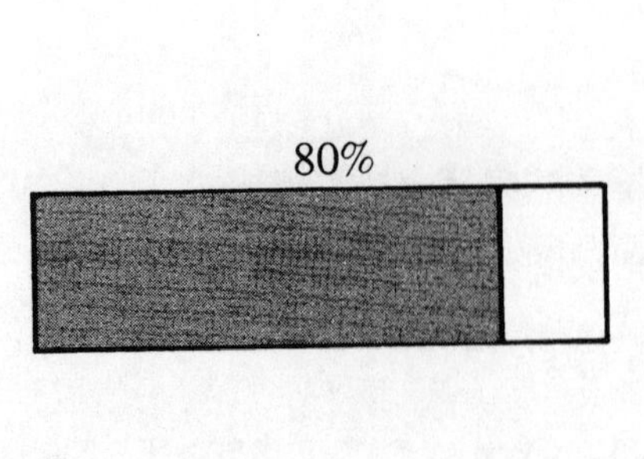

PROBLEM SET 4.4B

Shade the following to show an approximation of the percent given.

1. 20%

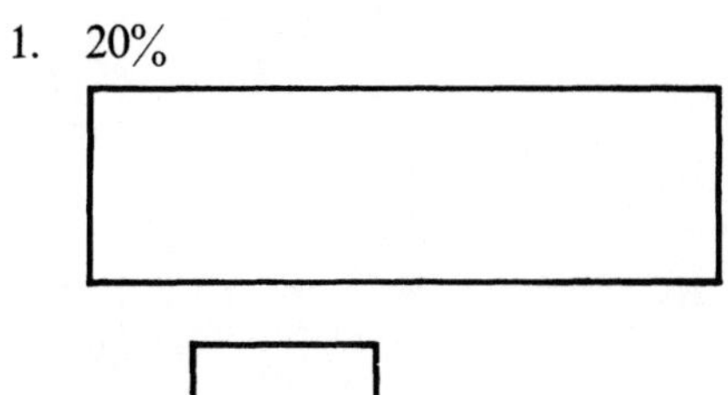

2. 25%

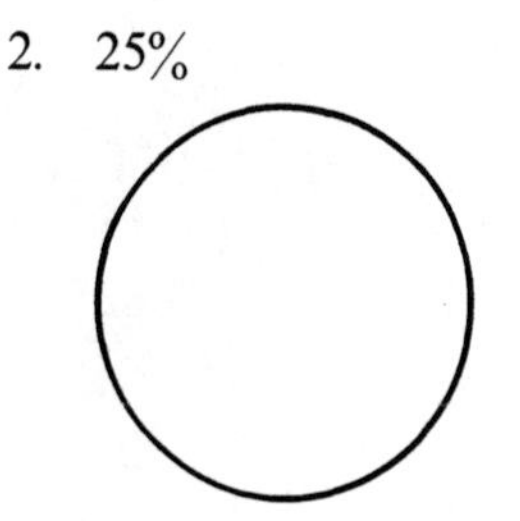

3. 3%

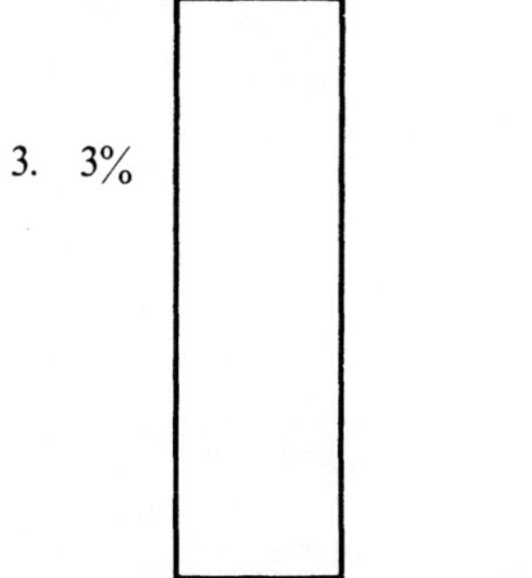

4. 60%

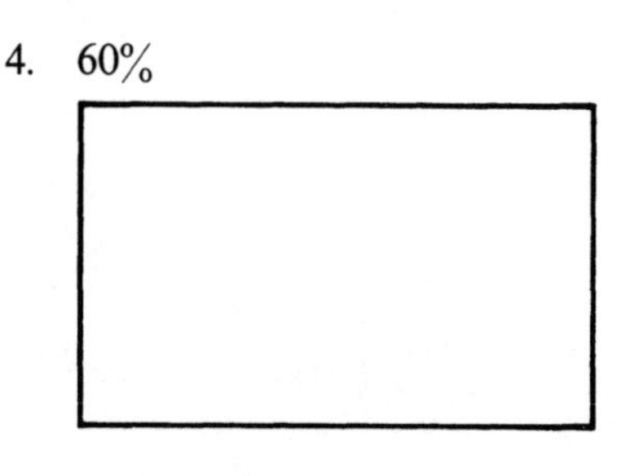

Remember, percent is a fraction. By itself, a figure like 2% represents a small part of something but whether it's a large or small amount depends on what you're taking part of. Two percent of a dollar is only 2 cents but 2% of the gross sales of General Motors just might keep you in hamburgers for a while. Before you can calculate 2% of the sales of General Motors or any other calculation with percent, it is necessary to rewrite the percentage as a decimal or a fraction.

Changing Percent Notation to Fraction Notation

Since percent means per 100, you can think of the percent sign % as $\div 100$.

% means $\div 100$. However, to divide, we multiply by the reciprocal. So practically, % means $\times \frac{1}{100}$. To convert a percent to a fraction, simply replace the % symbol with $\times \frac{1}{100}$ and simplify.

EXAMPLE 29 Write 25% as a fraction

$25\% = 25 \times \frac{1}{100}$ Replace % with $\times \frac{1}{100}$

$25\% = \frac{25}{100}$

$25\% = \frac{1}{4}$ After reducing

Bankers and advertisers like expressions such as $12\frac{1}{2}\%$. They can be converted to fractions.

EXAMPLE 30 Write $12\frac{1}{2}\%$ as a fraction.

$$12\frac{1}{2}\% = \left(12\frac{1}{2}\right) \times \frac{1}{100}$$

$$= \frac{25}{2} \times \frac{1}{100} \qquad \text{Rewrite } 12\frac{1}{2} \text{ as a pure fraction}$$

$$= \frac{\overset{1}{\cancel{25}}}{2} \times \frac{1}{\underset{4}{\cancel{100}}} \qquad \text{Simplify}$$

$$= \frac{1}{8}$$

PROBLEM SET 4.4C

Change each of the following percents to fractions reduced to lowest terms.

1. 50%	2. 70%	3. 12%	4. 40%
5. 33%	6. 13%	7. 15%	8. 68%
9. 77%	10. 35%	11. $33\frac{1}{3}\%$	12. $66\frac{2}{3}\%$
13. $16\frac{2}{3}\%$	14. $18\frac{3}{4}\%$	15. $37\frac{1}{2}\%$	16. $62\frac{1}{2}\%$
17. $\frac{1}{5}\%$	18. $\frac{1}{2}\%$	19. $\frac{3}{4}\%$	20. $\frac{1}{6}\%$
21. 100%	22. 120%	23. 150%	24. 200%

Changing Percent to Decimals

To change a percent to a decimal, we again rely on the idea that % means ÷ 100. Recall that the easy way to divide by 100 is to move the decimal point two places to the left.

$$234.5 \div 100 = 2.345$$

$$4.56 \div 100 = 0.0456$$

$$70 \div 100 = 0.70$$

We do the same to write a percent as a decimal fraction.

EXAMPLE 31 Write 25% as a decimal.

$25\% = 0.25$ Percent sign is removed and decimal is moved two places to the left

The decimal is at the end of a whole number.

EXAMPLE 32 Write $37\frac{1}{2}\%$ as a decimal.

$37\frac{1}{2}\% = 37.5\%$	First, replace the $\frac{1}{2}$ with 0.5
$= 0.375$	Remove the % sign and move the decimal two places to the left

PROBLEM SET 4.4D

Change the following percents to decimals.

1. 10%	2. 20%	3. 15%	4. 12%
5. 7%	6. 8%	7. 5.6%	8. 7.9%
9. 30.25%	10. 98.09%	11. $16\frac{1}{2}\%$	12. $34\frac{1}{4}\%$
13. $78\frac{3}{4}\%$	14. $92\frac{2}{5}\%$	15. $6\frac{1}{4}\%$	16. $18\frac{3}{4}\%$
17. $37\frac{1}{2}\%$	18. $62\frac{1}{2}\%$	19. $\frac{1}{2}\%$	20. $\frac{2}{5}\%$
21. $\frac{1}{10}\%$	22. $\frac{3}{4}\%$	23. $\frac{1}{4}\%$	24. $\frac{1}{5}\%$

Changing from Decimal Notation to Percent Notation

To change from a percent to a decimal, we moved the decimal point two places to the left. To go the other way, simply move the decimal in the other direction and add a percent sign. Here's an example to show why this rule is mathematically correct.

EXAMPLE 33 Write 0.625 as a percent.

Remember, you can multiply anything by 1 without changing its value. In this case, we will write 1 as $\frac{100}{1} \cdot \frac{1}{100}$.

$0.625 = 0.625 \times \frac{100}{1} \times \frac{1}{100}$	Multiply by 1
$= 62.5 \times \frac{1}{100}$	After multiplying 0.625 by $\frac{100}{1}$
$= 62.5\%$	Replace $\times\frac{1}{100}$ with %

EXAMPLE 34 Here are more examples of decimals and the equivalent percent notation.

$0.70 = 70\%$ $\quad 0.05 = 5\%$ $\quad 1.0 = 100\%$ $\quad 2 = 200\%$

PROBLEM SET 4.4E

Change the following decimals to percent notation.

1. 0.45 2. 0.20 3. 0.50 4. 0.51 5. 0.125

6. 0.165 7. 0.06 8. 0.08 9. 0.2 10. 0.9

11. 3.6 12. 5.8 13. 3.56 14. 9.02 15. 0.1625

16. 0.1875 17. 0.005 18. 0.0025 19. 5.0 20. 8.0

Review Your Skills

Reduce the following fractions to lowest terms.

21. $\frac{18}{51}$ 22. $\frac{15}{39}$ 23. $\frac{125}{100}$ 24. $\frac{175}{100}$

Changing from Fraction Notation to Percent Notation

If fractions have denominators of 100, it's easy to convert to a percent.

$$\frac{60}{100} = 60\% \qquad \frac{20}{100} = 20\% \qquad \frac{10}{100} = 10\%$$

For those fractions that don't have a denominator of 100, we multiply by the number 1, written as $\frac{100}{1} \times \frac{1}{100}$, to introduce a denominator of 100. Remember, anything multiplied by 1 yields the same thing.

EXAMPLE 35 Write $\frac{1}{4}$ as a percent.

$$\frac{1}{4} = \frac{1}{4} \times 100 \times \frac{1}{100}$$ ← This is equal to one

$$= \frac{1}{4} \times 100\%$$ Replace $\times \frac{1}{100}$ with %

$$= \frac{1}{\not{4}_{1}} \times \not{100}^{25}\%$$ Simplify

$$= 25\%$$

EXAMPLE 36 Write $\frac{3}{8}$ as a percent.

$$\frac{3}{8} = \frac{3}{8} \times 100 \times \frac{1}{100}$$ Multiply by 1

$$= \frac{3}{\not{8}_{2}} \times \not{100}^{25}\%$$ Replace $\times \frac{1}{100}$ with %

$$= \frac{75}{2}\%$$ To simplify $\frac{75}{2}$, we divide $75 \div 2$ and get a decimal result

$$= 37.5\%$$

EXAMPLE 37 Write $\frac{2}{3}$ as a percent.

$$\frac{2}{3} = \frac{2}{3} \times 100 \times \frac{1}{100} \qquad \text{Multiply by 1}$$

$$= \frac{200}{3}\% \qquad \text{Replace } \times \frac{1}{100} \text{ with } \%$$

$$= 66\frac{2}{3}\%$$

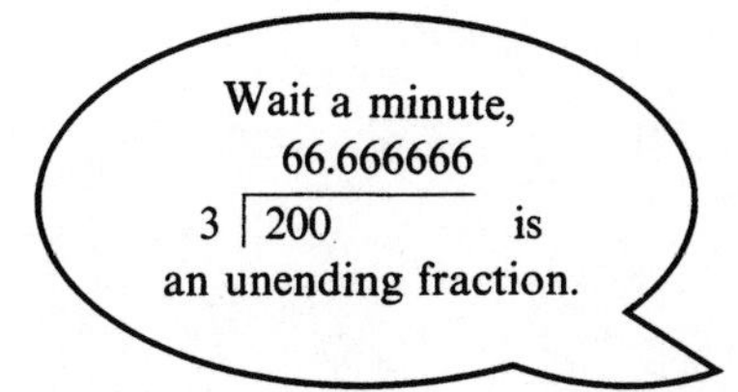

If you're using a calculator, use 66.667%, or you can write $66\frac{2}{3}\%$.

The easy way out of this is to memorize:

$$33\frac{1}{3}\% = \frac{1}{3} \qquad 66\frac{2}{3}\% = \frac{2}{3}$$

PROBLEM SET 4.4F

Change the following fractions to percents.

1. $\frac{1}{2}$
2. $\frac{3}{4}$
3. $\frac{1}{5}$
4. $\frac{2}{5}$
5. $\frac{4}{5}$
6. $\frac{3}{5}$
7. $\frac{1}{3}$
8. $\frac{5}{8}$
9. $\frac{5}{6}$
10. $\frac{3}{16}$
11. $\frac{5}{16}$
12. $\frac{3}{32}$
13. $\frac{7}{1000}$
14. $\frac{1}{6}$
15. $\frac{7}{8}$
16. $\frac{5}{32}$

Using Percentages

The basic formula for finding a percentage of something is the same formula used to find a fraction of something.

The basic formula used to find a fraction of a number is

Fraction × Whole = Amount

To find $\frac{3}{4}$ of 100, we apply this formula

$$\text{Fraction} \times \text{Whole} = \text{Amount}$$

$$\frac{3}{4} \cdot 100 = \text{Amount}$$

$$75 = \text{Amount}$$

Because percent is just another way to express a fraction, the same formula is used to find percentage. To work percent problems,

Fraction × Whole = Amount is written

$$\text{Rate} \times \text{Base} = \text{Amount}$$

$$R \cdot B = A$$

where R = Rate, which is the percent expressed as a fraction or decimal

B = Base or quantity that we are seeking a part of

A = Amount or portion resulting

EXAMPLE 38 What amount is 30% of 120?

Here are two ways to solve the problem.

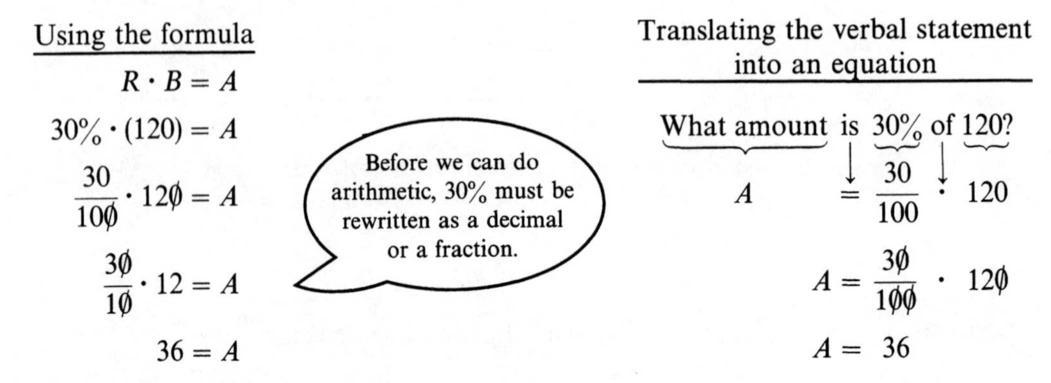

Using the formula

$$R \cdot B = A$$

$$30\% \cdot (120) = A$$

$$\frac{30}{10\cancel{0}} \cdot 12\cancel{0} = A$$

$$\frac{3\cancel{0}}{1\cancel{0}} \cdot 12 = A$$

$$36 = A$$

Translating the verbal statement into an equation

What amount is 30% of 120?

$$A = \frac{30}{100} \cdot 120$$

$$A = \frac{3\cancel{0}}{1\cancel{0}\cancel{0}} \cdot 12\cancel{0}$$

$$A = 36$$

EXAMPLE 39 What is 14.25% of $3,000?

This time it will be easier to express the percent as a decimal: 14.25% = 0.1425.

Using the formula

$$R \cdot B = A$$

$$14.25\% \cdot (\$3{,}000) = A$$

$$0.1425(\$3{,}000) = A$$

$$\$427.50 = A$$

Translating the verbal statement into an equation

What is 14.25% of $3,000?

$$A = 0.1425 \cdot (\$3{,}000)$$

$$A = \$427.50$$

This example tells you how much interest you would collect from $3,000 if you invested it for one year at 14.25% simple interest.

EXAMPLE 40 30% of what number is 120?

In this example, we are looking for the base. Again we show two solutions.

Using the formula

$$R \cdot B = A$$

$$30\% \cdot B = 120$$

$$\frac{30}{100} \cdot B = 120$$

$$\frac{\cancel{100}}{\cancel{30}} \cdot \frac{\cancel{30}}{\cancel{100}} \cdot B = 12\cancel{0} \cdot \frac{100}{3\cancel{0}}$$

$$B = \underline{\qquad}$$

Translating the verbal statement into an equation

30% of what number is 120?

$$\frac{30}{100} \cdot x = 120$$

$$\frac{\cancel{100}}{\cancel{30}} \cdot \frac{\cancel{30}}{\cancel{100}} \cdot x = 12\cancel{0} \cdot \frac{100}{3\cancel{0}}$$

$$x = \underline{\qquad}$$

400, 400

30% of 400 is 120.

EXAMPLE 41 What percent of 80 is 16?

$$\underbrace{\text{What percent}}_{x}\ \underbrace{\text{of}}_{\cdot}\ \underbrace{80}_{80}\ \underbrace{\text{is}}_{=}\ \underbrace{16}_{16}$$

$$x \cdot 80 = 16$$

$$\frac{x \cdot 80}{80} = \frac{16}{80}$$

$$\frac{x \cdot \overset{1}{\cancel{80}}}{\underset{1}{\cancel{80}}} = \frac{\overset{1}{\cancel{16}}}{\underset{5}{\cancel{80}}}$$

$$x = \frac{1}{5}$$

Expressing x as a percent,

$$5 \overline{)1.00} \quad 0.20$$
$$\underline{1\,0}$$
$$00$$

$$x = 20\%$$

Why 20% rather than $\frac{1}{5}$?

Because the problem asked "What percent?" 20% and $\frac{1}{5}$ mean the same thing.

PROBLEM SET 4.4G

Solve the following.

1. What is 28% of 150?
2. What is 15% of 40?
3. What percent of 160 is 40?
4. What percent of 120 is 30?
5. 40% of what number is 56?
6. $85\frac{1}{2}\%$ of 400 is what?
7. $37\frac{1}{2}\%$ of what number is 33?
8. 13 is what percent of 65?
9. What is 125% of 96?
10. $62\frac{1}{2}\%$ of what number is 85?
11. 99 is what percent of 72?
12. What is $\frac{1}{2}\%$ of 860?

4.5 APPLICATIONS OF EQUATIONS

There are many uses for equations in business and science. Percentage equations in business is a good example of the power of algebra to make complicated arithmetic simple.

Percents are simply fractions with a denominator of 100. Therefore, 25% means $\frac{25}{100}$.

The basic formula to find a percentage is:

Percent formula

Rate × Base = Amount

$$R \cdot B = A$$

where R = Rate expressed as a percent

B = Base or quantity that we are seeking a part of

A = Amount or portion resulting

To show how this formula works, let us use a practical problem. Most states charge a sales tax. While the rate of sales tax varies from state to state, it is frequently 6%.

EXAMPLE 42 Suppose a couple purchased a chair for $200 and the rate of sales tax was 6%. Find the amount of sales tax paid and the total purchase price.

Use the formula $R \cdot B = A$.

The Rate R is 6%.

The Base B is $200.

The tax paid is based on the size of the purchase; therefore, $B = \$200$.

Find A

$$R \cdot B = A$$

$$6\%(200) = A$$

$$\frac{6}{100}(200) = A$$

$$12 = A$$

The amount of sales tax is $12. The total paid is $200 + $12 = $212.

The problem of finding the sales tax could have been stated:

What amount is 6% of $200?

$$A = \frac{6}{100} \cdot \$200$$

$$A = 12$$

Why not use 0.06 instead of $\frac{6}{100}$ for 6%?

Use whatever is easier for you in the problem. If you were using a calculator, 0.06 would have been easier.

With the percent formula there are three variables: *Base*, *Rate*, and *Amount*. If we know values for any two of these variables, we can find the value of the third.

EXAMPLE 43 Using the same chair example above, suppose that the couple got home and looked at their bill. They saw a price of \$200 and a sales tax of \$12. Here is how they calculated the rate of tax while sitting in the new chair.

Use the formula $R \cdot B = A$.

The Rate R is unknown.

The Base B is \$200.

The Amount A is \$12.

Find R

$$R \cdot B = A$$
$$R(\$200) = \$12$$
$$\frac{200R}{200} = \frac{12}{200}$$
$$R = 0.06$$
$$R = 6\%$$

The problem of finding the rate could have been stated:

$$\underbrace{\text{What percent}}_{R}\ \underset{\cdot}{\text{of}}\ \underbrace{\$200}_{200}\ \underset{=}{\text{is}}\ \underbrace{\$12?}_{12}$$

The rate R is 6%.

A third use of the percent formula is when the rate and amount are known and we want to find the base.

EXAMPLE 44 An amusement park owner knows from past ticket sales that 8% of the people who visit will ride the Thriller before 10 A.M. The owner uses this information to get an early estimate of the daily crowd in the park. With the crowd estimate, extra employees can be called in to work if necessary.

If 1,152 people rode the Thriller by 10 A.M., what attendance can be expected?

We know

Rate = 8%

Amount = 1,152 Thriller riders

We want to find

Base = projected attendance for the day

$$\text{Rate} \cdot \text{Base} = \text{Amount}$$
$$8\% \cdot B = 1{,}152$$
$$0.08 \cdot B = 1{,}152$$
$$\frac{0.08}{0.08} \cdot B = \frac{1{,}152}{0.08}$$
$$B = 14{,}400$$

The owner can expect 14,400 visitors for the day.

This problem could have been set up this way:

$$\underbrace{8\%}_{0.08}\ \underset{\cdot}{\text{of}}\ \underbrace{\text{total attendance}}_{x}\ \underset{=}{\text{is}}\ \underbrace{1{,}152 \text{ Thriller riders}}_{1{,}152}$$

PROBLEM SET 4.5A

Solve the following problems.

1. After rebates and discounts, a car listed at $9,375 was bought for $7,500. What percent of the list price was paid?
2. On a test with 100 questions, a student received 84%. How many questions did he get correct?
3. A baseball player gets a hit 40% of the time he's at bat. If he went to bat 220 times during the season, how many hits did he get?
4. A baseball player gets a hit 30% of the time he's at bat. If he got 90 hits during the season, how many times was he at bat?
5. On a test with 50 questions, a student received 70%. How many questions did the student get correct?
6. A $680 sofa was bought on sale for $510. What percent of the original price was paid?
7. Mr. Salesman, who earned a 15% commission on all sales, sold $2,500 worth of TVs during one week. How much did he earn selling TVs?
8. An appliance saleswoman sold $3,200 worth of appliances in one week. Her earnings were $384. What was her rate of commission?
9. A woman paid $2.34 sales tax on a $39 dress. What was the rate of sales tax?
10. Teri paid $1.41 sales tax on a $23.50 item. What was the rate of sales tax?
11. A businessman made a $396 profit in a project that paid 11% yearly interest. How much money did he invest at 11%?
12. A businesswoman invested $2,000 and made a $300 profit. What was her rate of profit?
13. A student got 16 questions correct on a test and received 80%. How many questions were on the test?
14. A family must make a 20% down payment on a $96,000 home. How much of a down payment must they make?
15. A candidate received 3,000 votes out of the 24,000 votes cast. What percent of the votes did the candidate receive?
16. Poly High School played 25 basketball games and lost 6. What percent of the games did they lose?
17. Clerks at a music store are allowed a discount of 15%. What would a clerk pay for a $12 compact disk?
18. A store advertises towels at "$33\frac{1}{3}\%$ off." What will towels marked $7.50 sell for?
19. A man earning $150 a week got a $30-a-week raise. What was the percent increase?
20. A store paid 50¢ for melons and sold them for 63¢. Based on the store's cost, what was the percent of profit or increase?
21. A man bought a car for $6,000. At the end of one year, he sold the car but could only get $2,500 for it. What was the percent of decrease?
22. A 4-ounce sample of ground meat was tested and found to contain 1.3 ounces of fat. What percent of the meat was fat?
23. Thirty-five percent of a family's income is spent for food. If $532 a month is spent for food, how much is the income each month?
24. You work hard, so you get a 6% raise. That means you get $9 more a week. How much were you making each week before the raise?
25. John saves $17 a week, which is 10% of his pay. What is his regular pay?

26. Mama Lulu's recipe for pizza sauce calls for 11% herbs and spices. If 97.9 gallons of herbs and spices are used, how much pizza sauce will be made?

27. Jeff McNett's overtime pay is $17.46, which is 12% of his total pay. Find his total pay.

28. At Goldie's Delicatessen, 16% of all customers order a dill pickle. In a recent week, 372 pickles were sold. Find the total number of customers.

29. Susan Wright estimates that the total sales at her Magic Mart food store next month will be $175,000, with total advertising expenses of $3,500. What percent of the total sales will be spent on advertising?

30. Sid's Pharmacy has a total monthly payroll of $6,800, with $1,496 of this amount going for fringe benefits. What percent of the total payroll goes for fringe benefits?

31. Tom paid $1.68 sales tax on a $33.60 item. What was the rate of sales tax?

32. A businesswoman invested $2,000 and lost $300. What was her rate of loss?

Review Your Skills

Change the following percents to decimals.

33. 7% 34. 8% 35. $12\frac{1}{2}\%$ 36. $8\frac{1}{4}\%$

Calculate Interest

Simple interest method

A very basic method of calculating interest is called the *simple interest method.* If the simple interest method is used to calculate the interest, the interest is paid only on the principal amount for a specified period. Simple interest can be computed by using the formula $I = PRT$, where I is the interest, P is the principal, R is the rate, and T is the time in years.

EXAMPLE 45 How much interest is earned on a $500 investment at 6% interest per year for two years?

$$I = P \cdot R \cdot T$$

$$I = 500 \cdot 6\% \cdot 2$$

$\frac{6}{100}$

$$I = 500 \cdot \underline{\qquad} \cdot 2$$

$$I = 500 \cdot \frac{6}{\cancel{100}} \cdot 2$$

(with 500 reduced to 5 and 100 reduced to 1)

$60

$$I = \underline{\qquad}$$

EXAMPLE 46 What principal must be invested at 7% per year to earn $168 in one year?

$$I = P \cdot R \cdot T$$

$$168 = P \cdot 7\% \cdot 1$$

$$168 = P \cdot \frac{7}{100}$$

$\frac{100}{7}, \frac{100}{7}$

$$\left(\underline{\qquad}\right) \cdot 168 = P \cdot \frac{7}{100} \cdot \left(\underline{\qquad}\right)$$

$2,400

$$\underline{\qquad} = P$$

This could also have been solved by first multiplying both sides by 100 and then dividing both sides by 7.

Another application of equations is the formula used to calculate distance.

EXAMPLE 47 The distance traveled by a car is given by the formula

Distance formula $d = r \cdot t$

How fast would you have to drive to go 120 miles in $2\frac{1}{2}$ hours?

To solve this problem, substitute for the values that are known in the formula.

$$d = r \cdot t$$

$$120 = r \cdot \frac{5}{2} \qquad 2\frac{1}{2} \text{ hrs} = \frac{5}{2} \text{ hrs}$$

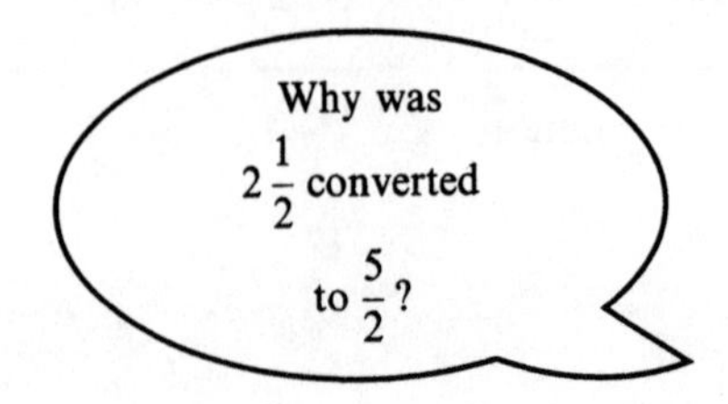

It avoids confusing notation. $2\frac{1}{2}$ means $2 + \frac{1}{2}$, but $2\frac{1}{x}$ means $2 \cdot \frac{1}{x}$. So in algebra, we stay away from mixed numbers.

To solve for r, multiply both sides by 2.

$$2 \cdot 120 = 2 \cdot r \cdot \frac{5}{2}$$

$240 = 5r$ 5 is multiplying r

$\frac{240}{5} = \frac{5r}{5}$ Divide both sides by 5

48

$\_\_\_\_\_ = r$

$$r = 48 \frac{\text{miles}}{\text{hour}}$$

There are many geometric uses for equations. One of these uses is the formula for the volume of a rectangular solid.

EXAMPLE 48 The volume of a rectangular solid is given by the formula

Volume formula $V = l \cdot w \cdot h$

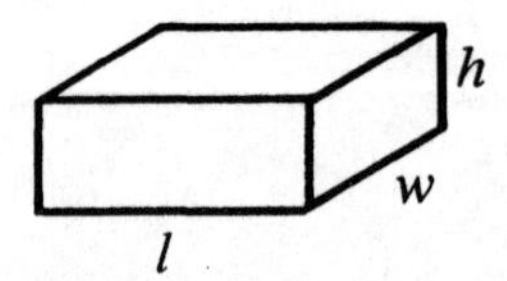

Find the volume of a box 20 centimeters long, 3 centimeters wide, and 4 centimeters high.

$V = l \cdot w \cdot h$

$V = (20)(3)(4)$ Substituting

240

$V = \_\_\_\_\_$ cubic centimeters

EXAMPLE 49 Find the length of a box 4 meters wide and 5 meters high, if its volume is to be 120 cubic meters.

Substituting known quantities in the formula

$$V = l \cdot w \cdot h$$

$$(120) = l \cdot (4) \cdot (5)$$

20

$$120 = l \cdot \_\_\_\_ \qquad \text{Simplifying}$$

$$\frac{120}{20} = l \cdot \frac{20}{20}$$

6

$$\_\_\_\_ = l$$

$$l = 6 \text{ meters}$$

PROBLEM SET 4.5B

Solve the following.

1. How much money must be invested at 9% per year to earn \$180 in one year?
2. How many years will it take a \$1,000 investment to earn \$400 at 8% per year?
3. If \$1,000 is invested at 8% per year and \$2,000 is invested at 9% per year, what is the total amount of interest earned in a year?
4. At what rate would \$3,000 have to be invested to earn \$720 in two years?
5. Using the distance formula $d = rt$, how long will a car traveling at 52 miles per hour take to travel 182 miles?
6. Find the height (h) of a rectangle solid if the volume (V) is 210 cubic meters, the length (l) 12 meters, and the width (w) 7 meters. Use the formula $V = lwh$.
7. $V = \pi r^2 h$ is the formula for the volume of a cylinder. Find the height (h) of a cylinder with volume (V) 628 cubic centimeters and radius (r) 10 centimeters. Use 3.14 for the value of π.
8. A useful physics formula is $E = IR$, which says that the voltage drop across a resistor is equal to the current times the resistance in ohms. Find the resistance (R) of an electric iron that draws 10 amperes of current (I) at a voltage (E) of 110 volts.
9. Find the current (I) in an electrical circuit if the resistance (R) is 15 ohms and the voltage (E) is 210 volts. Use the formula $E = IR$.
10. Use the formula $F = \frac{9}{5}C + 32$ to find the Celsius temperature if the Fahrenheit reading is 86.

Hint: Perimeter is the distance around the outside.

11. The formula for the perimeter of a square is $P = 4s$. Find the length of the side of a square with perimeter 25.24 meters.
12. The area of a triangle is given by the formula $A = \frac{bh}{2}$. Find the length of the base (b) of a triangle with an area of 27 square centimeters and a height (h) of 12 centimeters.
13. Use the formula $P = 2l + 2w$ to find the length (l) of a rectangle whose perimeter (P) is 20 centimeters and width (w) is $2\frac{1}{2}$ centimeters.
14. People's intelligence quotients, or IQ, are determined by multiplying their mental age by 100 and dividing this product by their chronological age, $Q = \frac{100 \cdot (m)}{c}$. Find the mental age ($m$) of a person who has an IQ (Q) of 120 and a chronological age (c) of 20 years.

15. When shopping for some furniture, Mr. and Mrs. Newlywed found a sofa and two chairs in their town for \$950. The sales tax in their state was 4%. Just across the state line they could buy the same pieces of furniture for \$935, but the sales tax was 6%. Which was the better buy?

16. Use the formula $C = \frac{5}{9}(F - 32)$ to find the temperature in Fahrenheit (F) if the Celsius reading (C) is 25°.

17. The "High Pressure 'em" used car lot offers two plans of earnings for its salespeople. Plan I calls for a guaranteed salary of \$100 per week plus 5% commission on all sales. Plan II calls for a guaranteed salary of \$140 per week plus 4% commission. A salesperson who averages \$6,500 in sales a week should adopt which plan?

Review 4

NOW THAT YOU HAVE COMPLETED UNIT 4, YOU SHOULD BE ABLE TO:

Use the following definitions in your working vocabulary.

Conditional equation—An equation that is true only for some values of the variable.

Solution set—The set of values that makes an equation true.

Equivalent equations—Equations that have the same solution set.

Apply the following rules and properties to solve equations.

The Addition and Subtraction Property of Equivalent Equations

If the same quantity is added to or subtracted from both sides of an equation, the new equation is equivalent to the original equation.

The Multiplication and Division Property of Equivalent Equations

If both sides of an equation are multiplied or divided by a nonzero quantity, the new equation is equivalent to the original equation.

Symmetric Property of Equality

If $a = b$, then $b = a$, where a and b represent any algebraic expression.

Use the following steps to solve first-degree equations.

1. *Simplify* by combining any like terms. Use the distributive property to *remove parentheses.*
2. Use addition and subtraction to *get all terms with the specified variable on one side* and all other terms on the other side.
3. *Combine any like terms.*
4. Use multiplication to *remove any fractions.*
5. Use multiplication or division to *make the coefficient of the variable 1.*

Solve equations for a specified variable in terms of the remaining variables.

Treat all the remaining variables and costants as though they were numbers. The answer will probably have letters on one side but the desired variable will be alone on the other side.

REVIEW TEST 4

Solve for the variable. *(4.1)*

1. $x + 6 = -18$ ____

2. $7b + 4 - 6b = 5 - 9$ ____

3. $2a - 4 = a - 12$ ____

4. $3x + 2(x - 3) = 4(x + 2)$ ____

Solve for the variable. *(4.2)*

5. $-x = 5$ ____

6. $5b - 1 = 2b + 3$ ____

7. $\frac{z}{6} = -3$ ____

8. $3x = 10 - (x - 2)$ ____

9. $2(x - 3) = 4(x - 3)$ ____

10. $5(x + 2) + x = 3(x - 1) - 2$ ____

11. $\frac{t}{3} + 1 = 3$ ____

12. $\frac{6y - 3y}{2} = -6$ ____

Solve for the indicated variable. *(4.3)*

13. $x + y = z$ for y ____

14. $P = 2l + 2w$ for l ____

15. $y = \frac{x}{k}$ for x ____

16. $V = lwh$ for w ____

17. $l = a + (n - 1)d$ for n ____

18. $s = 2r(r + h)$ for h ____

Change the following percents to fractions reduced to lowest terms. (4.4)

19. 18%

20. $87\frac{1}{2}\%$

21. $\frac{2}{5}\%$

22. 140%

Change the following percents to decimals. (4.4)

23. 18%

24. 6.2%

25. $8\frac{3}{4}\%$

26. $\frac{3}{5}\%$

Change the following decimals to percent notation. (4.4)

27. 0.39

28. 0.6

29. 0.03

30. 0.2475

Change the following fractions to percents. (4.4)

31. $\frac{4}{5}$

32. $\frac{3}{1000}$

33. $\frac{9}{16}$

34. $\frac{7}{32}$

35. What is $42\frac{1}{2}\%$ of 124?

36. 36 is what percent of 135?

37. $83\frac{1}{3}\%$ of what number is 210?

Solve the following. (4.5)

38. Use the formula $F = \frac{9}{5}C + 32$ to find the Celsius temperature (C) if the Fahrenheit (F) reading is 95.

39. Using the formula $I = P \cdot R \cdot T$, find the rate of interest (R) on a principal (P) of \$800 for 4 years (T) if the interest (I) is \$280.

40. $P = 2l + 2w$ is the formula for the perimeter of a rectangle. Find the width (w) of a rectangle with a perimeter (P) of 108 inches and a length (l) of 30 inches.

41. The area of a triangle is given by the formula $A = \frac{b \cdot h}{2}$. Find the height (h) of the triangle if the base (b) is 9 centimeters and the area (A) is 144 square centimeters.

42. In one year, the price of gasoline increased 15%. The new price was 129.9 cents per gallon. To the nearest tenth of a cent, what was the price of gasoline at the beginning of the year?

43. A computer store pays a straight commission of 12% to its salespeople. After the commission is paid, the management receives \$2,868.80 for a personal computer system. What was the original price of the system?

44. The enrollment at Johnson High School dropped from 2,050 to 1,722 during a 2-year period. What was the percent of decrease in enrollment?

REVIEW YOUR SKILLS

Evaluate the following.

45. $(-6 + 4)^3 - (-8) \cdot (3 - 5)$

46. $8^2 \div 4 - 2 + 6$

Multiply using the distributive property.

47. $-3x^2ya^4\,(-x^2y^3 + 5a^3 - 4x^3y^2a^0)$

Simplify.

48. $-6x + 3y - 8xy + 3x^2y - 3y - 5xy + 4x - 8x^2y$

Cumulative Review 1–4

Add or subtract the following. *(1.2)*

1. $-13-(-12)$
2. $-5-(4-6)$
3. $-3-8+(-2)$

Multiply or divide the following. *(1.3)*

4. $(-6)(12)$
5. $(-39)\div(-3)$

Evaluate.

6. $16\div(-4)\cdot 2$
7. $6^2\div 4-2+4$
8. $[8^2\cdot 2+4]\div 4$ *(1.4)*
9. $3a^2-2a+5$ if $a=-2$ *(2.1)*
10. $a^2(2a^2b^3-c^2)$ if $a=-2$, $b=-1$, and $c=2$ *(2.1)*
11. $3x-(8a+2)$ if $x=\frac{2}{3}$ and $a=-\frac{1}{4}$ *(3.4)*
12. $\frac{1}{3}a^3-12ax^2-6xy$ if $a=3$, $x=-\frac{1}{2}$, and $y=\frac{1}{3}$ *(3.4)*

Multiply the following. *(2.2)*

13. $(-5x^4)(2y^3)$
14. $(-4a^2b^3c^0)(-6a^0b^2c)$

Reduce to lowest terms. *(3.1)*

15. $\frac{-60}{-72}$
16. $-\frac{-18ay}{-15}$

Perform the indicated operations. Reduce all answers to lowest terms, and write the fractions in standard form. *(3.2)*

17. $\frac{24xy}{-42}\cdot\frac{56x^2}{32}$
18. $\frac{-36ab^2}{24}\div\frac{-15}{10a}$

Use the distributive property to simplify the following. Write the answer without parentheses.

19. $-2x^2(x^2 - 3x + 4)$ ***(2.3)***

20. $-4a^2b^3(-a^0b + 2a^3b^2c - 3a^3b^0c^2)$ ***(2.3)***

21. $-\frac{2}{3}(-15a^2 + 9x)$ ***(3.2)***

Add or subtract the following polynomials. (2.4)

22. $(-5x + 4) - (-8x - 2)$

23. $(4y^2 - 6y + 1) - (2y^2 + 4y - 3)$

24. $(a^2 + 3b - 4) + (-2a^2 - 4a + 2) - (3a^2 - 5a + 4b - 6)$

Simplify. (2.4)

25. $-4xy + 4x - x^2y + 4 - 5x + 2xy + 4x^2y$

Add or subtract the following. Reduce all answers to lowest terms and write the fractions in standard form. (3.3)

26. $\frac{-11xy}{12} + \frac{5xy}{12}$

27. $\frac{-5ab^2}{9} - \frac{-7ab^2}{15}$

Solve for the variable.

28. $9a + 3 - 4a = 2a - 12$ ***(4.1)***

29. $4x + 3(x - 2) = 3(x + 4)$ ***(4.1)***

30. $\frac{a}{4} + 2 = -1$ ***(4.2)***

31. $\frac{3x - 4}{4} = 3$ ***(4.3)***

Solve for the indicated variable. (4.3)

32. $V = lwh$ for h

33. $S = 2a + (n - 1)d$ for d

34. Change $62\frac{1}{2}\%$ to a fraction reduced to lowest terms. ***(4.4)***

35. Change $12\frac{1}{2}\%$ to a decimal. ***(4.4)***

36. Change 0.3475 to a percent. ***(4.4)***

37. $83\frac{1}{3}\%$ of what number is 135? ***(4.4)***

38. 17 is what percent of 85? ***(4.4)***

39. Use the formula $C = \frac{5}{9}(F - 32)$ to find the temperature on the Celsius scale for 68° Fahrenheit. ***(4.4)***

40. Find the width (w) of a rectangle with perimeter (P) of 112 inches and a length (l) of 24 inches. The formula for the perimeter of a rectangle is $P = 2l + 2w$. ***(4.5)***

UNIT 5

Word Problems

OBJECTIVES

After you have successfully completed this unit, you will be able to:

1. Translate English phrases involving numbers (5.1), ages (5.3), coins (5.4), and distance, rate, and time (5.5) into algebraic phrases using mathematical symbols
2. Set up an equation and solve it for problems involving numbers (5.1), consecutive integers (5.2), ages (5.3), coins (5.4), and distance, rate, and time (5.5)

5.1 NUMBER PROBLEMS

Mathematical Models

In daily life mathematical problems are first expressed as a question in English. Some typical questions that require mathematics are: "Which is the shortest way to work?" "What selling price will give the company maximum profit?" and "How much more weight can we add to the airplane?" But before any question can be answered mathematically, it must be translated from English into an equivalent algebraic expression.

While there are many English words to express an idea, relatively few algebraic symbols correspond to these words. The basic symbols used in translating English to algebra are

$$+, -, \cdot, \div, =$$

As addition and subtraction are closely related, so are the phrases that indicate these operations in both English and algebra.

Addition

English phrase	Algebra
A number plus two	$n + 2$
A number increased by two	$n + 2$
Two greater than a number	$n + 2$
Two more than a number	$n + 2$
The sum of a number and two	$n + 2$

Subtraction

English phrase	Algebra
A number minus two	$n - 2$
A number decreased by two	$n - 2$
Two smaller than a number	$n - 2$
Two less than a number	$n - 2$
The difference between a number and two	$n - 2$
The difference between two and a number	$2 - n$

Multiplication and division are also closely related. Therefore, the phrases that indicate them are similar.

Multiplication		*Division*	
English phrase	Algebra	English phrase	Algebra
Two times a number	$2n$	A number divided by two	$\frac{n}{2}$
Twice a number	$2n$	Half of a number	$\frac{n}{2}$
The product of two and a number	$2n$	The quotient of two and a number	$\frac{2}{n}$
The product of a number and two	$n \cdot 2$	The quotient of a number and two	$\frac{n}{2}$

Many English words can be replaced with an equal sign.

English word	Algebraic symbol
is, are, equal, makes, yields, gives, will be	$=$

We may use any letter as a variable. However, we usually try to pick a letter that reminds us of what the variable represents.

$t =$ time $\quad w =$ weight $\quad n =$ number $\quad d =$ distance

EXAMPLE 1 Use mathematical symbols to translate the following into algebraic expressions.

a plus $b = a + b$

a more than $b =$ ________ $\quad b + a$

a less than $b =$ ________ $\quad b - a$

x times $y =$ ________ $\quad x \cdot y$

Five more than the product of x and $y =$ ________ $\quad xy + 5$

Eight greater than $p =$ ________ $\quad p + 8$

One-eighth of $n =$ ________ $\quad \frac{1}{8}n$ or $\frac{n}{8}$

Seven less than twice $x =$ ________ $\quad 2x - 7$

Twice the quantity seven less than $x =$ ________ $\quad 2(x - 7)$

Twice the difference of x and seven $=$ ________ $\quad 2(x - 7)$

Three more than twice a number $=$ ________ $\quad 2x + 3$

The quotient of a and $b =$ ________ $\quad \frac{a}{b}$

PROBLEM SET 5.1A

Write the following English phrases as algebraic expressions. Let n represent the number.

1. Add ten to a number.
2. Add a number to five.
3. A number increased by six.
4. Five increased by a number.
5. Subtract a number from fifteen.
6. Subtract eight from a number.

7. Five times a number.
8. The product of a number and six.
9. A number divided by negative eight.
10. The sum of the square of a number and twelve.
11. The difference between two times a number and sixteen.
12. Eight minus three times a number.
13. Twice a number increased by seven.
14. Twelve decreased by three times a number.
15. The product of twice a number and negative eight.
16. Three times a number subtracted from one-third the same number.
17. Twice a number added to half the same number.
18. Multiply a number decreased by three by the number itself.
19. Divide twice a number plus seven by eight.
20. A number increased by six is subtracted from four times the same number.
21. Multiply a number increased by eight by twice the same number.
22. A number decreased by seven is divided by nine.
23. Three times the sum of a number and negative six.
24. Fourteen decreased by five times a number.

Review Your Skills

Solve for the variables.

25. $3n = n - 7$ 26. $n = 4n - 6$ 27. $5n = 6n - 9$ 28. $3n = 10 - 2n$

Number Problems

We can solve many problems by finding an algebraic equation that corresponds to the problem and then solving the equation.

EXAMPLE 2 Three times a number is 24. Find the number.

$3n$

Let n represent the number. Then three times the number is represented by ______.

Write an equation.

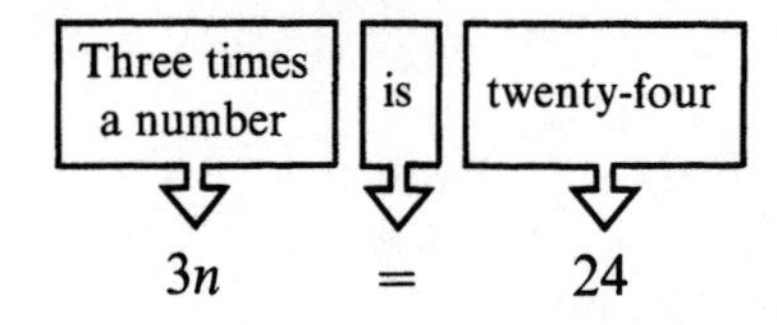

Solve the equation.

$$\frac{3n}{3} = \frac{24}{3}$$

$$n = 8$$

Check.

$$3(8) = 24$$

EXAMPLE 3 Twice a number is six more than the number.

Write an equation.

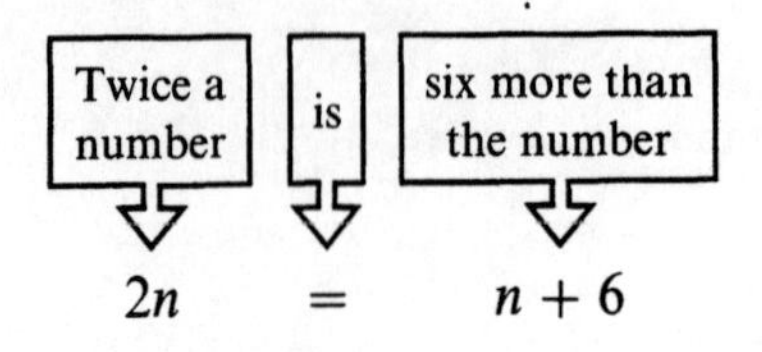

Solve the equation.

$$2n = n + 6$$

$$2n - n = n + 6 - n$$

6 $$n = \_\_\_\_$$

Check.

$$2(6) \stackrel{?}{=} (6) + 6$$

$$12 = 12$$

EXAMPLE 4 Negative ten increased by a number yields six times the number.

Write an equation

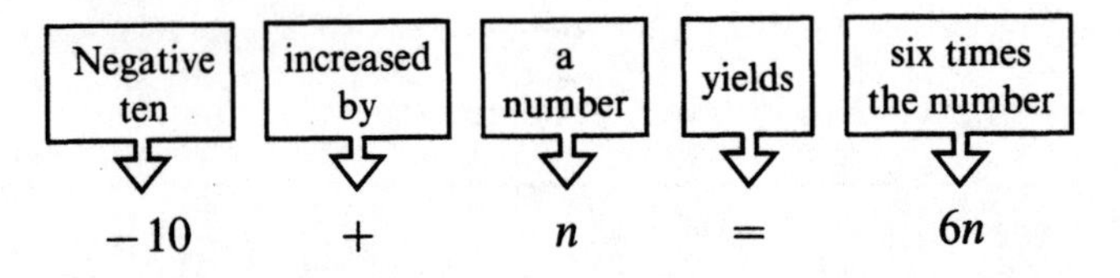

Solve the equation.

$$-10 + n = 6n$$

$$-10 + n - n = 6n - n$$

$5n$ $$-10 = \_\_\_\_$$

$$-2 = n$$

Check.

$$-10 + (-2) \stackrel{?}{=} 6 \cdot (-2)$$

$$-12 = -12$$

PROBLEM SET 5.1B

Write an equation, then solve it to find the number.

1. Four times a number is 28. Find the number.
2. One-third of a number is 12. Find the number.
3. Two-thirds of a number is 16. Find the number.
4. Three-fourths of a number is 15. Find the number.

5. If twice a number is diminished by 2, the result is 14. Find the number.
6. A number increased by 9 yields twice the number. Find the number.
7. The sum of 8 and a number increased by 12 is 32. Find the number.
8. The sum of 9 and twice a number is 23. Find the number.
9. If the sum of two numbers is 98 and one of them is 29, what is the other number?
10. The difference of a number and 21 is 16. Find the number.
11. Two less than three times a number is 37. Find the number.
12. The sum of a number and four times the same number is -35. Find the number.
13. One-half the sum of a number and 6 is 12. Find the number.
14. Two times the sum of a number and 19 is 60. Find the number.
15. Four increased by three times a number is equal to 14 diminished by twice the same number. Find the number.
16. Six decreased by eight times a number is equal to 18 increased by four times the same number. Find the number.
17. One-half a number added to 8 yields 26. Find the number.
18. One-third of a number subtracted from 4 yields -2. Find the number.
19. Two-fifths of a number subtracted from 6 yields -2. Find the number.
20. Three-fourths of a number added to 5 yields 17. Find the number.
21. One-fourth of the sum of a number and 9 yields 4. Find the number.
22. One-third of the difference of twice a number and 11 is 9. Find the number.
23. Two-thirds of the sum of three times a number and 15 yields the same result as four times the same number.
24. Three-fourths of the difference of two times a number and 18 yields the same result as three times the sum of the same number and 1.

5.2 CONSECUTIVE INTEGER PROBLEMS

Consecutive integers are one example of things that occur in a regular pattern. The skills learned in dealing with consecutive integers can be applied to many business, science, and engineering problems. Recall that the integers are the set of numbers

$$\{\ldots, -4, -3, -2, -1, 0, 1, 2, 3, 4, \ldots\}$$

Consecutive integers

Consecutive means following one right after the other. So, *consecutive integers* are integers that follow in order. 7, 8, 9 are consecutive integers. $-2, -1, 0, 1$ are also consecutive integers. Notice we could write the consecutive integers 9, 10, 11 as 9, $9 + 1$, and $9 + 2$ if we wished to make a point that 9 was the first integer in the group. We can use this same technique to represent consecutive integers when we don't know the first integer.

Let n = first integer

$n + 1$ = second integer

$n + 2$ = third integer

Notice that each integer is 1 greater than the previous integer. The difference between two consecutive integers is 1.

EXAMPLE 5 If $n = 6$, write the integer designated by each of the following in the blank.

7, 8, 9

$n + 1 =$ _____ $n + 2 =$ _____ $n + 3 =$ _____

10, 5, 3

$n + 4 =$ _____ $n - 1 =$ _____ $n - 3 =$ _____

EXAMPLE 6 Find the sum of three consecutive integers whose sum is 63.

Pick a variable and define its meaning.

Let $n =$ first integer

$n + 1 =$ second integer

$n + 2 =$ third integer

Make a model and express it as an equation.

[First integer] + [Second integer] + [Third integer] = 63

$n \quad + \; n + 1 \; + \; n + 2 \; = 63$

Solve the equation.

$$n + n + 1 + n + 2 = 63$$

$$3n + 3 = 63 \quad \text{Collect like terms on left side}$$

$$3n = 60 \quad \text{Subtract 3 from both sides}$$

$$n = 20 \quad \text{Divide both sides by 3}$$

At this point we have found the first integer. The problem asked for three consecutive integers. They are

$$n = 20$$

$$n + 1 = 21$$

$$n + 2 = 22$$

Check.

$$20 + 21 + 22 \stackrel{?}{=} 63$$

$$63 = 63$$

Odd integers

Odd integers are numbers from the set

$$\{\ldots, -5, -3, -1, 1, 3, 5, \ldots\}$$

Notice that each odd integer is separated from the next odd integer by 2. Therefore, if we let n equal an odd integer, the next odd integer is $n + 2$.

Four consecutive odd integers starting with 3 are 3, 5, 7, 9 or 3, 3 + 2, 3 + 4, 3 + 6.

EXAMPLE 7 If $n = 5$, write the odd integer named by each of the following in the blank.

7, 9, 3, 1

$n + 2 =$ _____ $n + 4 =$ _____ $n - 2 =$ _____ $n - 4 =$ _____

$k + 2$

EXAMPLE 8 If k is an odd integer, write an expression for the next odd integer after k. ________

Even integers

Even integers are numbers from the set

$$\{\ldots, -4, -2, 0, 2, 4, \ldots\}$$

Consecutive even integers are also separated by 2. Therefore, n, $n + 2$, $n + 4$ could represent three consecutive even integers if n represents an even integer.

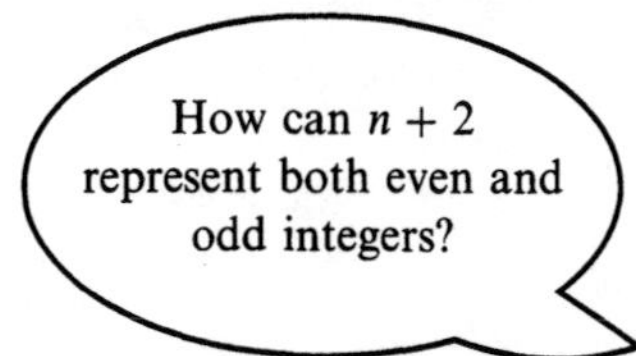

Notice that the value of $n + 2$ depends on the value of n. If n is even, $n + 2$ is even, and if n is odd, $n + 2$ is odd.

6, 2, 8, 0

EXAMPLE 9 If $n = 4$, write the even integer represented by the following expressions in the blank.

$n + 2 =$ ______ $n - 2 =$ ______ $n + 4 =$ ______ $n - 4 =$ ______

EXAMPLE 10 Find three consecutive odd integers whose sum is 57.

Pick a variable and define its meaning.

Let $n =$ first odd integer
$n + 2 =$ second odd integer
$n + 4 =$ third odd integer

Write an equation.

First odd integer + Second odd integer + Third odd integer = 57

$$n + n + 2 + n + 4 = 57$$

Solve the equation.

$$n + n + 2 + n + 4 = 57$$
$$3n + 6 = 57$$
$$3n = 51$$
$$n = 17$$

At this point, all we have found is the first odd integer. The problem asked for all three.

$$n = 17$$
$$n + 2 = 19$$
$$n + 4 = 21$$

Check.

$$17 + 19 + 21 \stackrel{?}{=} 57$$
$$57 = 57$$

EXAMPLE 11 Find three consecutive even integers so that the sum of the first two is three times the third.

Pick a variable.

Write an equation.

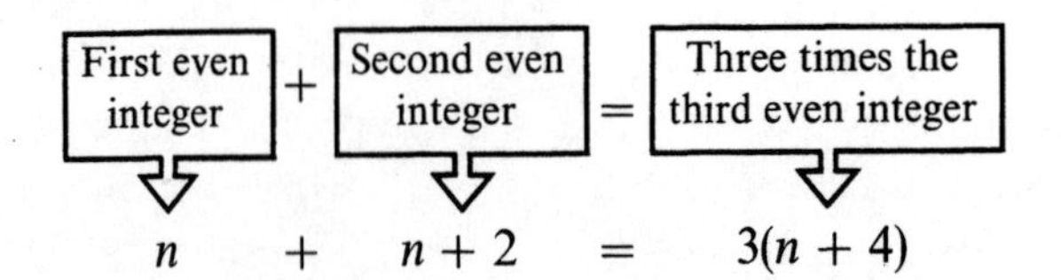

$$n + n+2 = 3(n+4)$$

Solve the equation.

$$n + n + 2 = 3(n + 4)$$
$$2n + 2 = 3n + 12$$
$$2 = n + 12$$
$$-10 = n$$

The three integers are

$$n = -10$$
$$n + 2 = -8$$
$$n + 4 = -6$$

Check.

$$(-10) + (-8) \stackrel{?}{=} 3(-6)$$
$$-18 = -18$$

PROBLEM SET 5.2

Solve the following problems involving consecutive integers. It is important that you clearly define your variable and write the equation you are using.

1. Find two consecutive integers whose sum is 15.
2. Find two consecutive even integers whose sum is 22.
3. Find two consecutive odd integers whose sum is 36.
4. Find two consecutive even integers whose sum is -26.
5. Find two consecutive odd integers whose sum is -32.
6. Find three consecutive integers whose sum is 21.
7. Find three consecutive odd integers whose sum is 21.
8. Find three consecutive even integers whose sum is -12.
9. Find three consecutive even integers if the sum of the first two minus the third is 16.
10. Find three consecutive integers if the sum of the first and last integer is equal to the second integer.
11. Find three consecutive integers if the sum of the first two integers is 6 more than the third integer.
12. Find three consecutive odd integers if four times the first is equal to the sum of the other two integers.
13. Find two consecutive integers whose sum is three times the first integer.

14. Find two consecutive even integers whose sum is three times the second integer.
15. Find three consecutive odd integers if four times the first is equal to the sum of the other two integers.
16. Find two consecutive integers if two times the first plus the second is 19.
17. Find two consecutive integers if two times the first integer minus the second integer is 19.
18. Find two consecutive odd integers if two times the first plus the second is 41.
19. Find two consecutive even integers if the first plus three times the second is 38.
20. Find two consecutive integers if two times the first integer minus three times the second is 19.
21. Find two consecutive integers if five times the first integer is 8 more than four times the second integer.
22. Find three consecutive odd integers if six times the first integer plus the second integer is five times the third integer.
23. Find three consecutive odd integers if five times the first integer minus three times the second integer is one more than seven times the third integer.
24. Find three consecutive integers if four times the first integer minus three times the second integer is equal to three more than two times the third integer.

5.3 AGE PROBLEMS

One of life's experiences no one can avoid is growing older. Since we all share this experience, it's a good one to use in extending our problem-solving skills.

If we represent a person's age as a, we can say the following:

$$a = \text{present age}$$
$$a + 5 = \text{age 5 years from now}$$
$$a - 5 = \text{age 5 years ago}$$

Use symbols to represent each of the following on the blanks below:

Answer	
$30 - 5$ or 25	The age of a 30-year-old man, 5 years ago. ________
$x - 5$	The age of a man who is x years old, 5 years ago. ________
$x + 5$	The age of a man who is x years old, 5 years from now. ________
$21 + n$	The age of a woman who is now 21, n years from now. ________
$30 - s$	The difference in age between a 30-year-old mother and her son who is s years old. ________
$2x$	The age of a father who is twice as old as his x-year-old son. ________
$12 - x$	The age of a girl who is x years younger than her 12-year-old brother. ________
$x + 20$	The age of a father who is 20 years older than his x-year-old son. ________
$x + 20 + 8$ or $x + 28$	Eight years from now, the father will be ________ years old.
$x - 6$	The age of a girl who is 6 years younger than her x-year-old friend. ________
$x - 6 + 8$ or $x + 2$	Eight years from now, the girl will be ________ years old.
$2x$	The age of a woman who is now twice as old as her x-year-old son. ________
$2x + 6$	Six years from now, she will be ________ years old.
$15 - x$	If a 15-year-old girl lived x years in Kansas, how many years did she live outside of Kansas? ________

PROBLEM SET 5.3A

Use symbols to represent each of the following.

1. The age of a 20-year-old woman 4 years from now.
2. The age of a 35-year-old man 6 years ago.
3. The age of a 10-year-old boy 6 years ago.
4. The age of a 39-year-old woman 21 years from now.
5. The age of a 25-year-old man x years from now.
6. The age of a 29-year-old woman n years ago.
7. The age of an x-year-old girl 5 years ago.
8. The age of an n-year-old man 5 years from now.
9. The sum of the ages of a 25-year-old mother and her 3-year-old daughter.
10. The sum of the ages of a 32-year-old father and his x-year-old son.
11. The difference in age of a 28-year-old father and his 6-year-old son.
12. The difference in age of a 39-year-old mother and her n-year-old daughter.
13. The difference in age of an n-year-old mother and her 10-year-old daughter.
14. The difference in ages of a father and his son if the father is twice as old as his x-year-old son.

Review Your Skills

Solve for x.

15. $x + 12 = 3(x - 4)$

16. $7x + 10 = 2(x + 10)$

Solving Age Problems

EXAMPLE 12 Ten years from now Mary will be twice as old as she was 3 years ago. How old is Mary now?

Let $x =$ Mary's present age

Then

$x + 10 =$ Mary's age 10 years from now

$x - 3 =$ Mary's age 3 years ago

Make a model.

[Mary's age 10 years from now] = 2 · [Mary's age 3 years ago]

Write an equation. $x + 10 = 2(x - 3)$

Solve.

$$x + 10 = 2x - 6$$
$$x + 16 = 2x$$
$$16 = x$$

Mary is now 16 years old.

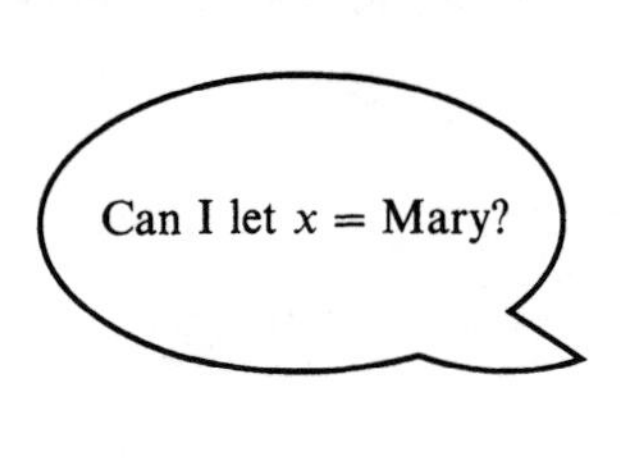

No! x can only stand for a number. Mary is a person. x could represent
Mary's age,
Mary's height,
Mary's weight.
x cannot represent Mary's car or Mary's cat.

In age problems involving more than one person, remember *everybody gets older at the same rate.* Therefore, in 10 years everybody will be 10 years older. Sometimes a table representing the important information at each time mentioned in the problem is useful.

EXAMPLE 13 George is three times as old as Robert. However, 10 years from now he will be only twice as old as Robert. Find their ages.

	Age now	Age 10 years from now
Robert	x	$x + 10$
George	$3x$	$3x + 10$

Make a model.

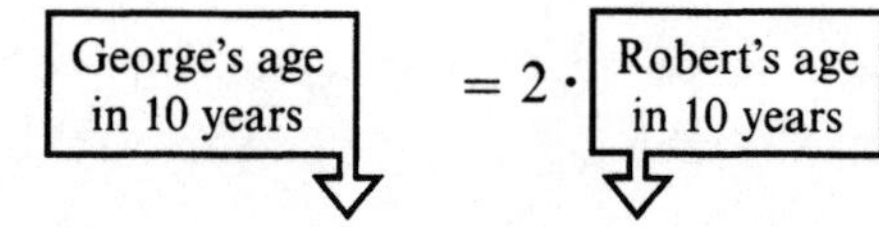

Write an equation. $3x + 10 = 2(x + 10)$

Solve.
$$3x + 10 = 2x + 20$$
$$x + 10 = 20$$
$$x = 10$$

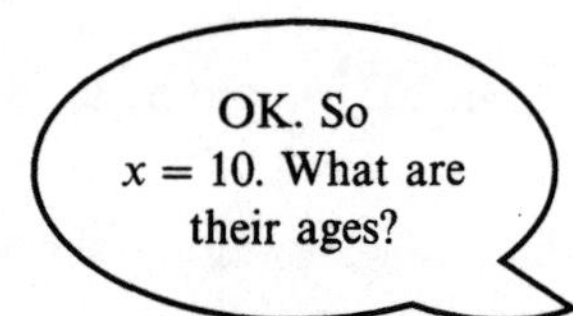

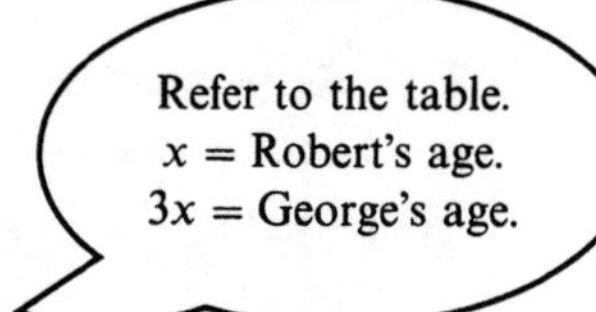

Robert is now 10, George is $3(10) = 30$.

EXAMPLE 14 A girl is 4 years older than her brother and 22 years younger than her father. Five years ago their father was twice as old as the sum of the ages of his children. Find the current ages of the father and his children.

Make a table.

	Age now	Age 5 years ago
Sister	x	$x - 5$
Brother	$x - 4$	$(x - 4) - 5 = x - 9$
Father	$x + 22$	$(x + 22) - 5 = x + 17$

Make a model.

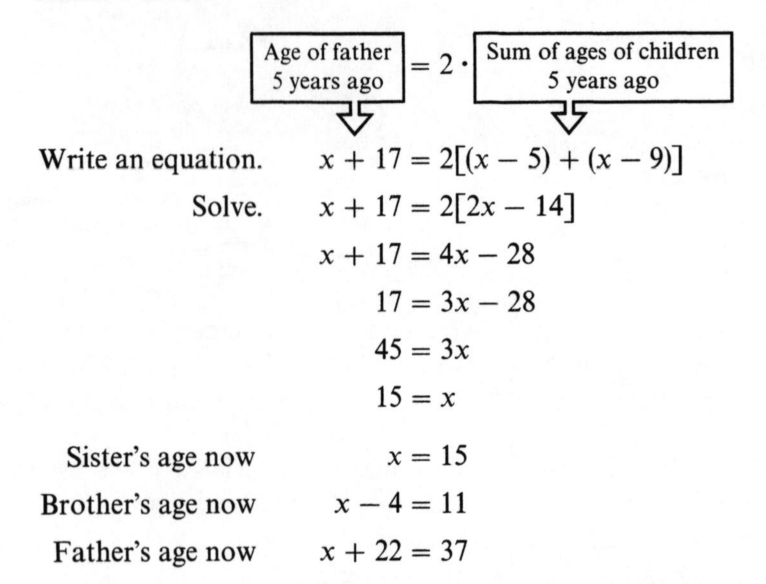

Write an equation. $x + 17 = 2[(x - 5) + (x - 9)]$

Solve.

$$x + 17 = 2[2x - 14]$$
$$x + 17 = 4x - 28$$
$$17 = 3x - 28$$
$$45 = 3x$$
$$15 = x$$

Sister's age now $x = 15$

Brother's age now $x - 4 = 11$

Father's age now $x + 22 = 37$

PROBLEM SET 5.3B

After identifying the variable and writing the equation, solve the following age problems.

1. Ten years from now, Clare will be twice as old as she is today. How old is she?
2. Four years from now, Ben will be three times as old as he is today. How old is he?
3. Dawn's age now is twice what it was 6 years ago. What is her present age?
4. Michael's present age is three times what it was 12 years ago. Find his present age.
5. In 3 years, a girl will be twice as old as she was 4 years ago. How old is she now?
6. Ten years ago a man was one-third as old as he will be in 12 years. How old is he now?
7. Eight years from now Harry will be three times as old as he was 2 years ago. How old is Harry now?
8. Five years ago Ricardo was one-fourth as old as he will be 7 years from now. How old is Ricardo now?
9. Twelve years ago Jeremy was one-half as old as he will be 6 years from now. How old is Jeremy today?
10. In 10 years Susanne will be three times as old as she was 6 years ago. How old is Susanne today?
11. Marcia is 5 years older than her brother. If the sum of their ages is 59, how old is Marcia?
12. Jack is now 6 years older than his brother. Last year the sum of their ages was 16. How old is Jack now?
13. Jim is now 8 years older than his sister. Two years ago he was twice as old as she was. How old is Jim today?
14. John is now three times as old as his sister Sally. In 4 years he will be only twice as old as she will be then. How old are they now?
15. Louise is four times as old as Maria. However, 6 years from now she will be only three times as old as Maria. Find their ages.
16. Leona is three times as old as her brother Don. Four years ago she was 10 years older than her brother. How old is Leona now?
17. Susan is 4 years older than her brother. Six years ago she was three times as old as her brother. Find their ages.

18. Duong is twice as old as his sister. Four years ago he was three times as old as his sister. Find their ages.

19. Jose is 6 years older than Manuel. Nine years ago Jose was twice as old as Manuel. Find their ages now.

20. Maria is 4 years older than her sister and 23 years younger than her mother. Six years ago the mother was three times as old as the sum of her daughters' ages. Find the present age of the mother and each of the children.

21. Tony is 8 years younger than his brother and 34 years younger than his father. Three years ago the father was three times as old as the sum of the sons' ages. Find the present age of the father and each son.

22. Lynn is 2 years older than her sister and 5 years younger than her brother. Five years ago her brother's age was twice that of the sum of Lynn's and her younger sister's ages. How old is Lynn now?

Review Your Skills Solve for the variable.

23. $0.1x + 0.25(x - 1) = 2.55$

24. $0.05(3x) + 0.1x = 3.75$

5.4 COIN PROBLEMS

Many useful algebra problems can be solved using the two basic principles that are applied in the solution of coin puzzles.

5.4A FIRST PRINCIPLE

The value of any number of identical coins is equal to the value of one coin times the number of coins.

One nickel is worth 5 cents.

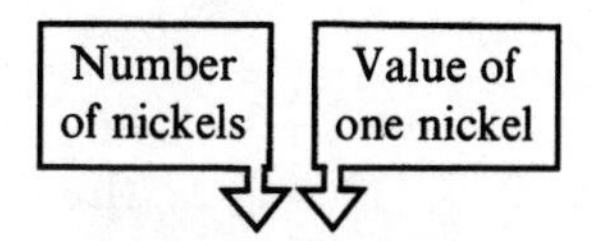

Two nickels are worth $2 \cdot 5$ cents.

12 nickels are worth $12 \cdot 5$ cents.

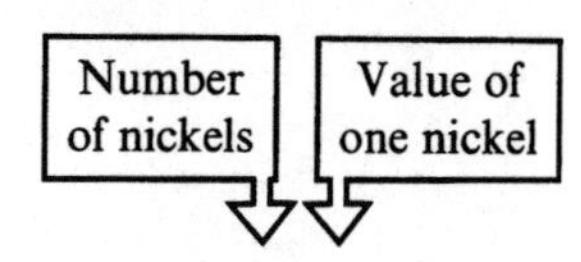

We prefer writing $5n$ to $n \cdot 5$. You may do it either way.

n nickels are worth $n \cdot 5$ cents.

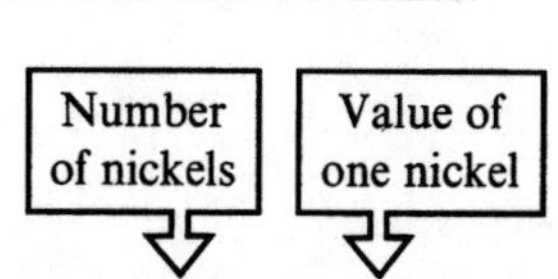

$x, 5$ — x nickels are worth ______ · ______ cents.

$5(x + 3)$ — $x + 3$ nickels are worth __________ cents.

$10n$ — n dimes are worth ______ cents.

$10 \cdot 2n$ — $2n$ dimes are worth __________ cents.

5.4B SECOND PRINCIPLE

The total value of any mixture is equal to the sum of the values of the parts.

EXAMPLE 15 A bank contains three times as many nickels as dimes. The total amount of money in the bank is $1.50. Find the number of each kind of coin in the bank.

$3x$

$10x$

$5 \cdot 3x$

The problem tells us there are three times as many nickels as dimes in the bank. If we let x equal the number of dimes in the bank, we can represent the number of nickels as ______.

The value of x dimes is ______.

value of $3x$ nickels is ________.

Make a table.

	Number of coins	Value of coins
Dimes	x	$10 \cdot x$
Nickels	$3x$	$5 \cdot 3x$

Make a model and express it as an equation.

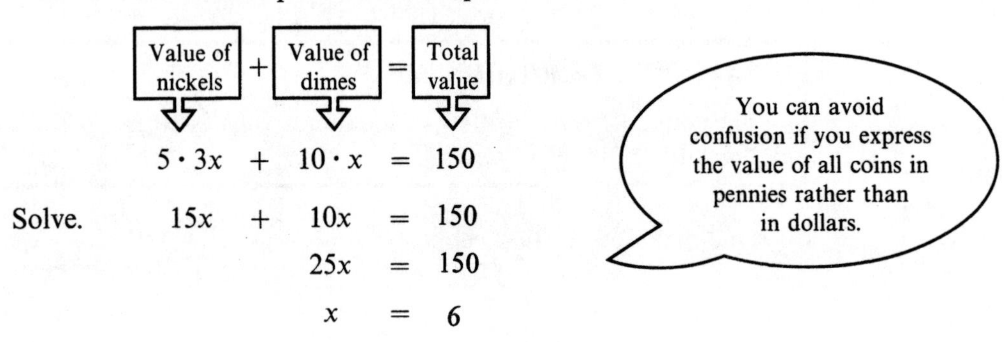

$$5 \cdot 3x + 10 \cdot x = 150$$

Solve.

$$15x + 10x = 150$$

$$25x = 150$$

$$x = 6$$

There are 6 dimes in the bank. Therefore, there are $3x$, or $3(6) = 18$, nickels in the bank.

Check.

6 dimes + 18 nickels = ?

$6 \cdot 10 + 18 \cdot 5 =$

60 cents + 90 cents = 150 cents.

EXAMPLE 16 A man had \$1.15 in change. He had 6 more pennies than quarters and 3 fewer nickels than quarters. How many of each coin did he have?

Notice that quantities of each kind of coin are described in terms of the number of quarters.

Let n = number of quarters.

$25n$

The value of the quarters is then ______.

The number of pennies is 6 more than the number of quarters; therefore, there are ________ pennies.

$n + 6$

$1(n + 6)$

The value of the pennies is ________.

$n - 3$

He had 3 fewer nickels than quarters. Therefore, the number of nickels is ________.

$5(n - 3)$

The value of the nickels is ________.

Make a table.

	Number of coins	Value of each coin	Value of all coins
Pennies	$n + 6$	1¢	$1 \cdot (n + 6)$
Nickels	$n - 3$	5¢	$5 \cdot (n - 3)$
Quarters	n	25¢	$25n$

Make a model and express it as an equation.

Value of quarters + Value of nickels + Value of pennies = Total value

$$25n + 5(n - 3) + 1(n + 6) = 115$$

Solve.

$$25n + 5(n - 3) + 1(n + 6) = 115$$
$$25n + 5n - 15 + n + 6 = 115$$
$$31n - 9 = 115$$
$$31n = 124$$
$$n = 4$$

There are 4 quarters and $n - 3 = 1$ nickel and $n + 6 = 10$ pennies.

Check.

4 quarters + 1 nickel + 10 pennies = ?

100 cents + 5 cents + 10 cents = 115 cents

EXAMPLE 17 A rock concert made \$6,250 from the sale of 325 tickets. If reserved tickets sold for \$25 each and general admission tickets sold for \$10 each, how many general admission tickets were sold?

Since we are looking for the number of general admission tickets, we will let n represent the number of general admission tickets. The total number of tickets sold was 325.

$325 - n$

Therefore, the number of reserved tickets sold was ________.

Make a table.

	Number of tickets	Value of each ticket	Value of tickets
General admission	n	\$10	$10n$
Reserved admission	$325 - n$	\$25	$25(325 - n)$

Make a model and express it as an equation.

Value of general admission tickets	+	Value of reserved tickets	=	Value of all tickets
⇩		⇩		⇩
$10 \cdot n$		$+ \; 25 \cdot (325 - n)$	$=$	6250

$$10 \cdot n + 25 \cdot (325 - n) = 6250$$
$$10n + 8125 - 25n = 6250$$
$$-15n = 6250 - 8125$$
$$n = 125 \quad \text{General admission tickets}$$

PROBLEM SET 5.4

Solve the following coin problems.

1. If you have 8 dimes, how much money do you have?
2. If you have 11 nickels, how much money do you have?
3. If you have 26 nickels, how much money do you have?
4. If you have 16 quarters, how much money do you have?
5. If you have 9 quarters, how much money do you have?
6. If you have 15 fifty-cent pieces, how much money do you have?
7. If you have n nickels, how much money do you have?
8. If you have d dimes, how much money do you have?
9. Find the total value of 6 nickels and 7 dimes.
10. Find the total value of 3 nickels and 11 quarters.
11. Find the total value of 7 pennies, 11 nickels, and 13 dimes.
12. Find the total value of 12 pennies, 4 dimes, and 7 quarters.
13. Find the total value of n nickels, d dimes, and q quarters.
14. Find the total value of x pennies, y dimes, and q quarters.

15a. If you have 12 coins and x of the coins are not nickels, how many nickels do you have?

15b. What is the value of $12 - x$ nickels?

15c. Jane has 12 coins in her purse that have a total value of $1. If these coins consist of nickels and dimes, how many of each does she have?

16. Jack has a handful of pennies and nickels. If he has a total of 19 coins worth 47 cents, how many of each does he have?

17. Debbie has three times as many nickels as dimes. If she has a total of $1.50, how many nickels and dimes does she have?

18. A coin collection is made up of dimes and quarters. If there are 4 more dimes than quarters resulting in a total of $3.55, how many of each coin is in the collection?

19. Ruth has $3.15 in nickels and quarters. If there are 3 more quarters than nickels, how many of each kind of coin does she have?

20. John has one-fourth as many quarters as dimes. If he has a total of $1.95, how many of each coin does he have?

21. Jose has one-third as many nickels as dimes, if he has a total of $2.45, how many of each coin does he have?

22. In a collection of coins made up of dimes and nickels, there are 3 more nickels than dimes. If there is a total of $1.95 in the collection, how many of each coin are there?

23. Jonalee has $1.65 in nickels, dimes, and quarters. If she has 4 more dimes than quarters and 1 more nickel than quarters, how many of each coin does she have?

24. Mary has $3.88 in pennies, dimes, and quarters. If she has 3 more dimes than quarters and 5 fewer pennies than dimes, how many of each coin does she have?

25. Michael has $2.08, consisting of pennies, nickels, and dimes. If he has one-half as many dimes as nickels and three times as many pennies as nickels, how many of each coin does he have?

26. Laura has $3.75, consisting of nickels, dimes, and quarters. If there are 3 more quarters than dimes and twice as many nickels as quarters, how many of each coin does Laura have?

27. A total of 1,020 tickets were sold to a football game. Adult tickets sold for $1.50 and children's for $1.00. If a total of $1,340 was collected, how many adults bought tickets?

28. Twice as many adults as children attended a football game. If adult tickets sold for $1.75 and children's tickets sold for 50¢, and a total of $640 was collected, how many of each ticket were sold?

29. Several neighbors went to a movie together. $22.30 was spent on a total of 14 tickets. If adult admission was $2.25 and children's admission was $1.10, how many children attended the movie?

30. Several Boy Scouts along with some fathers as sponsors spent an evening playing miniature golf. Each Scout paid $1.50 and each father paid $2.25 admission. If a total of $45 was paid for 26 tickets, how many of each played golf?

31. The total receipts for a concert were $3,280 for a total of 760 admissions. If reserved seats sold for $6 and general admission was $4, how many of each kind were sold?

32. The total receipts at a major league baseball game were $137,800 for a total paid attendance of 41,600 fans. There were twice as many reserved seats as box seats and five times as many general admission as box seats. If box seats sold for $6, reserved seats for $4, and general admission was $2.50, how many of each were sold?

POINTERS TOWARD CRITICAL THINKING

Critical does not necessarily mean unfavorable. You can use *critical* in the sense of critical thinking to mean:

careful—accurate, clearly defined

precise—specific

Above all, critical thought means clear, well-defined thinking.

Let's look at an idea we frequently gloss over.

The equal sign can be used only to compare the same kind of quantities.

You cannot say

6 apples = 6 oranges

In fact, you cannot even say

6 apples ≠ 4 oranges

= and ≠ are comparisons between like quantities.

You can say

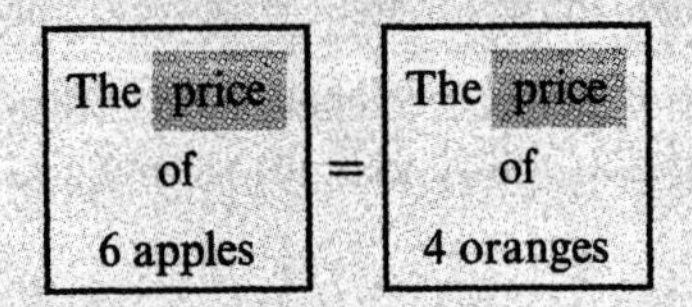

which is really a statement: dollars = dollars.

Or you can say

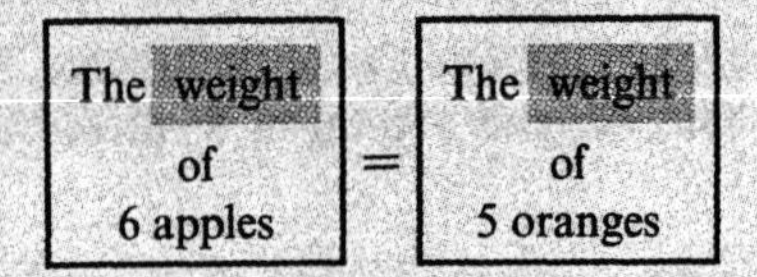

which is really a statement: ounces = ounces.

5.5 DISTANCE PROBLEMS

Problems involving distance, rate, and time occur in real life. The skills required to solve these problems can be used in planning trips to Disneyland for a vacation or they can be used to plan the trip of a rocket to Jupiter.

If you drive on an interstate highway for 2 hours at a steady speed of 65 miles per hour, you would expect to go $2 \cdot (65) = 130$ miles. This leads to the basic formula used in distance problems.

Distance formula

5.5 DISTANCE FORMULA

Distance = Rate · Time

$d = r \cdot t$

This formula allows us to directly calculate the distance covered by an object moving at a constant speed.

EXAMPLE 18 Find the distance covered by an airplane traveling at 300 miles per hour (mph) for $2\frac{1}{2}$ hours.

$$d = r \cdot t$$

$$d = 300\,\frac{\text{miles}}{\text{hour}} \cdot \left(2\frac{1}{2}\right) \text{hours}$$

$$d = 300\,\frac{\text{miles}}{\cancel{\text{hour}}} \cdot \frac{5}{2}\,\cancel{\text{hours}}$$

$$d = 750 \text{ miles}$$

In any physics equation, the units on both sides of the equation must be the same. In this case, we have miles = miles. This is one way scientists check to see that they have written the correct equation.

PROBLEM SET 5.5A

Solve the following distance problems.

1. Find the distance covered by a family traveling by car at a constant speed of 52 mph for $3\frac{1}{2}$ hours.
2. Find the total distance covered by a train traveling at a constant speed of 68 mph for $4\frac{1}{4}$ hours.
3. Find the total distance traveled by a salesman traveling at 48 mph for $3\frac{1}{2}$ hours and at 50 mph for the next $1\frac{1}{2}$ hours.
4. Find the total distance traveled by an airplane traveling at 320 mph for $2\frac{1}{4}$ hours and at 360 mph for the next $2\frac{3}{4}$ hours.
5. A family on a vacation traveled a total of 312 miles in $6\frac{1}{2}$ hours of driving time. What was the average speed of travel?
6. An airplane traveled 1,260 miles in $5\frac{1}{4}$ hours. What was the speed of the airplane?
7. An airplane traveled 1,027 miles at 316 mph. How long did the trip take?
8. How long would a family traveling at 45 mph take to travel 195 miles?

Here is some practice at expressing the elements of distance problems in algebraic notation.

EXAMPLE 19 d represents the distance in miles driven by a man.

$2d$ — Twice that distance is ________.

$d + 6$ — Six miles more than the distance is ________.

$d - 5$ — Five miles less than his distance is ________.

$\frac{d}{2}$ — Half his distance is ________.

$d + 3$ — Three miles farther than his distance is ______.

$100 - d$ — The remaining distance he must travel on a 100-mile trip after he traveled d miles is ________.

EXAMPLE 20 A woman drives at r mph.

$2r$ — A speed twice as fast as her speed is ________.

$r - 20$ — A rate 20 mph slower than her speed is ________.

$\frac{r}{2}$ — A speed half as fast as her speed is ________.

$r + 5$ — A speed 5 mph faster than her speed is ________.

EXAMPLE 21 A woman drives at an average rate of 50 mph.

50 — In 1 hour she would travel ________ miles.

100 — In 2 hours she would travel ________ miles.

$\frac{3}{2} \cdot 50 = 75$ — In $1\frac{1}{2}$ hours she would travel ________.

$50t$ — In t hours she would travel ________ miles.

$50(t + 1)$ — In $t + 1$ hours she would travel ________ miles.

$50(t + 2)$ — In 2 hours more than t hours she would travel ________ miles.

$50 \cdot 2 = 100$ — The speed of a plane flying twice as fast as the woman is driving would be ________ mph.

$50 - x$ — The speed of a car going x mph slower than the woman's speed would be ________ mph.

$50 + x$ — The speed of a car going x mph faster than the woman was driving would be ________ mph.

EXAMPLE 22 A total trip took 5 hours at 40 mph.

2; $40 \cdot 2 = 80$ — If the first part took 3 hours, the remaining portion took ________ hours. The distance covered on the second part was ________ miles.

$5 - t$; $40(5 - t)$ — If the last part of the trip took t hours, the first part took ________ hours. The distance covered on the first part of the trip was ________ miles.

$5 - 1 = 4$; $40 \cdot 4 = 160$ — If 1 hour was spent on rest stops, ________ hours were spent on driving. The distance covered while driving was ________ miles.

$2 \cdot 5 = 10$; $55 \cdot 10 = 550$ — Another trip that took twice as long took ________ hours. At 55 mph that trip was ________ miles.

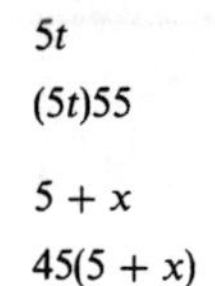

Another trip that took t times as long took ________ hours. At 55 mph that trip was ________ miles.

A trip that took x hours longer to make took ________ hours. At 45 mph that trip was ________ miles.

EXAMPLE 23 A car and a truck pass each other going in opposite directions on a highway. If the truck is traveling at 40 mph and the car is going 55 mph, how long will it be before they are 190 miles apart?

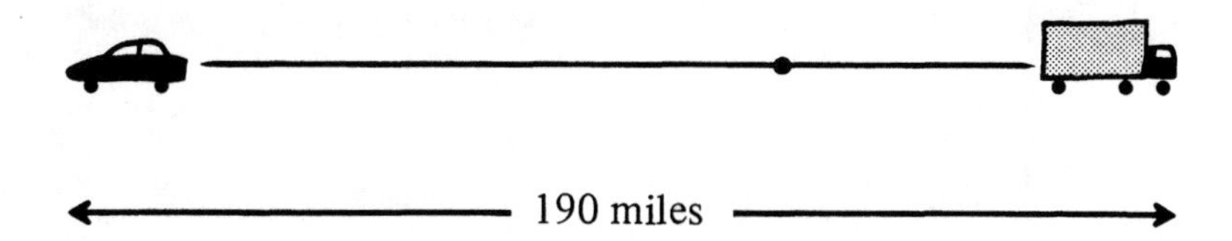

Make a model.

Distance by car	+	Distance by truck	=	Total distance

A chart can be helpful in distance problems to represent each element in the distance formula.

	r Rate	$\cdot$	t Time	$=$	d Distance
Car					
Truck					

We can fill in the first two columns on rate and time with the information in the word problem. Since $d = r \cdot t$, the third column, distance, is the product of whatever is in the first two columns.

In Example 23 the rate of the car is 55 mph and the rate of the truck is 40 mph. Notice that both have been traveling for the same time after they met. Since we don't know what the time period is, we will let t be the time for both the car and the truck.

Our chart looks like this.

	r	$\cdot$ t	$=$ d
Car	55	t	product ⇒ $55 \cdot t$
Truck	40	t	product ⇒ $40 \cdot t$

Our model tells us how to write the equation.

Distance by car	+	Distance by truck	=	Total distance

$$
\begin{aligned}
55t + 40t &= 190 \\
95t &= 190 \\
t &= 2
\end{aligned}
$$

It will take 2 hours for the car and the truck to get 190 miles apart.

POINTERS FOR CRITICAL THINKING

Denominate numbers

Occasionally we work with abstract numbers like

$3 + 5 = 8$

Usually we work with denominate numbers

3 miles + 5 miles = 8 miles

These are called dimensions. They tell us the type of quantity we are adding.

Whenever you use an equal sign, dimensions on one side of the equality must match the dimension on the other side.

3 hours = 2 hours + 1 hour   is legal

It is wrong to write

3 hours = 3 miles   This is illegal

When you add or subtract denominate numbers, the dimensions must be the same.

Legal

3 hours + 6 hours = 9 hours	is OK
80 miles − 50 miles = 30 miles	is OK

Illegal

3 hours + 3 miles	is wrong
$50 \dfrac{\text{miles}}{\text{hour}} + 20 \text{ miles}$	is wrong
55 minutes − 10 miles	is wrong

It is possible to multiply and divide denominate numbers. In this case we treat the dimensions like variables.

$$\frac{200 \text{ miles}}{4 \text{ hours}} = 50 \frac{\text{miles}}{\text{hour}}$$

$$\frac{55 \text{ miles}}{\text{hour}} \cdot 2 \text{ hours} = 110 \text{ miles}$$

Notice the dimensions on the left side of the equal sign still match the dimensions on the right side.

EXAMPLE 24 A private plane flying at 150 mph left Los Angeles headed for San Francisco. One hour later a jet flying at 600 mph left traveling the same route. How long will it take the jet to overtake the private plane?

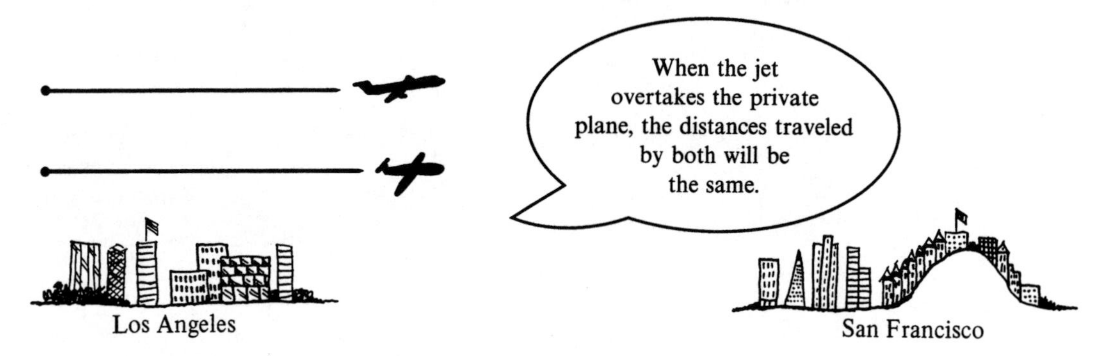

Make a chart for the time the jet overtakes the plane.

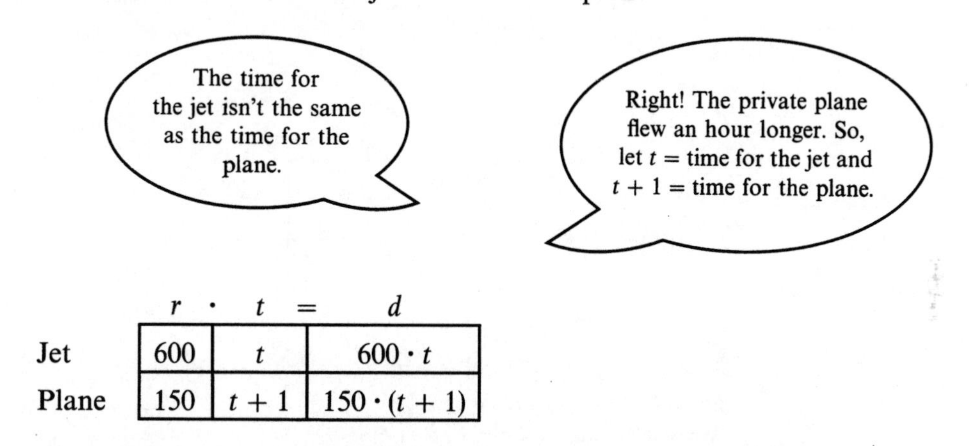

	r ·	t =	d
Jet	600	t	$600 \cdot t$
Plane	150	$t + 1$	$150 \cdot (t + 1)$

Make a model for the distances at the time the jet overtakes the plane.

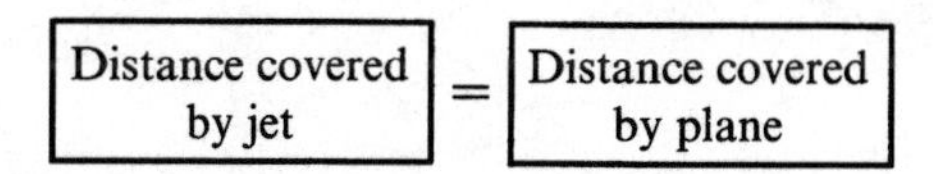

Write an equation.

$$600t = 150 \cdot (t + 1)$$

$$600t = 150t + 150$$

$$450t = 150$$

$$t = \frac{150}{450}$$

$$t = \frac{1}{3} \text{ hour}$$

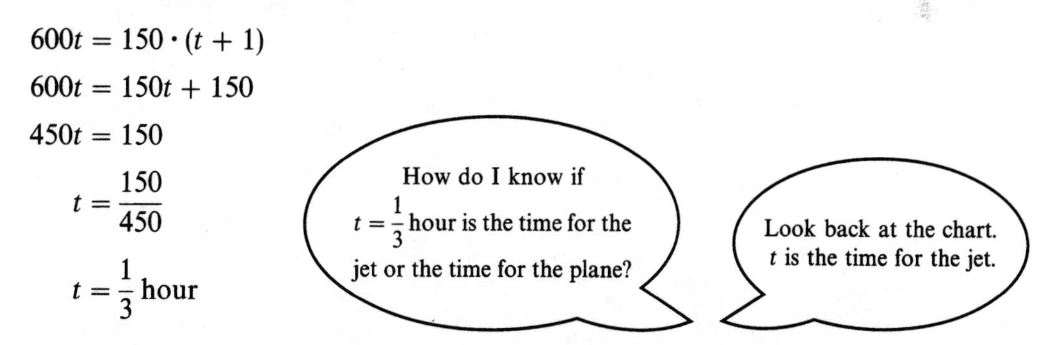

The jet will catch the private plane 20 minutes after the jet takes off.

Note: In solving word problems, you are frequently expected to provide an idea that is part of common knowledge. In Example 24 you were expected to realize that when the jet overtakes the small plane, they both will have traveled the same distance.

Be careful that you *don't overlook the obvious.* Frequently, the knowledge you must provide is so much a part of your basic assumptions that you may not realize it's there. See if you can identify the common knowledge in the next example before you read the bubble.

EXAMPLE 25 On a trip to see his girlfriend, Gilbert traveled at an average rate of 55 mph. However, on the return trip, due to heavy traffic and low motivation, Gilbert only averaged 45 mph. If his total driving time was 6 hours, how far away from his girlfriend did Gilbert live?

True, but it's common knowledge that "the distance going is equal to the distance returning."

Make a chart.

	r ·	t =	d
Going	55	t	________
Returning	45	$6 - t$	________

$55t$

$45(6 - t)$

I can see why the rates are 55 and 45, but how did you know what to put in the time column?

Well, I don't know the time spent going or the time spent returning, but I do know that he drove 6 hours in all. So I let t = time going and $6 - t$ = time returning.

Make a model of the distances.

Distance going	=	Distance returning

Write an equation.

$$55 \cdot t = 45 \cdot (6 - t)$$
$$55t = 270 - 45t$$
$$100t = 270$$
$$t = 2.7 \text{ hours}$$

Now we have the time, $t = 2.7$ hours, that it took him to drive to his girlfriend's house but the problem asked for the distance to his girlfriend's house. To get the distance, we must multiply that time by the rate.

$$d = r \cdot t$$
$$= 55 \cdot (2.7)$$
$$= 148.5 \text{ miles}$$

Always reread the problem to make sure you have answered the question.

EXAMPLE 26 An aircraft carrier and a tanker are 240 miles apart when they start toward each other. If the carrier travels 12 mph faster than the tanker and they meet in 4 hours, how fast does each ship travel?

If they are going to meet, then together they must travel the 240 miles between them.

They both travel for 4 hours, but the rate of the aircraft carrier is 12 mph faster than the tanker.

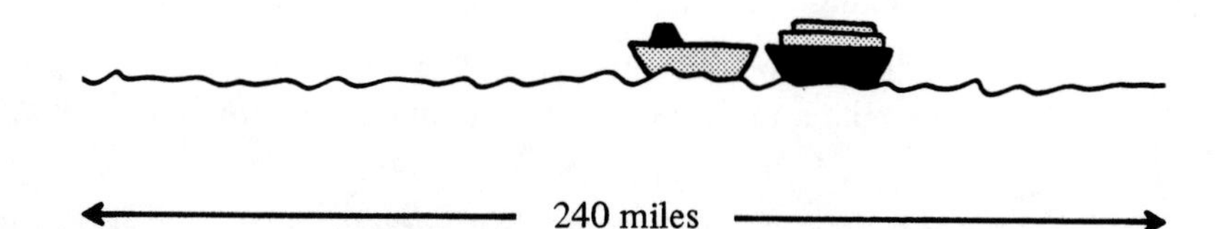

Make a chart.

	r	$\cdot\ t =$	d
Tanker	r	4	______
Carrier	$r + 12$	4	______

$r \cdot 4$

$(r + 12)4$

Make a model.

Distance by carrier	+	Distance by tanker	=	Total distance

Write an equation.

$$(r + 12) \cdot 4 + r \cdot 4 = 240$$
$$4r + 48 + 4r = 240$$
$$8r + 48 = 240$$
$$8r = 192$$
$$r = 24$$

Referring back to the chart,

the speech the tanker travels is $r =$ ________

and the speed the aircraft carrier travels is $r + 12 =$ ________

24 mph

36 mph

PROBLEM SET 5.5B

Solve the following distance problems.

1. Willis and Don left the same airport in light planes at the same time but in opposite directions. Willis traveled at 120 mph and Don traveled at 150 mph. How long will it be before they are 810 miles apart?

r ·	t =	d

2. Ken and John left Los Angeles on separate bicycles traveling in opposite directions. Ken rode 3 mph faster than John and in 4 hours they were 81 miles apart. How fast was each traveling?

3. Jim left Denver on a bicycle, riding at 12 mph. Four hours later Sally left by car, driving at 52 mph over the same route. How long will it take Sally to overtake Jim?

4. John and Glorya each rented a light airplane. They left the same airport at the same time in the same direction. John's plane traveled at 160 mph, whereas Glorya traveled at 120 mph. After $3\frac{1}{4}$ hours how far apart were the two planes?

5. Jack and Jill left Los Angeles in separate cars traveling in opposite directions. Jack drove 10 mph faster than Jill and in 3 hours they were 270 miles part. How fast was each traveling?

6. Bill and Bob left Chicago at the same time driving in opposite directions. Bill drove at 40 mph, and at the end of 4 hours they were 380 miles apart. How much faster was Bob driving than Bill?

7. While moving, Albert and Maria rented a truck to carry their furniture. Albert left for their new home with the truckload of furniture traveling at 40 mph. One hour later Maria left with the children in the family car traveling at 50 mph. How long will it take Maria to overtake Albert?

8. Ben left Miami for Chicago in a single-engine plane flying at 120 mph. One hour later Jim flew out at 200 mph over the same route. How long did it take Jim to overtake Ben?

9. Terry rode his bicycle into the country riding at 12 mph. Due to a flat tire he had to push his bicycle home, going at 4 mph. If he was gone a total of 2 hours, how far out into the country had he ridden?

10. Paul took his family out into the country for a picnic lunch. Since it was near lunchtime, he drove at 50 mph going to the picnic area. A leisurely lunch and recreation took $2\frac{1}{2}$ hours, after which they drove home through heavy traffic at 25 mph. If the total time spent was 4 hours, how far into the country was the picnic area?

11. The Jones family left San Diego for San Francisco by car traveling at 50 mph. Seven hours later the Smith family left by plane flying at 120 mph. How long after the Smith family left did they pass the Jones family?

12. A salesperson drove from New York to Philadelphia in 2 hours. Due to heavy traffic it took him 1 hour longer, traveling 18 mph slower, to return to New York. How fast did he drive going to Philadelphia?

13. A long-distance trucker left Chicago for Dallas, Texas, driving at 40 mph. The trucker left some important papers in the truck terminal. A half hour after the trucker left, the dispatcher drove in the company pickup at 55 mph to deliver the papers to the truck driver. How long did it take the pickup to overtake the truck?

14. A salesperson flew the company airplane to a customer across the state at an average speed of 150 mph. Due to a headwind her average speed on the return trip was 120 mph. How far did she travel going to the customer if the return trip took 1 hour longer than the flight out?

Review 5

NOW THAT YOU HAVE COMPLETED UNIT 5, YOU SHOULD BE ABLE TO:

Use the following definitions and formulas to build models for word problems.

Consecutive integers—Integers that follow in order. Each is separated from the preceding integer by 1. If n represents an integer, then the next two consecutive integers are $n + 1$ and $n + 2$.

Even integers—Integers from the set $\{\ldots, -4, -2, 0, 2, 4, \ldots\}$. If n represents an even integer, then the next two consecutive even integers are $n + 2$ and $n + 4$.

Odd integers—Integers from the set $\{\ldots, -3, -1, 1, 3, \ldots\}$. If n represents an odd integer, then the next two consecutive odd integers are $n + 2$ and $n + 4$.

Distance formula

$$\text{Rate} \cdot \text{Time} = \text{Distance}$$
$$r \cdot t = d$$

Translate common English phrases into algebraic expressions.

Solve word problems by applying the following suggestions.

1. Read the problem carefully. Determine what is asked for and what is given.
2. Look for the sentences that are statements of equality.
3. Identify elements from common knowledge that may help solve the problem.
4. List any formulas that might apply to the problem.
5. Make a model using boxes and English phrases to describe the problem.
6. Select a variable to represent the key unknown quantity.
7. Translate the model into an algebraic equation. Express all quantities in the model in terms of the chosen variable.
8. Solve the resulting equation.
9. Answer the question asked by the problem.
10. Check the solution in the original problem.

Review Test 5

Represent the following English phrases by an algebraic phrase. (5.1 through 5.5)

1. Two less than a number ______
2. The difference between three times a number and 25 ______
3. The sum of twice a number and 12 ______
4. Divide the sum of a number and 8 by 7 ______
5. Six decreased by three times a number ______
6. The sum of two consecutive integers ______
7. The age of a 35-year-old man x years ago ______
8. The age of an n-year-old woman 6 years from now ______
9. The sum of the ages of an x-year-old mother and her y-year-old son ______
10. The difference in age of a 42-year-old father and his x-year-old son ______
11. The difference in age of a mother and her x-year-old daughter if the mother is three times as old as her daughter ______
12. The value of x dimes ______
13. If you have n nickels, how much money do you have? ______
14. The value of y nickels and n quarters ______
15. The value of n dimes and twice as many nickels ______
16. The speed of a car traveling x mph faster than a bicycle traveling 20 mph ______
17. The speed of a car traveling twice as fast as one traveling x mph ______

Select a letter to represent the unknown quantity. Set up an equation and solve it. (5.1 through 5.5)

18. The sum of a number and four times the same number is 85. Find the number.

19. Find three consecutive odd integers whose sum is 69.

20. Twenty-eight years from now, John will be three times as old as he is now. How old is John now?

21. Laura is 4 years older than her brother. Two years ago Laura was three times as old as her brother. How old are they now?

22. Dick has \$5.20 in nickels and quarters. If he has three times as many nickels as quarters, how many of each coin does he have?

23. The total receipts for a high school football game were \$5,160 for a total of 2,220 admissions. If student tickets sold for \$2 and adult tickets for \$3, how many of each were sold?

24. A family on a trip traveled a total of 390 miles in $7\frac{1}{2}$ hours of driving time. What was the average driving speed?

25. Ted and Joe, flying light planes, left Chicago's O'Hare Airport at the same time. Ted headed due east toward New York and Joe headed due west toward Los Angeles. After 4 hours of flying time, they were 1,320 miles apart. If Ted's plane flew 30 mph faster than Joe's, what was the speed of each plane?

REVIEW YOUR SKILLS

Evaluate the following if $a = 3$, $b = -2$, $c = -4$.

26. $a^2(ab^3 - c)$

Multiply using the distributive property.

27. $-3a^2x^3y(4xy^3 - 6a^2x^3y - 5a^3y^3)$

Perform the indicated operations. Reduce all answers to lowest terms, and write the answers in standard form.

28. $\frac{-36x^2}{10} \div \frac{-8}{25}$

29. $\frac{-7x^2}{12} - \frac{-8x^2}{15}$

UNIT 6

Functions and Relations—Graphing Linear Equations

OBJECTIVES

After you have successfully completed this unit, you will be able to:

1. Find the value of a function for the given value of x (6.1)
2. Represent ordered pairs on a graph and name the coordinates of points on a graph (6.2)
3. Graph a first-degree equation using a table of points (6.2)
4. Find the slope of a line through two specified points (6.3)
5. Find the equation of a line given its slope and the y-intercept (6.4)
6. Find the equation of a line given its slope and a point (6.4)
7. Find the equation of a line given any two points (6.4)
8. Sketch graphs of lines using the slope-intercept method or x- and y-intercepts method (6.5)

6.1 FUNCTIONS

Some quantities are so directly related that if you are given the value of one variable, you can supply the value of the second. The cost of hamburger at $1.20 per pound is an example. If I tell you the number of pounds, you can tell me the cost.

Function

6.1 DEFINITION

y is said to be a *function* of x, if for each value of x there is one and only one value for y.

One way to tell the value of y to be paired with each permissible value of x is to use a set of ordered pairs.

Ordered pair

(3, 5) is called an *ordered pair*. It is a pair because there are two numbers inside the parentheses. It is an ordered pair because (3, 5) is different from (5, 3). The order of the numbers is important since $(3, 5) \neq (5, 3)$.

Components

In the ordered pair (3, 5), the numbers 3 and 5 are called *components* of the ordered pair. 3 is the first component and 5 is the second component.

Some other ordered pairs are

$$(7, 2) \quad (5, -46) \quad (-5, -12) \quad (0, 0) \quad (0, 4) \quad (x, y)$$

EXAMPLE 1 Ordered pairs can be used to show how much to pay a person.

Hours worked	Pay earned	Ordered pairs
1	6.85	(1, 6.85)
2	13.70	(2, 13.70)
5	34.25	(5, 34.25)
10	68.50	(10, 68.50)
20	137.00	(20, 137.00)

Functional notation

If you were an employer, you would find the set of ordered pairs in Example 1 incomplete. Another way to indicate the pay earned is with *functional notation.*

$$P = f(n)$$

Read $P = f(n)$ as "P equals f of n."

This notation says that the pay earned, represented by the variable P, depends on the number of hours worked, represented by the variable n.

Exactly how the pay earned depends on the hours worked is explained by the equation

$$f(n) = 6.85 \cdot n$$

This function can be evaluated for each permissible replacement of n. $f(3)$ means the value of $f(n)$ when $n = 3$.

EXAMPLE 2 Find $f(3)$, if $f(n) = 6.85n$.

$$f(n) = 6.85 \cdot n$$
$$f(3) = 6.85 \cdot 3$$
$$= 20.55$$

The expression $f(x)$ is read "f of x." This expression in functional notation indicates that the value of some other quantity depends on the replacement value selected for x. Suppose that the letter y is chosen to represent the quantity that is functionally related to x. Then we can write

$$y = f(x)$$

This expression tells us only that y is related to x. If we wish to specify exactly how y is related to x, we must specify $f(x)$. One possible way to specify $f(x)$ is with an equation:

$$f(x) = 3x - 1$$

This is equivalent to saying $y = 3x - 1$. The function can be evaluated for each permissible replacement of x. The expression $f(5)$ means the value of $f(x)$ when $x = 5$.

EXAMPLE 3 Find $f(5)$, if $f(x) = 3x - 1$.

$$f(x) = 3x - 1$$
$$f(5) = 3(5) - 1 \quad \text{Replace each occurrence of } x \text{ with } 5$$
$$= 15 - 1$$
$$= 14$$

EXAMPLE 4 Find $f(-4)$, if $f(x) = 3x - 1$.

$$f(x) = 3x - 1$$
$$f(-4) = 3(-4) - 1$$
$$= -12 - 1$$
$$= -13$$

The value of $f(x)$ is -13 when x is -4.

EXAMPLE 5 For $f(x) = 2x + 3$, evaluate each of the following: $f(1)$, $f(2)$, $f(-2)$, $f(0)$.

$f(x)$	$= 2x$	$+ 3$
$f(1)$	$= 2(1)$	$+ 3 = 5$
$f(2)$	$= 2(2)$	$+ 3 = 7$
$f(-2)$	$= 2(-2)$	$+ 3 =$ ____
$f(0)$	$= 2(0)$	$+ 3 =$ ____

-1

3

If the value of another quantity also depends on the value of x, it might be expressed this way:

$$g(x) = 7x - 5$$

This is read "g of x equals seven x minus five."

EXAMPLE 6 Complete the following table for $g(x) = 7x - 5$.

x	$g(x)$	$= 7x$	$- 5$
1	$g(1)$	$= 7(1)$	$- 5 = 2$
3	$g(3)$	$= 7(3)$	$- 5 = 16$
-3	$g($____$)$	$= 7($____$)$	$- 5 = -26$
0	$g($____$)$	$= 7($____$)$	$- 5 = -5$
5	$g($____$)$	$= 7($____$)$	$- 5 = 30$

We frequently write $y = f(x)$. This simply says that the value that depends on x is called y. In Example 4, we may replace $f(x)$ with y.

The equation

$$f(x) = 3x - 1$$

now reads

$$y = 3x - 1$$

The value of y still depends on the value used to replace x.

EXAMPLE 7 The value of y that is given by $f(x) = 3x - 1$ is shown for several values of x in the table below.

x	$f(x)$	$= 3x$	$- 1 = y$
1	$f(1)$	$= 3(1)$	$- 1 = 2$
2	$f(2)$	$= 3(2)$	$- 1 = 5$
0	$f(0)$	$= 3(0)$	$- 1 = -1$
-2	$f(-2)$	$= 3(-2)$	$- 1 = -7$
5	$f(5)$	$= 3(5)$	$- 1 = 14$

Value of x — Resulting value of $f(x)$ or y

The table shows how each value of $y = f(x)$ is calculated for the various replacements for x. The important idea, however, is that for each x value that goes into the function exactly one $f(x) = y$ value comes out. We can use ordered pairs to express this important information in shorter form.

From the table above

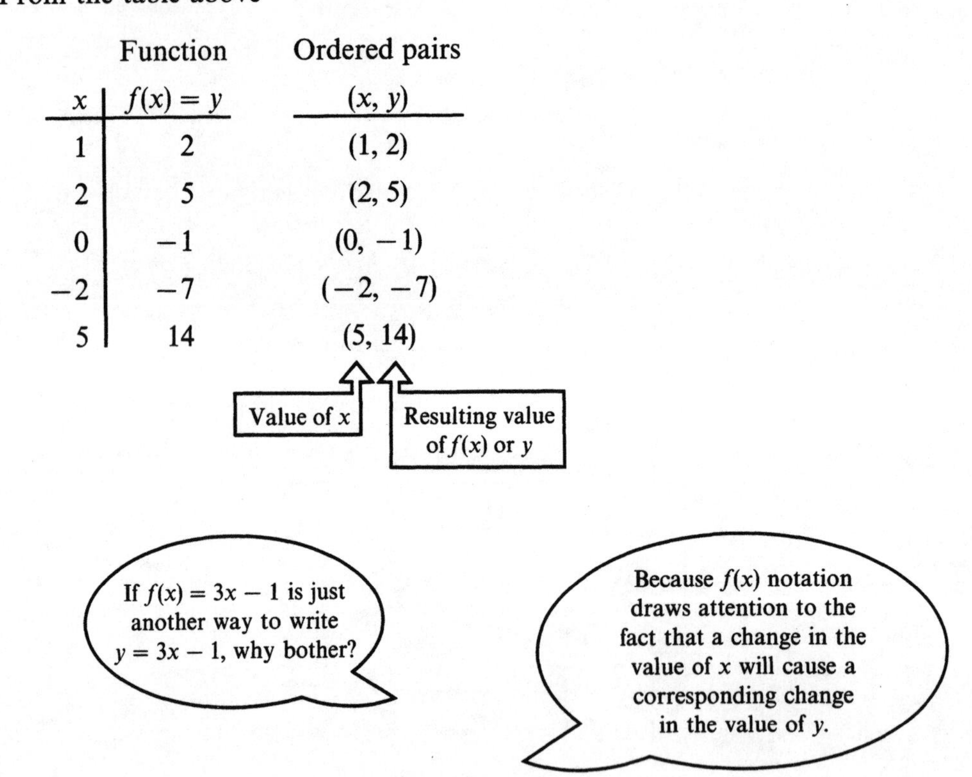

	Function	Ordered pairs
x	$f(x) = y$	(x, y)
1	2	(1, 2)
2	5	(2, 5)
0	-1	$(0, -1)$
-2	-7	$(-2, -7)$
5	14	(5, 14)

EXAMPLE 8 Complete the following table for $f(x) = 6 - 2x$.

		Function		Ordered pairs
x	$f(x)$	$= 6 - 2x$	$= y$	(x, y)
1	$f(1)$	$= 6 - 2(1)$	$= 4$	$(1, 4)$
-1	$f(-1)$	$= 6 - 2(-1)$	$= 8$	$(-1, 8)$
0	$f(0)$	$= 6 - 2(0)$	$=$ ____	$(0,$ ____$)$
3	$f($____$)$	$= 6 - 2($____$)$	$=$ ____	$($____, ____$)$
-4	$f($____$)$	$= 6 - 2($____$)$	$=$ ____	$($____, ____$)$

6 6

3, 3, 0 3, 0

$-4, -4, 14$ $-4, 14$

Independent variable

Dependent variable

When one variable may assume any one of a set of values, it is referred to as an *independent variable.* The quantity that changes as a result of a change in the independent variable is called the *dependent variable.* The value of the dependent variable depends on the value of the independent variable. The equation specifies how the value of the dependent variable depends on the value of the independent variable.

In functional notation, the independent variable is enclosed in parentheses.

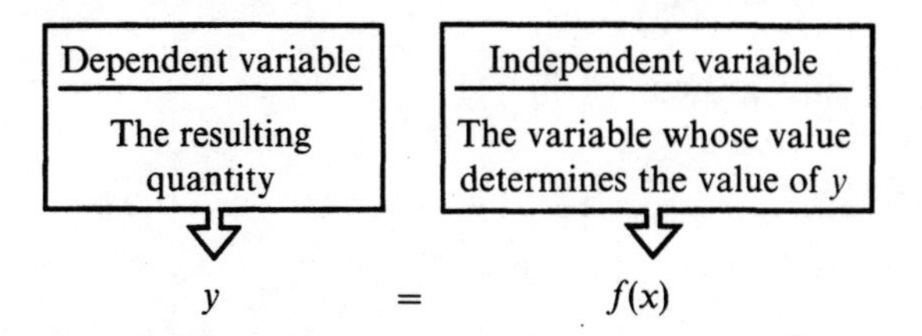

In the function $f(x) = 3x - 1$

Since $\quad y = f(x) \quad$ and $\quad f(x) = 3x - 1$

then $\quad y = 3x - 1$

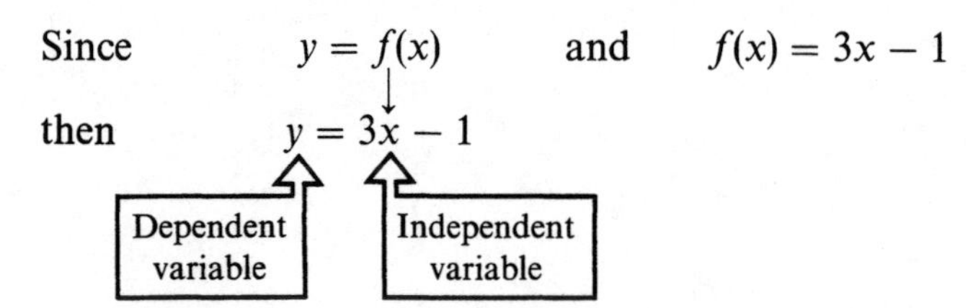

When ordered pairs are used to express a function, the independent variable is always written first.

In an earlier example we expressed the pay earned as a function of n, the number of hours worked. In that example, the number of hours worked was the independent variable and the pay earned was the dependent variable.

Notice in the pay example that there are some values the independent variable n cannot assume. For instance, n cannot be a negative number of hours.

Domain

The *domain* of a function is the set of possible replacements for the independent variable. The independent variable is written first in an ordered pair. Therefore, it is possible to define the domain as the set of first components in the set of ordered pairs.

Range

The *range* of a function is the set of possible values for the dependent variable. If the function is defined by a set of ordered pairs, the range of the function is the set of all possible second components in the set of ordered pairs.

PROBLEM SET 6.1

Given $f(x) = 3x - 2$, find the following.

1. $f(2)$ 2. $f(3)$ 3. $f(0)$ 4. $f(5)$

Given $t(x) = 4x - 5$, find the following.

5. $t(-1)$ 6. $t(-4)$ 7. $t(-3)$ 8. $t(0)$

Given $g(x) = -4x + 6$, find the following.

9. $g(-3)$ 10. $g(-1)$ 11. $g(0)$ 12. $g(2)$

Given $h(x) = \frac{3}{4}x - 3$, find the following.

13. $h(-8)$ 14. $h(-4)$ 15. $h(4)$ 16. $h(0)$

Given $p(x) = \frac{-2}{3}x + 5$, find the following.

17. $p(-6)$ 18. $p(0)$ 19. $p(-3)$ 20. $p(3)$

Complete the following tables.

21.

x	$f(x) = 4x + 2 = y$	(x, y)
3	$f(3) = 4(3) + 2 = 14$	(3, 14)
−4		
0		
−6		
2		
1		

22.

x	$f(x) = -2x + 5 = y$	(x, y)
−3		
4		
2		
0		
−1		
6		

6.2 GRAPHING ORDERED PAIRS AND FIRST-DEGREE EQUATIONS

Graphing Ordered Pairs

Two ways to represent the ordered pairs that belong to a function are a table or a set of ordered pairs. Both methods are accurate. However, for a large set of ordered pairs, we need a more convenient method to represent the ordered pairs and to show any pattern that may exist. A graph is the most commonly used method.

x-axis
y-axis

On the next page is a *rectangular graph.* The horizontal line, called the *x-axis*, is placed at right angles to the vertical line, called the *y-axis.*

Origin

The point where the two lines cross is called the *origin.*

By agreement, the *x*-axis is marked with positive numbers to the right of the origin and negative numbers to the left of the origin, with zero at the origin. The *y*-axis is marked with positive numbers going up from the origin and negative numbers going down from the origin. Zero is at the origin.

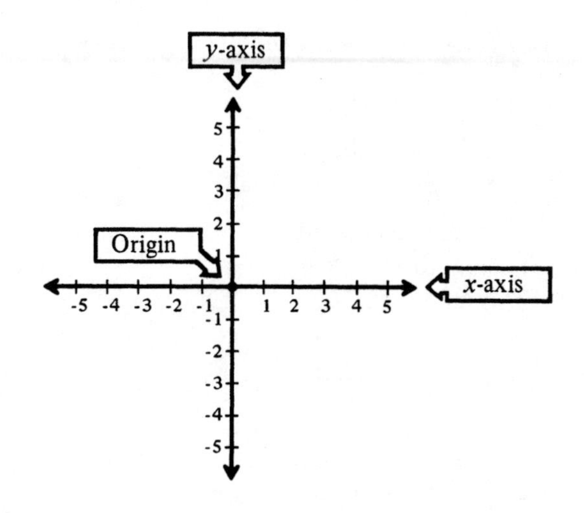

We use points to represent each ordered pair. The first component of the ordered pair tells how far right or left along the x-axis to place the point. If the first component is a positive number, the point will be to the right along the x-axis; if it is a negative number, the point will be to the left along the x-axis. The second component tells how far up or down the y-axis to place the point. If the second component is a positive number, the point will be above the x-axis; and if the second component is a negative number, the point will be below the x-axis.

EXAMPLE 9 Plot the point (4, 3).

Start at the origin,
go 4 units right,
then 3 units up.

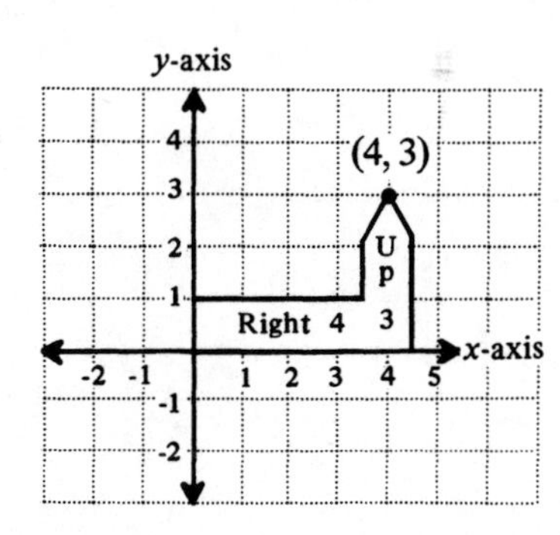

Coordinates

Each ordered pair of the form (x, y) is referred to as the *coordinates* of a point. For every ordered pair, there is exactly one point on the graph. On the other hand, for every point of the graph, there is exactly one ordered pair.

EXAMPLE 10 Plot the point $(4, -3)$.

Start at the origin.

The first coordinate, 4, says go 4 units right.

The second coordinate, -3, tells us to then move 3 units down because it is negative.

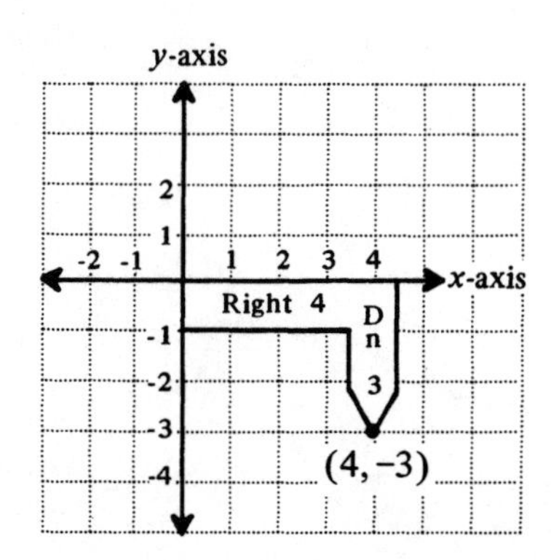

EXAMPLE 11 Plot the point $(-2, -3)$.

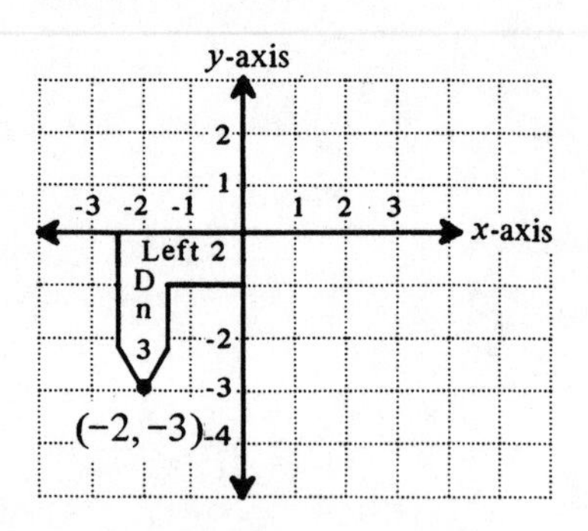

Start at the origin.

The first coordinate, -2, says go 2 units to the left.

The second coordinate, -3, tells us to then move 3 units down.

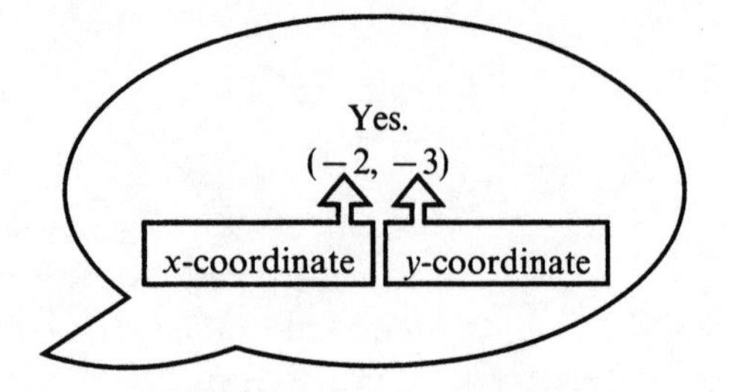

EXAMPLE 12 Plot the point $(0, 3)$.

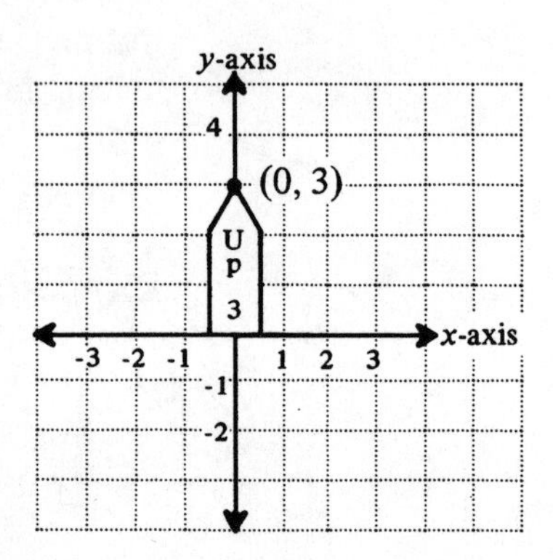

Start at the origin.

The first coordinate, 0, says stay at the origin.

The second coordinate, 3, says move up 3 units.

EXAMPLE 13 Give the coordinates of each point labeled on the graph.

$A(4, 2)$
$B(4, 4)$
$C(-1, 4)$
$D(-5, 2)$
$E(-4, -3)$
$F(-1, -1)$
$G(1, -1)$
$H(5, 0)$
$I(0, -5)$

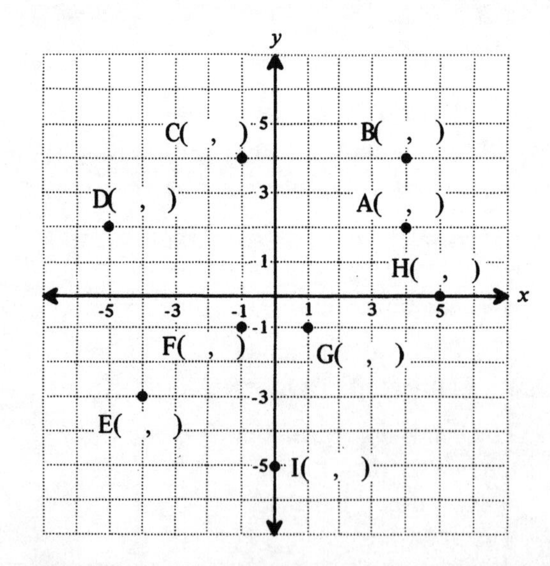

PROBLEM SET 6.2A

1. Plot the following points on the same graph.

   $(-2, 0)$, $(3, -1)$, $(5, 2)$, $(-3, 1)$, $(-4, -2)$, $(3, 4)$, $(5, -6)$, $(0, 2)$

2. Plot the following points on the same graph.

   $(4, 0)$, $(3, 5)$, $(-6, 1)$, $(-3, -2)$, $(0, 4)$, $(7, -2)$, $(0, -4)$, $(2, 1)$

Give the coordinates of each point labeled on the graphs below.

3. 4.

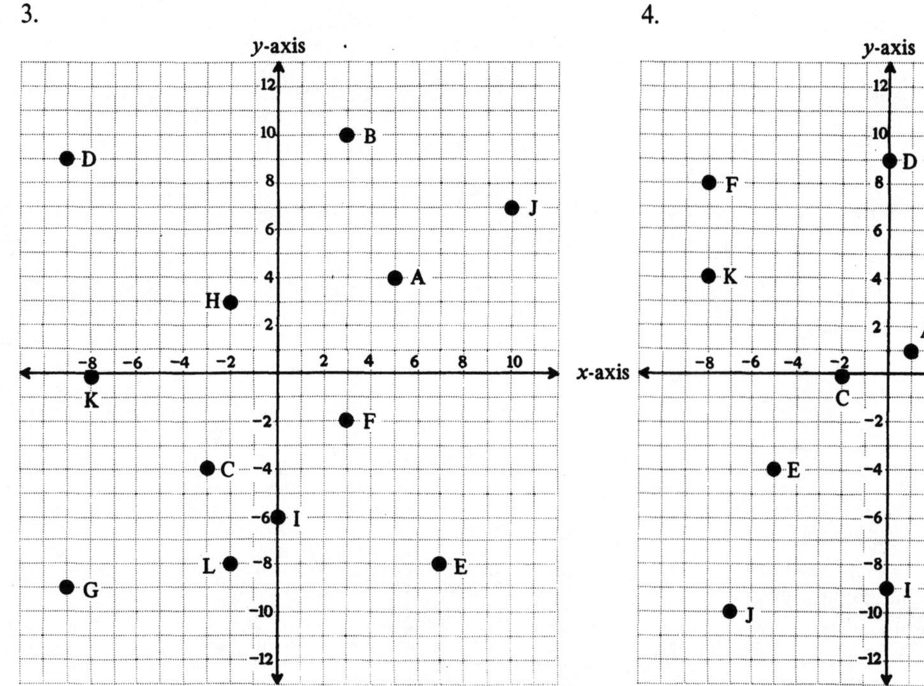

Graphing First-Degree Equations

First-degree equation

An equation of the form $y = f(x)$, where $f(x)$ is an expression involving a single variable raised to the first power only, is a *first-degree equation.*

$y = 3x + 2$ and $y = \dfrac{-2}{3}x - 7$ are two different first-degree equations.

To draw a graph of a first-degree equation, we need to evaluate $f(x)$ for several values of x.

EXAMPLE 14 Graph $y = 2x - 3$.

x	Function $2x - 3 = y$	Ordered pairs (x, y)
1	$2(1) - 3 = -1$	$(1, -1)$
0	$2(0) - 3 = -3$	$(0, -3)$
2	$2(2) - 3 = 1$	$(2, 1)$
4	$2(4) - 3 = 5$	$(4, 5)$

Why select those values for x?

I could have picked any values, so I chose ones that were easy to work with.

Then we plot the points that correspond to the ordered pairs.

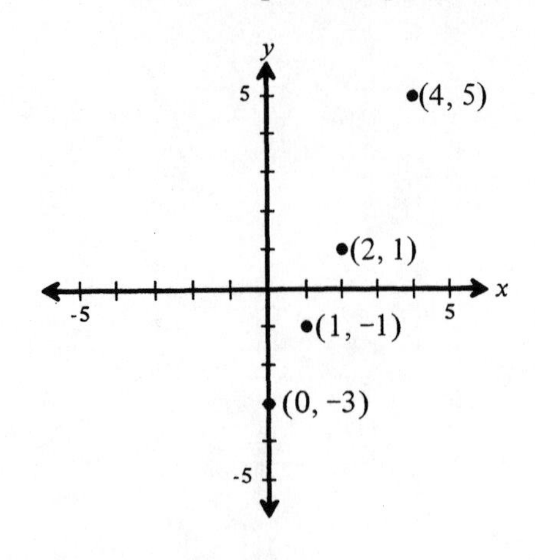

Hey, these points all lie on a straight line! Can I draw a line through them?

Yes, but read the following to see what a solid line means.

When we draw a solid line, we say that the coordinates of every point on the line satisfy the function. Notice the line also passes through the point $(-1, -5)$ and the point $\left(\frac{3}{2}, 0\right)$.

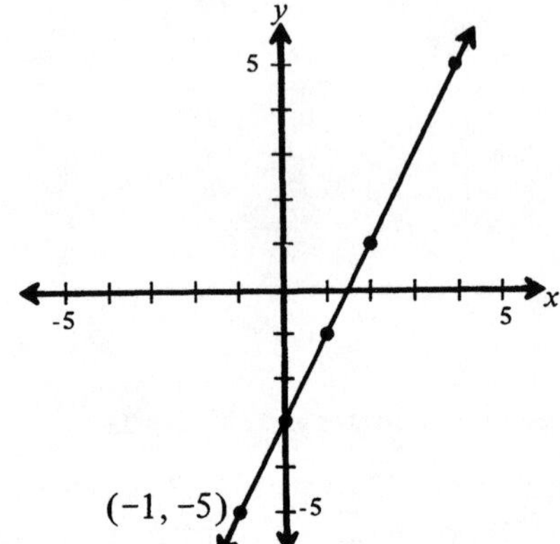

Test the point $(-1, -5)$

$$y = 2x - 3$$
$$-5 \overset{?}{=} 2(-1) - 3$$
$$-5 \overset{?}{=} -2 - 3$$
$$-5 = -5$$

It works.

The point $(-1, -5)$ is one of the points of the function. You may wish to try the coordinates of a few other points that are on the line of the function. You will find that the coordinates of any point on the line satisfy the equation $y = 2x - 3$. You will also find that the coordinates of any point that is not on the line will not satisfy the equation.

Notice that the equation $y = 2x - 3$ says that the value of y is twice the value of x less three. The points along the line are the set of points whose second coordinate is twice the first coordinate less three. The line is therefore said to be the graph of the equation $y = 2x - 3$. A graph is a way to picture the solution set of an equation.

EXAMPLE 15 Graph $y = \frac{-2}{3}x + 1$.

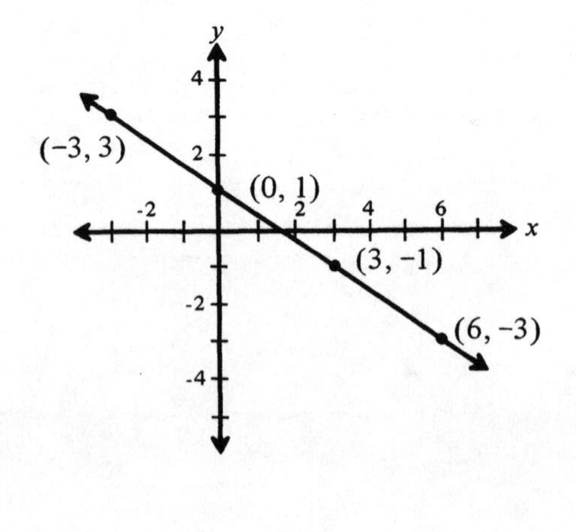

	Function	Ordered pairs
x	$\frac{-2}{3}x \quad + 1 = y$	(x, y)
0	$\frac{-2}{3}(0) \quad + 1 = 1$	(0, 1)
3	$\frac{-2}{3}(3) \quad + 1 = -1$	(3, −1)
6	$\frac{-2}{3}(6) \quad + 1 = -3$	(_____, _____)
−3	$\frac{-2}{3}(-3) + 1 = 3$	(_____, _____)

6, −3

−3, 3

How come all the x values are multiples of 3?

You can pick any x you wish. Therefore, you might as well pick the ones that make it easy to calculate y. When the denominator of the fraction is 3, multiples of 3 are easy.

EXAMPLE 16 Graph $-3x + 2y = 8$.

Before we graph this equation, we will solve it for y.

$$-3x + 2y = 8$$

$$2y = 3x + 8$$

$$\frac{2y}{2} = \frac{3x}{2} + \frac{8}{2}$$

$$y = \frac{3}{2}x + 4$$

Complete the following table.

			Function	Ordered pairs
		x	y	(x, y)
		2	$\frac{3}{2}(2) + 4 = 7$	(2, 7)
10	10	4	$\frac{3}{2}(4) + 4 =$ ______	(4, ______)
0, 4	0, 4	0	$\frac{3}{2}($______$) + 4 =$ ______	(______, ______)
$\frac{3}{2}(-2) + 4$, 1	-2, 1	-2	______ = ______	(______, ______)

Notice that multiples of 2 were picked because the denominator of the fraction is 2.

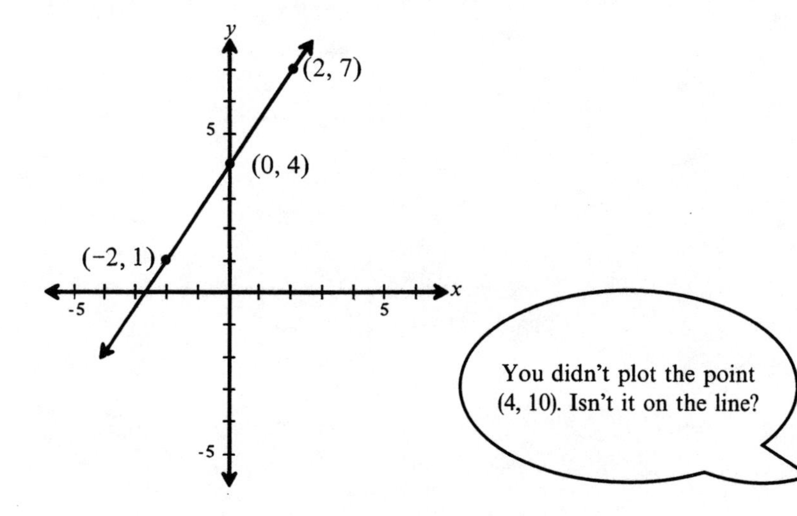

You didn't plot the point (4, 10). Isn't it on the line?

Yes, it is, but my graph isn't large enough to include it. Either extend the graph or plot another point.

Two points are enough to draw a line but it's good practice to use at least three points so that if you make a mistake you can tell.

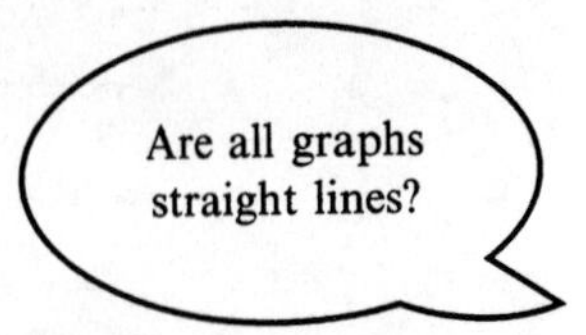

No, only the graphs of first-degree equations are straight lines. That's why they are sometimes called *linear* equations.

PROBLEM SET 6.2B

Complete the following tables and make a set of ordered pairs for the following equations. Sketch the graphs.

1. $y = 3x - 4$

x	$3x - 4$	(x, y)
-2		
0		
1		
3		
4		

2. $y = -2x + 3$

x	$-2x + 3$	(x, y)
-2		
-1		
0		
1		
2		

3. $y = \frac{-1}{2}x - 3$

x	$\frac{-1}{2}x - 3$	(x, y)
-4		
-2		
0		
2		
-6		

4. $y = \frac{1}{3}x - 3$

x	$\frac{1}{3}x - 3$	(x, y)
-6		
-3		
0		
3		
6		

5. $2x + 3y = 6$ (Solve for y first.)

x	$-\frac{2}{3}x + 2$	(x, y)
-6		
-3		
0		
3		
6		

6. $3x - 2y = 6$ (Solve for y first.)

x		(x, y)
-4		
-2		
0		
2		
6		

Make a table of ordered pairs and sketch the graph for the following equations.

7. $y = x + 2$
8. $y = -x + 2$
9. $y = 2x - 4$
10. $y = -3x - 1$
11. $y = \frac{3}{4}x - 2$
12. $y = \frac{2}{3}x + 2$
13. $y = -\frac{2}{3}x + 3$
14. $y = -\frac{3}{4}x - 2$
15. $x + y = 4$
16. $x - y = 3$
17. $4x - y = 5$
18. $2x + y = 3$
19. $4x + 3y = 9$
20. $5x - 2y = 6$
21. $2x - 4y = -12$
22. $3x + 3y = -6$

Review Your Skills

Evaluate the following. Reduce to the lowest terms. Write your answer in the standard form of a fraction.

23. $\frac{8 - 12}{7 - 13}$

24. $\frac{17 - 11}{4 - 13}$

6.3 SLOPE

Think for a minute about the minimum information required to locate a line. Two points will do it. You could also locate a line if you knew one point on the line and the proper angle to draw the line through that point. *The angle or inclination of a line is called its slope.*

EXAMPLE 17

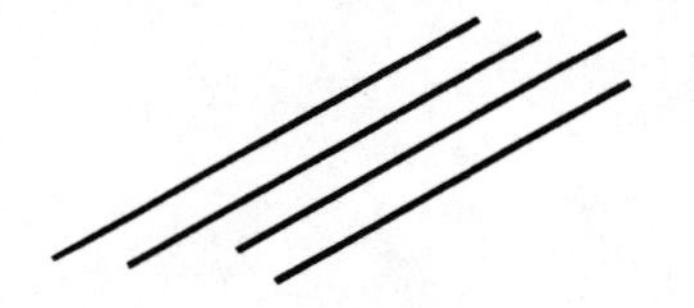
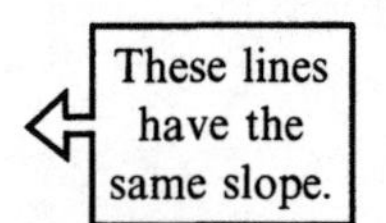

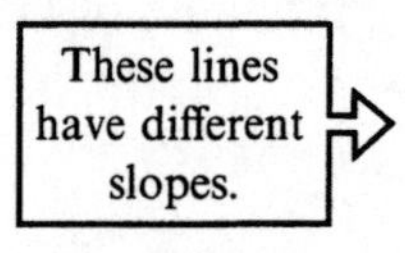

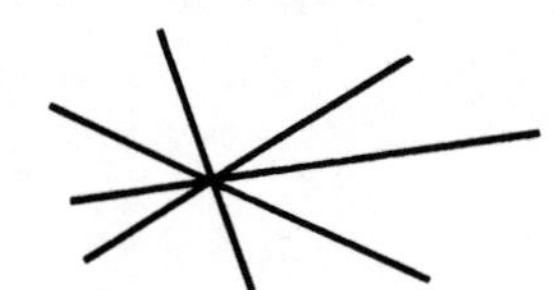

Before you can draw a line with a particular slope, you need to know how to measure slope.

EXAMPLE 18 Consider the equation $y = 2x$.

A table of values for the equation is

x	$y = 2x$
0	0
1	2
2	4
3	6
4	8

Using the table of values above, answer these questions.

4, 1
2

As x varies from 1 to 2, y varies from 2 to ______. The change in x is ______ and the corresponding change in y is ______.

2, 6, 2
4

As x varies from 1 to 3, y varies from ______ to ______. The change in x is ______ and the corresponding change in y is ______.

2, 8, 3
6

As x varies from 1 to 4, y varies from ______ to ______. The change in x is ______ and the corresponding change in y is ______.

2

If x increases by 1, y increases by ______.

4

If x increases by 2, y increases by ______.

6

If x increases by 3, y increases by ______.

2

The change in y is always ______ times the change in x.

Using the table of values, sketch the graph of $y = 2x$.

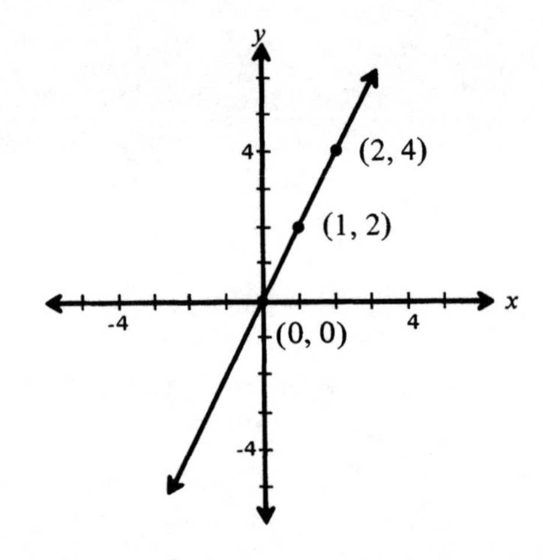

EXAMPLE 19 Consider the equation $y = 4x$.

A table of values for the equation is

x	$y = 4x$
0	0
1	4
2	8
3	12
4	16

Use the table of values to answer these questions.

8, 1
4

As x varies from 1 to 2, y varies from 4 to ______. The change in x is ______ and the corresponding change in y is ______.

4, 12, 2
8

As x varies from 1 to 3, y varies from ______ to ______. The change in x is ______ and the corresponding change in y is ______.

4, 16, 3
12

As x varies from 1 to 4, y varies from ______ to ______. The change in x is ______ and the corresponding change in y is ______.

4

If x increases by 1, y increases by ______.

8

If x increases by 2, y increases by ______.

16

If x increases by 4, y increases by ______.

4

The change in y is always ______ times the change in x.

Using the table of values, sketch the graph of $y = 4x$.

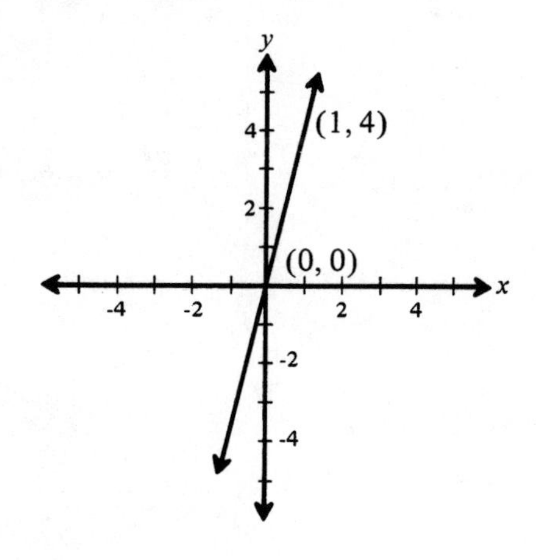

Yes

Is the graph of $y = 4x$ steeper than the graph of $y = 2x$? ______

We can see that the steepness or inclination of a line depends on its slope.

Slope

6.3 DEFINITION

The *slope* of a line between two points with coordinates (x_1, y_1) and (x_2, y_2) is the ratio of the change in the y-coordinates of the two points to the change in the x-coordinates of the points.

$$\text{Slope} = \frac{\text{change in } y}{\text{change in } x} = \frac{\text{rise}}{\text{run}} = \frac{y_2 - y_1}{x_2 - x_1}$$

What do the little 1's and 2's on the x's and y's mean?

The little numbers are called subscripts. y_2 is read "y sub 2." When we want to use the same variable letter for two different quantities, we use subscripts to show that they are different.

Then y_1 and y_2 are different quantities?

Yes. In the above definition, we are talking about two points each with an x- and y-coordinate, so we say the first point has coordinates (x_1, y_1) and the second point has coordinates (x_2, y_2).

EXAMPLE 20 The letter m is used to represent the slope.

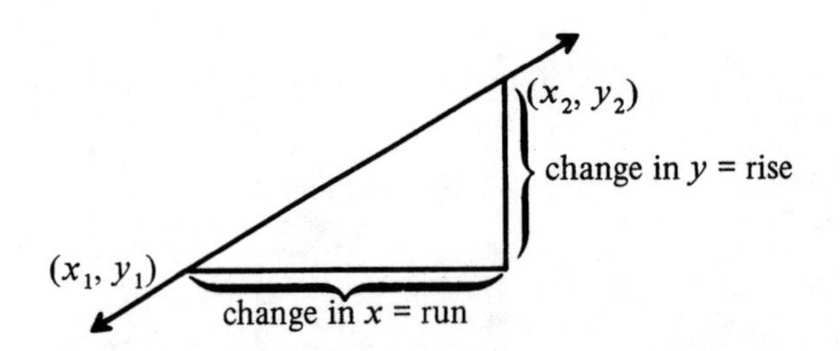

$$m = \frac{\text{change in } y}{\text{change in } x} = \frac{\Delta y}{\Delta x}$$

$$= \frac{\text{rise}}{\text{run}}$$

$$m = \frac{y_2 - y_1}{x_2 - x_1}$$

Δ is used as shorthand for "change in," thus Δx means change in x.

EXAMPLE 21 Find the slope of the line below.

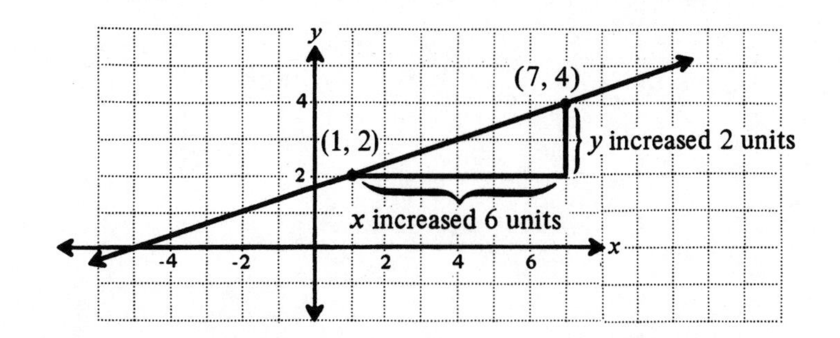

To find the change in y, either count squares or notice that the y-coordinate of the second point (4) minus the y-coordinate of the first point (2) is $4 - 2 = 2$ units.

The change in x is $7 - 1 = 6$ units.

$$\text{Slope} = m = \frac{\Delta y}{\Delta x} = \frac{y_2 - y_1}{x_2 - x_1} = \frac{4 - 2}{7 - 1} = \frac{2}{6} = \frac{1}{3}$$

A slope of $\frac{1}{3}$ tells us that the change in the value of y will be $\frac{1}{3}$ the change in the value of x.

If x is increased 6 units, y will increase $\frac{1}{3} \cdot (6) = 2$ units.

EXAMPLE 22 Find the slope of the line below.

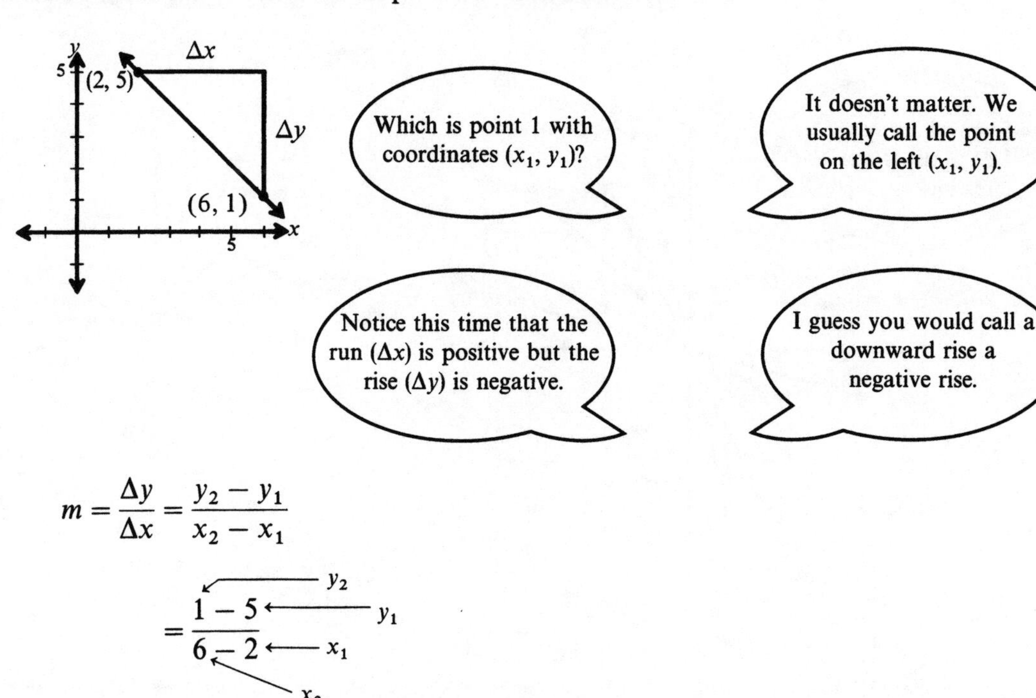

$$m = \frac{\Delta y}{\Delta x} = \frac{y_2 - y_1}{x_2 - x_1}$$

$$= \frac{1 - 5}{6 - 2} \quad \begin{matrix} \leftarrow y_2 = 1 \text{ (first term)},\; y_1 = 5 \\ \leftarrow x_2 = 6,\; x_1 = 2 \end{matrix}$$

$$= \frac{-4}{4} = -1$$

Try it the other way around; call (6, 1) point 1. Then

$$m = \frac{y_2 - y_1}{x_2 - x_1} = \frac{5 - 1}{2 - 6} = \frac{+4}{-4} = -1$$

EXAMPLE 23 Complete the table below showing the slope between selected pairs of points on the line below.

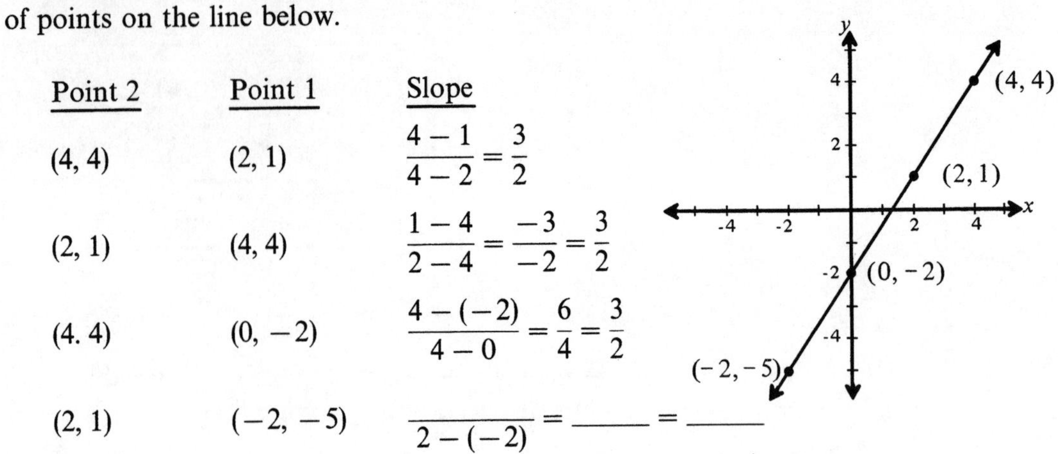

Point 2	Point 1	Slope
(4, 4)	(2, 1)	$\frac{4-1}{4-2} = \frac{3}{2}$
(2, 1)	(4, 4)	$\frac{1-4}{2-4} = \frac{-3}{-2} = \frac{3}{2}$
(4. 4)	(0, −2)	$\frac{4-(-2)}{4-0} = \frac{6}{4} = \frac{3}{2}$
(2, 1)	(−2, −5)	$\frac{\_\_\_\_}{2-(-2)} = \_\_\_\_ = \_\_\_\_$
(−2, −5)	(0, −2)	$\_\_\_\_ = \_\_\_\_ = \_\_\_\_$

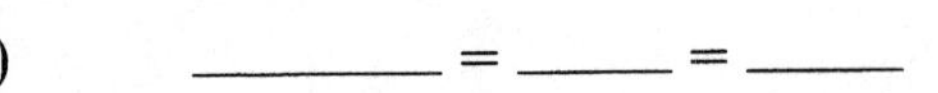

$\frac{1-(-5)}{}, \frac{6}{4}, \frac{3}{2}$

$\frac{-5-(-2)}{-2-0}, \frac{-3}{-2}, \frac{3}{2}$

EXAMPLE 24 The following graphs are lines passing through the same point but having different slopes. Note the direction of each line with its given slope.

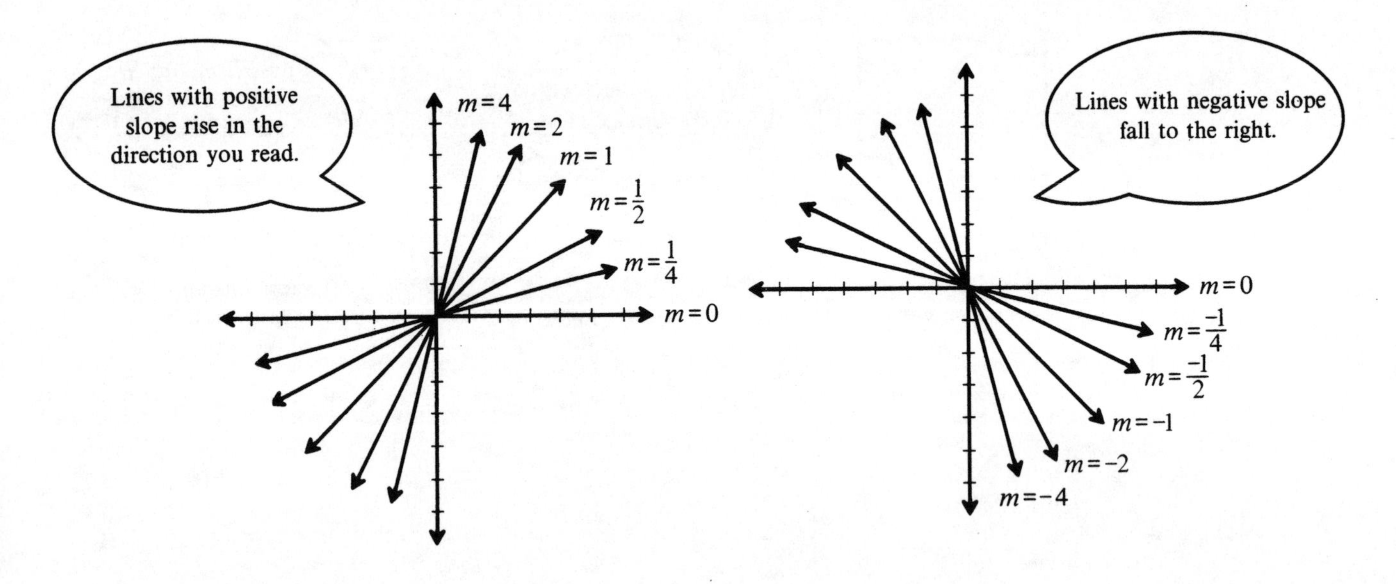

C

B

A

EXAMPLE 25 Study the lines on the right.

Which line has a negative slope? ______

Which line has a positive slope? ______

Which line has zero slope? ______

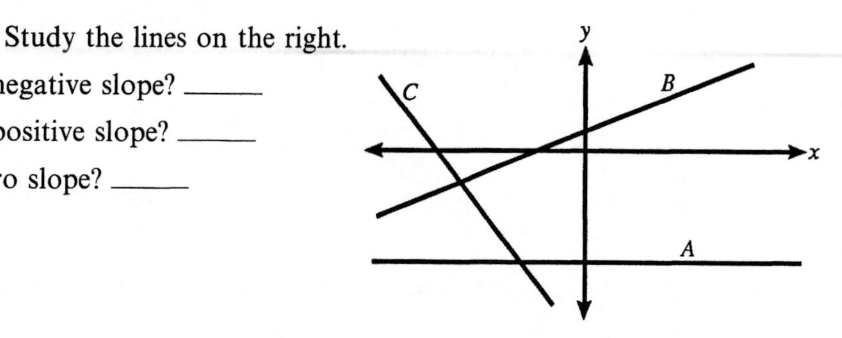

PROBLEM SET 6.3

Sketch a line through each pair of points and find the slope of the line through the points.

1. (3, 4), (8, 6)
2. (6, 7), (2, 9)
3. (2, 6), (−3, 4)
4. (−6, −4), (4, 3)
5. (4, −6), (2, 3)
6. (−5, 3), (2, 7)
7. (−4, 6), (3, 2)
8. (−3, 5), (6, 1)
9. (−6, −1), (2, 3)
10. (−3, −2), (4, 5)
11. (−4, −2), (3, −4)
12. (−5, −3), (2, −7)
13. (−3, 4), (2, −4)
14. (−5, 4), (4, −6)
15. (−7, 2), (−3, 1)
16. (−8, 5), (−6, −3)
17. (5, 2), (7, −6)
18. (8, 4), (2, −2)
19. (−2, 7), (−6, −4)
20. (6, 7), (−2, −4)

6.4 WRITING THE EQUATION OF A LINE

Slope-Intercept Form

y-Intercept

Every straight line (except a vertical one) crosses the y-axis somewhere. *Where the line crosses the y-axis, the x-coordinate is zero.* Therefore, we will say that in general a straight line crosses the y-axis at $(0, b)$. The y-coordinate, b, is called the *y-intercept*.

The slope of any straight line is

$$m = \frac{y_2 - y_1}{x_2 - x_1}$$

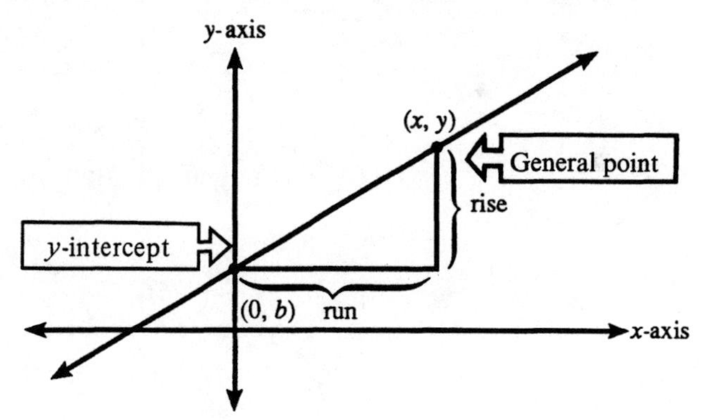

A general point on a line can be represented by the ordered pair (x, y). The slope of a line passing through the general point (x, y) and the y-intercept $(0, b)$ is

$$m = \frac{y - b}{x - 0} \qquad \text{Using the points } (0, b) \text{ and } (x, y)$$

$$m = \frac{y - b}{x}$$

$$mx = y - b \qquad \text{Multiplying both sides by } x$$

$$mx + b = y \qquad \text{Adding } b \text{ to both sides}$$

$$y = mx + b \qquad \text{Symmetric property of equality}$$

Slope-intercept

6.4A DEFINITION

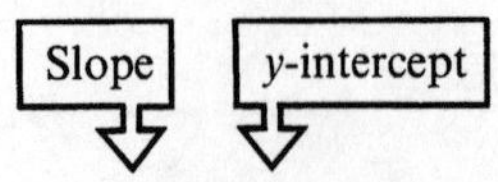

$y = mx + b$ is called the *slope-intercept* form of the equation of a line.

The slope-intercept form is one of the most useful forms of the equation of a straight line, because we can graph the line very easily from it.

In the equation $y = 3x - 4$, the slope is 3 and the y-intercept is -4.

$y = 3x - 4$

Slope | y-intercept

EXAMPLE 26 Give m, b, and the coordinates of the y-intercept for each of the equations below.

Equation	m	b	Coordinates of y-intercept
$y = 2x + 3$	2	3	(0, 3)
$y = 2x - 3$	2	-3	____
$y = 5x + 4$	5	____	____
$y = -5x + 4$	____	____	____
$y = \frac{2}{3}x + 3$	____	____	____
$y = -4x$	____	____	____
$y = x$	____	____	____

$(0, -3)$

4 $(0, 4)$

-5 4 $(0, 4)$

$\frac{2}{3}$ 3 $(0, 3)$

-4 0 $(0, 0)$

1 0 $(0, 0)$

EXAMPLE 27 Find the equation of the line with slope -3 and y-intercept -4.

Start with the general equation of a line in slope-intercept form.

$$y = mx + b$$
$$y = -3x + (-4)$$
$$y = -3x - 4$$

Just substitute the proper values for m and b.

EXAMPLE 28 Find the equation of the line with slope $\frac{2}{3}$ and y-intercept 4.

$$y = mx + b$$

$y =$ ______ $x +$ ______

$\frac{2}{3}$, 4

PROBLEM SET 6.4A

Find the equation of the line with the given slope and y-intercept.

1. Slope 2, y-intercept 4
2. Slope 3, y-intercept 7
3. Slope -5, y-intercept 6
4. Slope -6, y-intercept 8
5. Slope -3, y-intercept -4
6. Slope -4, y-intercept -3
7. Slope $\frac{3}{4}$, y-intercept 4
8. Slope $\frac{5}{6}$, y-intercept -3
9. Slope $\frac{-4}{7}$, y-intercept -5
10. Slope $\frac{-3}{5}$, y-intercept 2
11. Slope $\frac{-3}{2}$, y-intercept 6
12. Slope $\frac{-5}{2}$, y-intercept -7
13. Slope 4, y-intercept 0
14. Slope -3, y-intercept 0
15. Slope 0, y-intercept 5
16. Slope 0, y-intercept -5
17. Slope -2, passing through (0, 3)
18. Slope $\frac{1}{2}$, passing through (0, -1)
19. Slope $\frac{-2}{3}$, passing through (0, 0)
20. Slope $\frac{5}{6}$, passing through the origin

(17) Hint: Since (0, 3) is on the y-axis, the y-intercept is 3.

Review Your Skills

Evaluate the following.

21. $mx + b$, if $m = \frac{2}{3}$, $x = 6$, $b = -4$
22. $mx + b$, if $m = \frac{-3}{4}$, $x = -8$, $b = 4$

Equation of a Line Given the Slope and a Point

If we know the slope and any point on a line, it is possible to write the equation of the line. To do this, we must first find b, the y-intercept.

EXAMPLE 29 Find the equation of a line with a slope of -3 and passing through the point (5, 7).

$y = mx + b$ is the general equation of a line. To find the equation of a specific line, we substitute the value of the slope for m and the coordinates of the known point for x and y. Then we solve for b, which is the y-intercept.

$$y = mx + b$$

becomes

$$7 = (-3)(5) + b$$
$$7 = -15 + b$$
$$22 = b$$

Now that b is known, the specific equation of a line with slope -3 and y-intercept 22 can be written.

$y = -3x + 22$ ⇐ is the equation of a line passing through the point (5, 7) with a slope of -3

EXAMPLE 30 Find the equation of a line with a slope of $\frac{2}{5}$ and passing through the point $(5, -2)$.

Find the y-intercept, b

$$y = mx + b$$

becomes

$$-2 = \left(\frac{2}{5}\right)(5) + b \qquad \text{Substitute } (5, -2) \text{ for } x \text{ and } y \text{ and } \frac{2}{5} \text{ for } m$$

$$-2 = 2 + b$$

$$-4 = b$$

Use the slope and the y-intercept to write the equation.

Substituting $m = \frac{2}{5}$ and $b = -4$, the equation

$$y = mx + b$$

becomes

$$y = \frac{2}{5}x - 4$$

PROBLEM SET 6.4B

Find the equation of a line with the given slope and passing through the given point.

1. Slope $= 2$, passing through $(3, 7)$
2. Slope $= 3$, passing through $(2, 10)$
3. Slope $= -2$, passing through $(-1, 4)$
4. Slope $= -3$, passing through $(5, -2)$
5. Slope $= \frac{2}{3}$, passing through $(3, 7)$
6. Slope $= \frac{3}{4}$, passing through $(4, 6)$
7. Slope $= \frac{-1}{2}$, passing through $(4, 3)$
8. Slope $= \frac{5}{3}$, passing through $(3, 2)$
9. Slope $= \frac{-4}{7}$, passing through $(-7, 2)$
10. Slope $= 0$, passing through $(-1, 4)$
11. Slope $= \frac{-4}{5}$, passing through $(0, 0)$
12. Slope $= \frac{-6}{5}$, passing through $(-5, 0)$
13. Slope $= \frac{2}{5}$, passing through $(2, 3)$
14. Slope $= \frac{-2}{3}$, passing through $(5, -2)$
15. Slope $= \frac{5}{4}$, passing through $(-6, 2)$
16. Slope $= \frac{-4}{3}$, passing through $(2, 3)$

Review Your Skills

Find the slope of the line through the given points.

17. $(-1, 2), (8, 6)$
18. $(-2, -6), (3, -4)$

Equation of a Line Through Two Points

If we only know two points on the line, we must find the slope before we can write the equation.

EXAMPLE 31 Find the equation of a line through the points (5, 6) and (−5, 0).

Find the slope of the line through the two points.

$$m = \frac{y_2 - y_1}{x_2 - x_1} = \frac{6 - 0}{5 - (-5)} = \frac{6}{10} = \frac{3}{5}$$

We have found $m = \frac{3}{5}$ and can calculate the value of b using either point, (5, 6) or (−5, 0).

To find the y-intercept, use the point (5, 6). Substitute $x = 5$, $y = 6$, and $m = \frac{3}{5}$ in the slope-intercept form of the line.

$$y = mx + b$$

becomes

$$6 = \frac{3}{5}(5) + b$$

$$6 = 3 + b$$

$$3 = b$$

Use the slope and the y-intercept to write the equation.

Substitute $m = \frac{3}{5}$ and $b = 3$.

$$y = mx + b$$

becomes

$$y = \frac{3}{5}x + 3$$

This is the equation of the line through the points (5, 6) and (−5, 0).

We used (5, 6) to substitute in the general equation. Next, substitute the other point (−5, 0) in the specific equation to see that it also satisfies the equation.

$$y = \frac{3}{5}x + 3$$

$$0 \stackrel{?}{=} \frac{3}{5}(-5) + 3$$

$$0 \stackrel{?}{=} -3 + 3$$

$$0 = 0$$

It checks.

EXAMPLE 32 Find the equation of the line through the points $(-4, 3)$ and $(6, 7)$.

Find the slope of the line through the points.

$$m = \frac{y_2 - y_1}{x_2 - x_1}$$

$$= \frac{7 - 3}{6 - (-4)}$$

$$= \frac{4}{10}$$

$$= \frac{2}{5}$$

Find the y-intercept.

Substitute the point $(6, 7)$ and the slope $\frac{2}{5}$ in the slope-intercept form of the line.

$$y = mx + b$$

becomes

$$7 = \frac{2}{5}(6) + b \qquad \text{Substitute } y = 7,\ x = 6,\ m = \frac{2}{5}$$

$$7 = \frac{12}{5} + b$$

$$7 - \frac{12}{5} = b \qquad \text{Subtract } \frac{12}{5}$$

$$\frac{35}{5} - \frac{12}{5} = b \qquad \text{Common denominator}$$

$$\frac{23}{5} = b$$

Using the slope and the y-intercept, write the equation.

Substitute $m = \frac{2}{5}$ and $b = \frac{23}{5}$.

$$y = mx + b$$

becomes

$$y = \frac{2}{5}x + \frac{23}{5}$$

This is the equation of the line through the points $(-4, 3)$ and $(6, 7)$.

What would happen if I used the point $(-4, 3)$ instead of $(6, 7)$?

You would have the same equation for an answer. Try it.

PROBLEM SET 6.4C

Find the equation of the line passing through the given points.

1. (0, 3), (4, 0)
2. (−3, 0), (0, −5)
3. (0, 4), (5, 3)
4. (0, 2), (4, −1)
5. (0, −3), (−4, 6)
6. (0, −4), (−5, 2)
7. (0, 6), (−5, −6)
8. (0, 3), (−4, 3)
9. (−3, 4), (5, 8)
10. (−2, 8), (6, 7)
11. (−1, −6), (−5, 3)
12. (−4, −3), (−1, 7)
13. (−6, 3), (4, −5)
14. (−2, 7), (2, 7)
15. (−2, −3), (−6, −7)
16. (−4, −3), (−7, −6)
17. (2, 3), (1, 4)
18. (2, 3), (−1, −4)
19. (2, 3), (1, −4)
20. (1, 4), (2, 3)
21. (2, 3), (4, 1)
22. (3, 2), (4, 1)

Review Your Skills

Use the distributive property to write the product as a sum.

23. $\frac{-3}{4} \cdot (x - 2)$

24. $\frac{-3}{7} \cdot (x + 4)$

Another Method for Finding the Equation of a Line Given the Slope and a Point

We have learned that the slope of a line is the same everywhere.

The ordered pair (x, y) is a general point on the line. The slope of a line passing through the general point (x, y) and the fixed point (x_1, y_1) is:

$m = \frac{y - y_1}{x - x_1}$	Using points (x_1, y_1) and (x, y)
$\frac{y - y_1}{x - x_1} = m$	Symmetric property of equality
$(x - x_1)\frac{y - y_1}{x - x_1} = m(x - x_1)$	Multiply both sides by $(x - x_1)$
$y - y_1 = m(x - x_1)$	

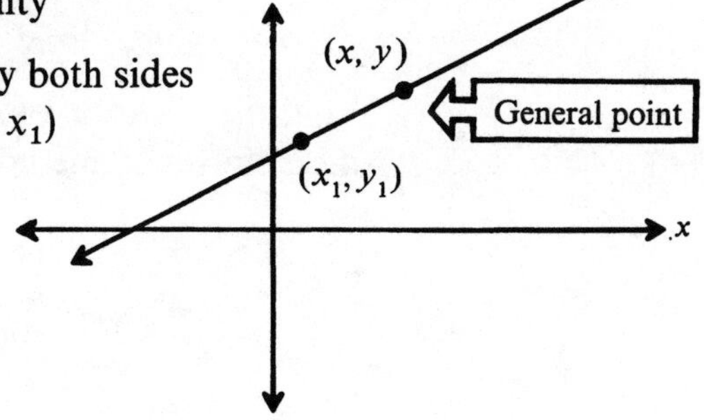

Point-slope

> **6.4B DEFINITION**
> $y - y_1 = m(x - x_1)$ is called the *point-slope* form of the equation of a line.

To use the point-slope form of the equation of a line, we need to know the slope of the line and the coordinates of one point on the line.

EXAMPLE 33 Find the equation of a line with slope -2 and passing through the point $(-4, 3)$.

$y - y_1 = m(x - x_1)$ is the point-slope form of the equation of a line

$y - 3 = -2[x - (-4)]$ Substitute values in the formula

$y - 3 = -2(x + 4)$ Simplify the right side

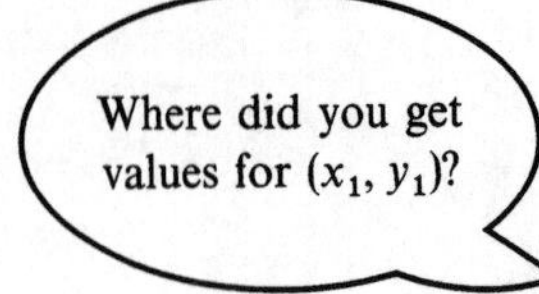

The problem gives a point on the line: $(-4, 3)$. This tells us $x_1 = -4$, and $y_1 = 3$.

$y - 3 = -2x - 8$ Use the distributive property

$y = -2x - 5$ Add 3 to both sides

EXAMPLE 34 Find the equation of a line with slope $\frac{2}{3}$ and passing through the point $(4, 5)$. $y_1 = 5$, $x_1 = 4$, $m = \frac{2}{3}$

$y - y_1 = m(x - x_1)$

$y - 5 = \frac{2}{3}(x - 4)$ Substitute values in the formula

$y - 5 = \frac{2}{3}x - \frac{8}{3}$ Use the distributive property

$y = \frac{2}{3}x - \frac{8}{3} + 5$ Add 5 to both sides

$y = \frac{2}{3}x + \frac{7}{3}$ Combine the $\frac{-8}{3}$ and 5

PROBLEM SET 6.4D

Find the equation of a line with given slope and passing through the given point. Use the point-slope method.

1. Slope $= 2$, passing through $(3, 2)$
2. Slope $= 3$, passing through $(2, 6)$
3. Slope $= -2$, passing through $(4, 5)$
4. Slope $= -4$, passing through $(1, 0)$
5. Slope $= \frac{2}{3}$, passing through $(3, 2)$
6. Slope $= \frac{3}{4}$, passing through $(4, 6)$
7. Slope $= \frac{3}{5}$, passing through $(-2, 4)$
8. Slope $= \frac{4}{5}$, passing through $(-3, 5)$
9. Slope $= \frac{-3}{2}$, passing through $(4, -2)$
10. Slope $= \frac{-4}{3}$, passing through $(6, -7)$
11. Slope $= \frac{-3}{7}$, passing through $(-1, 5)$
12. Slope $= \frac{-4}{7}$, passing through $(-2, 3)$
13. Slope $= 0$, passing through $(4, -7)$
14. Slope $= \frac{6}{5}$, passing through $(-2, -5)$

PROBLEM SET 6.4E

Find the equation of a line with given slope and passing through the given point. Use either method.

1. Slope = 2, passing through $(1, -1)$
2. Slope = 5, passing through $(-1, -1)$
3. Slope = 3, passing through $(-6, 1)$
4. Slope = 4, passing through $(-5, 2)$
5. Slope = -4, passing through $(5, 6)$
6. Slope = -5, passing through $(7, 2)$
7. Slope = $\frac{3}{5}$, passing through $(6, -2)$
8. Slope = $\frac{4}{7}$, passing through $(2, -8)$
9. Slope = $\frac{-2}{7}$, passing through $(3, 4)$
10. Slope = $\frac{-3}{8}$, passing through $(5, 7)$
11. Slope = $\frac{7}{3}$, passing through $(-4, 5)$
12. Slope = $\frac{5}{3}$, passing through $(6, -4)$
13. Slope = $\frac{-4}{3}$, passing through $(-2, 6)$
14. Slope = 0, passing through $(3, -8)$

6.5 GRAPHING LINES

Graphing Lines by the Slope-Intercept Method

We can use the slope-intercept form of an equation to graph its line.

EXAMPLE 35 Sketch a graph of $y = \frac{4}{3}x - 2$.

-2 The y-intercept in this equation is ______.

$\frac{4}{3}$ The slope of this equation is $\frac{\Delta y}{\Delta x} =$ ______.

Step 1
Plot the y-intercept.

Step 2
Starting from the y-intercept point, $(0, -2)$, draw Δx and Δy.

Step 3
Draw a line.

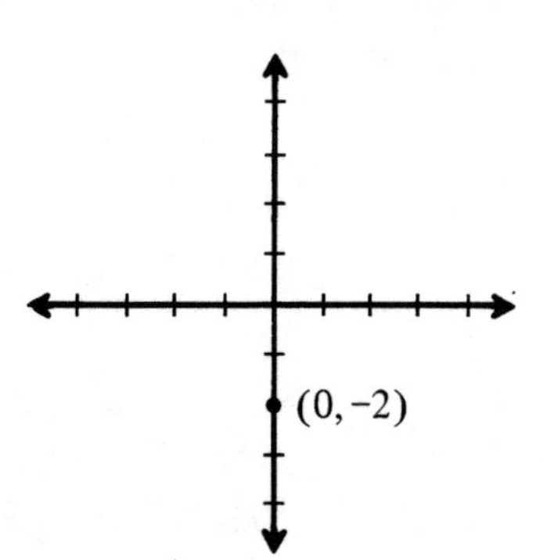

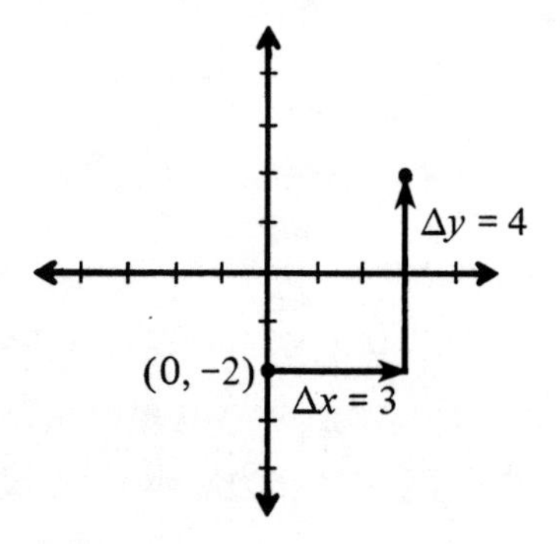

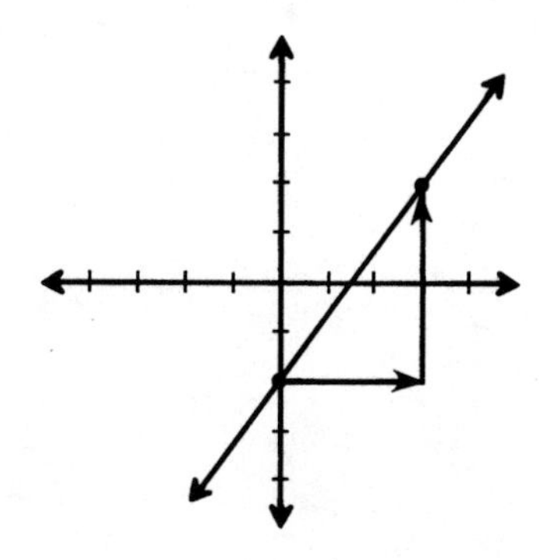

EXAMPLE 36 Sketch a graph of $y = -3x + 4$.

4

The y-intercept of this equation is ______.

The slope of this equation is $\dfrac{\Delta y}{\Delta x} = -3$ or $\dfrac{-3}{1}$.

A slope of -3 means that for every unit x increases y will decrease by 3 units.

Step 1
Plot the y-intercept.

Step 2
Starting from the y-intercept point, $(0, 4)$, draw Δx and Δy.

Step 3
Draw a line.

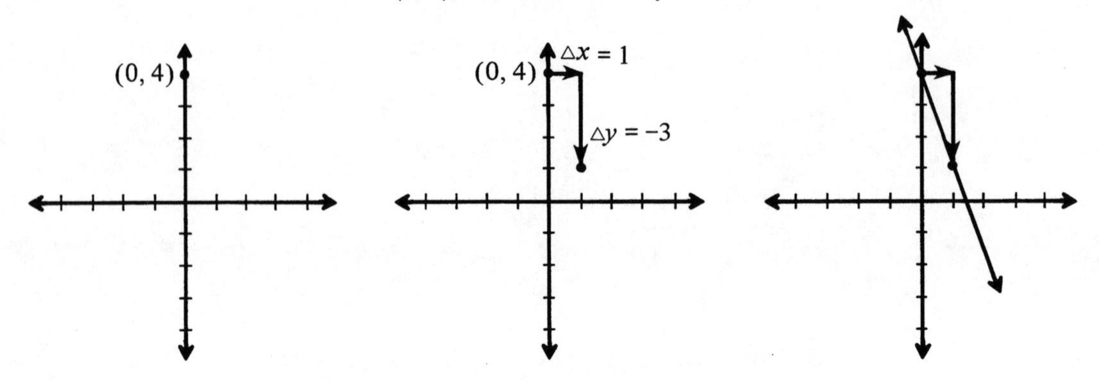

EXAMPLE 37 Sketch a graph of $3x + 4y = 8$.

First rewrite the equation in slope-intercept form by solving for y, then graph using the slope and y-intercept.

$$3x + 4y = 8$$

$$4y = -3x + 8$$

$$\frac{4}{4}y = \frac{-3}{4}x + \frac{8}{4}$$

$$y = \frac{-3}{4}x + 2$$

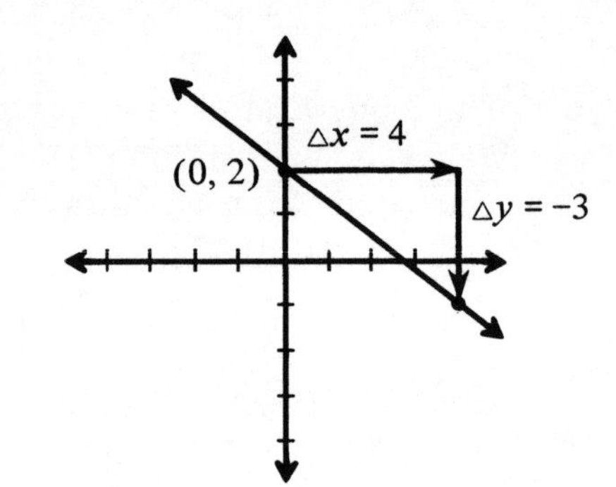

The y-intercept is $(0, 2)$.

The slope is $\dfrac{-3}{4}$.

EXAMPLE 38 Sketch $y = 2x - 2$ and $y = 2x + 3$.

slope

What is alike in the two equations? ______________

y-intercept

What is different in the two equations? ______________

direction

What is alike in the two graphs? ______________

where they cross the y-axis

What is different in the two graphs? ______________

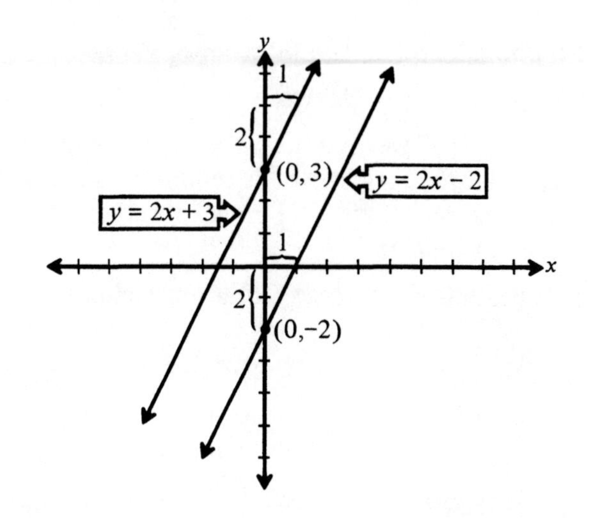

Parallel

Example 38 illustrates the fact that lines with the same slope have the same inclination. They are said to be *parallel.*

Where the line crosses the y-axis is determined by the y-intercept.

PROBLEM SET 6.5A

Sketch the following graphs using the slope and y-intercept method.

1. $y = x + 2$
2. $y = x - 3$
3. $y = 2x + 1$
4. $y = 2x - 1$
5. $y = -2x - 1$
6. $y = -2x + 1$
7. $y = \frac{2}{3}x - 2$
8. $y = \frac{3}{4}x + 3$
9. $y = \frac{-3}{5}x + 4$
10. $y = \frac{-2}{3}x + 5$
11. $y = \frac{1}{2}x - 4$
12. $2x + y = 3$
13. $-3x + y = 1$
14. $-2x + 3y = 6$
15. $-3x + 2y = 6$
16. $-3x + 2y = -6$
17. $4x + 3y = 12$
18. $4x - 3y = 12$
19. $x - 2y = -8$
20. $y = \frac{2}{3}x$
21. $y = -3x$
22. $y = x$
23. $y = \frac{2}{3}x + 2$
24. $3x - y = 6$

Review Your Skills

Evaluate the following.

If $f(x) = \frac{3}{4}x - 4$, find

25. $f(-4)$
26. $f(0)$

If $g(x) = \frac{-2}{3}x + 5$, find

27. $g(-6)$
28. $g(0)$

Graphing Lines by the Intercepts Method

EXAMPLE 39 Sketch a graph of $3x + 4y = 8$ using the x- and y-intercepts.

In Example 37 we used the slope-intercept form of the line to graph $3x + 4y = 8$. Another way to graph $3x + 4y = 8$ is by finding its x- and y-intercepts, that is, the points where it crosses each axis. We have observed that when the line crosses the y-axis, the x-coordinate is 0.

Substituting $x = 0$ in the equation yields

$$3(0) + 4y = 8$$
$$4y = 8$$
$$y = 2 \quad \text{is the } y\text{-intercept}$$

Where the line crosses the x-axis, the y-coordinate is 0.

We find the x-intercept by substituting $y = 0$ in the equation.

$$3x + 4(0) = 8$$
$$3x = 8$$
$$x = \frac{8}{3} \quad \text{is the } x\text{-intercept} \quad \left(\frac{8}{3} = 2\frac{2}{3}\right)$$

Step 1
Plot the intercepts.

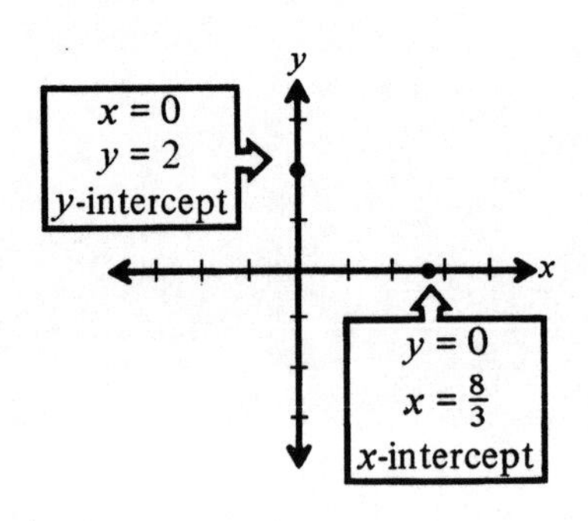

Step 2
Draw a line.

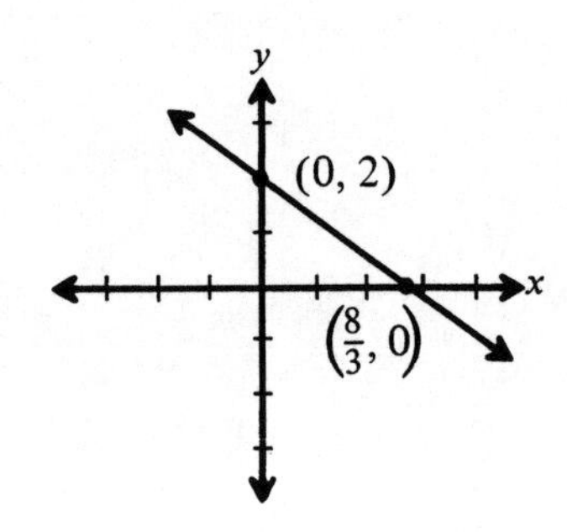

EXAMPLE 40 Sketch a graph of $2x - y = -6$ by the intercepts method.

x

To find the y-intercept, we substitute ______ = 0.

Substituting in the equation yields

$$2(0) - y = -6$$
$$-y = -6$$

6

$y =$ ______ is the y-intercept

y

To find the x-intercept, we substitute ______ = 0.

$$2x - (0) = -6$$
$$2x = -6$$

-3

$x =$ ______ is the x-intercept

Step 1
Plot the intercepts.

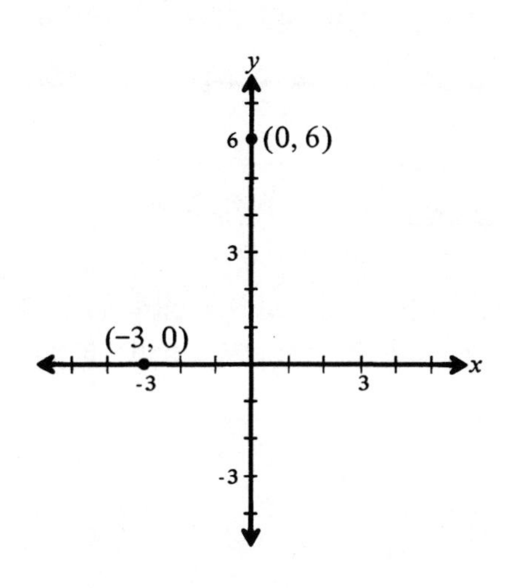

Step 2
Draw a line.

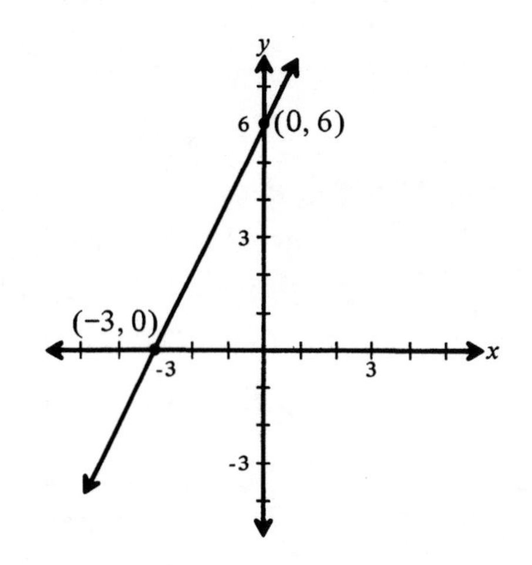

Two Special Cases

Case 1 **EXAMPLE 41** Sketch the graph of $y = 2$.

Since there is no x term in this equation, we consider the coefficient of the x term to be zero. The equation can be written $0x + y = 2$.

Make a table. Ordered pairs

x	y	(x, y)
-2	2	$(-2, 2)$
0	2	$(0, 2)$
3	______	(________)
5	______	(________)

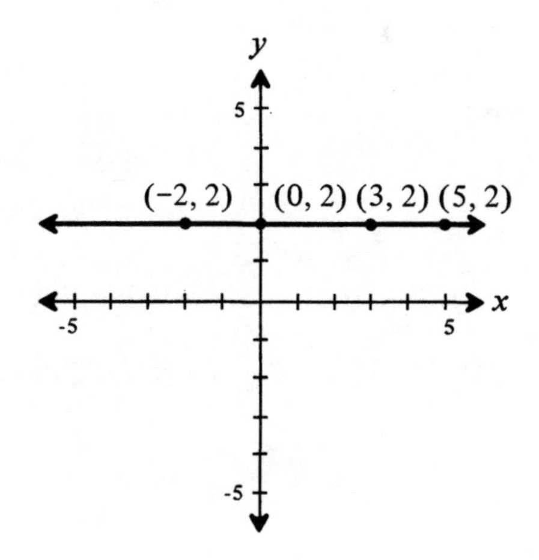

2 3, 2

2 5, 2

What is the change in y when x changes

from -2 to 0? ______

from 0 to 3? ______

from any point to another? ______

0

0

0

Using the coordinates (0, 2) and (5, 2)

$$m = \frac{y_2 - y_1}{x_2 - x_1} = \frac{2 - 2}{5 - 0} = \frac{0}{5} = \text{\_\_\_\_\_\_}$$

0

Intuitively, $y = 2$ means all points that are up 2 from the x-axis.

Graph of $y = b$

> **6.5A GRAPH OF $y = b$**
>
> In general, the graph of $y = b$ is a straight line parallel to the x-axis with zero slope. It crosses the y-axis at $y = b$.

Case 2 **EXAMPLE 42** Sketch the graph of $x = 2$.

Since there is no y term in the equation $x = 2$, the coefficient of the y term may be considered to be zero. The equation becomes $x + 0y = 2$.

In the equation $x + 0y = 2$, the only value that can be assigned to x is 2. Notice, however, you can assign any value to y and the equation will be true as long as x is 2.

Make a table. Ordered pairs

x	y	(x, y)
2	-2	$(2, -2)$
2	0	$(2, 0)$
____	2	(________)
____	3	(________)

2 2, 2

2 2, 3

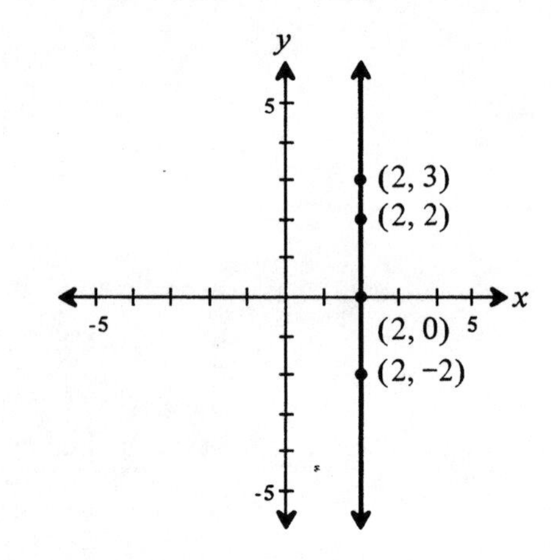

The graph of $x = 2$ is a vertical line that crosses the x-axis at $x = 2$.

Intuitively, $x = 2$ means all points that are $+2$ units from the y-axis.

By definition, slope $= \dfrac{\text{change in } y}{\text{change in } x}$. For the equation $x = 2$, x does not change ($\Delta x = 0$).

Therefore, the denominator of our fraction will always be zero. But division by zero is illegal. Therefore, the slope of the vertical line $x = 2$ is undefined.

Graph of $x = a$

> **6.5B GRAPH OF $x = a$**
>
> In general, the graph of $x = a$ is a straight line parallel to the y-axis. It crosses the x-axis at $x = a$.

PROBLEM SET 6.5B

Sketch the following graphs using the intercepts method.

1. $x + y = 4$
2. $x - y = 3$
3. $2x + 3y = 6$
4. $3x + 2y = 6$
5. $3x - 2y = 6$
6. $2x + 3y = 12$
7. $6x - 3y = 6$
8. $2x - 5y = 10$
9. $-3x + 4y = 12$
10. $-2x + 2y = 8$
11. $-4x + 2y = -8$
12. $2x + y = 10$
13. $3x - y = 9$
14. $x - 4y = -8$
15. $2x - y = 8$
16. $-x - 3y = 6$
17. $x + y = 7$
18. $x - y = 7$
19. $-x + y = -7$
20. $x = 3$
21. $x = -5$
22. $y = 4$
23. $y = 1$
24. $y = 0$
25. $x = 0$
26. $x - 2y = 4$
27. $2x - y = 6$
28. $x = -3$

Graphing Lines (Summary)

Generally, sketching graphs of lines can be done by one of the following methods:

1. A table of values (ordered pairs)
2. The slope-intercept method
3. The intercepts method

Regardless of the method used, the same graph is obtained. With practice, it can be determined which method is best in a given situation.

EXAMPLE 43 Sketch the graph of $3x + 4y = 12$ using the table of values method.

Solve for y.

$$3x + 4y = 12$$
$$4y = -3x + 12$$
$$y = \frac{-3}{4}x + 3$$

Make a table.

	Function	Ordered pairs
x	$y = \frac{-3}{4}x + 3$	(x, y)
-4	$\frac{-3}{4}(-4) + 3 = 6$	$(-4, 6)$
0	$\frac{-3}{4}(0) + 3 = 3$	$(0, 3)$
3	$\frac{-3}{4}(3) + 3 = \frac{3}{4}$	$\left(3, \frac{3}{4}\right)$
4	$\frac{-3}{4}(4) + 3 = 0$	$(4, 0)$

By using multiples of 4, fractions can be avoided.

Make a graph.

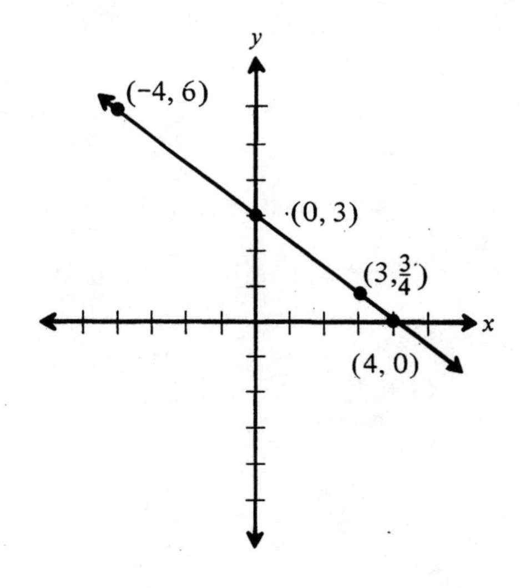

EXAMPLE 44 Sketch the same graph of $3x + 4y = 12$ using the intercepts method. Substituting $x = 0$ in the equation yields

$$3(0) + 4y = 12$$
$$4y = 12$$
$$y = 3 \qquad \text{is the } y\text{-intercept}$$

Substituting $y = 0$ in the equation yields

$$3x + 4(0) = 12$$
$$3x = 12$$
$$x = 4 \qquad \text{is the } x\text{-intercept}$$

A table using the intercepts method can easily be made.

x	y
0	
	0

Filling in the values yields

x	y
0	3
4	0

Sketch the graph.

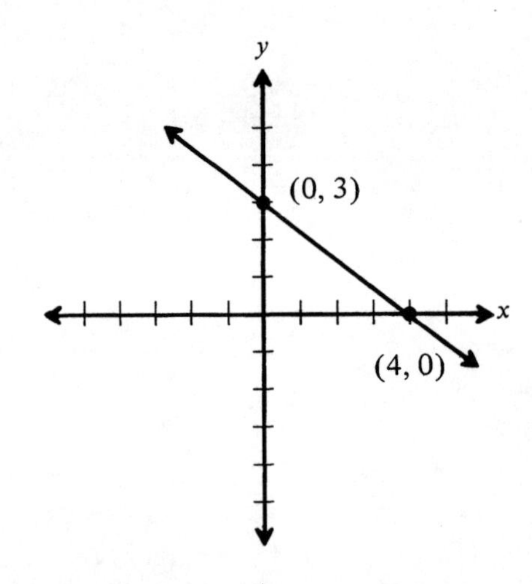

This is the same graph as Example 43.

EXAMPLE 45 Sketch the graph of $3x + 4y = 12$ using the slope-intercept method.

Solve for y.

$$3x + 4y = 12$$
$$4y = -3x + 12$$
$$y = \frac{-3}{4}x + 3$$

The y-intercept is (0, 3), the slope is $\frac{-3}{4} = \frac{\Delta y}{\Delta x}$

$\Delta x = 4$

$\Delta y = -3$

A slope of $\frac{-3}{4}$ means that for every increase of 1 unit in x, y will decrease by $\frac{3}{4}$ of a unit.

You can also say that a slope of $\frac{-3}{4}$ means that if you increase x by 4 units, y will decrease by 3 units.

Sketch the graph.

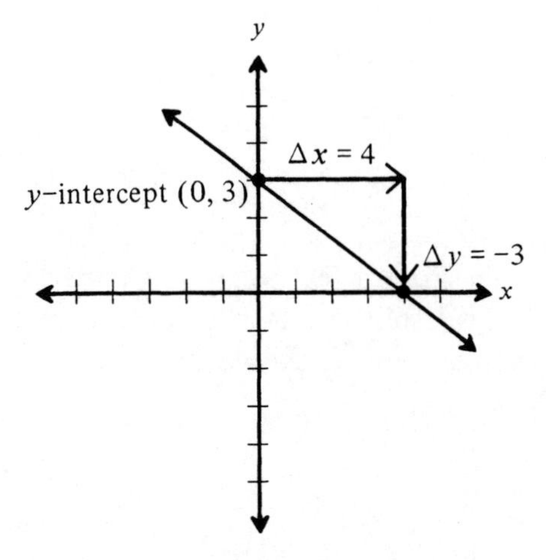

Examples 43, 44, and 45 show three ways to graph $3x + 4y = 12$. Use the method that seems easiest to you.

POINTERS FOR CHOOSING AN EASY METHOD TO GRAPH A LINE

Sometimes a particular method is impossible. For example, sketch the graph of $2y = 3x$. Try the intercepts method.

Substitute $x = 0$

$2y = 3(0)$

$2y = 0$

$y = 0$

x	y
0	0

Substitute $y = 0$

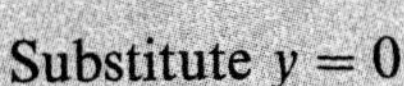

$2(0) = 3x$

$0 = 3x$

$0 = x$

x	y
0	0

Both intercepts give us the same point.

To draw a line, we need at least two points. The intercepts method cannot be used in this problem, nor is it a very good choice whenever the two intercepts are close together.

Next we will illustrate what happens when the two intercepts are close together by trying to graph a different equation.

Now try sketching the graph of $2x + 3y = 1$ using the intercepts method.

Substitute $x = 0$

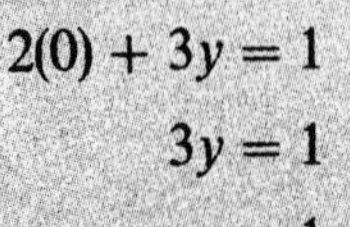

$$2(0) + 3y = 1$$
$$3y = 1$$
$$y = \frac{1}{3}$$

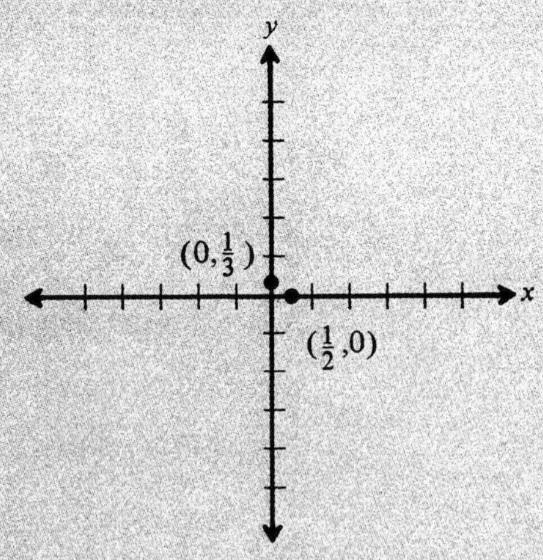

Substitute $y = 0$

$$2x + 3(0) = 1$$
$$2x = 1$$
$$x = \frac{1}{2}$$

x	y
0	$\frac{1}{3}$
$\frac{1}{2}$	0

The points are fractions, which are difficult to plot, and they are so close together that it is difficult to sketch the graph using the intercepts method. Therefore, another method should be used.

Try the slope-intercept method.

$$2x + 3y = 1$$
$$3y = -2x + 1$$
$$y = \frac{-2}{3}x + \frac{1}{3} \quad \text{Solve for } y$$

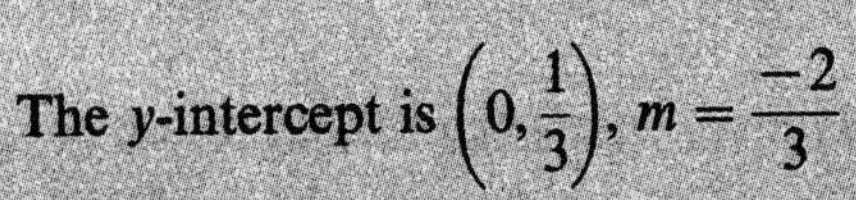

The y-intercept is $\left(0, \frac{1}{3}\right)$, $m = \frac{-2}{3}$

$$\Delta x = 3$$
$$\Delta y = -2$$

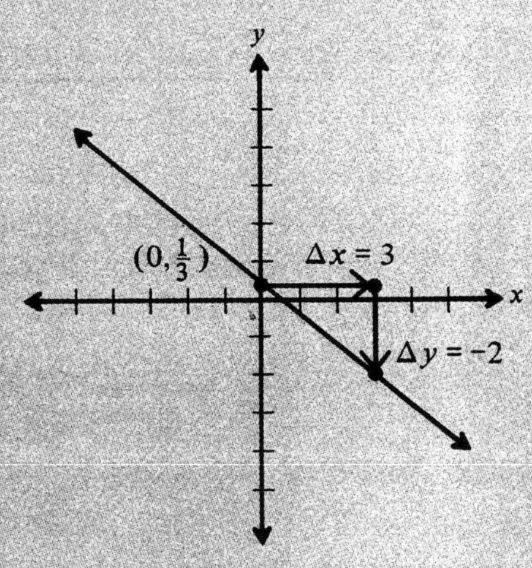

Using the slope-intercept method was better, but we still had to deal with a fraction. Let's try the table of values method.

$$2x + 3y = 1$$

Solving for y we have

$$2x + 3y = 1$$
$$3y = -2x + 1$$
$$y = \frac{-2}{3}x + \frac{1}{3}$$

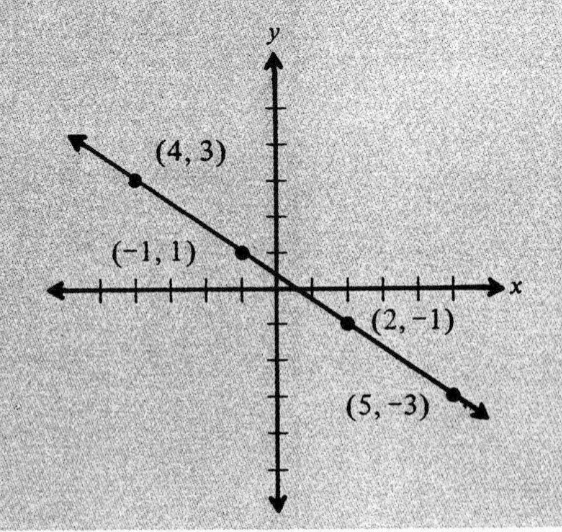

x	Function $\frac{-2}{3}x + \frac{1}{3} = y$	Ordered pairs (x, y)
-4	$\frac{-2}{3}(-4) + \frac{1}{3} = 3$	$(-4, 3)$
-1	$\frac{-2}{3}(-1) + \frac{1}{3} = 1$	$(-1, 1)$
2	$\frac{-2}{3}(2) + \frac{1}{3} = -1$	$(2, -1)$
4	$\frac{-2}{3}(4) + \frac{1}{3} = -2\frac{1}{3}$	$\left(4, -2\frac{1}{3}\right)$
5	$\frac{-2}{3}(5) + \frac{1}{3} = -3$	$(5, -3)$

Why didn't you plot $x = 4$?

Because $x = 4$ gave me a messy value, so I tried $x = 5$, which gave me a nicer number.

How do I know which method to use?

Choose the method with which you feel most comfortable, or the one you like best.

PROBLEM SET 6.5C

Sketch the following graphs using any method.

1. $x + y = 5$
2. $x + y = 4$
3. $x - y = 4$
4. $4x - y = 8$
5. $x + 2y = 5$
6. $x - 2y = 7$
7. $2x + 4y = 5$
8. $3x + 2y = 7$
9. $-x + 3y = 4$
10. $x - 3y = 5$
11. $3x + 4y = 24$
12. $3x - 4y = 24$
13. $4x - 5y = 7$
14. $6x - 7y = 5$
15. $5x - 2y = 3$
16. $-3x + 4y = 6$
17. $-x + y = 5$
18. $-x + y = 3$
19. $x = 0$
20. $x = -2$
21. $y = 4$
22. $y = 0$
23. $x = 5$
24. $y = -3$

6.6 PARALLEL AND PERPENDICULAR LINES (OPTIONAL)

EXAMPLE 46 Sketch $y = -3x - 3$ and $y = -3x + 2$.

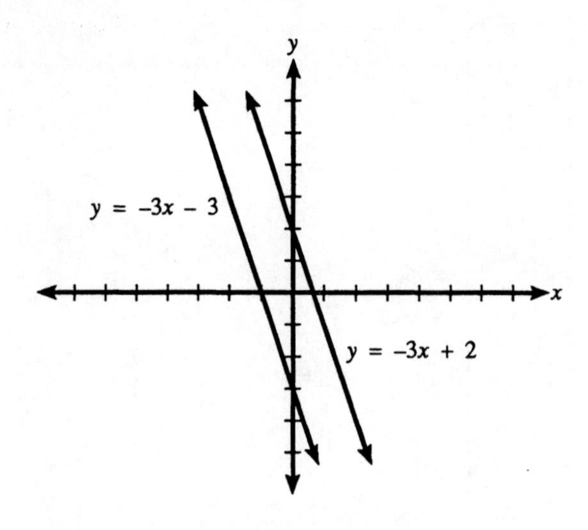

slope

y-intercept

direction

where they cross the y-axis

What is alike in the two equations? ______________

What is different in the two equations? ______________

What is alike in the two graphs? ______________

What is different in the two graphs? ______________

Parallel

Example 46 illustrates the fact that lines with the same slope are *parallel.*

PROBLEM SET 6.6A

Sketch the graphs of each pair of lines on the same coordinate system.

1. $y = \frac{2}{3}x + 5$ and $y = \frac{2}{3}x - 5$
2. $y = \frac{-2}{3}x + 5$ and $y = \frac{-2}{3}x - 5$
3. $y = 3x - 3$ and $y = 3x$
4. $y = -2x + 6$ and $y = -2x + 1$
5. $y = \frac{3}{5}x$ and $y = \frac{3}{5}x + 3$
6. $y = \frac{-3}{5}x + 2$ and $y = \frac{-3}{5}x - 2$
7. $y = -3x - 3$ and $y = -3x + 1$
8. $y = \frac{1}{3}x - 6$ and $y = \frac{1}{3}x + 2$

It is possible to sketch lines parallel to a given line using the slope-intercept method of sketching graphs.

EXAMPLE 47 Sketch a line parallel to the line $y = \frac{3}{4}x - 4$ with a y-intercept 5.

To sketch the line $y = \frac{3}{4}x - 4$, we see that the slope is $\frac{3}{4}$ and the y-intercept is -4.

Locate the point $(0, -4)$, the y-intercept. The line has a rise of 3 and a run of 4. Connect the points.

To sketch the line parallel to the given line with y-intercept 5, locate the point $(0, 5)$. This line also has a rise of 3 and a run of 4. Connect the points.

Since the lines have the same slope, they are parallel.

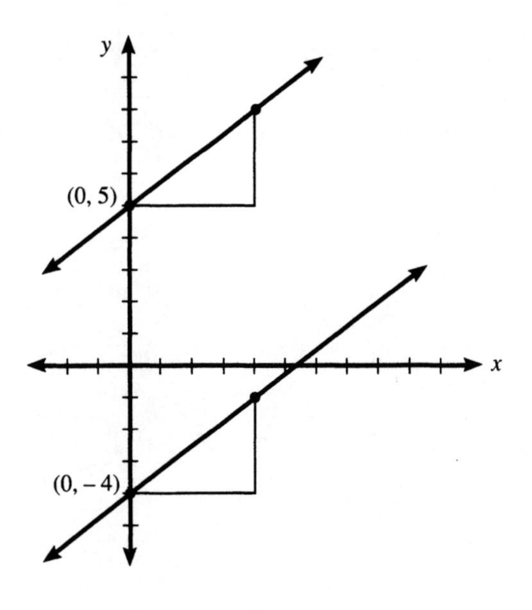

PROBLEM SET 6.6B

Sketch a second line parallel to the given line. The second line should cross the y-axis as specified.

1. Parallel to $y = 3x - 4$, passing through $(0, 3)$
2. Parallel to $y = -3x$, passing through $(0, -4)$
3. Parallel to $y = \frac{2}{5}x + 3$, y-intercept of -1
4. Parallel to $y = \frac{2}{5}x + 3$, crossing the y-axis at the origin
5. Parallel to $y = \frac{-3}{5}x + 4$, passing through $(0, -2)$
6. Parallel to $y = \frac{3}{5}x - 3$, passing through $(0, 4)$
7. Parallel to $y = \frac{7}{4}x + 2$, y-intercept of 3
8. Parallel to $y = \frac{-7}{4}x - 3$, crossing the y-axis at $y = 4$

It is possible to write the equation of a line parallel to another line passing through any point we specify.

EXAMPLE 48 Write the equation of a line passing through the point $(-5, 4)$ and parallel to the line $y = \frac{2}{5}x - 4$.

Since the desired line is to be parallel to $y = \frac{2}{5}x - 4$, we know its slope must be $\frac{2}{5}$.

Therefore this is really a problem of writing the equation of a line with slope $\frac{2}{5}$ passing through the point $(-5, 4)$.

We will use one of the methods we used earlier to do this.

$y = mx + b$ General equation of a line

Since we know values for x, y, and m, we can substitute them to find the value of b.

$$4 = \frac{2}{5}(-5) + b$$

$$4 = -2 + b$$

$$4 + 2 = b$$

$$6 = b$$

Therefore, the equation we want is

$$y = \frac{2}{5}x + 6$$

It is parallel to $y = \frac{2}{5}x - 4$ because it has the same slope.

We can establish that $y = \frac{2}{5}x + 6$ contains the point $(-5, 4)$ by substituting these values in the equation.

$$y = \frac{2}{5}x + 6$$

$4 = \frac{2}{5}(-5) + 6$ Substituting

$$4 = -2 + 6$$

$$4 = 4$$

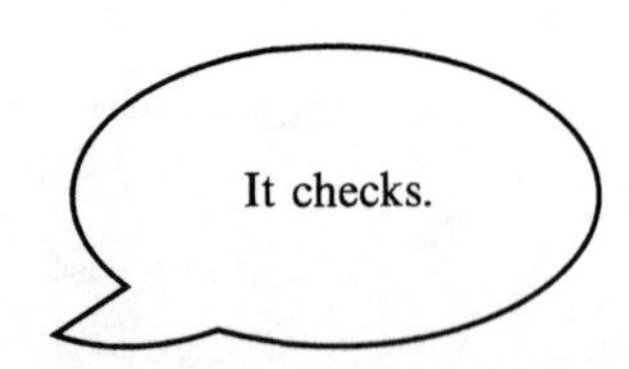

In higher math classes, it is possible to prove the following theorem.

Perpendicular lines

> **6.6 THEOREM: PERPENDICULAR LINES**
>
> Two lines are perpendicular if their slopes are negative reciprocals; that is, a line with slope m_1 is perpendicular to line 2 with slope m_2 if
>
> $$m_2 = \frac{-1}{m_1} \quad \text{or} \quad m_1 m_2 = -1$$

EXAMPLE 49 Use the theorem for perpendicular lines to sketch a line perpendicular to $y = \frac{3}{2}x - 4$ with y-intercept 3.

To sketch the given line, locate the point $(0, -4)$, the y-intercept, on the graph; then use the slope to find the direction.

The theorem states that the slope of the line perpendicular to the given line is given by

$$m_2 = \frac{-1}{m_1}$$

The desired slope is

$$m_2 = \frac{-1}{\frac{3}{2}}$$

$$m_2 = \frac{-2}{3}$$

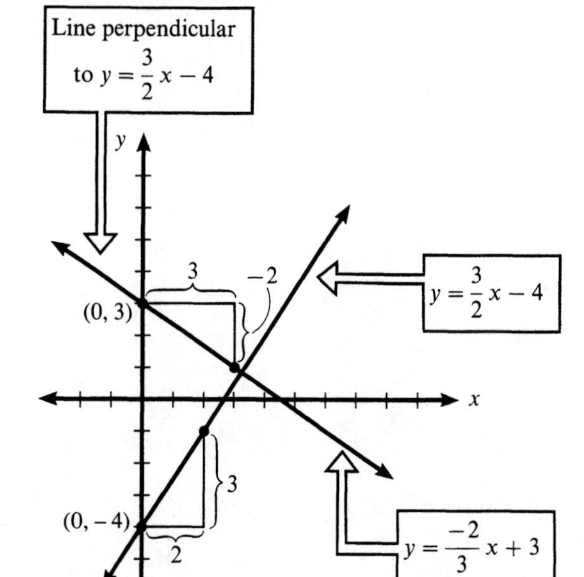

With y-intercept 3 and slope $\frac{-2}{3}$, we can sketch the graph.

PROBLEM SET 6.6C

Use the method of Example 49 to sketch a line perpendicular to the given line with the specified intercept.

1. Perpendicular to $y = 3x - 2$, y-intercept 2
2. Perpendicular to $y = -3x$, passing through $(0, -3)$
3. Perpendicular to $y = \frac{2}{3}x - 3$, passing through $(0, 4)$
4. Perpendicular to $y = \frac{-3}{5}x + 2$, crossing the y-axis at $y = 4$
5. Perpendicular to $y = -4x + 2$, passing through $(0, 6)$
6. Perpendicular to $y = 4x - 2$, passing through $(0, -3)$
7. Perpendicular to $y = \frac{4}{3}x + 3$, crossing the y-axis at $y = -4$
8. Perpendicular to $y = \frac{6}{5}x - 2$, y-intercept 3

We can also use Theorem 6.6 to write the equation of a line perpendicular to a given line passing through any desired point.

EXAMPLE 50 Find the equation of a line perpendicular to $y = \frac{2}{3}x + 4$, passing through the point $(-4, 4)$.

Since the slope of the desired line m_2 is perpendicular to a line with slope $\frac{2}{3}$, the desired slope is

$$m_2 = \frac{-1}{m_1}$$

$$m_2 = \frac{-1}{\frac{2}{3}}$$

$$= \frac{-3}{2}$$

Now write the equation of a line passing through $(-4, 4)$ with slope $\frac{-3}{2}$.

Using the point-slope form of a line

$$y - y_1 = m(x - x_1)$$

$$y - (4) = \frac{-3}{2}(x - (-4))$$

$$y - 4 = \frac{-3}{2}(x + 4)$$

$$y - 4 = \frac{-3}{2}x - 6$$

$$y = \frac{-3}{2}x - 2$$

Plot both lines.

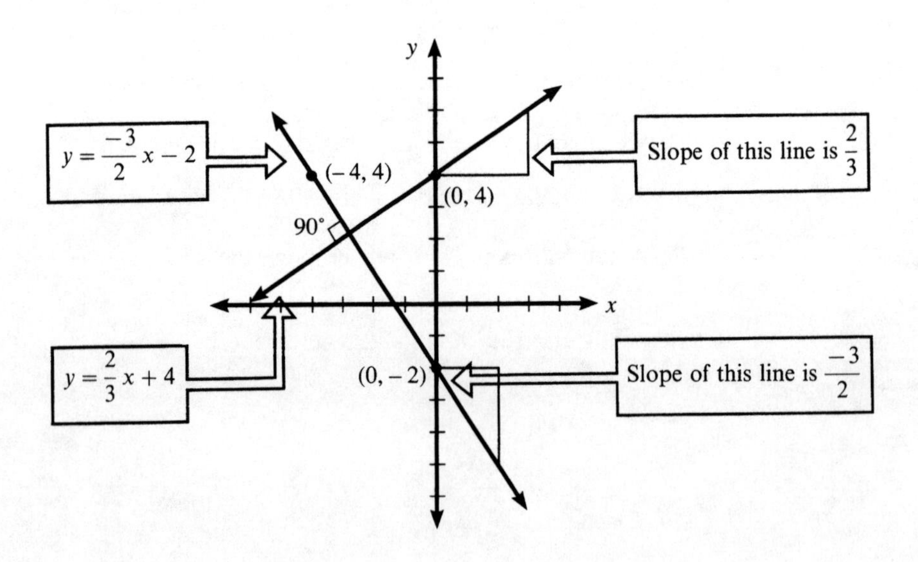

EXAMPLE 51 Find the equation of a line through $(-1, 3)$ and perpendicular to the line $3x + 5y = -10$. Sketch the lines.

We must find the slope of the given equation before we can find the desired slope of the required line.

Hence, solve the given equation for y.

$$3x + 5y = -10$$
$$5y = -3x - 10$$
$$y = \frac{-3}{5}x - 2$$

Since the slope of the desired line m_2 is perpendicular to the line with slope $\frac{-3}{5}$, the desired slope is

$$m_2 = \frac{-1}{m_1}$$
$$= \frac{-1}{\frac{-3}{5}}$$
$$= \frac{5}{3}$$

Now we write the equation of a line passing through $(-1, 3)$ with slope $\frac{5}{3}$.

Use the point-slope form of a line.

$$y - y_1 = m(x - x_1)$$
$$y - (3) = \frac{5}{3}(x - (-1))$$
$$y - 3 = \frac{5}{3}(x + 1)$$
$$y - 3 = \frac{5}{3}x + \frac{5}{3}$$
$$y = \frac{5}{3}x + \frac{5}{3} + 3$$
$$y = \frac{5}{3}x + \frac{5}{3} + \frac{9}{3}$$
$$y = \frac{5}{3}x + \frac{14}{3}$$

Plot both lines.

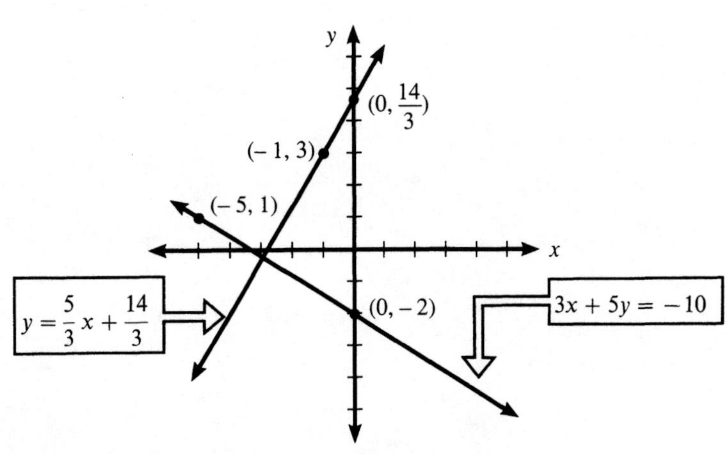

PROBLEM SET 6.6D

a. Find the equation of a line that is parallel to each given line and contains the given point.
b. Find the equation of the line that is perpendicular to each given line and contains the given point.

1. $y = \frac{2}{3}x + 5$, $(-6, 4)$
2. $y = \frac{-2}{5}x - 6$ $(-2, -3)$
3. $x - 2y = 3$, $(2, -3)$
4. $3x + 2y = 4$, $(4, 5)$
5. $y = \frac{-2}{3}x - 5$, $(4, -6)$
6. $y = \frac{2}{5}x + 6$, $(-3, -2)$
7. $2x - 3y = 4$, $(1, 4)$
8. $2x - 5y = 4$, $(-3, 5)$

Find the equation of the line with the following conditions.

9. Parallel to $2x - 3y = 6$, crossing the x-axis at -2
10. Perpendicular to $2x - 3y = 9$, with x-intercept -2
11. Perpendicular to $3x + 4y = 12$, passing through $(0, 3)$
12. Parallel to $3x + 4y = 8$, crossing the y-axis at 3
13. Parallel to $3x - 4y = 8$, passing through $(-2, 0)$
14. Perpendicular to $4x - 3y = 6$, passing through $(-3, 0)$
15. Perpendicular to $5x + 2y = 4$, crossing the x-axis at -3
16. Parallel to $3x + 5y = 5$, with x-intercept 4
17. Parallel to $x = 2$, passing through $(-3, 4)$
18. Perpendicular to $x = -3$, crossing the y-axis at 3
19. Perpendicular to $y = 4$, with x-intercept $(-5, 0)$
20. Parallel to $y = -2$, passing through $(-4, 5)$
21. Parallel to $y = -2$, crossing the y-axis at -2
22. Perpendicular to $y = -3$, with x-intercept -2
23. Perpendicular to $x = -6$, passing through $(-5, 4)$
24. Parallel to $x = 5$, passing through $(2, -5)$

Review 6

NOW THAT YOU HAVE COMPLETED UNIT 6, YOU SHOULD BE ABLE TO:

Use the following definitions in your working vocabulary.

Function—y is said to be a function of x if for each value of x there is one and only one value for y.

Independent variable—A variable that may assume any one of a set of values.

Dependent variable—The quantity that changes as a result of a change in the independent variable.

Domain of a function—Set of possible replacements for the independent variable.

Range of a function—Set of possible values for the dependent variable.

x-axis—Horizontal line in a rectangular graph.

y-axis—Vertical line in a rectangular graph.

Origin—Point where horizontal and vertical axes meet.

Slope of a line—The angle or inclination of a line.

x-intercept—The point where a line crosses the x-axis.

y-intercept—The point where a line crosses the y-axis.

Apply the following rules and definitions.

To evaluate a function, substitute the value given for the variable and follow the rules for order of operations.

To draw a graph of a first-degree equation, use one of the following methods.

1. Table of values—Plot the points and connect them with a straight line.
2. Slope-intercept method—Plot the y-intercept on the graph. Use the slope to locate a second point by drawing the change in y resulting from the change in x, starting at the y-intercept. Draw a straight line through the y-intercept and the second point determined from rise and run.
3. Intercepts method—Plot the y-intercept found by substituting zero for the x-value in the equation. Plot the x-intercept found by substituting zero for the y-value in the equation. Draw a straight line through the two intercepts.

To find the slope of a line passing through two points, use the formula

$$m = \frac{y_2 - y_1}{x_2 - x_1}$$

To write the equation of a line given the slope and y-intercept, substitute the values of m and b in the equation

$$y = mx + b$$

To write the equation of a line given the slope and one point, substitute the slope for m and substitute the coordinates of the known point (x, y) into $y = mx + b$, then solve for the value of b. Finally, substitute the values of m and b in the equation $y = mx + b$.

To write the equation of a line if two points are known, use the formula

$$m = \frac{y_2 - y_1}{x_2 - x_1}$$

to find the slope. Then apply the method used when one point and the slope are known.

To write the equation of a line parallel to a given line and passing through a given point, find the slope of the given line and then apply the method used when one point and the slope are known.

To write the equation of a line perpendicular to a given line and passing through a given point, use the negative reciprocal of the slope of the given line as the slope of the line perpendicular to it. Then apply the method used when one point and the slope are known.

REVIEW TEST 6

1. Complete the following table for $f(x) = -3x + 4$. ***(6.1)***

x	$f(x) = -3x + 4$
-2	
0	
3	
-4	
2	

2. Given $f(x) = 2x - 3$, fill in the blanks. ***(6.1)***

$f(0) =$ ____________

$f(1) =$ ____________

$f(-6) =$ ____________

$f(-4) =$ ____________

$f(3) =$ ____________

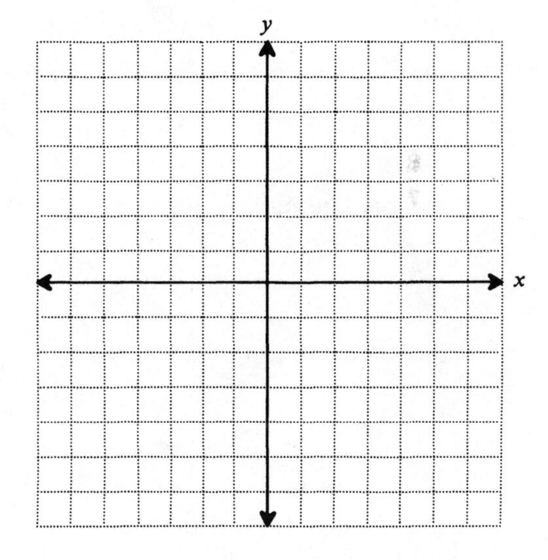

3. Plot the following points on a graph:
$(3, 2), (3, -4), (-6, 1), (-2, -4), (-5, 2)$
Label your points. ***(6.2)***

Complete the following tables and sketch the graphs. (6.2)

4. $y = \frac{1}{2}x - 3$

x	$\frac{1}{2}x - 3$
0	
2	
-4	

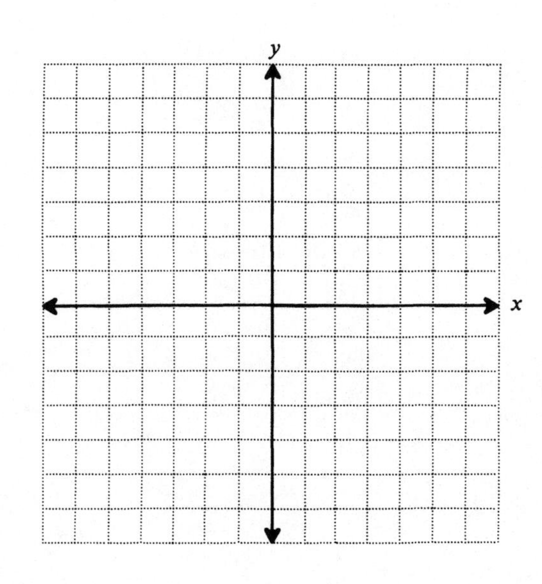

5. $y = -2x + 3$

x	$-2x + 3$
0	
3	
-1	

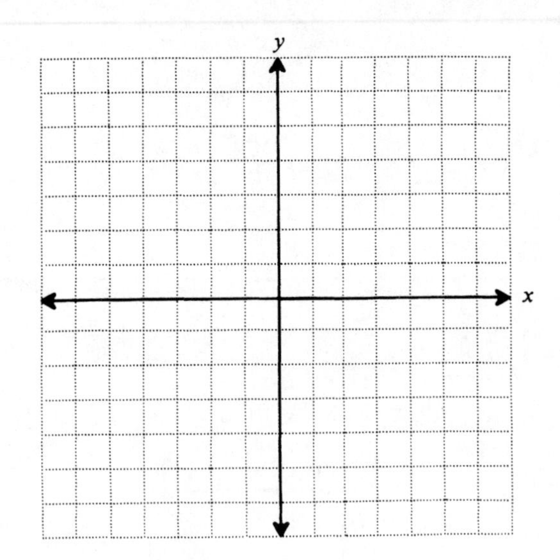

6. Find the slope of the line through the two points $(-2, -4)$ and $(3, 6)$. *(6.3)*

7. Find the slope of the line through the two points $(-3, 4)$ and $(1, -3)$. *(6.3)*

8. Find the equation of the line with slope -4 and y-intercept -3. *(6.4)*

9. Find the equation of the line with slope $\frac{2}{3}$ and y-intercept 4. *(6.4)*

10. Find the equation of the line with slope $\frac{2}{3}$ and through the point $(3, 5)$. *(6.4)*

11. Find the equation of the line with slope $\frac{-3}{4}$ and through the point $(2, -3)$. *(6.4)*

12. Use the slope-intercept method to find the equation of the line through the points $(-2, 3)$ and $(4, 7)$. *(6.4)*

13. Use the point-slope method to find the equation of the line through the points $(-4, -1)$ and $(3, -2)$. *(6.4)*

Sketch the following graphs using the slope and y-intercept method. Be sure to label the scale on each axis. Also label the y-intercept. *(6.5)*

14. $y = \frac{3}{4}x - 1$

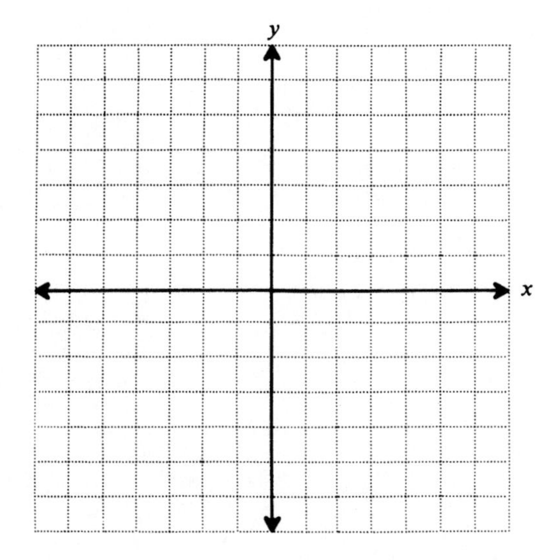

15. $y = -3x + 2$

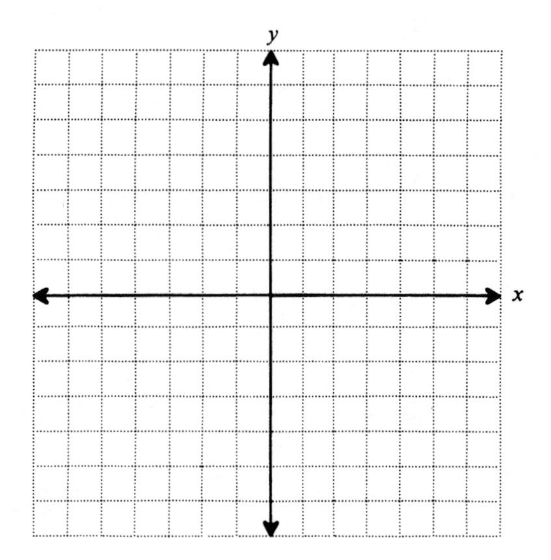

16. $y = \frac{-3}{4}x + 5$

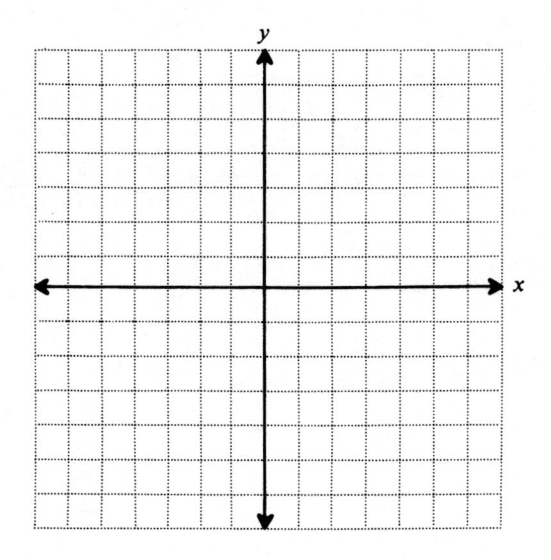

17. $y = \frac{7}{2}x - 3$

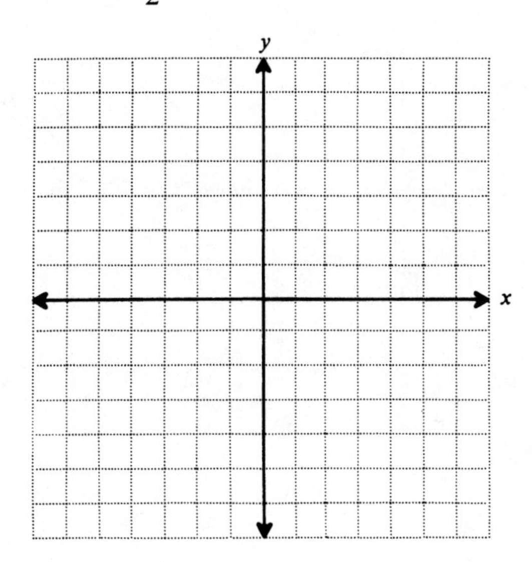

Sketch the following graphs using the intercepts method. Label each intercept.* *(6.5)

18. $3x + 2y = 12$

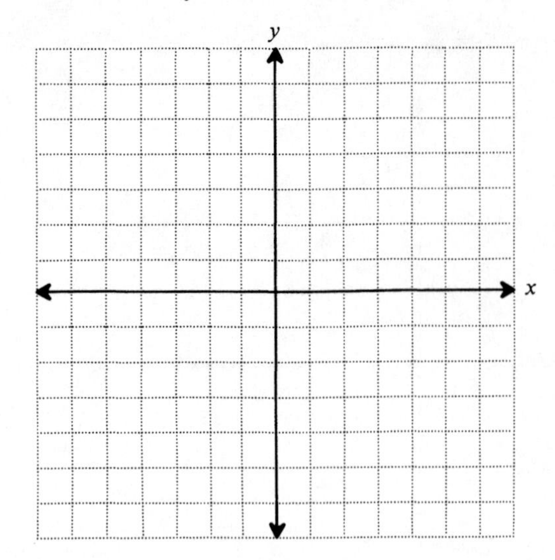

19. $-2x + 5y = -10$

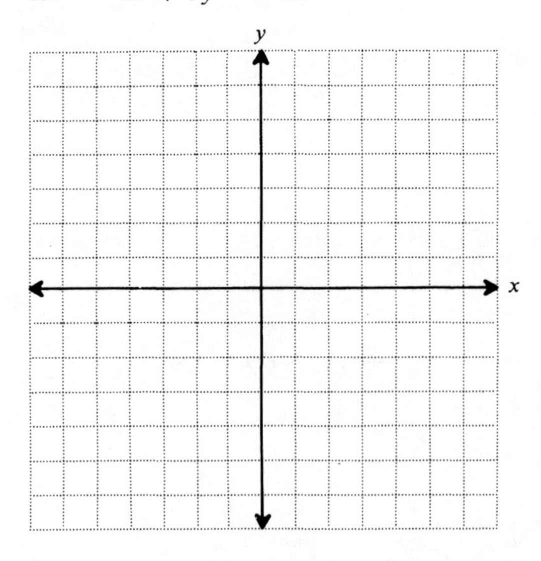

20. Sketch the graph of $x = -2$. ***(6.5)***

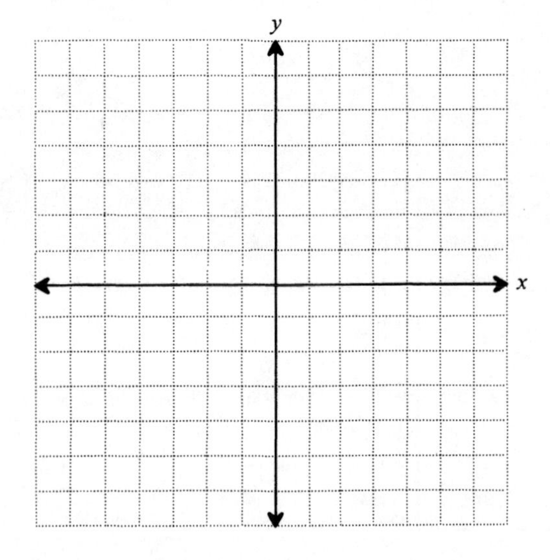

21. Sketch the graph of $y = 4$. ***(6.5)***

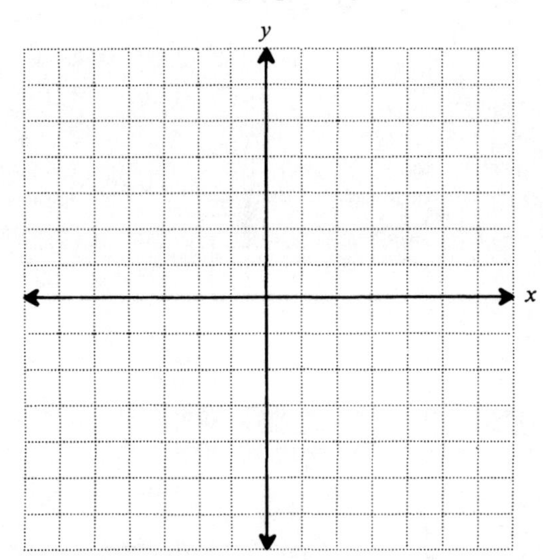

Sketch the following graphs using any method. Show the y-axis intercept and label units on each axis. ***(6.5)***

22. $3x - 4y = 12$

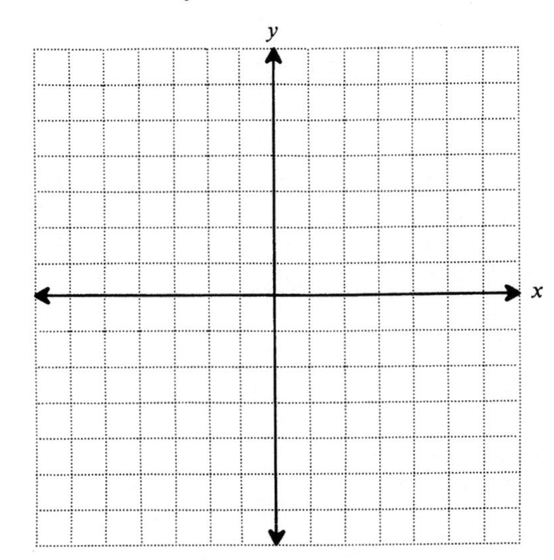

23. $2x - y = 6$

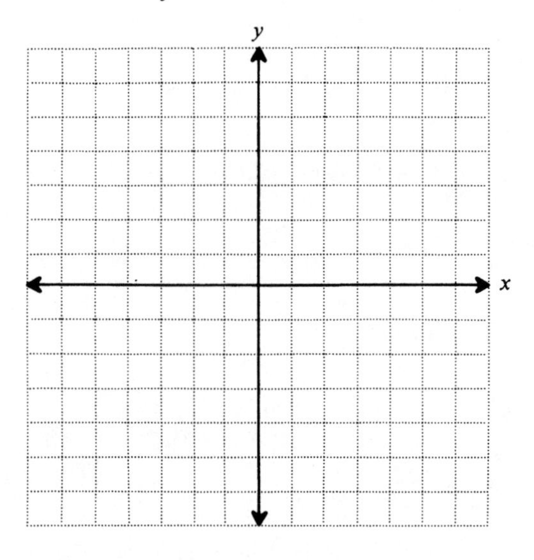

24. $x - 3y = 7$

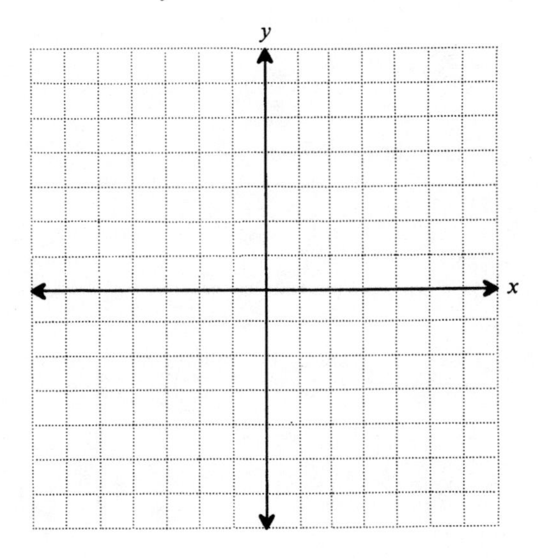

25. $3x - 5y = 9$

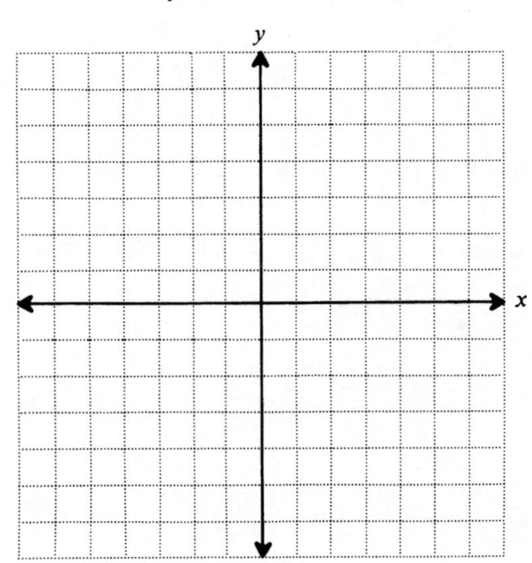

REVIEW YOUR SKILLS

Solve for the indicated variable.

26. $S = \frac{n}{2}(n + l)$ for l

27. $l = a + (n - 1)d$ for d

Change the following percents to fractions reduced to lowest terms.

28. $66\frac{2}{3}\%$

29. $12\frac{1}{2}\%$

Evaluate the following.

30. $3(a^2 + b - c) - \frac{3}{4}(abc)$ if $a = -2$, $b = \frac{2}{3}$, $c = -1$

Questions 31–34 are optional.

31. Find the equation of the line parallel to the line $2x - 5y = 6$, passing through the point $(-1, 2)$. ***(6.6)***

32. Find the equation of the line perpendicular to the line $-5x + 3y = 6$, passing through the point $(-3, -4)$. ***(6.6)***

33. Sketch a line parallel to the given line with the specified y-intercept. ***(6.6)***

    $y = \frac{-2}{7}x + 3$ $\quad$ y-intercept -2

34. Sketch a line perpendicular to the given line with the specified y-intercept. ***(6.6)***

    $y = \frac{3}{4}x + 2$ $\quad$ passing through $(0, -3)$

UNIT 7

Systems of Equations

OBJECTIVES

After you have successfully completed this unit, you will be able to:

1. Estimate the simultaneous solution of two linear equations by graphing the equations (7.1)
2. Find the simultaneous solution of two linear equations by the addition method (7.2)
3. Find the simultaneous solution of two linear equations by the substitution method (7.3)
4. Solve word problems by setting up two simultaneous equations and solving them (7.4)

7.1 SOLVING SYSTEMS OF EQUATIONS BY GRAPHING

The graphs of first-degree equations with two variables are straight lines; therefore, they are called *linear equations.*

Standard form of a linear equation

$Ax + By = C$ is called the *standard form of a linear equation,* where A, B, and C are constants.

That means once A, B, and C are given, we will have a specific equation of a line.

EXAMPLE 1 For the general equation $Ax + By = C$, give the specific equation if $A = 1$, $B = 2$, and $C = 3$.

$$Ax + By = C$$
$$(1)x + (2)y = (3)$$
$$x + 2y = 3$$

EXAMPLE 2 For the general equation $Ax + By = C$, give the specific equation if $A = 3$, $B = -2$, $C = 4$.

$$Ax + By = C$$

$3, -2, 4$

$$(\_\_\_\_)x + (\_\_\_\_)y = (\_\_\_\_)$$

$3x - 2y, 4$

$$\_\_\_\_\_\_ = \_\_\_\_$$

To graph the equation $Ax + By = C$, we could write it in slope-intercept form by solving for y.

$$Ax + By = C$$
$$By = C - Ax$$
$$y = \frac{C}{B} - \frac{A}{B}x$$
$$y = \frac{-A}{B}x + \frac{C}{B} \quad \text{where} \quad m = \frac{-A}{B} \quad \text{and} \quad b = \frac{C}{B}$$

Frequently, it is easier to graph equations written in standard form by finding their x- and y-intercepts.

EXAMPLE 3 Graph both equations below on the same graph.

$$3x - y = 9 \quad \text{and} \quad x + 2y = -4$$

Since a line crosses the y-axis at $x = 0$, we will find the y-intercept of the first equation by substituting $x = 0$ for x.

$$3x - y = 9$$
$$3(0) - y = 9$$
$$y = -9 \qquad (0, -9) \text{ is the } y\text{-intercept of the first equation}$$

To find the x-intercept of the first equation replace y with zero.

$$3x - y = 9$$
$$3x - (0) = 9$$
$$3x = 9$$
$$x = 3 \qquad (3, 0) \text{ is the } x\text{-intercept}$$

Now we can graph the line $3x - y = 9$.

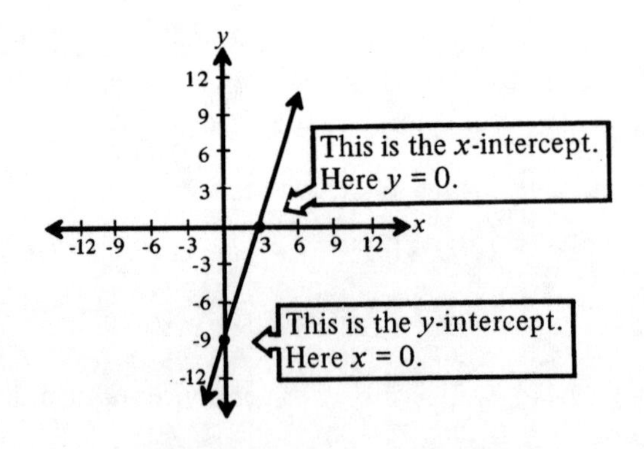

Similarly, find the x- and y-intercepts of the second equation, $x + 2y = -4$.

y-intercept	x-intercept
$x + 2y = -4$	$x + 2y = -4$
$(0) + 2y = -4$	$x + 2(0) = -4$
$y = -2$	$x = -4$

Plot both equations on the same graph:

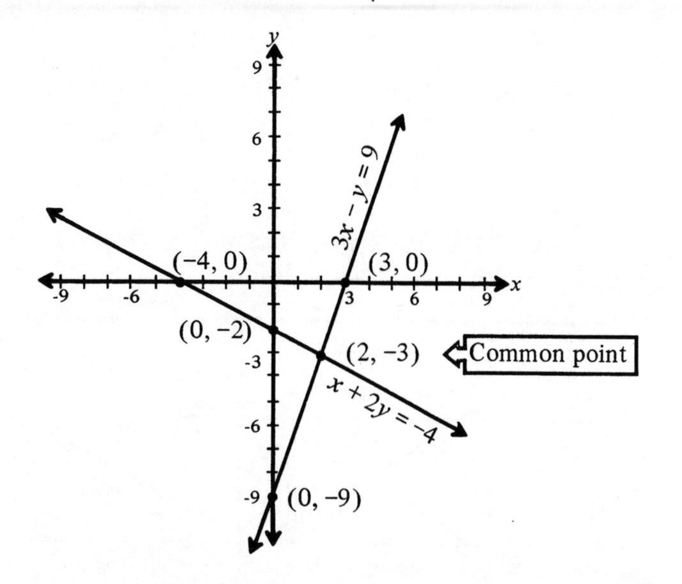

Simultaneous solution

In Example 3, notice the point where the two lines cross. $(2, -3)$ is on both lines. When this point is substituted in either equation, it makes the equation true. The point that is common to both lines is called the *simultaneous solution* to the system of equations. $(2, -3)$ is the simultaneous solution to Example 3.

Test the solution by substituting $(2, -3)$ in both equations.

$3x - y = 9$	$x + 2y = -4$
$3(2) - (-3) \stackrel{?}{=} 9$	$(2) + 2(-3) \stackrel{?}{=} -4$
$6 + 3 \stackrel{?}{=} 9$	$2 + (-6) \stackrel{?}{=} -4$
$9 = 9$	$-4 = -4$

(2, −3) checks in both equations.

EXAMPLE 4 Find the simultaneous solution to the system.

$$x + 4y = -8$$
$$3x + 2y = 6$$

The solution is found by graphing both equations on the same axis and finding the common point.

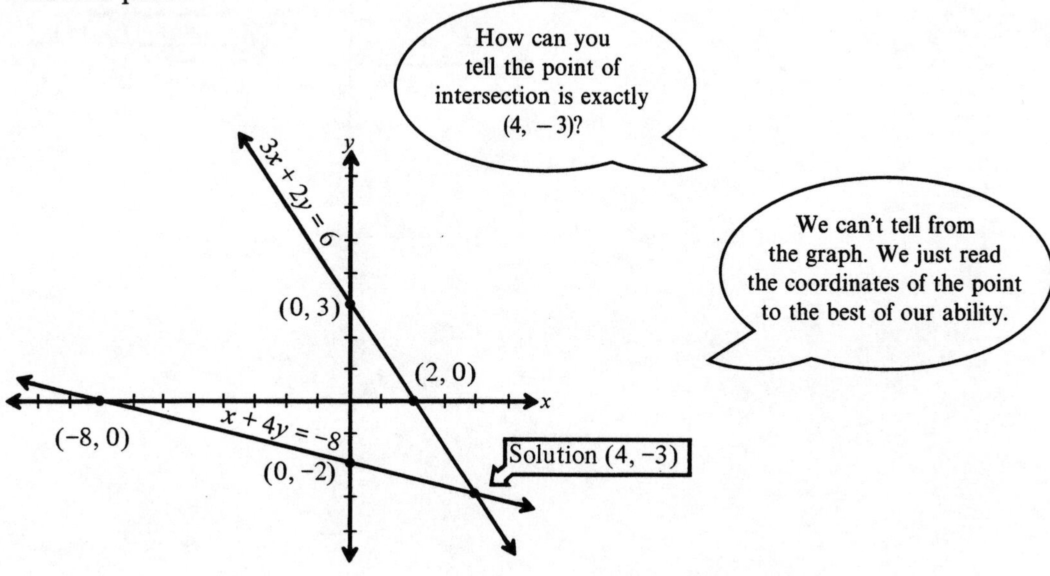

Test the solution $(4, -3)$ by substituting in both equations.

$$\begin{aligned} x + 4y &= -8 \\ (4) + 4(-3) &\stackrel{?}{=} -8 \\ 4 + (-12) &\stackrel{?}{=} -8 \\ -8 &= -8 \end{aligned} \qquad \begin{aligned} 3x + 2y &= 6 \\ 3(4) + 2(-3) &\stackrel{?}{=} 6 \\ 12 + (-6) &\stackrel{?}{=} 6 \\ 6 &= 6 \end{aligned}$$

As you search for a simultaneous solution to a system of equations by graphical methods, your result will fall into one of the three categories below.

Typical system of equations	$5x - 6y = -4$ $4x + 5y = 5$	$x - 2y = 3$ $x - 2y = 5$	$x - 2y = 3$ $2x - 4y = 6$
Typical graph			
Descriptive name	Intersecting lines Independent equations	Parallel lines Inconsistent equations	Same line Dependent equations
Number of solutions	One	None	Since both equations are actually the same line, any ordered pair that satisfies one equation satisfies the other.

It is only possible to find a meaningful solution for the case with intersecting lines.

EXAMPLE 5 Find the simultaneous solution to the system.

$$y = -4$$
$$3x + y = 5$$

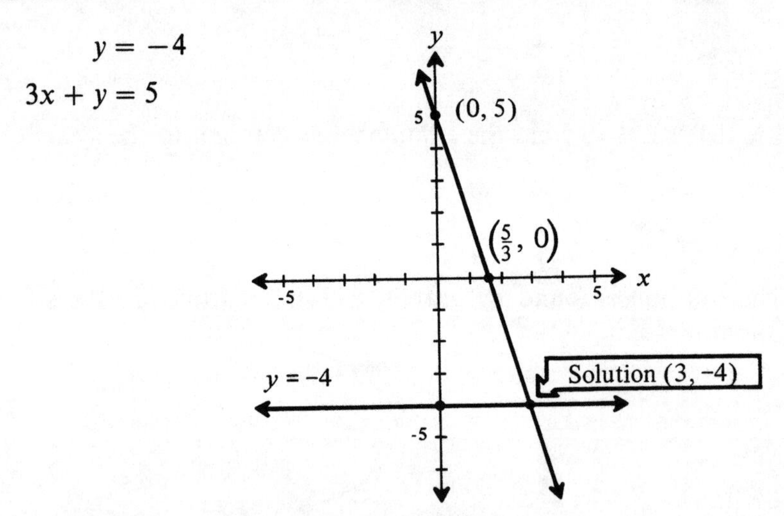

Test the solution $(3, -4)$ by substitution.

$$\begin{aligned} y &= -4 \\ (-4) &\stackrel{?}{=} -4 \\ -4 &= -4 \end{aligned} \qquad \begin{aligned} 3x + y &= 5 \\ 3(3) + (-4) &\stackrel{?}{=} 5 \\ 9 + (-4) &\stackrel{?}{=} 5 \\ 5 &= 5 \end{aligned}$$

PROBLEM SET 7.1

Substitute the ordered pair directly into each equation of the system to decide whether that ordered pair is a simultaneous solution of the given system of equations.

1. $x + y = 5$ $(4, 1)$
   $x - y = 3$

2. $2x - y = 6$ $(4, 2)$
   $x + 2y = 8$

3. $x - 2y = 1$ $(5, -1)$
   $2x + 3y = 7$

4. $2x + 3y = 5$ $(-2, 3)$
   $4x + y = -6$

5. $2x + y = 2$ $(-1, 4)$
   $-3x + 4y = 6$

6. $x - 2y = -7$ $(-3, 2)$
   $-2x - 3y = 0$

Graph both equations on the same graph. Estimate the simultaneous solution of the given system, if there is one.

7. $y = x + 2$
   $y = -x + 8$

8. $3x - 2y = 12$
   $x + 4y = -10$

9. $3x + y = -2$
   $x + 2y = 6$

10. $3x + 2y = -3$
    $-5x + y = 5$

11. $x + 3y = -2$
    $2x - y = 10$

12. $4x - y = 0$
    $6x + 2y = 7$

13. $2x - y = 3$
    $4x - 2y = -6$

14. $x - 3y = -3$
    $2x - 6y = 6$

15. $3x + 4y = -4$
    $-2x + 3y = -3$

16. $-3x + 2y = 10$
    $2x - 3y = -10$

17. $-3x + y = -6$
    $6x - 2y = 12$

18. $x + y = -4$
    $2x + 4y = -6$

19. $2x + y = -2$
    $3x + 2y = -6$

20. $2x + y = -2$
    $-x + 2y = 11$

21. $3x - y = 6$
    $6x - 2y = 12$

22. $-x + 2y = 4$
    $2x - 4y = -8$

23. $4x - y = -6$
    $-2x + 3y = 8$

24. $-3x + y = 9$
    $2x - y = -7$

25. $-x + 2y = 4$
    $2x - 4y = -8$

26. $-2x + y = -5$
    $3x + 2y = -3$

27. $x + 2y = 9$
    $y = 4$

28. $2x - 3y = -3$
    $x = -3$

29. $x = 3$
    $y = -2$

30. $x = 0$
    $y = 5$

31. $x = 5$
    $x = 2$

32. $y = 0$
    $y = -2$

33. $x = 2$
    $y = 0$

7.2 SOLVING SYSTEMS OF EQUATIONS BY ADDITION

Graphical methods of solving equations work, but they are time consuming. Also, it is hard to get precise answers with graphical methods. As you might expect, mathematicians have easier ways of finding the simultaneous solution to two equations. One method rests on the addition property of equality.

EXAMPLE 6 Find the solution to the system of equations.

$2x - y = 1$

$2x + y = 7$

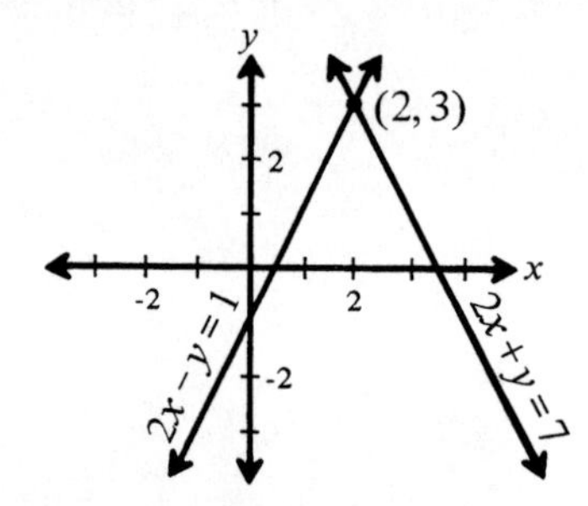

Graphing tells us that (2, 3) is the simultaneous solution.

Since we can't solve an equation with two variables, we need a technique to eliminate a variable from one equation of the system.

If you take the second equation of the system quite literally, it says

$$2x + y = 7 \quad \text{or} \quad 7 = 7$$

The addition property of equality says that you can add the same thing to both sides of an equation. We use that property to add 7 to both sides of the first equation.

$$\begin{array}{r} 2x - y = 1 \\ + \qquad\quad 7 = 7 \\ \hline 2x - y + 7 = 1 + 7 \end{array}$$

This result is true but not very useful.

However, if we add 7 to equation 1 in the form of equation 2, we get a very useful result.

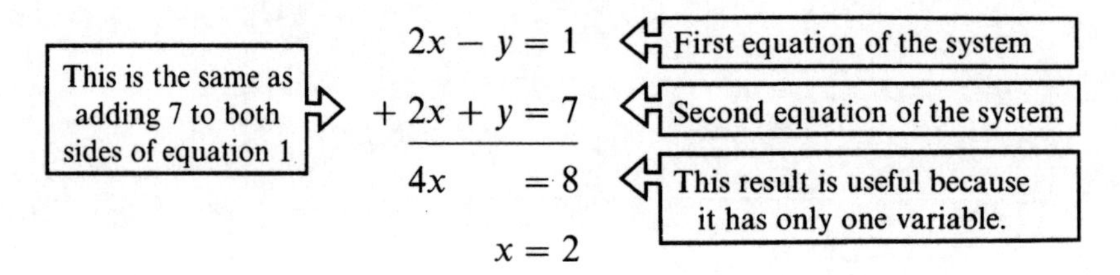

Now substitute $x = 2$ into either equation to find the value for y.

Using the second equation

$$\begin{aligned} 2x + y &= 7 \\ 2(2) + y &= 7 \\ 4 + y &= 7 \\ y &= 3 \end{aligned}$$

Again we find that (2, 3) is the simultaneous solution to both equations.

EXAMPLE 7 Find the simultaneous solution to the system.

$y + 4x = 8$

$5x - y = 1$

First rewrite the system so the x's and y's are lined up in columns.

$$\begin{array}{rl} 4x + y = 8 & \\ 5x - y = 1 & \\ \hline 9x \qquad = 9 & \quad \text{Adding both equations} \\ x = \underline{\qquad} & \quad \text{Solving for } x \end{array}$$

Next substitute $x = 1$ in either equation to find y.

1

$$4x + y = 8$$
$$4(\_\_\_\_\_) + y = 8$$
$$4 + y = 8$$
$$y = 4$$

(1, 4) is the simultaneous solution to both equations.

We should check our solution (1, 4) in *both* equations.

1, 4

$4x + y = 8$	$5x - y = 1$
$4(1) + (4) \stackrel{?}{=} 8$	$5(\_\_\_\_\_) - (\_\_\_\_\_) \stackrel{?}{=} 1$
$4 + 4 \stackrel{?}{=} 8$	$5 - 4 \stackrel{?}{=} 1$
$8 = 8$	$1 = 1$

(1, 4) satisfies both equations.

In both of the earlier examples, it was possible to eliminate one variable by simply adding the two equations. Sometimes we need to manipulate one or both equations before we add.

EXAMPLE 8 Solve the system.

$$4x + y = 5$$
$$3x - 4y = -1$$

Notice if we multiply both sides of the first equation by 4, the y terms in the system of equations will be eliminated when we add the equations.

$4x + y = 5$	Multiply by 4 ⇨	$16x + 4y = 20$
$3x - 4y = -1$	No change ⇨	$3x - 4y = -1$
	Add ⇨	$19x = 19$
		$x = 1$

Substitute $x = 1$ into one of the original equations.

$$4x + y = 5$$
$$4(1) + y = 5$$
$$4 + y = 5$$
$$y = 1$$

Why use the first equation?

Because it looks easier to solve for y.

1, 1

(_____, _____) is the simultaneous solution.

Check the solution in both original equations.

$4x + y = 5$	$3x - 4y = -1$
$4(1) + (1) \stackrel{?}{=} 5$	$3(1) - 4(1) \stackrel{?}{=} -1$
$4 + 1 \stackrel{?}{=} 5$	$3 - 4 \stackrel{?}{=} -1$
$5 = 5$	$-1 = -1$

(1, 1) satisfies both equations.

EXAMPLE 9 Solve the system.

$$3x + 4y = 20$$
$$2x - 5y = -2$$

In this system, multiplying only one equation by an integer will not eliminate a variable. However, if we multiply the first equation by 2, and the second equation by -3, the x terms will be eliminated when we add because $6x + (-6x) = 0$.

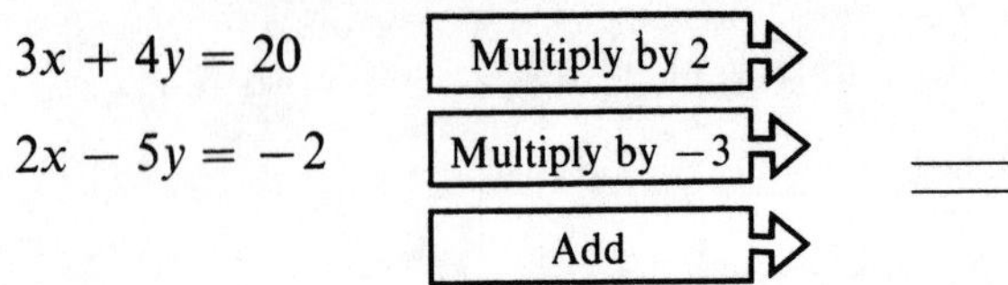

$$6x + \quad 8y = 40$$
$$\underline{\quad\quad} x + \underline{\quad\quad} y = \underline{\quad\quad}$$
$$23y = 46$$
$$y = 2$$

−6, 15, 6

Substituting $y = 2$ into one of the original equations,

$$3x + 4y = 20$$
$$3x + 4(2) = 20$$
$$3x + 8 = 20$$
$$3x = 12$$
$$x = 4$$

4, 2

The solution is (_____, _____).

Check the solution by substituting in both original equations.

$3x + 4y = 20$	$2x - 5y = -2$
$3(4) + 4(2) \stackrel{?}{=} 20$	$2(4) - 5(2) \stackrel{?}{=} -2$
$12 + 8 \stackrel{?}{=} 20$	$8 - 10 \stackrel{?}{=} -2$
$20 = 20$	$-2 = -2$

(4, 2) checks.

EXAMPLE 10 Solve the system.

$$2x + 3y = 3$$
$$4x - 6y = -2$$

$2x + 3y = 3$	Multiply by 2	$4x + 6y = 6$
$4x - 6y = -2$	No change	$4x - 6y = -2$
	Add	$8x = 4$

$$x = \frac{4}{8}$$

$$x = \frac{1}{2}$$

Substitute $x = \frac{1}{2}$ into one of the original equations.

$$2x + 3y = 3$$
$$2\left(\frac{1}{2}\right) + 3y = 3$$
$$1 + 3y = 3$$
$$3y = 2$$
$$y = \frac{2}{3}$$

$\frac{1}{2}, \frac{2}{3}$

The solution is (______, ______).

Check the solution by substituting in both original equations.

$$2x + 3y = 3 \qquad 4x - 6y = -2$$
$$2\left(\frac{1}{2}\right) + 3\left(\frac{2}{3}\right) \stackrel{?}{=} 3 \qquad 4\left(\frac{1}{2}\right) - 6\left(\frac{2}{3}\right) \stackrel{?}{=} -2$$
$$1 + 2 \stackrel{?}{=} 3 \qquad 2 - 4 \stackrel{?}{=} -2$$
$$3 = 3 \qquad -2 = -2$$

$\left(\frac{1}{2}, \frac{2}{3}\right)$ checks.

Dependent Equations

Sometimes the addition method yields puzzling results.

EXAMPLE 11 Solve the system.

$$3x + 4y = 8$$
$$-6x - 8y = -16$$

2

We can eliminate the x terms if we multiply the first equation by ______.

$3x + 4y = 8$	Multiply by 2	$6x + 8y = 16$
$-6x - 8y = -16$	No change	$-6x - 8y = -16$
	Add	$0 = 0$

That's hard to argue with.

The reason for this result is that the second original equation is just the first equation multiplied by -2.

First equation		Second equation
$3x + 4y = 8$	Multiply by -2	$-6x - 8y = -16$

Dependent equations

We really don't have two independent equations. We have two equivalent equations. This case is sometimes called *dependent equations.* In this text, the answers will say "same line" for this case.

Dependent equations
Same line

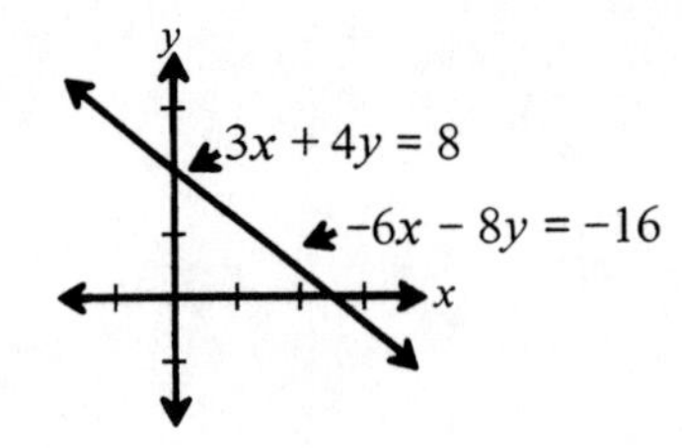

Inconsistent Equations

Here is another case you should be aware of.

EXAMPLE 12 Solve the system.

$$x - 2y = 2$$
$$2x - 4y = 8$$

−2

We can eliminate the x terms if we multiply the first equation by ______.

$x - 2y = 2$	Multiply by −2 ⇨	$-2x + 4y = -4$
$2x - 4y = 8$	No change ⇨	$2x - 4y = 8$
	Add ⇨	______ $= 4$

0

A graph of both equations shows the problem. Since the two lines are parallel, they do not intersect. There is no common solution. Equations whose graphs are parallel lines are sometimes called *inconsistent.* The answers in this text will call them "parallel lines."

Inconsistent equations

Inconsistent equations
Parallel lines
No solution

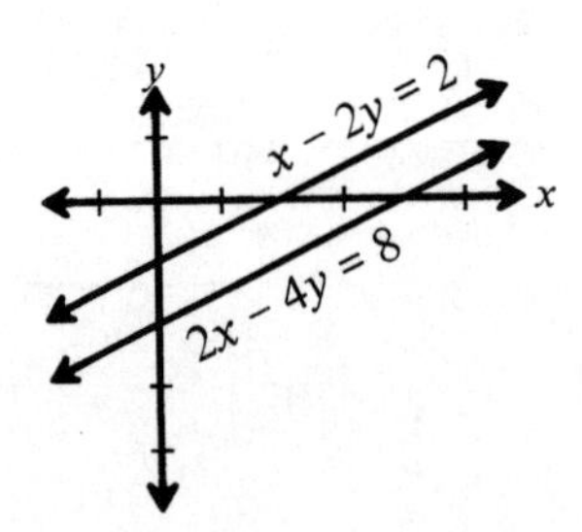

No solution means that there is no ordered pair (x, y) that makes both equations true.

SUMMARY

Follow these steps to solve a system of equations:

1. If necessary, multiply one or both of the equations by a constant to eliminate one of the variables.
2. Add the equations and solve for the remaining variable.
3. Substitute the value of this variable into one of the original equations and solve for the other variable.
4. If after using the addition method you get a result that is $0 = 0$, the two apparently distinct equations really represent only one line.
5. If the addition method yields a false equation like $0 = -10$, the system of equations is two parallel lines, and there is no solution.

PROBLEM SET 7.2

Use the addition method to find the simultaneous solution, if one exists, for the following equations.

1. $x + y = 6$
   $x - y = 4$
2. $x + y = 10$
   $x - y = -6$
3. $x + y = -4$
   $x - y = -4$
4. $x + y = 1$
   $-x + y = 5$
5. $x + y = 5$
   $x + y = -5$
6. $2x + y = -2$
   $x + y = 1$
7. $2x + y = -1$
   $3x - y = -9$
8. $x - 2y = -6$
   $-2x + 4y = 12$
9. $-x + 2y = 10$
   $x + y = 5$
10. $2x - y = 8$
    $x + 3y = 4$
11. $-4x + 2y = -8$
    $2x - y = 4$
12. $3x - y = 6$
    $-6x + 2y = -12$
13. $2x + 5y = 1$
    $-x + 2y = 4$
14. $2x + 5y = 1$
    $x + 2y = 4$
15. $2x + 3y = 6$
    $3x - y = -13$
16. $3x + 4y = 7$
    $x - 2y = 9$
17. $2x + 3y = 1$
    $3x + 2y = -6$
18. $-3x + 2y = -12$
    $5x - 4y = 20$
19. $3x + 4y = 10$
    $2x - 2y = -12$
20. $3x - 2y = 4$
    $-6x + 4y = -8$
21. $2x + 5y = 11$
    $3x + 8y = 16$
22. $3x + 4y = 10$
    $6x - 3y = -2$
23. $5x - 2y = 8$
    $-10x + 4y = 16$
24. $4x - 3y = -4$
    $-2x + 5y = 16$
25. $5x + 2y = 3$
    $-2x + 3y = 14$
26. $3x + 2y = 8$
    $4x + 3y = 13$
27. $3x - 5y = 10$
    $-9x + 15y = -20$
28. $3x + 4y = -8$
    $-5x + 8y = 17$
29. $3y - x = 12$
    $x + 2y = 8$
30. $3x - 4y = -2$
    $y - 2x = 3$
31. $4x + 3y = 4$
    $5y - 6x = -6$
32. $2x + y = 2$
    $4x - y = -8$
33. $2x + 5y = 2$
    $-4x + 5y = 5$
34. $4x - 2y = -6$
    $y - 2x = 3$
35. $3x + 4y = -1$
    $6x - 12y = 13$
36. $3x + 4y = 2$
    $-6x + 5y = 9$
37. $2y - 3x = 6$
    $6x - 4y = -12$
38. $2x + 4y = 7$
    $-x - 2y = 8$
39. $2x + 3y = 10$
    $-3x + 4y = -15$

Review Your Skills Find the value of the following.

40. Evaluate $3x + 4y$ if $x = -2$, $y = 3$

41. Evaluate $-5x + 3y$ if $x = \frac{-2}{5}$, $y = \frac{1}{3}$

42. Evaluate $6x - 5y$ if $x = \frac{1}{2}$, $y = 0$

7.3 SOLVING SYSTEMS OF EQUATIONS BY SUBSTITUTION

The addition method can be used to solve any system of linear equations. However, for some systems the substitution method is easier. The substitution method is easiest when it is possible to solve one equation for a variable in terms of the other variable, without fractions.

EXAMPLE 13 Solve the system.

$$y = 2x - 4$$
$$x + 3y = 16$$

In the system above, the first equation neatly specifies the value of y in terms of x. If we substitute this value of y in the second equation, we will have an equation in one variable that we can solve for x.

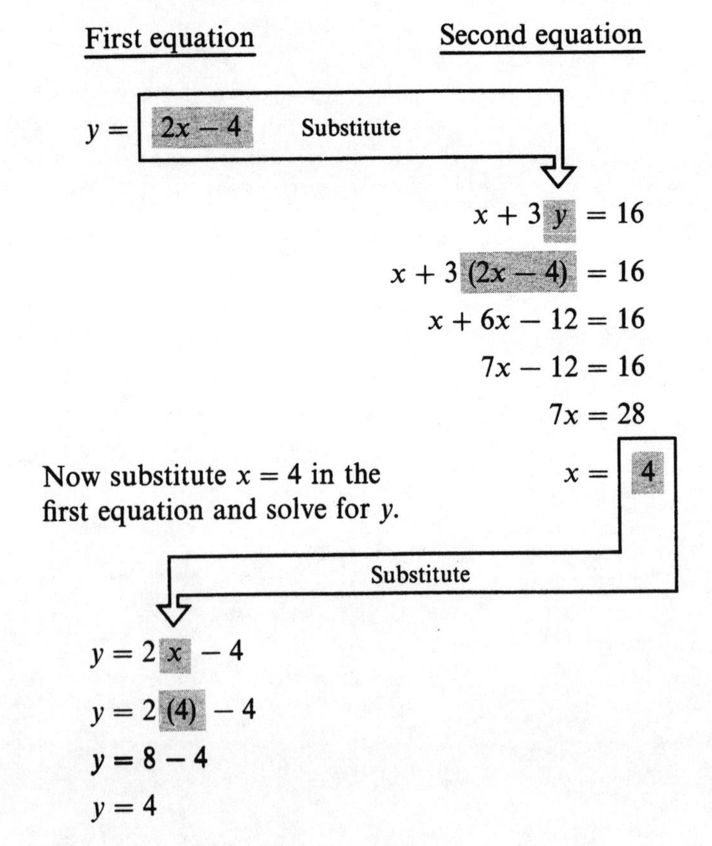

The simultaneous solution is (4, 4).

Check the solution (4, 4) in both equations.

$$\begin{array}{rlcrl} y &= 2x - 4 & \qquad & x + 3y &= 16 \\ (4) &\stackrel{?}{=} 2(4) - 4 & & (4) + 3(4) &\stackrel{?}{=} 16 \\ 4 &\stackrel{?}{=} 8 - 4 & & 4 + 12 &\stackrel{?}{=} 16 \\ 4 &= 4 & & 16 &= 16 \end{array}$$

EXAMPLE 14 Solve the system.

$$-3x + y = -2$$
$$6x - 3y = 15$$

The first equation can be conveniently solved for y. Substitute this expression for y in the second equation.

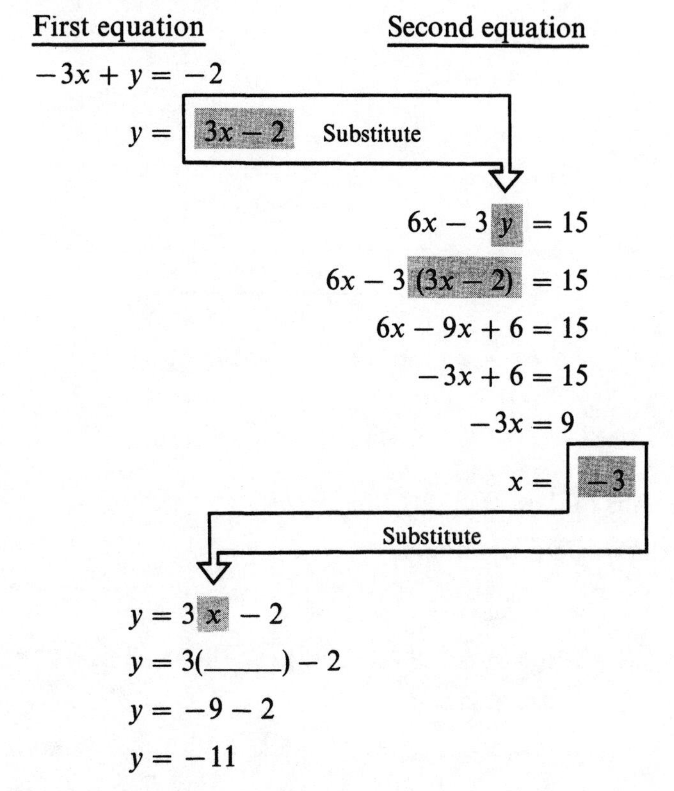

(−3, −11)

The simultaneous solution is __________.

Check the solution in the original equations.

$$\begin{array}{rlcrl} -3x + y &= -2 & \qquad & 6x - 3y &= 15 \\ -3(-3) + (-11) &\stackrel{?}{=} -2 & & 6(-3) - 3(-11) &\stackrel{?}{=} 15 \\ 9 + (-11) &\stackrel{?}{=} -2 & & -18 - (-33) &\stackrel{?}{=} 15 \\ -2 &= -2 & & 15 &= 15 \end{array}$$

EXAMPLE 15 Solve the system.

$$2x + 3y = -4$$
$$x - 5y = -2$$

Look at the second equation. The coefficient of the x term is 1; therefore, it's easy to solve this equation for x in terms of y. Then substitute this solution for x into the first equation and solve for y.

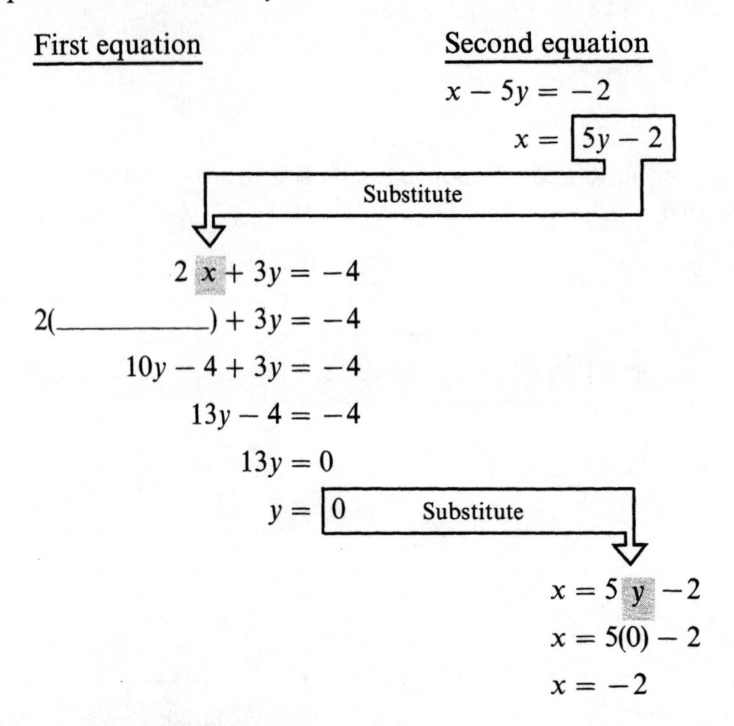

5y − 2

The solution is $(-2, 0)$.

Check in the original equations.

$$2x + 3y = -4 \qquad x - 5y = -2$$
$$2(-2) + 3(0) \stackrel{?}{=} -4 \qquad (-2) - 5(0) \stackrel{?}{=} -2$$
$$-4 + 0 \stackrel{?}{=} -4 \qquad -2 - 0 \stackrel{?}{=} -2$$
$$-4 = -4 \qquad -2 = -2$$

It checks.

The substitution method generally works best when the coefficient of one of the variables is 1 in either equation.

EXAMPLE 16 Solve the system.

$$2x + 3y = 3$$
$$x - 9y = 5$$

In the second equation, the coefficient of the x term is 1. It is easy to solve this equation for x in terms of y. Then substitute this representation of x into the first equation and solve for y.

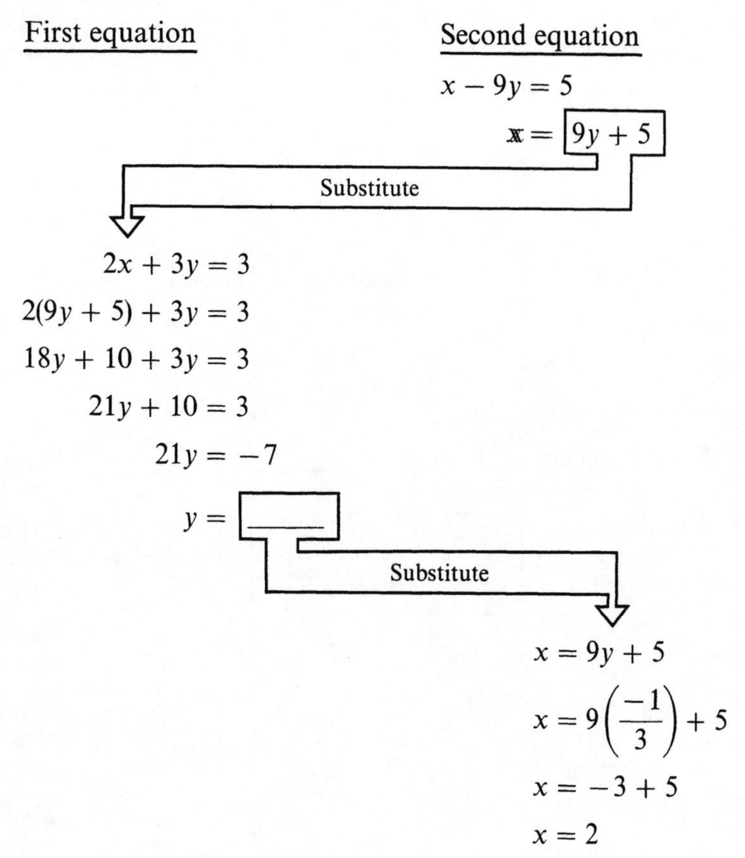

$\frac{-1}{3}$

Solution set is $\left(2, \frac{-1}{3}\right)$.

Check in the original equations.

$$2x + 3y = 3 \qquad\qquad x - 9y = 5$$
$$2(2) + 3\left(\frac{-1}{3}\right) \stackrel{?}{=} 3 \qquad\qquad 2 - 9\left(\frac{-1}{3}\right) \stackrel{?}{=} 5$$
$$4 - 1 \stackrel{?}{=} 3 \qquad\qquad 2 + 3 \stackrel{?}{=} 5$$
$$3 = 3 \qquad\qquad 5 = 5$$

It checks.

SOME POINTERS FOR BETTER UNDERSTANDING

Here's how to tell what kind of a system you're dealing with. When you solve your system of equations, your answers will be one of three types.

Typical answer	**Graphical representation**	**Meaning**
I. $x = 4$ $y = -5$, a valid ordered pair.	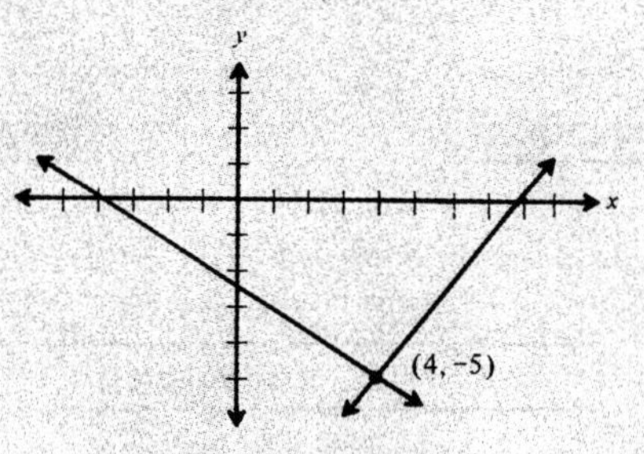 Intersecting lines Independent equations	Two independent equations that intersect at the point $(4, -5)$.
II. When solving for the variable, the system of equations reduces to a true statement like $4 = 4$ or $0 = 0$.	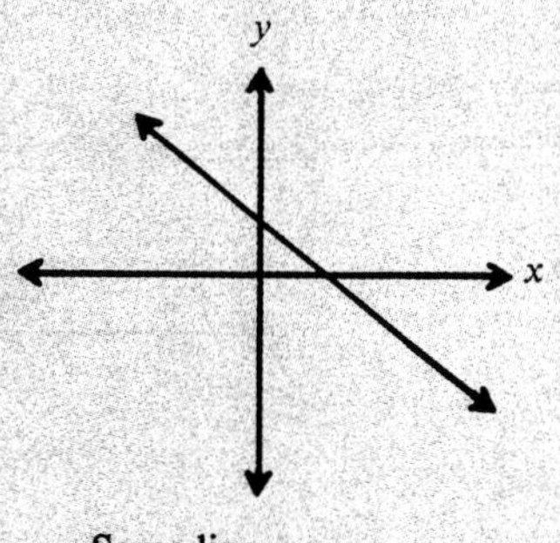 Same line Dependent equations	We tried to find a solution assuming there were two independent equations that intersected at one point. What we have is two representations of the same line. The two equations are called dependent because they are really equivalent equations.
III. When solving for a variable, the system of equations reduces to a false statement like $4 = 7$ or $0 = -5$.	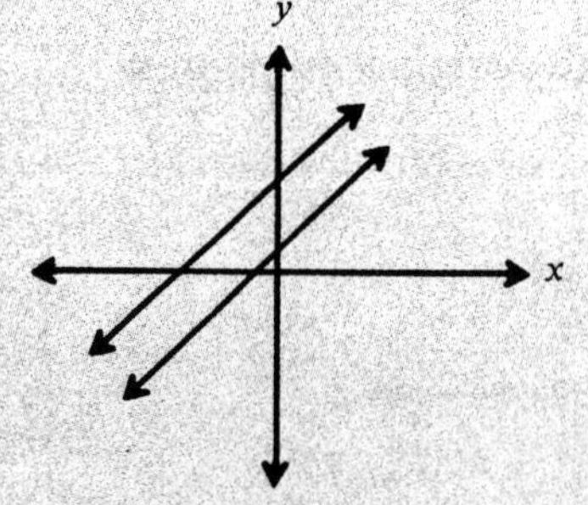 Parallel lines Inconsistent equations	The attempted solution led to an absurdity because it was based on the false assumption that the two lines intersected.

PROBLEM SET 7.3

Use the substitution method to find the simultaneous solution, if one exists, for the following equations.

1. $x + y = 6$
   $y = x - 2$

2. $x - y = 6$
   $y = 2x + 3$

3. $2x + y = 1$
   $y = 3 - x$

4. $2x + y = 9$
   $y = 2x - 7$

5. $x + 2y = 4$
   $x = 3y - 6$

6. $x - 2y = 7$
   $x = 3y + 9$

7. $x + 2y = 6$
   $x = 6 - 2y$

8. $5x + y = 8$
   $y = 9 - 5x$

9. $2x + y = 4$
   $y = 4x + 1$

10. $x + 3y = 3$
    $x = 3y + 1$

11. $2x + 3y = -8$
    $y = 3x + 12$

12. $-2x + 3y = -1$
    $x = 3y + 5$

13. $2x - 3y = 1$
    $x + 2y = 4$

14. $x - 2y = -3$
    $3x + y = 5$

15. $2x + 5y = 2$
    $x + 2y = 2$

16. $3x + 2y = 1$
    $2x + y = -1$

17. $3x - y = 6$
    $6x - 2y = 10$

18. $3x + y = -4$
    $5x - 4y = -1$

19. $3x + y = -1$
    $6x - 2y = 10$

20. $x + 8y = -1$
    $2x + 10y = 1$

21. $-x + 3y = 7$
    $3x + 5y = -7$

22. $-x + 5y = 4$
    $3x - 15y = -12$

23. $2x + 3y = 4$
    $x + 2y = 5$

24. $-2x + y = 8$
    $3x - 4y = -12$

25. $2x - y = 4$
    $-6x + 3y = 6$

26. $x - 3y = 4$
    $3x + 6y = 2$

27. $x - 3y = 3$
    $3x + 2y = -13$

28. $2x - y = 11$
    $3x + 7y = 8$

29. $x - 3y = 1$
    $3x - 6y = 4$

30. $4x - y = -5$
    $2x - 3y = 8$

31. $-x + 4y = -8$
    $2x - 8y = 16$

32. $-x + 4y = -6$
    $3x - 8y = 15$

33. $4x + 2y = 1$
    $8x - y = 7$

34. $2x - 4y = 7$
    $x = 2y + 3$

35. $-x + 3y = -7$
    $3x - 4y = 6$

36. $3x - 2y = -9$
    $2x - y = -6$

SOME POINTERS FOR FASTER SOLUTIONS

I've learned three ways to solve systems: graphing, addition, substitution. Which is best?

That depends on the system of equations. Generally, we use the method we find easiest.

Graphing is a good choice

1. If a visual display is desired, as when you want to illustrate that there is no solution.
2. If accuracy is not too important.

Addition is a good choice

1. If one of the variables has the same coefficient in both equations, as in

$$3x + 7y = 10$$
$$3x - 9y = 15$$

2. If one of the coefficients of a variable in one equation is a multiple of the coefficient of the same variable in the other equation, as in

$$3x - 11y = 5$$
$$6x + 4y = 7$$

3. If all coefficients are different, for example,

$$3x - 5y = 15$$
$$2x + 7y = 12$$

4. If you are writing a computer program to solve systems of equations, addition is usually easiest to program.

Substitution is a good choice

1. If one equation in the system is already solved for a variable, for example,

$$3x + 6y = 19$$
$$y = 2x - 5$$

2. If it is easy to solve an equation for one of the variables.

For example,

$$x - 7y = 10$$
$$y - 3x = -2$$

When is it easy?

When the coefficient of any variable in the system is 1.

Frequently, substitution lets us take advantage of an especially easy case. Any method will work. Use the one that seems easiest at the moment.

7.4 APPLICATIONS

The ability to solve equations in two variables makes it easier to solve many applications problems.

EXAMPLE 17 The difference of two numbers is 8. The sum of the numbers is 16. Find the numbers.

Let

$x =$ larger number

$y =$ smaller number

$x - y = 8$ The difference of the numbers is 8.

$x + y = 16$ The sum of the numbers is 16.

$2x = 24$ Add the two equations to eliminate y.

12

$x =$ _____

To find the other number, substitute in either equation.

$$x - y = 8$$
$$(12) - y = 8$$
$$-y = -4$$

4

$y =$ _____

The numbers are 12 and 4.

Check.

$x - y = 8$	$x + y = 16$
$12 - 4 \stackrel{?}{=} 8$	$12 + 4 \stackrel{?}{=} 16$
$8 = 8$	$16 = 16$

EXAMPLE 18 A bank contains twice as many dimes as nickels. The value of the dimes less the value of the nickels is $1.35. Find the number of each kind of coin.

Let $n =$ number of nickels

$d =$ number of dimes

$d = 2n$ There are twice as many dimes as nickels.

$10 \cdot d - 5 \cdot n = 135$ The value of the dimes less the value of the nickels is 135 cents.

Substitute $d = 2n$ in the second equation.

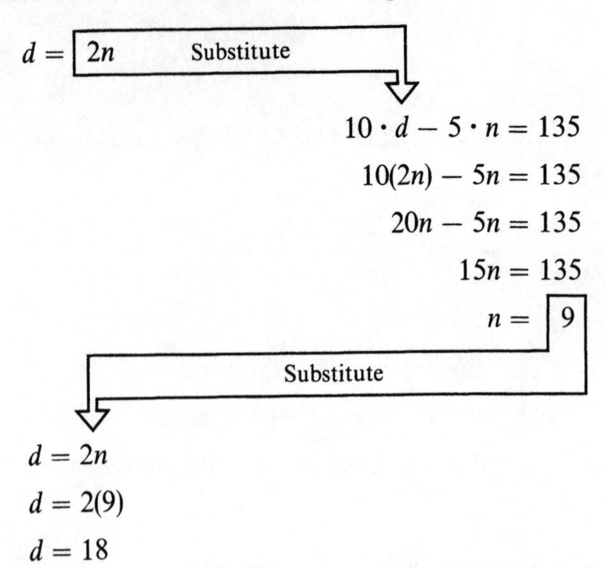

$$d = 2n$$
$$d = 2(9)$$
$$d = 18$$

18, 9

There are ______ dimes and ______ nickels in the bank.

Two equations are often useful in problems dealing with investments. First, let's look at how interest is paid on an investment.

A 12% rate of return on an investment means that each year 12% interest is paid for the use of the investor's money. Therefore, a person who invests \$3,000 for one year at 12% could expect to collect

$$\$3{,}000 \cdot 12\% =$$

$$\overset{30}{\cancel{3000}} \cdot \frac{12}{\underset{1}{\cancel{100}}} = \$360$$

Recall

$$12\% = \frac{12}{100}$$

That is, each year the investor would collect an income of \$360 as payment for the use of the money.

EXAMPLE 19 Find the income from an investment of \$4,000 at 16% interest for one year.

$$\$4{,}000 \cdot 16\% = \text{income for one year}$$

$$\overset{40}{\cancel{4000}} \cdot \frac{16}{\underset{1}{\cancel{100}}} = \$640$$

The investor will receive \$640 per year return on the investment of \$4,000.

EXAMPLE 20 Give the yearly income from the following investments.

$2,000 at 12% returns ________ $240

$500 at 15% returns ________ $75

$2,200 at 10% returns ________ $220

$\$x$ at 13% returns ________ $\$\frac{13}{100}x$

$\$y$ at 9% returns ________ $\$\frac{9}{100}y$

EXAMPLE 21 Jane Sharp, an ambitious car salesperson, received a bonus of $2,000 at the end of the year for her outstanding sales record. She invested part in a long-term savings certificate paying 14% interest and the remainder in a short-term savings certificate paying 12% interest. If her total income at the end of the year was $264, how much was invested at each rate?

First build a model for each major idea in the problem.

Model I   [Part invested at 14%] + [Part invested at 12%] = [Total Investment]

Model II   [Income from 14% investment] + [Income from 12% investment] = [Total Income]

Select letters to make equations from the models.

Let x = part invested at 14%

y = part invested at 12%

Make an equation for each model.

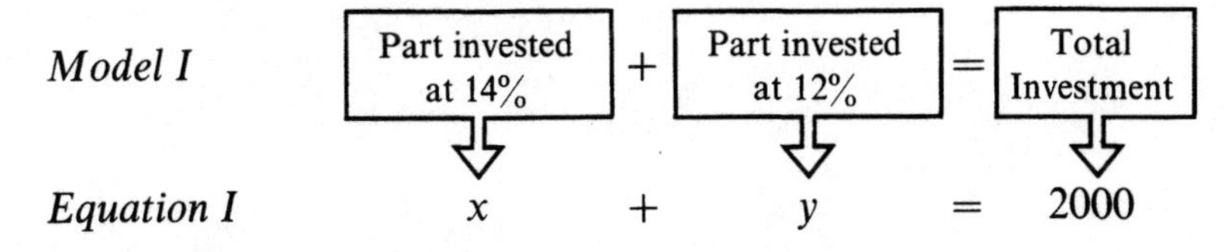

Total investment was $2,000.

Model II   [Income from 14% investment] + [Income from 12% investment] = [Total Income]

Equation II $\frac{14}{100} \cdot x + \frac{12}{100} \cdot y = 264$

Income from 14% investment plus income from 12% investment totals 264 dollars.

Multiply both sides of the second equation by 100 to eliminate the 100 in the denominator. Multiply the first equation by -12 to eliminate the y terms.

$x + y = 2000$ [Multiply by -12] ⇨ $-12x - 12y = -24{,}000$

$\frac{14}{100}x + \frac{12}{100}y = 264$ [Multiply by 100] ⇨ $14x + 12y = 26{,}400$

[Add] ⇨

$$2x = 2{,}400$$

$$x = 1{,}200$$

Substitute $x = 1200$ in the first equation.

$$x + y = 2000$$
$$(1200) + y = 2000$$
$$y = 800$$

$x = \$1{,}200$ invested at 14%. $y = \$800$ invested at 12%.

EXAMPLE 22 A chef wishes to make a special blend using two kinds of coffee beans for a banquet. Type A bean costs \$2.00 a pound and type B bean costs \$3.20 a pound. He figures that he will need a total of 12 pounds of coffee for the occasion. The manager wants to keep the price down in order to stay within the budget. In order to do so, the bookkeeper said the total cost was to be \$30.00. How many pounds of each type should the chef use to meet all the requirements?

Make a table.

Type	Number of pounds	Price per pound	Cost of coffee in mixture
A	x	\$2.00	$2.00x$
B	y	\$3.20	$3.20y$

Make models and express as equations.

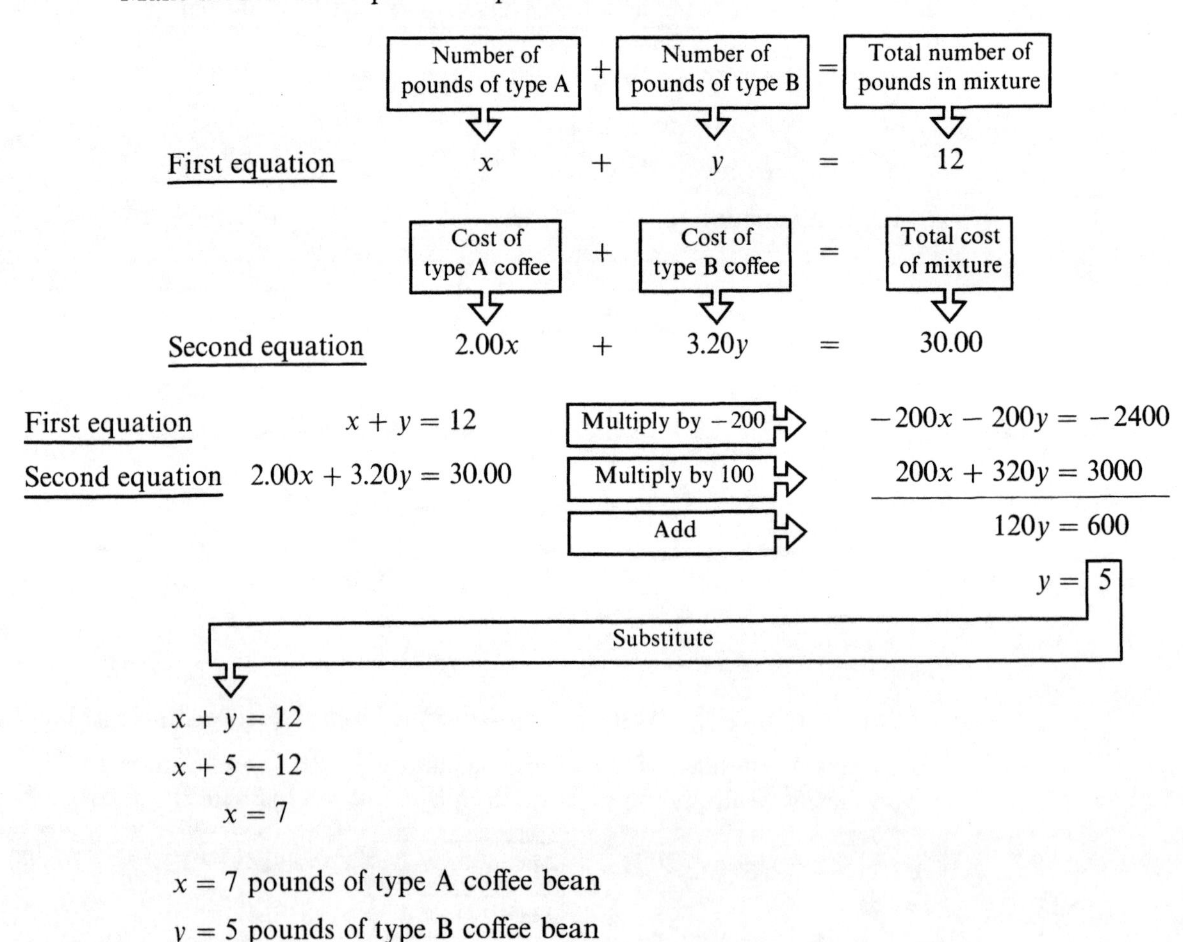

$$x + y = 12$$
$$x + 5 = 12$$
$$x = 7$$

$x = 7$ pounds of type A coffee bean

$y = 5$ pounds of type B coffee bean

PROBLEM SET 7.4

Solve the following problems by using simultaneous equations. State in words exactly what your variables represent. Set up two equations and solve them.

1. The sum of two numbers is 18, and their difference is 4. Find the numbers.
2. The sum of two numbers is 14, and their difference is 2. Find the numbers.
3. The sum of two numbers is 48, and one number is twice the other. Find the numbers.
4. The sum of two numbers is 69, and one of the numbers is twice the other. Find the numbers.
5. Find two numbers whose sum is 68 if one number is three times the other number.
6. The difference of two numbers is 8, and their sum is -32. Find the numbers.
7. The sum of two numbers is 72, and one number is 8 more than the other. Find the numbers.
8. Find two numbers whose sum is 6 if twice the first number minus the second number is 12.
9. Find two integers whose sum is -3 if three times the first integer plus the second integer is -1.
10. One number is 9 less than another and their sum is 61. Find the numbers.
11. A collection of coins contains 7 more nickels than dimes. If the collection is worth \$2.30, how many of each coin are there?
12. A small bank contains 22 coins consisting of dimes and quarters. If the bank contains \$3.10, how many of each coin are in the bank?
13. A girl selling papers has a pocket full of coins consisting of dimes and quarters. If she has three times as many dimes as quarters and a total of \$8.80 worth of coins, how many of each coin does she have?
14. A man just sold his old car for \$360 and was paid in cash consisting of \$5 bills and \$20 bills. If he received 3 more \$20 bills than \$5 bills, how many of each did he receive?
15. The total receipts at a concert were \$3,800. General admission seats sold for \$5.00 and reserved seats sold for \$7.50. If 160 more general admission seats were sold than reserved seats, how many of each were sold?
16. A child's bank contains several dimes and quarters worth \$5.40. If there were twice as many quarters as dimes, how many of each were in the bank?
17. The gate receipts at a ball game amounted to \$1,500. Adult tickets sold for \$2.25, and student tickets sold for \$1.00. If a total of 950 tickets were sold, how many of each were sold?
18. The total receipts at a concert were \$2,295. Reserved seats sold for \$6.50 and general admission was \$3.50. If twice as many general admission tickets were sold as reserved seat tickets, how many of each were sold?
19. A real estate firm invested \$18,000 worth of its profits in second mortgages. Part of the money was invested at 12% and the remainder at 10%. If twice as much was invested at 12% as at 10%, how much was invested at each rate?
20. A midwestern farmer sold his farm for \$80,000 cash. He invested \$20,000 more at 10% than he did at 8%. How much did he invest at each rate?
21. A retiree has \$45,000 in investments. One investment is at 16% and the other one is at 12%. If her annual income from both accounts is \$6,400, how much is in each account?
22. A broker has twice as much money invested at 8% as he has invested at $6\frac{1}{2}$%. If his total income from the two investments is \$2,700, how much does he have invested at each rate?

23. A woman spent \$19.00 for some walnuts and pecans. The walnuts cost \$2.50 a pound and the pecans cost \$3.00 a pound. If she has a total of 7 pounds of nuts, how many pounds of each does she have?

24. Joe bought peanuts and cashews for a party. He bought three times as many pounds of peanuts as cashews. If he had a total of 6 pounds of nuts, how many pounds of each did he buy?

Review Your Skills Write in exponential form.

25. $(2)(2)(2)(2)(a)(a)(b)(b)(b)$

26. $(-3)(-3)(-3)(a)(b)(b)$

Evaluate the following for $a = -2$, $b = 3$.

27. $2a^2b - a$

28. $2ab^2 - 3ab$

Review 7

NOW THAT YOU HAVE COMPLETED UNIT 7, YOU SHOULD BE ABLE TO:

Use the following definitions in your working vocabulary.

Linear equations—First-degree equations with two variables

Standard form of a linear equation—$Ax + By = C$

Independent equations—Equations whose graphs intersect at a point

Parallel lines—Lines that have the same slope and do not intersect

Inconsistent equations—Equations whose graphs are parallel

Dependent equations—Equations that represent the same line

Simultaneous solution—The point that is common to the graphs of two linear equations

Solve systems of equations by graphing.

To use the graphing method to solve systems of equations:

1. Graph both lines on the same axis.
2. a. If the lines intersect, determine the point of intersection (intersecting lines).
   b. If the lines do not intersect, determine whether the lines represent inconsistent equations (parallel lines) or dependent equations (same line).

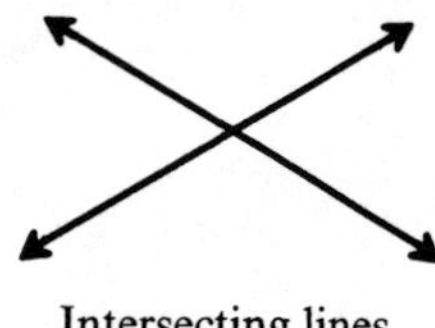

Intersecting lines
Independent equations

Parallel lines
Inconsistent equations

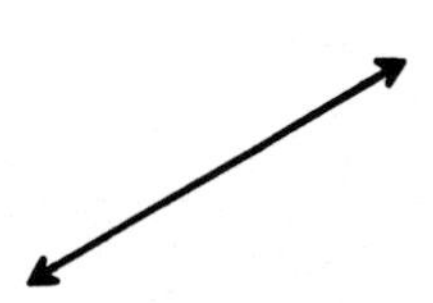

Same line
Dependent equations

Solve systems of equations using the addition method.

To use the addition method to solve systems of equations:

1. Write one equation underneath the other.
2. If necessary, multiply one or both equations by a constant so the coefficients of one variable have the same numerical value but have opposite signs.
3. Add the two equations so that one variable is eliminated.
4. Solve for the value of the remaining variable.
5. Substitute this value in either original equation and solve for the other variable.
6. Check your solution by substituting in the other equation.

Solve systems of equations using the substitution method.

To use the substitution method of solving systems of equations:

1. Solve the easiest equation for one of the variables.
2. Substitute the expression you have found for that variable in the other equation and solve for the remaining variable.
3. Substitute this value in the original equation and solve for the other variable.
4. Check your solution by substituting the values for both variables in both equations.

Solve word problems using two equations and two variables.

To solve word problems:

1. Read the problem carefully to determine what is asked for and what is given. Also look for sentences that are statements of equality.
2. List any formulas that might apply to the problem.
3. Make models using boxes and English phrases to describe the problem.
4. Select variables to represent the quantities in the boxes.
5. Translate your models into algebraic equations.
6. Solve the resulting equations.
7. Answer the question asked by the problem.
8. Check the solution in the original problem.

Review Test 7

Graph both equations on the same graph and estimate the simultaneous solution of each system. (7.1)

1. $3x - 2y = -12$
   $x + 2y = 4$ ____

2. $2x + 3y = 3$
   $x - 3y = 6$ ____

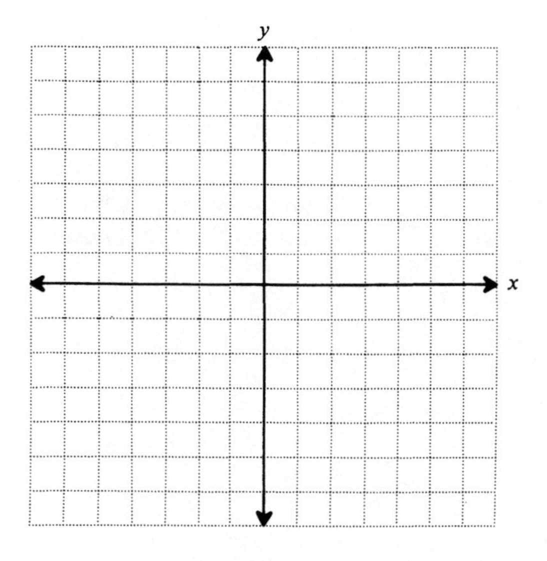

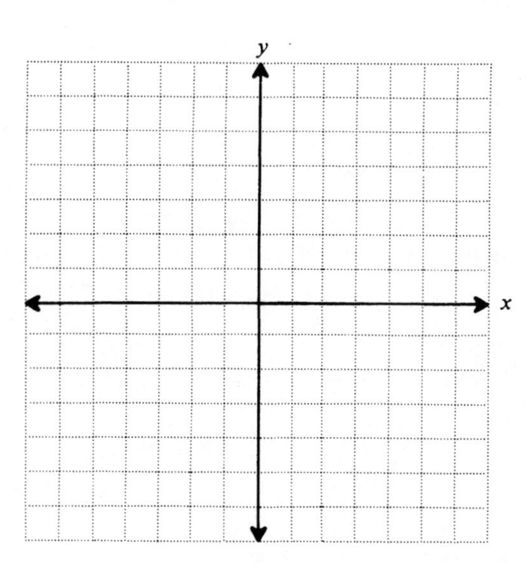

Use the addition method to find the simultaneous solution, if one exists, for the following equations. (7.2)

3. $x + 3y = 5$
   $-2x + 7y = -10$ ____

4. $5x + 2y = -13$
   $2x - 3y = 10$

5. $2x - 3y = 4$
   $4x + 9y = 3$ ____

6. $-2x + 5y = 6$
   $4x - 10y = -12$ ____

Use the substitution method to find the simultaneous solution, if one exists, for the following equations. (7.3)

7. $3x + 4y = -1$
   $y = x + 5$ ____

8. $x + 2y = -6$
   $2x - 3y = 9$ ____

9. $12x - 2y = 8$
   $y - 6x = 5$ ____

10. $5x + 2y = -1$
    $x + 4y = 2$ ____

For problems 11, 12, 13, and 14, use any method to find the simultaneous solution, if one exists. (7.2 and 7.3)

11. $2x + 3y = -1$
    $3x - 4y = 24$ ______

12. $3x + y = -3$
    $6x + 2y = 10$ ______

13. $y = 2x + 7$
    $4x - 2y = -14$ ______

14. $5x + y = 1$
    $x + 2y = 5$ ______

Solve problems 15, 16, and 17 by using simultaneous equations. State in words exactly what your variable represents. Set up two equations and solve them. (7.4)

15. A collection of 33 coins is made up of nickels and dimes. If the total value is $2.40, how many of each is in the collection?

16. A 72-foot rope is cut into two pieces so that one piece is 18 feet longer than the other. How long is each piece?

17. Twenty pounds of Christmas candy was made by mixing filled candy with nonfilled candy. The filled candy sold for $4.25 a pound and the nonfilled candy sold for $3.60 a pound. The total cost of the 20 pounds of candy was $79.80. How many pounds of each kind was used?

REVIEW YOUR SKILLS

Solve for the variable.

18. $6(x - 2) - 5(x + 4) = 2 - x$

19. $\frac{6y - 7}{3} = 9$

20. Jim and Jane leave Chicago at the same time flying light airplanes. They are flying in opposite directions. If Jim's plane flies 20 mph faster than Jane's and they are 750 miles apart after $2\frac{1}{2}$ hours' flying time, how fast is each plane flying?

UNIT 8

Laws of Exponents

OBJECTIVES

After you have successfully completed this unit, you will be able to:

1. Multiply and divide powers using positive, negative, and zero exponents (8.1)
2. Raise a power to a power (8.2)
3. Find the power of a product (8.2)
4. Find the power of a quotient (8.2)
5. Evaluate an expression involving integer exponents for a given replacement set (8.1, 8.2)
6. Convert decimal numerals to numerals written in scientific notation (8.3)
7. Convert numerals written in scientific notation to decimal numerals (8.3)

8.1 BASIC LAWS OF EXPONENTS

In earlier units some of the laws of exponents were introduced. As a convenience they are repeated here.

8.1A REVIEW OF EXPONENTS

Exponential form

Exponential form: $a^n = \underbrace{a \cdot a \cdot a \ldots a}_{n \text{ factors}}$

Multiplication of powers

Multiplication of powers: $a^x \cdot a^y = a^{x+y}$

Zero exponent

Zero exponent: $a^0 = 1$ where $a \neq 0$

Now let's examine division of powers.

To divide different powers of the same base, first write out what the numerator and denominator mean.

$$\frac{a^5}{a^3} \text{ means } \frac{\overbrace{a \cdot a \cdot a \cdot a \cdot a}^{5\ a\text{'s}}}{\underbrace{a \cdot a \cdot a}_{3\ a\text{'s}}}$$

Dividing out common factors yields

$$\frac{a^5}{a^3} = \frac{\overset{1}{\cancel{a}} \cdot \overset{1}{\cancel{a}} \cdot \overset{1}{\cancel{a}} \cdot a \cdot a}{\underset{1}{\cancel{a}} \cdot \underset{1}{\cancel{a}} \cdot \underset{1}{\cancel{a}}}$$

$$= \frac{a \cdot a}{1}$$

$$= a^2$$

a^5 divided by a^3 yields a^{5-3} or a^2.

Division of powers

8.1B DIVISION OF POWERS

To find the quotient of two powers of the same base subtract the exponent of the denominator from the exponent of the numerator.

$$\frac{a^x}{a^y} = a^{x-y}$$

for any $a \neq 0$.

EXAMPLE 1 Divide $\frac{p^7}{p^3}$.

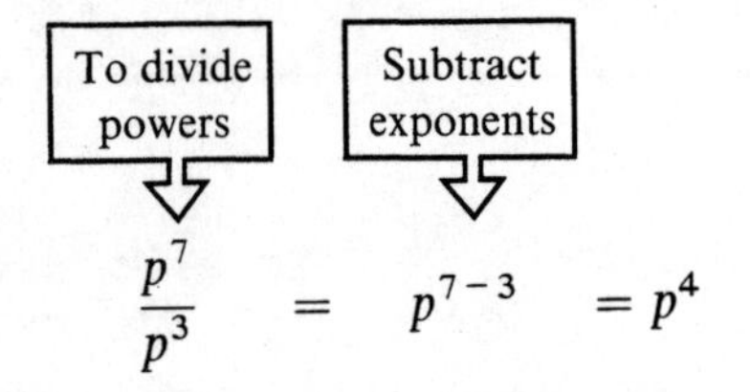

$$\frac{p^7}{p^3} = p^{7-3} = p^4$$

Notice the base, p, remains unchanged.

EXAMPLE 2 Evaluate $\frac{2^5}{2^3}$.

$$\frac{2^5}{2^3} = \frac{2 \cdot 2 \cdot 2 \cdot 2 \cdot 2}{2 \cdot 2 \cdot 2} = \frac{32}{8} = 4$$

or

$$\frac{2^5}{2^3} = 2^{5-3} = 2^2 = 4$$

Now what happens if the exponent of the denominator is greater than the exponent of the numerator?

$$\frac{a^3}{a^5} = \frac{\overset{1}{\cancel{a}} \cdot \overset{1}{\cancel{a}} \cdot \overset{1}{\cancel{a}}}{\underset{1}{\cancel{a}} \cdot \underset{1}{\cancel{a}} \cdot \underset{1}{\cancel{a}} \cdot a \cdot a} = \frac{1}{a \cdot a} = \frac{1}{a^2}$$

But if the law for division of powers is followed,

$$\frac{a^3}{a^5} = a^{3-5} = a^{-2}$$

Therefore, we will define $a^{-2} = \frac{1}{a^2}$.

Negative exponents in the numerator

8.1C DEFINITION: NEGATIVE EXPONENTS IN THE NUMERATOR

Any power in the numerator with a negative exponent can be written in the denominator with a positive exponent.

$$a^{-x} = \frac{1}{a^x}$$

for any $a \neq 0$.

The law for multiplication of powers still holds for negative exponents.

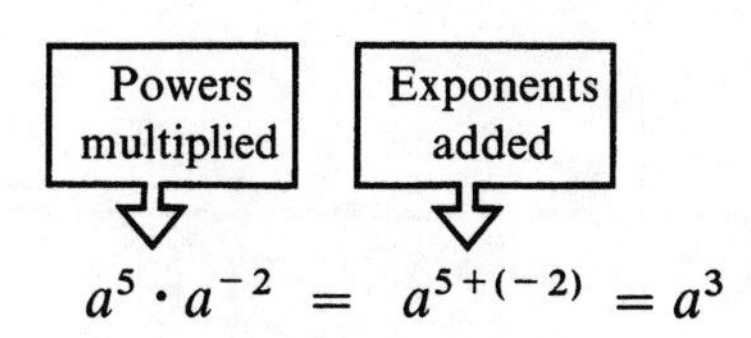

$$a^5 \cdot a^{-2} = a^{5+(-2)} = a^3$$

Using the definition of negative exponents yields the same result.

$$a^5 \cdot a^{-2} = a^5 \cdot \frac{1}{a^2}$$ Definition of negative exponents

$$= a^{5-2}$$ Division of powers

$$= a^3$$

EXAMPLE 3 Write the following with positive exponents.

a. $p^{-3} = \dfrac{1}{p^3}$

b. $x^{-5} =$ _____ $\quad \dfrac{1}{x^5}$

c. $p^{-2} \cdot q^3 =$

$\dfrac{1}{p^2} \cdot q^3 = \dfrac{q^3}{p^2}$ Only p has a negative exponent, therefore only p moves to the denominator.

d. $x^{-3} \cdot y^{-5} =$

$\dfrac{1}{x^3} \cdot \dfrac{1}{y^5} =$ _______ Both x and y have negative exponents, therefore both x and y move to the denominator $\quad \dfrac{1}{x^3y^5}$

e. $(2x)^{-4} =$

$\dfrac{1}{(2x)^4} = \dfrac{1}{16x^4}$ Negative exponent moves entire parentheses to denominator

f. $2x^{-4} = \dfrac{2}{x^4}$ Negative exponent only operates on x

Why did the 2 in the last example stay in the numerator?

Because the exponent only acts on the symbol it follows. If you want it to act on more than one symbol, you must enclose them in parentheses.

Denominators with Negative Exponents

Negative exponents can also appear in the denominator.

$$\frac{1}{x^{-2}} = \frac{1}{\frac{1}{x^2}}$$ Definition of negative exponents

$$= 1 \div \frac{1}{x^2}$$

$$= 1 \cdot \frac{x^2}{1}$$ Definition of division

$$= x^2$$

In general,

Negative exponents in the denominator

> **8.1D DEFINITION: NEGATIVE EXPONENTS IN THE DENOMINATOR**
>
> Any power in the denominator with a negative exponent can be written with a positive exponent in the numerator.
>
> $$\frac{1}{x^{-n}} = x^n$$
>
> where $x \neq 0$.

If a power with a negative exponent appears in the *denominator*, it can be placed in the *numerator* with a positive exponent.

Also, if a power with a negative exponent is in the *numerator*, it can be placed in the *denominator* with a positive exponent.

EXAMPLE 4 Write without negative exponents.

$\frac{1}{x^3}, 4x^3, \frac{x^4y^3}{9}$

$x^{-3} =$ ______ $\frac{4}{x^{-3}} =$ ______ $\frac{x^{-2}}{y^{-3}} = \frac{y^3}{x^2}$ $\frac{x^4}{9y^{-3}} =$ ______

PROBLEM SET 8.1A

Write without negative exponents.

1. a^{-3}
2. a^{-2}
3. x^{-4}
4. $\frac{1}{x^{-4}}$
5. $\frac{1}{y^{-5}}$
6. y^{-5}
7. $x^{-3}y^4$
8. x^3y^{-4}
9. $x^{-3}y^{-4}$
10. $\frac{x^{-3}}{y^4}$
11. $\frac{x^3}{y^{-4}}$
12. $\frac{x^{-3}}{y^{-4}}$
13. $xy^{-2}z^{-1}$
14. $x^{-2}yz^{-1}$
15. $\frac{x^2y^{-3}}{z^{-4}}$
16. $\frac{x^{-2}z^{-3}}{y^{-5}}$

EXAMPLE 5 Evaluate $a^{-3}b^2$ when $a = -3$ and $b = 4$.

$$a^{-3}b^2 = \frac{b^2}{a^3}$$

$$= \frac{(4)^2}{(-3)^3}$$

$$= \frac{16}{-27}$$

$$= \frac{-16}{27}$$

Rewrite with positive exponents, then substitute the values for a and b.

PROBLEM SET 8.1B

Evaluate for $a = 2$, $b = -3$, and $c = -1$.

1. a^2
2. b^2
3. a^3
4. a^{-3}
5. b^2
6. b^{-2}
7. b^{-3}
8. c^5
9. c^6
10. c^{-1}
11. ab^{-1}
12. $a^{-1}b$
13. $a^{-1}b^2$
14. a^2b^{-1}
15. cb^{-2}
16. $c^{-3}b^{-3}$
17. $a^2b^{-1}c$
18. $a^{-2}bc^{-3}$
19. $a^{-3}b^2c^{-1}$
20. $ab^{-3}c^{-4}$
21. $\frac{a}{b}$
22. $\frac{a^2}{b^2}$
23. $\frac{c^3}{b^2}$
24. $\frac{c^4}{a^3}$
25. $\frac{b^3}{c^5}$
26. $\frac{a^4}{c^2}$
27. $\frac{a^3b^2}{c^4}$
28. $\frac{c^5a^3}{b^3}$
29. $\frac{b^3c^2}{a^2}$
30. $\frac{a^3c^3}{b^2}$

Review Your Skills

Perform the indicated operations.

31. $3 + (-4)$
32. $-2 + (-5)$
33. $-6 + 2$
34. $5 - 7$
35. $3 - 2$

The laws of exponents can be used to simplify multiplication and division problems involving positive and negative exponents.

EXAMPLE 6 Simplify $x^{-4} \cdot x^{-3}$.

Powers multiplied ⇩ Exponents added ⇩

$$x^{-4} \cdot x^{-3} = x^{-4+(-3)}$$ Multiplication of powers

$$= x^{-7}$$

$$= \frac{1}{x^7}$$ Definition of negative exponents

EXAMPLE 7 Simplify $\frac{x^3}{x^4}$.

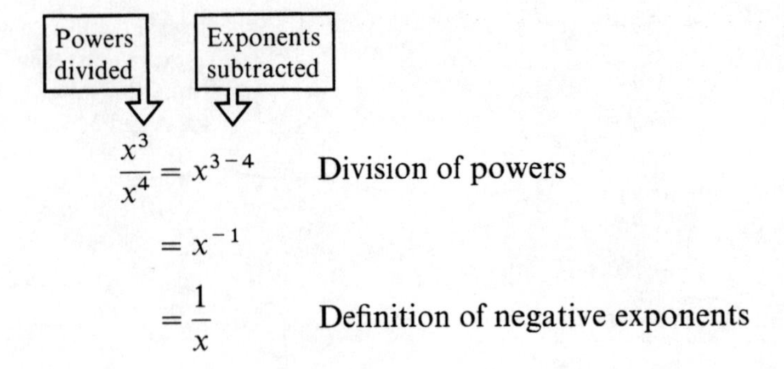

$$\frac{x^3}{x^4} = x^{3-4} \quad \text{Division of powers}$$

$$= x^{-1}$$

$$= \frac{1}{x} \quad \text{Definition of negative exponents}$$

Must the answer be written using positive exponents only?

No, negative exponents are mathematically correct. It's only convenient to write expressions without them.

EXAMPLE 8 Simplify $(a^4b^{-3})(a^2b^2)$.

$$(a^4b^{-3})(a^2b^2) = a^{4+2} \cdot b^{-3+2} \quad \text{Multiplication of powers for like bases}$$

$$= a^6 \cdot b^{-1}$$

$$= \frac{a^6}{b}$$

EXAMPLE 9 Simplify $(2a^3b^{-2})(-3a^{-3}b)$.

$$(2a^3b^{-2})(-3a^{-3}b) = 2 \cdot (-3) \cdot a^{3-3} \cdot b^{-2+1} \quad \text{Multiplication of powers}$$

$$= -6a^0b^{-1}$$

$$= \frac{-6a^0}{b}$$

$$= \frac{-6}{b}$$

$a^0 = 1$

To find the quotient of two monomials divide any numerical coefficients, and then use the rules of exponents for division.

EXAMPLE 10 Simplify $\frac{14x^3y}{-2x^2}$.

$$\frac{14x^3y}{-2x^2} = \frac{14}{-2} \cdot \frac{x^3}{x^2} \cdot \frac{y}{1}$$

$$= -7 \cdot x^{3-2} \cdot y$$

$$= -7xy$$

EXAMPLE 11 Simplify $\frac{-8a^{10}b}{-32a^7b^8}$.

> Remember $b = b^1$

$$\frac{-8a^{10}b}{-32a^7b^8} = \frac{-8}{-32} \cdot \frac{a^{10}}{a^7} \cdot \frac{b^1}{b^8}$$

$$= \frac{1}{4} \cdot a^{10-7} \cdot b^{1-8}$$ Divide the coefficients and subtract the exponents

$$= \frac{1}{4} \cdot a^3b^{-7}$$

$$= \frac{a^3b^{-7}}{4}$$

$$= \frac{a^3}{4b^7}$$ Rewrite using positive exponents

PROBLEM SET 8.1C

Simplify using the laws of exponents. Write the results without negative exponents.

1. $x \cdot x^2$
2. $x^2 \cdot x^{-1}$
3. $x^3 \cdot x^2$
4. $x^3 \cdot x^{-2}$
5. $x^{-3} \cdot x^2$
6. $x^{-3} \cdot x^{-2}$
7. $\frac{x^3}{x^2}$
8. $\frac{x^2}{x^3}$
9. $\frac{x}{x^4}$
10. $\frac{x^4}{x}$
11. $\frac{a}{a^5}$
12. $\frac{b^{10}}{b^3}$
13. $\frac{y^4}{y^4}$
14. $\frac{z^5}{z^6}$
15. $\frac{x^2y}{x^4}$
16. $\frac{x^3y^3}{y^4}$
17. $\frac{x^4y^3}{x^2}$
18. $\frac{a^3b^2}{a^2b}$
19. $\frac{r^4t}{r^3t^5}$
20. $\frac{x^2y^3}{x^5y^4}$
21. $\frac{a^6b^3}{a^3b^2}$
22. $\frac{m^7n^4}{m^2n^3}$
23. $\frac{x^3y}{x^5y^3}$
24. $\frac{ab^2}{a^6b^2}$
25. $\frac{2x^2y^5}{xy^2}$
26. $\frac{4ab^2}{12b^6}$
27. $\frac{21x^4}{-3x^4y}$
28. $\frac{-8a^5b^3}{-16a^3b^7}$
29. $\frac{15a^3b^2}{25a^3b^5}$
30. $\frac{-36b^7c^3}{12b^5c^2}$
31. $\frac{-4r^2t^2}{-6r^4t^2}$
32. $\frac{9x^2y^3}{-6xy^6}$
33. $(xy^2)(y^{-3})$
34. $(x^{-3})(x^4y^{-2})$
35. $(x^2y^3)(x^3y^{-2})$
36. $(a^{-3}b^2)(a^5b^{-5})$
37. $(a^{-2}b^6)(a^{-2}b)$
38. $(x^{-3}y^{-5})(x^{-1}z)$
39. $(2a^3)(a^2b)$
40. $(2x^2y)(3x^3y)$
41. $(-2r^4t^{-3})(3r^6t^3)$
42. $(-4b^5c^6)(-5b^{-5})$
43. $(ab^{-3})(4a^{-2}b^4)$
44. $(-3ab^{-3})(3b^3c)$
45. $(-2x^{-2}y)(-3x^{-3}y)$
46. $(-5s^{-1}t^{-2})(-s^{-4}t^4)$

8.2 POWERS RAISED TO POWERS

$(a^5)^3$ means $(a^5)(a^5)(a^5) = a^{5+5+5} = a^{15}$

also $5 \cdot 3 = 15$

$(p^4)^2$ means $(p^4)(p^4) = p^{4+4} = p^8$

also $4 \cdot 2 = 8$

These observations lead to the following property.

Power to a power

> **8.2A POWER RAISED TO A POWER**
>
> To raise a power to a power, multiply the exponents.
>
> $(a^x)^y = a^{x \cdot y}$

EXAMPLE 12 Simplify the following.

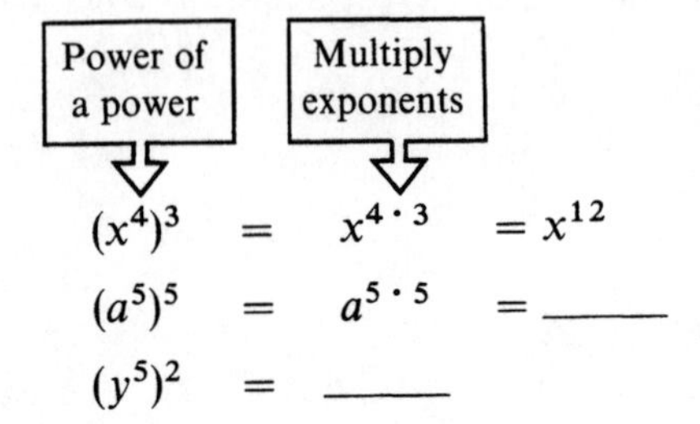

$(x^4)^3 = x^{4 \cdot 3} = x^{12}$

a^{25}

$(a^5)^5 = a^{5 \cdot 5} =$ ______

y^{10}

$(y^5)^2 =$ ______

Power of a Product

To see what happens when a product like (ab^2) is raised to a power, use the basic definition of exponents as repeated multiplication.

$(ab^2)^3$ means $(ab^2)(ab^2)(ab^2)$

Therefore,

$$(ab^2)^3 = (ab^2)(ab^2)(ab^2)$$

Since multiplication is commutative,

$$= a \cdot a \cdot a \cdot b^2 \cdot b^2 \cdot b^2$$
$$= a^3 \cdot b^6$$

Notice this is the same result as if we had multiplied the exponents.

$$(ab^2)^3 = (a^{1 \cdot 3})(b^{2 \cdot 3}) = a^3 \cdot b^6$$

Power of a product

> **8.2B POWER OF A PRODUCT**
>
> $(a^x \cdot b^y)^z = a^{x \cdot z} \cdot b^{y \cdot z}$

EXAMPLE 13 Simplify each of the following.

p^9q^3

(a) $(x^2y^4)^2 = x^{2 \cdot 2} \cdot y^{4 \cdot 2}$
$= x^4y^8$

(b) $(p^3q)^3 =$ ________

EXAMPLE 14 Simplify $(3a^4)^2$.

$(3a^4)^2 = 3^2a^8$ Power of a product

$= 9a^8$ Evaluate 3^2

EXAMPLE 15 Simplify $(-3a^4)^2$.

Be sure to enclose the negative sign of the -3 in the parentheses.

$(-3a^4)^2 = (-3)^2a^8$

$= 9a^8$ Evaluate $(-3)^2$

EXAMPLE 16 Evaluate the following for $a = 2$ and $b = -3$.

$(a^2b)^2 = [(2)^2(-3)]^2$ or $(a^2b)^2 = a^4b^2$

$= [4(-3)]^2$ $= (2)^4(-3)^2$

9 $= [-12]^2$ $= 16 \cdot$ _____

$= 144$ $= 144$

EXAMPLE 17 Evaluate the following for $a = 2$ and $b = -3$.

$(ab)^2 = [(2)(-3)]^2$ or $(ab)^2 = a^2b^2$

$= [-6]^2$ $= (2)^2(-3)^2$

36, 9 $=$ _____ $= 4 \cdot$ _____

36 $=$ _____

EXAMPLE 18 Evaluate the following for $a = 2$ and $b = -3$.

$(a + b)^2 = [(2) + (-3)]^2$ or

$= [-1]^2$

1 $=$ _____

Nope! There is no other easy way to evaluate this one. Raising a sum to a power comes later.

EXAMPLE 19 Evaluate the following for $a = 2$ and $b = -3$.

$a^2 + b^2 = (2)^2 + (-3)^2$

$= 4 + 9$

13 $=$ _____

Why don't Examples 18 and 19 have the same answer?

Because $(a + b)^2$ is not the same as $a^2 + b^2$. You'll learn how to evaluate $(a + b)^2$ in Unit 9.

PROBLEM SET 8.2A

Simplify the following.

1. $(a^2)^3$ 2. $(b^3)^2$ 3. $(x^4)^5$ 4. $(x^5)^2$ 5. $(b^3)^4$

6. $(a^2)^4$ 7. $(x^2y)^3$ 8. $(xy^2)^3$ 9. $(x^3y^2)^2$ 10. $(x^2y^3)^2$

11. $(a^5bc^2)^3$ 12. $(ab^2c^5)^3$ 13. $(a^4b^2c)^2$ 14. $(a^3b^4c^2)^3$ 15. $(2x)^3$

16. $(2x^2)^3$ 17. $(2x^4)^2$ 18. $(3x^5)^2$ 19. $(-x^2)^3$ 20. $(-x^3)^2$

21. $(-2x)^3$ 22. $(-2x^2)^3$ 23. $(-2x)^2$ 24. $(-2x^5)^2$ 25. $(4x^2y)^2$

26. $(5x^3y^2)^2$ 27. $(-4n^4p^2)^3$ 28. $(-2p^6q)^3$ 29. $(-6m^6n^3)^2$ 30. $(-7ab^7)^2$

31. $(5s^5t)^3$ 32. $(2x^4y^2)^4$ 33. $(-3x^2y^3)^3$ 34. $(-2a^3b^2)^4$

Evaluate the following for $a = -3$, $b = 2$, $c = -2$.

35. $(a^2b)^2$ 36. $(ab^2)^2$ 37. $(ab)^2$ 38. $(ab)^3$ 39. $(bc^2)^2$

40. $(b^2c)^2$ 41. $(ac^2)^2$ 42. $(a^2c)^2$ 43. $(bc)^3$ 44. $(b^2c)^3$

45. $(bc^2)^3$ 46. $(abc)^2$ 47. $(a + b)^2$ 48. $a^2 + b^2$ 49. $(b + c)^2$

50. $b^2 + c^2$ 51. $(a + b)^3$ 52. $a^3 + b^3$ 53. $(a - b)^3$ 54. $a^3 - b^3$

55. $(b + c)^3$ 56. $b^3 + c^3$ 57. $(b - c)^3$ 58. $b^3 - c^3$

59. $(a + b + c)^2$ 60. $a^2 + b^2 + c^2$ 61. $a^3 - b^3 + c^3$ 62. $a^3 + b^3 - c^3$

63. $(a^2 + b^2 - c^2)^2$ 64. $a^3 - b^3 - c^3$

Power of Quotients

Quotients may also be raised to a power.

$$\left(\frac{a}{b}\right)^2 \quad \text{means} \quad \frac{a}{b} \cdot \frac{a}{b} = \frac{a^2}{b^2}$$

and

$$\left(\frac{x^2}{y^3}\right)^2 \quad \text{means} \quad \frac{x^2}{y^3} \cdot \frac{x^2}{y^3} = \frac{x^4}{y^6}$$

In general,

Power of a quotient

8.2C POWER OF A QUOTIENT

$$\left(\frac{a^x}{b^y}\right)^z = \frac{a^{xz}}{b^{yz}}$$

where $b \neq 0$.

EXAMPLE 20 Simplify the following.

(a) $\left(\frac{a^3}{b^4}\right)^2 = \frac{a^{3\cdot 2}}{b^{4\cdot 2}} = \frac{a^6}{b^8}$

(b) $\left(\frac{a^2b^3}{c}\right)^3 = \frac{a^6b^9}{c^3}$

$\frac{9a^{10}}{16b^6}$

(c) $\left(\frac{-3a^5}{4b^3}\right)^2 = \frac{(-3)^2a^{10}}{(4)^2b^6} =$ ________

EXAMPLE 21 Evaluate each of the following for $a = 2$ and $b = -3$.

(a) $\left(\frac{a}{b}\right)^2 = \frac{a^2}{b^2}$

$= \frac{(2)^2}{(-3)^2}$

$= \frac{4}{9}$

(b) $\left(\frac{3a^2}{b}\right)^3 = \frac{3^3a^6}{b^3}$

$= \frac{3^3(2)^6}{(-3)^3}$

$= \frac{27\cdot 64}{-27}$

$= -64$

SUMMARY OF THE LAWS OF EXPONENTS

1. Zero exponent	$a^0 = 1$	$a \neq 0$
2. Negative exponent	$a^{-x} = \frac{1}{a^x}$	$a \neq 0$
3. Multiplication of powers	$a^x \cdot a^y = a^{x+y}$	
4. Division of powers	$\frac{a^x}{a^y} = a^{x-y}$	$a \neq 0$
5. Power to a power	$(a^x)^y = a^{xy}$	
6. Power of a product	$(a^xb^y)^z = a^{xz}b^{yz}$	
7. Power of a quotient	$\left(\frac{a^x}{b^y}\right)^z = \frac{a^{xz}}{b^{yz}}$	$b \neq 0$

PROBLEM SET 8.2B

Simplify each of the following.

1. $\left(\frac{a}{b}\right)^2$
2. $\left(\frac{a}{b^2}\right)^3$
3. $\left(\frac{a}{b}\right)^3$
4. $\left(\frac{a^2}{b}\right)^3$
5. $\left(\frac{a}{b^3}\right)^2$
6. $\left(\frac{a^3}{b^2}\right)^4$
7. $\left(\frac{xy}{z}\right)^2$
8. $\left(\frac{x^2y^3}{zw^0}\right)^2$
9. $\left(\frac{xy^3}{z^2}\right)^4$
10. $\left(\frac{x^2y}{z^6}\right)^5$
11. $\left(\frac{a^3x^0}{bc^2}\right)^6$
12. $\left(\frac{p}{q^3r^4}\right)^4$
13. $\left(\frac{a^2b^3}{c^5d}\right)^5$
14. $\left(\frac{xy^2}{a^3b}\right)^3$
15. $\left(\frac{2x}{y}\right)^2$
16. $\left(\frac{3x^3}{a^0y}\right)^2$
17. $\left(\frac{-2x^6}{y^3}\right)^0$
18. $\left(\frac{-2x^3}{b^0}\right)^2$
19. $\left(\frac{-3a^4}{b^2}\right)^3$
20. $\left(\frac{-2x^2}{y^4a^0}\right)^2$

21. $\left(\frac{-2x^3}{y^2}\right)^3$ 22. $\left(\frac{2x}{y^4}\right)^3$ 23. $\left(\frac{2a}{3b}\right)^2$ 24. $\left(\frac{3a}{2b^2}\right)^3$ 25. $\left(\frac{-3a^3}{4b^4}\right)^0$

26. $\left(\frac{-4a^2}{5b^3}\right)^2$ 27. $\left(\frac{-3x^2}{2y^3}\right)^3$ 28. $\left(\frac{-2x^4}{3y^3}\right)^4$ 29. $\left(\frac{-3a^4}{2b^4}\right)^0$ 30. $\left(\frac{-4a^2}{3b^3}\right)^3$

Evaluate the following for $a = 2$, $b = 3$, $x = -1$, $y = -2$.

31. $\left(\frac{a}{b}\right)^2$ 32. $\left(\frac{y}{b}\right)^2$ 33. $\left(\frac{y}{b}\right)^3$ 34. $\left(\frac{xy^0}{b}\right)^2$ 35. $\left(\frac{x}{a}\right)^3$

36. $\left(\frac{x}{a}\right)^4$ 37. $\left(\frac{a}{y}\right)^2$ 38. $\left(\frac{y}{a}\right)^5$ 39. $\left(\frac{x^2}{a}\right)^3$ 40. $\left(\frac{y^2}{x^3}\right)^2$

41. $\left(\frac{ab}{y}\right)^2$ 42. $\left(\frac{xy}{b}\right)^3$ 43. $\left(\frac{4y^2}{b^3}\right)^0$ 44. $\left(\frac{5a}{b}\right)^2$ 45. $\left(\frac{2a^2}{b^2}\right)^2$

8.3 SCIENTIFIC NOTATION

Scientists use the laws of exponents to express very big or very small numbers.

The mass of a proton is 0.00000000000000000000000167 kg.

The speed of light is 299,790,000 meters per second.

Neither of these figures is convenient to work with. However, we can use the properties of exponents and the way the decimal moves when you multiply or divide by a power of ten to simplify things.

$$1 = 1 = 10^0$$
$$10 = 10 = 10^1$$
$$100 = 10 \cdot 10 = 10^2$$
$$1000 = 10 \cdot 10 \cdot 10 = 10^3$$

$$\underbrace{10{,}000}_{\text{4 zeros}} = \underbrace{10 \cdot 10 \cdot 10 \cdot 10}_{\text{4 tens}} = 10^4 \quad (\text{Exponent } 4)$$

Notice that the number of zeros in the power of ten is equal to the number of times ten is used as a factor. The exponent also gives us the number of times ten is used as a factor.

To multiply by a power of ten, move the decimal point as many places to the right as there are zeros in the power of ten.

EXAMPLE 22

(a) $2.18 \times 10 = 21.8$

(b) $2.18 \times 100 = 218$

2,180 — (c) $2.18 \times 1{,}000 =$ ________

21,800 — (d) $2.18 \times 10{,}000 =$ ________

The exponent on the base (10) is the same as the number of zeros in the power of ten. When multiplying by a power of ten written with exponents, move the decimal point to the right the number of places indicated by the exponent. Add zeros if necessary.

EXAMPLE 23

(a) $2.18 \times 10^1 = 21.8$

(b) $2.18 \times 10^2 = 218$

2,180 — (c) $2.18 \times 10^3 =$ ________

21,800 — (d) $2.18 \times 10^4 =$ ________

PROBLEM SET 8.3A

Move the decimal point the proper number of places to complete the multiplications.

1. 4.0×10
2. 7.2×10^2
3. 3.654×100
4. 5.6×10
5. 8.62×10^3
6. 5.646×10^2
7. 8.265×10^1
8. 2.3×10^2
9. $3.86 \times 10{,}000$
10. $1.4 \times 1{,}000$
11. 5.6×10^4
12. 2.3×10^3
13. 4×100
14. $6 \times 1{,}000$
15. 9.23×10^3

Supply the missing powers of ten so each statement is true.

16. $3.26 \times$ ______ $= 32.6$
17. $5.6 \times$ ______ $= 560$
18. $2.34 \times$ ______ $= 23.4$
19. $8.65 \times$ ______ $= 86{,}500$
20. $9.674 \times$ ______ $= 96.74$
21. $3.2 \times$ ______ $= 3{,}200$
22. $1.0 \times$ ______ $= 100$
23. $6.0 \times$ ______ $= 600$
24. $7.32 \times$ ______ $= 732$
25. $5 \times$ ______ $= 5{,}000$
26. $6.74 \times$ ______ $= 67{,}400$
27. $3.26 \times$ ______ $= 326$
28. $6.5 \times$ ______ $= 6{,}500$
29. $6.86 \times$ ______ $= 68.6$
30. $9 \times$ ______ $= 900{,}000$

Scientific notation

8.3A DEFINITION

To write a number in *scientific notation*, write it as a number with a magnitude between 1 and 10 multiplied by a power of ten.

EXAMPLE 24

Number	Scientific notation
56	5.6×10^1
8,674	8.674×10^3
−92,300	-9.23×10^4
8,600,000	8.6×10^6

In Example 24, the magnitudes of all the numbers were greater than 1 and all exponents were positive. However, scientific notation can also be used to represent numbers with magnitudes less than 1. To express numbers with magnitudes between 0 and 1 in scientific notation, use a negative exponent.

When a number is multiplied by a power of ten with a positive exponent, the decimal moves to the right. Multiplying by a power of ten with a negative exponent moves the decimal to the left the number of places the exponent indicates.

EXAMPLE 25 Write as a decimal numeral without using exponents.

(a) $3.86 \times 10^{-2} = 0.0386$

(b) $-5.86 \times 10^{-1} = -0.586$

(c) $9.47 \times 10^{-3} = 0.00947$

(d) $5.92 \times 10^{-4} =$ ________ 0.000592

(e) $-2.1 \times 10^{-2} =$ ________ −0.021

8.3B RULE

To convert numbers from scientific notation to decimal notation, move the decimal point the number of places indicated by the exponent.

Positive exponents: Move the decimal right.

Negative exponents: Move the decimal left.

EXAMPLE 26 Write as a decimal numeral.

(a) $3.92 \times 10^4 = 39{,}200$

(b) $5.21 \times 10^{-3} = 0.00521$

(c) $-8.6 \times 10^2 =$ ________ −860

(d) $5.3 \times 10^{-1} =$ ________ 0.53

Converting decimals to scientific notation

8.3C RULE

To convert decimals to scientific notation:

1. Move the decimal point so that it is immediately after the first nonzero digit.
2. Multiply the number by the power of ten that would move the decimal to its original position.

Note: Positive exponents move the decimal to the right and negative exponents move the decimal to the left.

EXAMPLE 27 Write the following in scientific notation.

$5320 = 5.32 \times 10^3$

$0.0684 = 6.84 \times 10^{-2}$

$48.6 = 4.86 \times 10^1$

$-0.35 = -3.5 \times 10^{-1}$

10^2 — $865.6 = 8.656 \times$ ________

10^{-4} — $0.00046 = 4.6 \times$ ________

10^0 — $-9.6 = -9.6 \times$ ________

10^2 — $423 = 4.23 \times$ ________

In chemistry, physics, biology, and other scientific fields, scientific notation is frequently used to simplify expressions containing large or small numbers.

EXAMPLE 28 Use scientific notation to find the value of 0.00024×0.0000013.

Write in scientific notation.

$$\begin{aligned} 0.00024 \times 0.0000013 &= 2.4 \times 10^{-4} \times 1.3 \times 10^{-6} \\ &= 2.4 \times 1.3 \times 10^{-4} \times 10^{-6} \\ &= 3.12 \times 10^{-10} \\ &= 0.000000000312 \end{aligned}$$

EXAMPLE 29 Use scientific notation to find the value of $\dfrac{0.0042}{12{,}000}$.

Write in scientific notation.

$$\frac{0.0042}{12{,}000} = \frac{4.2 \times 10^{-3}}{1.2 \times 10^4}$$

$$= \frac{4.2}{1.2} \times \frac{10^{-3}}{10^4}$$

$$1.2\overline{)4.2_{\wedge}0} = 3.5; \quad 36,\ 60$$

$= 3.5 \times 10^{-3-4}$ Subtract the exponents

$= 3.5 \times 10^{-7}$

0.00000035 — $=$ ________

EXAMPLE 30
It is estimated that since the beginning of time all the natural gas resources in the United States amount to 1,075,000,000,000 cubic feet of natural gas. It is further estimated that by the time our country was 200 years old, we had already consumed 538,500,000,000 cubic feet of this resource. Use scientific notation to determine the fraction of the natural gas that has been used.

$$\frac{\text{natural gas used}}{\text{natural gas ever available}} = \frac{5.385 \times 10^{11} \text{ cubic ft}}{1.075 \times 10^{12} \text{ cubic ft}}$$

$$= 5 \times 10^{11-12} \qquad 1.075\overline{)5.385_\wedge}\;\; 5.$$

$$= 5 \times 10^{-1}$$

$$= 0.5$$

We have used over half of all the natural gas that was ever available to us.

PROBLEM SET 8.3B

Convert the following numbers in scientific notation to decimal numbers.

1. 3.2×10^3
2. 5.48×10
3. 7.67×10^4
4. 6.793×10^2
5. 5.6×10^2
6. 8.65×10^3
7. 8.1×10^{-4}
8. 1.2×10^{-1}
9. 3.678×10^1
10. 4.9×10^4
11. -4×10^{-5}
12. -9×10^2
13. 1.2×10^4
14. -3.678×10^2
15. 2.679×10^{-3}
16. 6.789×10^{-4}

Convert the following decimal numbers to numbers in scientific notation.

17. 0.0654
18. 0.962
19. 946
20. 3,800
21. 34.6
22. 256.8
23. 0.76
24. 0.000468
25. 958,000
26. 21,000,000
27. 0.000056
28. 0.0000000867
29. −218
30. 80,000,650
31. −0.00086
32. −0.0457
33. −20,000
34. 10,000,000
35. 0.000001
36. −0.001

Use scientific notation to simplify the following.

37. 0.0007×0.0000056
38. $390{,}000 \times 0.00048$
39. $17{,}800{,}000 \times 23{,}000$
40. $49{,}000 \times 0.000016$
41. $0.025 \times 0.0000000018$
42. 250×0.00000000018
43. $0.075 \times 240{,}000$
44. $0.004 \times 32{,}000$
45. $\dfrac{300}{25{,}000}$
46. $\dfrac{0.24}{60{,}000}$
47. $\dfrac{450{,}000}{2{,}500}$
48. $\dfrac{0.000045}{2{,}500}$
49. $\dfrac{0.00214}{350}$
50. $\dfrac{8{,}840}{260{,}000}$

51. The present rate of consumption of natural gas is 19,300,000,000 cubic feet per year. Our country has natural gas reserves of 540,400,000,000 cubic feet. How long can we expect these reserves to last at our present rate of consumption?

52. The federal standard for clean air says the maximum permissible level of pollution from particulate matter over our cities is 2.6×10^{-3} grams of particles per cubic meter. An area of a city that is 2 miles wide by 5 miles long has about 8×10^9 cubic meters of air in the first 1,000 feet above ground level. Estimate the total weight of the particle pollution in this amount of air at the maximum permissible level of pollution.

53. The sun's mass is approximately 1.98×10^{30} kilograms. The mass of the earth is approximately 6×10^{24} kilograms. How many times greater is the mass of the sun than the mass of the earth?

54. Astronomers measure large distances in light-years. One light-year is the distance traveled by a beam of light in one year. The speed of light is 300,000,000 meters per second. In a year there are 3.15×10^7 seconds. Use scientific notation to calculate the distance in one light-year.

Review Your Skills Write the following products as sums.

55. $a(x + 2)$

56. $b(y - 2)$

57. $2a(3a - b)$

58. $-3a(5a - 7)$

8 REVIEW

NOW THAT YOU HAVE COMPLETED UNIT 8, YOU SHOULD BE ABLE TO USE THE FOLLOWING RULES AND DEFINITIONS:

Laws of exponents

$$a^{-x} = \frac{1}{a^x} \qquad \frac{1}{a^{-x}} = a^x$$

$$a^x \cdot a^y = a^{x+y} \qquad (a^x b^y)^z = a^{x \cdot z} b^{y \cdot z}$$

$$\frac{a^x}{a^y} = a^{x-y} \qquad \left(\frac{a^x}{b^y}\right)^z = \frac{a^{x \cdot z}}{b^{y \cdot z}}$$

$$(a^x)^y = a^{x \cdot y} \qquad a^0 = 1$$

To evaluate exponential expressions, substitute the given values for the variables and follow the rules of order of operations.

To convert numbers from scientific notation to decimal numerals, move the decimal point the number of places indicated by the exponent.

Positive exponents: Move the decimal right.

Negative exponents: Move the decimal left.

To convert decimal numerals to scientific notation:

1. Move the decimal point so that it is immediately after the first nonzero digit.
2. Multiply the number by the power of ten that would return the decimal to its original position.

Review Test 8

Simplify the following using the laws of exponents. Write the results without negative exponents. (8.1 and 8.2)

1. $x^{-4} \cdot x^2 =$ _____
2. $(-2x)^2 =$ _____
3. $(x^4)^3 =$ _____
4. $\frac{x^3}{x^7} =$ _____
5. $\left(\frac{x^5}{y^2}\right)^4 =$ _____
6. $\frac{m^5n^2}{mn^4} =$ _____
7. $(ab^3)^2 =$ _____
8. $(x^{-3}y^5)(x^3y^{-2}) =$ _____
9. $(4a)(a^{-2}b^2) =$ _____
10. $\frac{-9a^3b^3}{12a^6b^2} =$ _____
11. $\left(\frac{x^3y}{z^2}\right)^5 =$ _____
12. $\left(\frac{-2a^2}{3b^3}\right)^3 =$ _____
13. $(4a^4b^2)^2 =$ _____
14. $(2x^{-5}y^{-1})(-3xy^{-3}) =$ _____

Evaluate the following for a = 3, b = −2, and c = −1. (8.1 and 8.2)

15. $a^{-2}b =$ _____
16. $(ac^3)^3 =$ _____
17. $(a + c)^4 =$ _____
18. $\frac{a^2}{b^3} =$ _____
19. $\left(\frac{b}{c^3}\right)^2 =$ _____
20. $\left(\frac{4a}{b^3}\right)^2 =$ _____

Convert the following numerals in scientific notation to decimal numerals. (8.3)

21. $5.67 \times 10^{-3} =$ _____
22. $8.146 \times 10^4 =$ _____

Convert the following decimal numerals to numerals in scientific notation. (8.3)

23. 2,678,000 = _____
24. 0.00743 = _____

Use scientific notation to simplify. (8.3)

25. $48{,}800 \times 0.0000035 =$ _____

REVIEW YOUR SKILLS

26. Find the slope of the line through the two points $(-3, 2)$ and $(5, -7)$.

27. Find the equation of the line with slope $\frac{-3}{4}$ and y-intercept -5.

28. Find the equation of the line with slope $\frac{-5}{4}$, passing through the point $(4, -5)$.

29. Use the point-slope method to find the equation of the line through the points $(-3, -6)$ and $(-5, 4)$.

30. Truong is three times as old as her sister. In 12 years, she will be twice as old as her sister. How old is Truong?

Cumulative Review 1–8

Evaluate.

1. $-2-(3-4)$ ***(1.4)***
2. $-2-3+(-4)$ ***(1.4)***
3. $5^2-2\cdot 5$ ***(1.4)***
4. $12\div 3\cdot(-2)$ ***(1.4)***
5. $3x^2-2x-1$ if $x=2$ ***(2.1)***
6. $3x^2-2x-1$ if $x=-2$ ***(2.1)***
7. $5x-3y$ if $x=\dfrac{-3}{5}$ and $y=\dfrac{2}{3}$ ***(3.4)***
8. $(ab^2)^3$ if $a=2$ and $b=-1$ ***(8.2)***

Perform the indicated operation and simplify.

9. $x^3(x^2+6x)$ ***(2.3)***
10. $-2x(x^2-4x+5)$ ***(2.3)***
11. $(2x^2+5x-7)+(x^2-3x-4)$ ***(2.4)***
12. $(3y^2-2y-4)-(2y^2-y+4)$ ***(2.4)***
13. $(4x+5)+(3-x)-(x-3)$ ***(2.4)***
14. $\dfrac{3b}{2}\cdot\dfrac{8}{9}$ ***(3.2A)***
15. $\dfrac{-15}{2}\div\dfrac{-5}{3}$ ***(3.2C)***
16. $\dfrac{3a}{4}+\dfrac{1}{6}$ ***(3.3C)***
17. $\dfrac{-2x}{7}-\dfrac{3x}{14}$ ***(3.3C)***

Solve for the variable.

18. $4x-1=3x-5$ ***(4.1)***
19. $6y-2=2y+6$ ***(4.2B)***
20. $3(x+3)=5(x-1)$ ***(4.2D)***
21. $2a=7-(a+1)$ ***(4.2D)***
22. $\dfrac{t}{3}+4=-2$ ***(4.2C)***
23. $x=4y-2$ for y ***(4.3)***
24. The area of a triangle is given by the formula $A=\dfrac{b\cdot h}{2}$. Find the height (h) if the base (b) is 8 inches and the area (A) is 12 square inches. ***(4.4)***

Represent the following English phrases by an algebraic expression. (5.1A)

25. Four more than twice a number.
26. The age of an x-year-old girl 10 years ago.
27. The value of d dimes and q quarters.

Select a letter to represent the unknown quantity. Set up an equation and solve it.

28. The sum of two consecutive integers is -85. Find the integers. ***(5.2)***

29. Sue is 9 years older than her brother. Three years ago she was four times as old as her brother. How old are they now? ***(5.3)***

30. Jane and Debbie left Chicago in separate cars, traveling in opposite directions. Jane drove 10 mph faster than Debbie, and in 5 hours they were 450 miles apart. How fast was each traveling? ***(5.5)***

31. If $f(x) = 4x - 2$, find $f(-3)$ ***(6.1)***

On a rectangular graph, sketch the graph of the following.

32. $y = \frac{2}{3}x - 4$ ***(6.5A)***

33. $2x - 5y = -10$ ***(6.5B)***

34. $x = 1$ ***(6.5B)***

35. Find the slope of a line through the points $(4, -1)$ and $(-2, -3)$. ***(6.3)***

36. Find the equation of a line with slope 3 and y-intercept -4. ***(6.4A)***

37. Find the equation of a line with slope $\frac{-2}{3}$ through the point $(6, 1)$. ***(6.4B)***

38. Find the equation of a line through the points $(8, 0)$ and $(4, -1)$. ***(6.4C)***

Solve the following systems of equations by graphing. (7.1)

39. $\begin{aligned} 3x - y &= 3 \\ x + 2y &= 8 \end{aligned}$

40. $\begin{aligned} 4x - y &= -8 \\ 2x - 3y &= 6 \end{aligned}$

Solve the following systems of equations by the addition method. (7.2)

41. $\begin{aligned} 3x + y &= 7 \\ 4x - 2y &= 6 \end{aligned}$

42. $\begin{aligned} 7x + 2y &= 10 \\ 2x + 3y &= -2 \end{aligned}$

Solve the following systems of equations by the substitution method. (7.3)

43. $\begin{aligned} 2x + 3y &= -1 \\ y &= x + 3 \end{aligned}$

44. $\begin{aligned} x + 4y &= 1 \\ 2x - y &= -7 \end{aligned}$

Solve the following by using two equations. State in words exactly what your variables represent. Set up two equations and solve them. (7.4)

45. Wendy, a 2-year-old, has three times as many nickels as dimes. If the total value of the coins is \$3, how many of each coin does she have in her piggy bank?

46. A grocer blends peanuts worth \$2.60 a pound and cashews worth \$3.50 a pound to obtain a 30-pound mixture worth \$88.80. How many pounds of each type of nut were used?

Simplify the following. Write the results without negative exponents.

47. $(5x^{-5}) \cdot (4x^4)$ ***(8.1)***

48. $(3ab^3)^2$ ***(8.2)***

49. $\dfrac{8xy^3}{2x^4y^3}$ ***(8.1)***

50. $\left(\dfrac{-2m^4}{3n^3}\right)^3$ ***(8.2)***

51. Write 0.000421 in scientific notation. ***(8.3)***

52. Simplify using scientific notation: $0.0031 \times 52{,}000$. ***(8.3)***

UNIT 9

Multiplication and Division of Polynomials

OBJECTIVES

After you have successfully completed this unit, you will be able to:

1. Find the product of two polynomials using the distributive property or the vertical method (9.1)
2. Find the product of two binomials using the FOIL method (9.2)
3. Find the product of the sum and difference of two terms (9.2)
4. Find the square of a binomial (9.2)
5. Divide a polynomial by a monomial (9.3)
6. Divide a polynomial by a binomial using long division (9.4)

9.1 MULTIPLICATION OF POLYNOMIALS

Polynomial

In Unit 2, we defined a *polynomial* as a term or a sum of terms, each term consisting of a numeral or the product of a numeral and one or more variables. The variables may be raised to any whole number power.

As you recall, the purpose of defining a polynomial was to avoid problems with division by zero along with some other problems to be mentioned later.

In Section 2.2, multiplication of monomials (single terms) was covered. Here is an example.

EXAMPLE 1 Multiply $-4a^3 \cdot 5a^2b^2$.

$$-4a^3 \cdot 5a^2b^2 = -20a^5b^2$$

Product of the numerical coefficients

Product of the variables

In Section 2.3, we explained that to multiply a monomial times a polynomial we apply the distributive law.

EXAMPLE 2 Multiply $6a^2(4a^2 - a + 2)$.

$$6a^2(4a^2 - a + 2) = 6a^2 \cdot 4a^2 - 6a^2 \cdot a + 6a^2 \cdot 2$$
$$= 24a^4 - 6a^3 + 12a^2$$

Now we will use the distributive property to multiply a polynomial by a polynomial.

Distributive property

9.1 THE DISTRIBUTIVE PROPERTY

$$a(b + c) = a \cdot b + a \cdot c \quad \text{and} \quad (b + c)a = b \cdot a + c \cdot a$$

where a, b, and c stand for any algebraic expression

Since a, b, and c may stand for any algebraic expression, we can use the distributive property to multiply polynomials.

EXAMPLE 3 Multiply $(x + 3)(2x + 4)$.

Think of $(x + 3)$ as Z.

Then

$$(x + 3)(2x + 4) = Z(2x + 4)$$

and

$$Z \cdot (2x + 4) = Z \cdot 2x + Z \cdot 4 \qquad \text{Using the distributive property}$$
$$= (x + 3)2x + (x + 3) \cdot 4 \qquad \text{Restoring } (x + 3) \text{ for } Z$$

Now distribute the $2x$ and the 4

$$= (x + 3)2x + (x + 3) \cdot 4$$
$$= x(2x) + 3(2x) + x(4) + 3(4)$$
$$= 2x^2 + 6x + 4x + 12$$
$$= 2x^2 + 10x + 12 \qquad \text{Collecting like terms}$$

EXAMPLE 4 Multiply $(a - 5)(a - 4)$.

Think of the first factor as a single quantity.

Let

$$(a - 5) = Z$$

Why Z?

Any symbol will do. The idea is to emphasize that $(a - 5)$ is a single quantity.

Then

$$(a - 5)(a - 4) =$$

becomes

$$Z \cdot (a - 4) = Za + Z(-4)$$

How come the subtraction sign on the left became an addition sign on the right?

You caught us, we skipped a step.

$Z(a - 4) = Z[a + (-4)]$

If we write it as an addition problem, the next step will be easier.

Now restoring Z with $a - 5$

$$(a-5)(a-4) = (a-5)(a) + (a-5)(-4)$$

Applying the distributive property

$$= a(a) - 5(a) + a(-4) - 5(-4)$$
$$= a^2 - 5a - 4a + 20$$
$$= a^2 - 9a + 20$$

Notice: The effect of this line is to multiply every term of the first binomial by every term of the second.

This same process can be done vertically in a method similar to multiplication in arithmetic. To multiply two polynomials, *multiply every term of the first polynomial by every term of the second polynomial*, and then collect the like terms.

$$\begin{array}{ll}
a \ -5 & \\
a \ -4 & \\
\hline
a^2 - 5a & \Leftarrow \text{Product of } a(a-5) \\
\quad - 4a + 20 & \Leftarrow \text{Product of } -4(a-5) \\
\hline
a^2 - 9a + 20 & \quad \text{Collect like terms}
\end{array}$$

EXAMPLE 5 Use the distributive property to multiply $(2x + y)(3x - y)$.

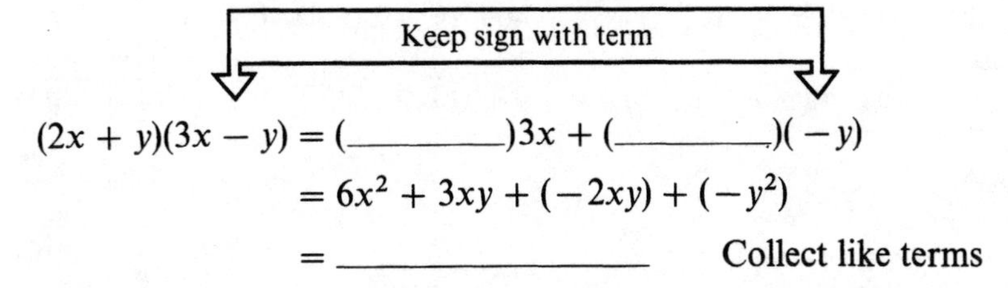

$2x + y,\ 2x + y$

$$(2x + y)(3x - y) = (\underline{\qquad\qquad})3x + (\underline{\qquad\qquad})(-y)$$
$$= 6x^2 + 3xy + (-2xy) + (-y^2)$$
$$= \underline{\qquad\qquad\qquad} \quad \text{Collect like terms}$$

$6x^2 + xy - y^2$

EXAMPLE 6 Use the vertical method to multiply $(2x + y)(3x - y)$.

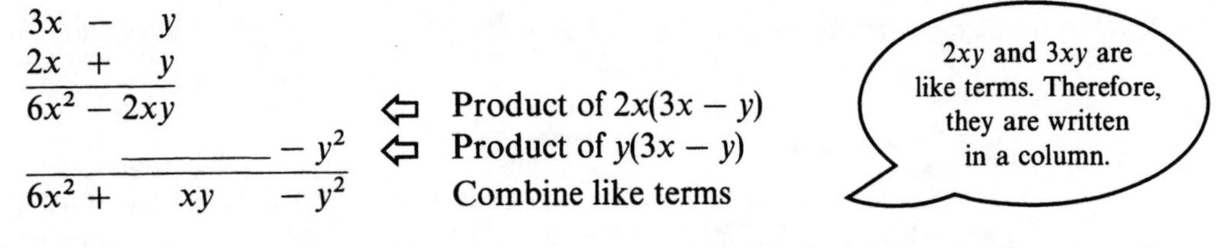

$$\begin{array}{ll}
3x \ - \ y & \\
2x \ + \ y & \\
\hline
6x^2 - 2xy & \Leftarrow \text{Product of } 2x(3x-y) \\
\quad \underline{\qquad\quad} - y^2 & \Leftarrow \text{Product of } y(3x-y) \\
\hline
6x^2 + \quad xy \quad - y^2 & \quad \text{Combine like terms}
\end{array}$$

$+3xy$

EXAMPLE 7 Use the vertical method to multiply $(2a^2 + 3a - 4)(5a^2 - 6a - 7)$.

$$\begin{array}{ll}
2a^2 + 3a - 4 & \\
5a^2 - 6a - 7 & \\
\hline
10a^4 + 15a^3 - 20a^2 & \Leftarrow \text{Product of } 5a^2(2a^2 + 3a - 4) \\
\quad - 12a^3 - 18a^2 + 24a & \Leftarrow \text{Product of } -6a(2a^2 + 3a - 4) \\
\quad\quad - 14a^2 - 21a + 28 & \Leftarrow \text{Product of } -7(2a^2 + 3a - 4) \\
\hline
10a^4 + 3a^3 - 52a^2 + 3a + 28 & \quad \text{Collect like terms}
\end{array}$$

PROBLEM SET 9.1

Find the following products.

1. $3a \cdot 4ab$
2. $-2a^2b \cdot ab$
3. $(-6x^2y)(-7xy^2)$
4. $(-5ax)(6a^3x)$
5. $8ax^2 \cdot (-7ay)$
6. $(9a^2b)(3a^2x)$
7. $(3x^3y)(-x^2y^2)$
8. $(-ab^2)(-2ab)$

Use the distributive property to find the following products.

9. $x(3x + 4)$
10. $y(-2y + 5)$
11. $5x(2x + 3)$
12. $6x(m^2 - 5m)$
13. $-2y(3y^2 + 4y)$
14. $4a(2a^2 + 1)$
15. $7x(3x^2 - 2x)$
16. $x^2(6x - 5)$
17. $2a^2(a^2 + 8a)$
18. $4ab(3a + 5b)$
19. $-2xy(6x^2 - 4y^2)$
20. $(2x + 3)(x + 4)$
21. $(2x + 3)(3x + 2)$
22. $(4x - 5)(x + 4)$
23. $(x - 2y)(x + 3y)$
24. $(3x + 2)(x - 3)$
25. $(5x + y)(4x - 3y)$
26. $(4t - 7)(3t - 2)$
27. $(3a - 2b)(6a - 2b)$
28. $(8a - 7x)(4a + 3x)$
29. $(3a - 4x)(-2a + 5x)$

Line up the following problems vertically, then multiply.

30. $(x - 2)(x + 4)$
31. $(x + 3)(x - 4)$
32. $(2x - 3)(x + 5)$
33. $(3x - 4)(4x + 7)$
34. $(5a - 6)(3a + 4)$
35. $(2x - 5y)(x + y)$
36. $(3a - 4b)(a + 3b)$
37. $(3x + 7y)(4x - 3y)$
38. $(8x - 9y)(7x + y)$
39. $(x - 3)(x^2 - 2x + 1)$
40. $(x + 2)(x^2 + 5x - 6)$
41. $(2x - 1)(x^2 + 5x - 3)$
42. $(3x + 4)(x^2 - 6x + 5)$
43. $(5x + 6)(3x^2 + 2x - 6)$
44. $(6x - 7)(2x^2 + 7x - 6)$
45. $(2x - 3)(5x^2 + 7x - 9)$
46. $(5y - 2)(3y^2 - 4y + 7)$
47. $(x^2 + x - 2)(x^2 - x + 2)$
48. $(x^2 + 2x - 4)(x^2 - 3x + 4)$
49. $(x^2 + 2x - 3)(x^2 + 4x - 5)$
50. $(x^2 + 3x - 4)(x^2 + 9x - 5)$
51. $(2a^2 - 6a + 3)(3a^2 + 7a - 7)$
52. $(5a^2 + 5a - 6)(3a^2 - 7a + 8)$

9.2 PRODUCT OF TWO BINOMIALS

The product of a binomial times a binomial occurs often enough that learning how to find it quickly is worthwhile.

Let's examine the product of $(3x + 2)(4x + 5)$. Notice that in this problem the first terms from each binomial are like terms, and the second terms from each binomial are like terms.

Like terms

$(3x + 2)(4x + 5)$

Like terms

EXAMPLE 8 Multiply $(3x + 2)(4x + 5)$ using the distributive property.

$$(3x + 2)(4x + 5) = (3x + 2)4x + (3x + 2)5$$

Applying the distributive property again,

$$= 3x(4x) + 2(4x) + (3x)(5) + 2(5)$$
$$+ \underbrace{8x + 15x}$$
$$= 12x^2 + 23x + 10$$

Now let's multiply vertically and notice which factors produce each term of the result.

$$\begin{array}{r} 3x + 2 \\ 4x + 5 \\ \hline 12x^2 + 8x \quad \quad \\ + 15x + 10 \\ \hline 12x^2 + 23x + 10 \end{array}$$

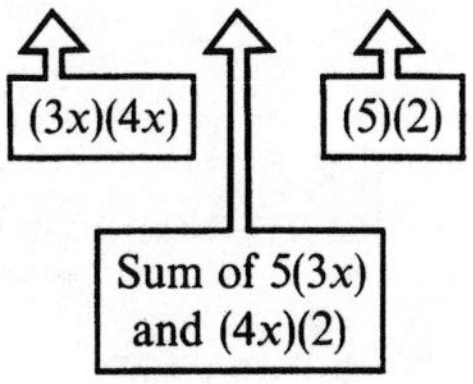

(3x)(4x) — (5)(2) — Sum of 5(3x) and (4x)(2)

Finally, let's write the result directly.

First, find the product of the *first* two terms of each binomial.

$$(3x + 2)(4x + 5) = 12x^2 +$$

Second, add the product of the *outer* factors plus the product of the two *inner* factors.

outer

inner

$$(3x + 2)(4x + 5) = 12x^2 + 23x +$$

$$\begin{array}{r} 8x \\ + 15x \\ \hline 23x \end{array}$$

Third, add the product of the *last* terms in each binomial.

$$(3x + 2)(4x + 5) = 12x^2 + 23x + 10$$

A device many people use to remember how to multiply two binomials is FOIL.

FOIL

F—Product of the First terms	$(3x \quad)(4x \quad)$	⇨	$12x^2$
O—Product of the Outside terms	$(3x \quad)(\quad 5)$	⇨	$+ 15x$
I—Product of the Inside terms	$(\quad 2)(4x \quad)$	⇨	$+ 8x$
L—Product of the Last terms	$(\quad 2)(\quad 5)$	⇨	$+ 10$
	$(3x + 2)(4x + 5)$	=	$12x^2 + 23x + 10$

EXAMPLE 9 Multiply $(x-3)(x-2)$ using the FOIL method.

First	$(x\quad)(x\quad)$	⇨	x^2
Outside	$(x\quad)(\;-2)$	⇨	$-2x$
Inside	$(\;-3)(x\quad)$	⇨	$-3x$
Last	$(\;-3)(\;-2)$	⇨	$+6$
	$(x-3)(x-2)$	$=$	x^2-5x+6

EXAMPLE 10 Multiply $(4x-2)(3x+1)$.

$12x^2$ The product of the first terms is ______.

$4x$ The product of the outside terms is ______.

$-6x$ The product of the inside terms is ______.

$-2x$ The sum of the outside product plus the inside product is ______.

-2 The product of the last terms is ______.

$12x^2-2x-2$ The product $(4x-2)(3x+1)$ is ______________.

EXAMPLE 11 Multiply $(3x+y)(4x-y)$ directly using FOIL.

$12x^2$, xy, $(-y^2)$ $(3x+y)(4x-y) =$ ________ + ________ + ________

Think $(3x)(4x)$ | Think $3x(-y) + 4x(y)$ | Think $y(-y)$

PROBLEM SET 9.2A

Use the FOIL method to multiply the following binomials.

1. $(x+2)(x-3)$
2. $(x+4)(x-6)$
3. $(a-2)(a-3)$
4. $(a-1)(a-5)$
5. $(y+2)(y+4)$
6. $(x+3)(x+4)$
7. $(x+3)(x-4)$
8. $(x-3)(x+4)$
9. $(x-3)(x-4)$
10. $(x-3)(2x-1)$
11. $(3b+1)(b-4)$
12. $(4a-1)(a-4)$
13. $(2y+3)(3y+1)$
14. $(2x-1)(4x-3)$
15. $(8x-7)(2x+1)$
16. $(6z+7)(z-2)$
17. $(3x+4)(6x-5)$
18. $(7a-2)(5a+3)$
19. $(7a+3)(4a-5)$
20. $(5x-2)(6x-7)$
21. $(8t+9)(5t+4)$
22. $(7x-3)(4x+7)$
23. $(x-y)(2x+y)$
24. $(x+y)(3x-y)$
25. $(2x+y)(4x+y)$
26. $(4a-b)(5a+b)$
27. $(2x-3y)(3x-8y)$
28. $(4x+3y)(5x-4y)$
29. $(5x-6y)(3x+4y)$
30. $(6r+7s)(3r-5s)$
31. $(2a+3b)(3a+4b)$
32. $(2a-3b)(3a+4b)$
33. $(2a+3b)(3a-4b)$
34. $(2a-3b)(3a-4b)$
35. $(4s-3t)(4s-3t)$
36. $(5s+4t)(5s+4t)$

Two Special Products

Square of a binomial

To find the *square of a binomial*, we can use the FOIL method; however, let's study the product using the vertical method.

EXAMPLE 12 Find the square $(a + b)^2$.

$$(a + b)^2 = (a + b)(a + b) = \begin{array}{r} a + b \\ a + b \\ \hline a^2 + ab \quad \\ + ab + b^2 \\ \hline a^2 + 2ab + b^2 \end{array}$$

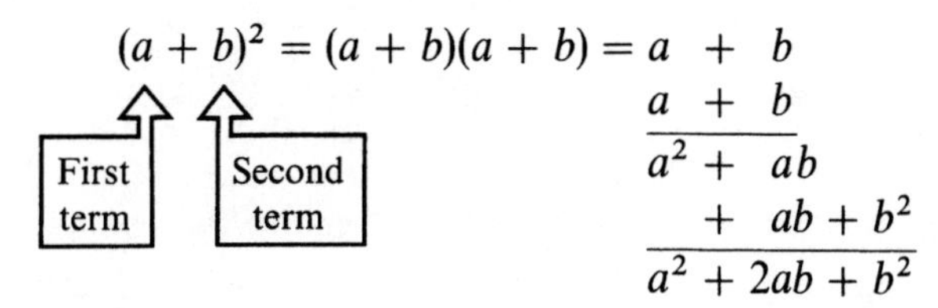

Notice that the result $(a + b)^2 = a^2 + 2ab + b^2$

is

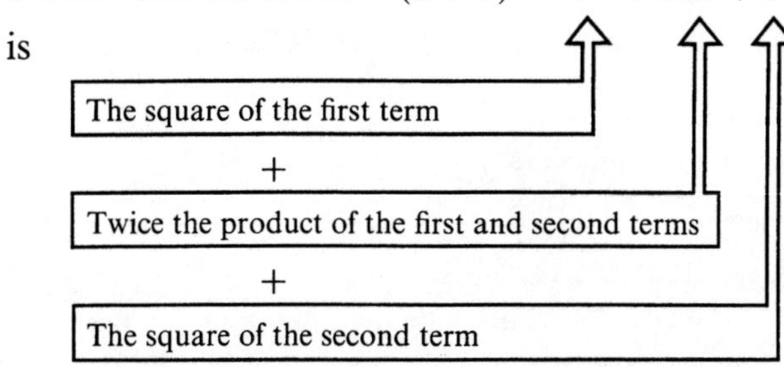

EXAMPLE 13 Find the square $(3x - y)^2$.

$$(3x - y)^2 = 9x^2 - 6xy + y^2$$

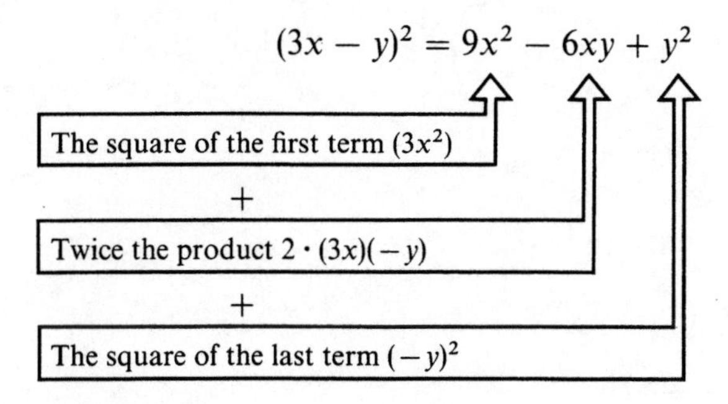

EXAMPLE 14 Find the square $(5x - 1)^2$.

The square of the first term is ______. ($25x^2$)

The product of the two terms is ______. ($-5x$)

Twice the product of the terms is ______. ($-10x$)

The square of the second term is ______. ($+1$)

The product $(5x - 1)^2 =$ ______________. ($25x^2 - 10x + 1$)

Notice that squaring a binomial *always* gives us three terms.

The middle term of the result is twice the product of the first and second terms of the binomial.

Product of a sum and difference

A binomial product like $(a + b)(a - b)$ is called the *product of a sum and difference* of two terms because in the first binomial the terms are added and in the second binomial the same terms are subtracted. These are called *conjugate* factors.

EXAMPLE 15 Use the vertical method to multiply $(a + b)(a - b)$.

$$\begin{array}{rrr} a & + b & \\ a & - b & \\ \hline a^2 & + ab & \\ & - ab & - b^2 \\ \hline a^2 & - & b^2 \end{array}$$

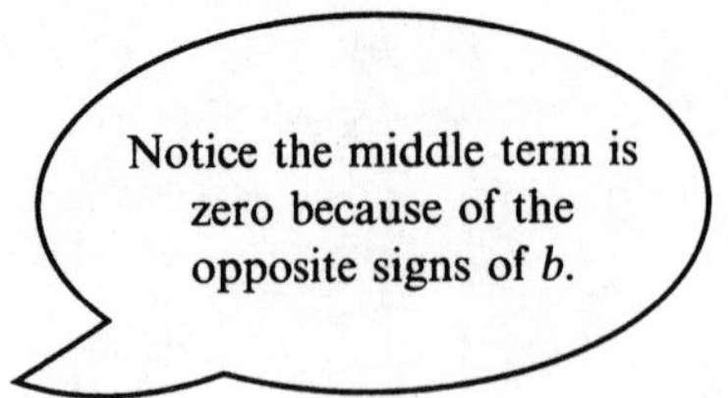

The product of the sum and difference of two terms

$$(a + b)(a - b) = a^2 - b^2$$

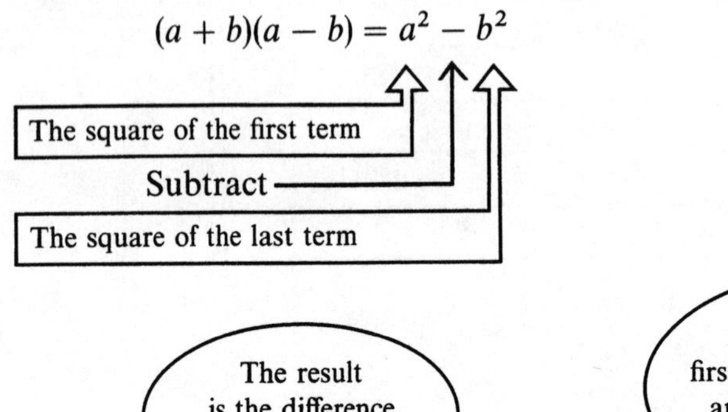

The result is the difference of two squares.

That is because the first terms of the binomials are the same and so are the last terms.

EXAMPLE 16 Multiply $(3x - 2y)(3x + 2y)$.

$$(3x - 2y)(3x + 2y) = 9x^2 - 4y^2$$

Square of the first term $(3x)^2$

Subtract

Square of the second term $(2y)^2$

EXAMPLE 17 Multiply $(1 - a)(1 + a)$.

The first term is ______. — 1

The square of the first term is ______. — 1

The second term is ______. — a

The square of the second term is ______. — a^2

The product $(1 - a)(1 + a) =$ ______. — $1 - a^2$

EXAMPLE 18 Multiply $(3p^2 - 5)(3p^2 + 5)$.

$3p^2$	The first term is ______.
$9p^4$	The square of the first term is ______.
5	The second term is ______.
25	The square of the second term is ______.
$9p^4 - 25$	The difference of the squares is ________.
$9p^4 - 25$	The product $(3p^2 - 5)(3p^2 + 5) =$ ________.

PROBLEM SET 9.2B

Find the following products using the short method.

1. $(a + 1)^2$
2. $(a - 2)^2$
3. $(x + 3)^2$
4. $(x - 3)^2$
5. $(x + 3)(x - 3)$
6. $(2a + 1)^2$
7. $(2a - 1)^2$
8. $(2a - 1)(2a + 1)$
9. $(3x + 2)(3x - 2)$
10. $(4x + 3)^2$
11. $(5y + 4)(5y - 4)$
12. $(7a - 1)(7a + 1)$
13. $(6a + 5)(6a - 5)$
14. $(x + y)^2$
15. $(x - y)^2$
16. $(x - y)(x + y)$
17. $(a - 2b)^2$
18. $(2a - b)^2$
19. $(3r - s)(3r + s)$
20. $(3x - 2y)^2$
21. $(3x + 2y)^2$
22. $(3x + 2y)(3x - 2y)$
23. $(x - 7y)^2$
24. $(6x + y)^2$
25. $(a - 2b)(a + 2b)$
26. $(2x - y)(2x + y)$
27. $(4x - 5y)(4x + 5y)$
28. $(3a + 4b)(3a - 4b)$
29. $(x - 3y)(x + 3y)$
30. $(2x + 5y)(2x - 5y)$
31. $(4x + y)^2$
32. $(x + 7y)^2$
33. $(x^2 - 3)^2$
34. $(3x^2 - y)^2$
35. $(3x^2 + 2)(3x^2 - 2)$
36. $(4x^2 + 3y)^2$
37. $(4x^2 - 3y)(4x^2 + 3y)$
38. $(3x^3 - 4y)^2$
39. $(5x + 3y^3)^2$

Review Your Skills

Use the distributive property to change a product to a sum or difference.

40. $\frac{1}{2}(2x^2 + 3x - 4)$
41. $(3x^3 + 6x^2 - 9x) \cdot \frac{1}{3x}$
42. $(4b^4 - 5b^3 - 6b) \cdot \frac{1}{2b}$

9.3 DIVISION OF A POLYNOMIAL BY A MONOMIAL

To divide a polynomial by a monomial recall the definition of division.

Division

> **9.3 DEFINITION: DIVISION**
>
> $a \div b = a \cdot \frac{1}{b}$ where $b \neq 0$

EXAMPLE 19 Divide $\frac{4x^2 + 8x}{2x}$.

$$\frac{4x^2 + 8x}{2x} = (4x^2 + 8x) \div 2x$$

$$= (4x^2 + 8x) \cdot \frac{1}{2x} \quad \text{Definition of division}$$

$$= \frac{4x^2}{2x} + \frac{8x}{2x} \quad \text{Distributive property}$$

$$= 2x + 4$$

Are the results the same if you divide each term of the polynomial by $2x$ and add the results?

Yes, but be careful that you divide *every* term by the divisor $2x$.

EXAMPLE 20 Divide $\frac{3a^3 + 18a^2 - 12a}{3a}$.

$$\frac{3a^3 + 18a^2 - 12a}{3a} = \frac{1}{3a} \cdot (3a^3 + 18a^2 - 12a)$$

$$= \frac{3a^3}{3a} + \frac{18a^2}{3a} - \frac{12a}{3a}$$

$$= \underline{\qquad\qquad}$$

$a^2 + 6a - 4$

EXAMPLE 21 Divide $\frac{5b^3 - b^2 + 5}{5b}$.

$$\frac{5b^3 - b^2 + 5}{5b} = \underline{\qquad} \cdot (5b^3 - b^2 + 5)$$

$\frac{1}{5b}$

$$= \frac{5b^3}{5b} - \frac{b^2}{5b} + \frac{5}{5b}$$

$$= b^2 - \frac{b}{5} + \frac{1}{b}$$

When dividing, we usually write the results with positive exponents.

EXAMPLE 22 Divide $\frac{8a^3b^3 - 12a^2b^2 + 2ab}{2ab^2}$.

$$\frac{8a^3b^3 - 12a^2b^2 + 2ab}{2ab^2} = \frac{8a^3b^3}{2ab^2} - \frac{12a^2b^2}{2ab^2} + \frac{2ab}{2ab^2}$$

$$= \underline{\qquad\qquad}$$

$4a^2b - 6a + \frac{1}{b}$

PROBLEM SET 9.3

Divide. Write your answers with positive exponents.

1. $\frac{3x + 3}{3}$
2. $\frac{4x + 8y}{4}$
3. $\frac{12a + 16b}{4}$
4. $\frac{15a + 24b}{3}$
5. $\frac{5a^2 + 10a}{a}$
6. $\frac{5a^2 + 10a}{5a}$
7. $\frac{6x^3 - 9x^2}{3x}$
8. $\frac{12b^3 - 2b^2 + 6b}{2b}$
9. $\frac{4x^3 + 12x^2}{x^2}$
10. $\frac{4x^3 + 12x^2}{4x^2}$
11. $\frac{4x^3 + 12x^2}{4x}$
12. $\frac{12a^2 - 6a + 6}{6a}$
13. $\frac{16x^2 - 10x + 4}{2x}$
14. $\frac{15y^3 + 25y^2 - 40y}{5y^2}$
15. $\frac{15x^2 - 3x}{3x}$
16. $\frac{15x^2y - 3xy}{3x}$
17. $\frac{15x^2y - 3xy}{3xy}$
18. $\frac{18a^2b^2 + 6a^2b}{2ab}$
19. $\frac{18a^2b^2 - 12a^2b}{6ab^2}$
20. $\frac{15a^2b^3 - 12ab^2}{-3ab^2}$
21. $\frac{-18a^3b^4 + 27a^2b^2}{-3a^2b}$
22. $\frac{2x^4y^2 + 3x^3y + x^2}{x^2y}$
23. $\frac{6a^3b^2 - 8a^2b^3 + 4ab^4}{2ab^2}$
24. $\frac{7a^4b^2 - 9a^3b^3 - 8a^2b^4}{a^2b^2}$
25. $\frac{16a^3 - 12a^4 + 24a^5}{4a^3}$
26. $\frac{-6a^3b - 9a^3b + 18ab}{-3ab}$
27. $\frac{9a^3b - 12ab^2 + 15a^2b}{3a^2b}$
28. $\frac{a^2b^3 - 5ab^2 - 10a^3b}{5ab^3}$
29. $\frac{a^2b^4 - 3ab^2 + 6a^3b^3}{-ab^2}$
30. $\frac{-8a^4b^3 + 12a^2b^2 - 16a^3b^4}{-4a^2b^2}$
31. $\frac{x^4y - 7x^2 + 4x^3y^3}{x^2y}$
32. $\frac{3x^3y - 6x^4a + 9x^2y^2}{3x^2y^2}$
33. $\frac{4x^2y^3 - 12x^3y^2 + 16xy^4}{-4x^2y^2}$
34. $\frac{-6ay^2 + 36a^2y^3 - 42a^3y}{-6ay^2}$
35. $\frac{-7x^4 + 14x^5y^3 - 21xy^4}{7xy^3}$
36. $\frac{-5ab^4 + 15ab^2 - b^4}{-5a^2b^2}$
37. $\frac{-30ab - 15b^3 + 12a^4}{-3ab}$
38. $\frac{-6x^2y^2 + 9x^3y^3 - 12xy^4}{-6x^2y^3}$
39. $\frac{-18a^2b + 9a^4b^2 - 21a^3b^3}{-9a^2b^2}$

9.4 LONG DIVISION OF POLYNOMIALS

Degree

The *degree* of a polynomial is the highest power used in any of its variables.

$5x^3 - 1$ is a 3rd degree polynomial

x^4 is a 4th degree polynomial

$x^2 + 2x + 1$ is a 2nd degree polynomial

When we write a polynomial, we usually order it so that the variable is written in decreasing powers.

$2x - x^3 + 14 + 7x^2$ is usually written $-x^3 + 7x^2 + 2x + 14$

$x^2 + x^4 + 2$ is usually written $x^4 + x^2 + 2$

To divide polynomials, it is necessary to write them not only in order of decreasing powers but to supply zero coefficients for any missing powers.

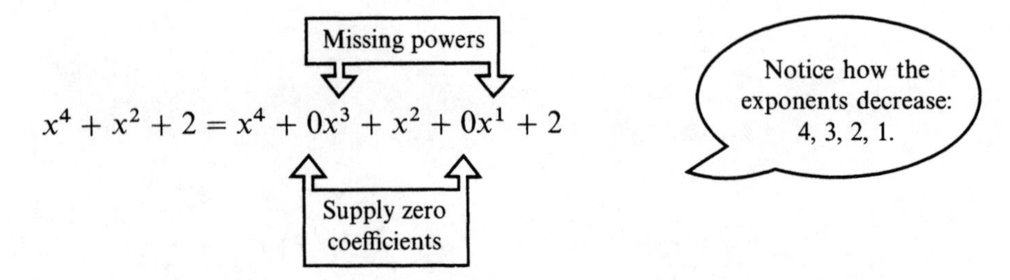

The process for long division of polynomials is very similar to long division in arithmetic. It works because division can be thought of as repeated subtraction.

EXAMPLE 23 $(2x^2 + 10x + 12) \div (x + 3)$

To divide $(2x^2 + 10x + 12) \div (x + 3) =$?

Dividend Divisor Quotient

set it up as a long division problem

$$\begin{array}{r} ? \\ x + 3 \overline{) 2x^2 + 10x + 12} \end{array}$$

Divisor → ; ← Quotient ; ← Dividend

2x

Step 1 Ask "What must I multiply x by to get $2x^2$?" or $2x^2 \div x =$ ______.

$$x + 3 \overline{) 2x^2 + 10x + 12}$$

$2x^2 \div x = ?$

Step 2 Write the $2x$ above the $10x$.

$$\begin{array}{r} 2x \\ x + 3 \overline{) 2x^2 + 10x + 12} \end{array}$$

Step 3 Multiply this $2x$ by $x + 3$.

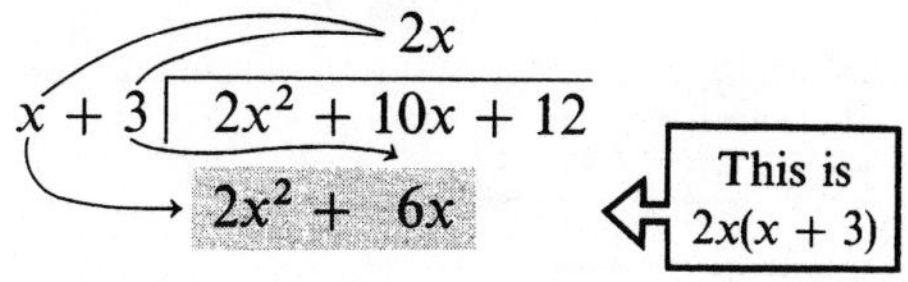

$$\begin{array}{r} 2x \\ x+3 \overline{) 2x^2 + 10x + 12} \\ 2x^2 + 6x \end{array}$$

This is $2x(x + 3)$

Step 4 Subtract $2x^2 + 6x$ from $2x^2 + 10x + 12$.

$$\begin{array}{r} 2x \\ x+3 \overline{) 2x^2 + 10x + 12} \\ \underline{2x^2 + 6x } \\ 4x + 12 \end{array}$$

4

Step 5 Ask "What must I multiply x by to get $4x$?" or $4x \div x =$ ______.

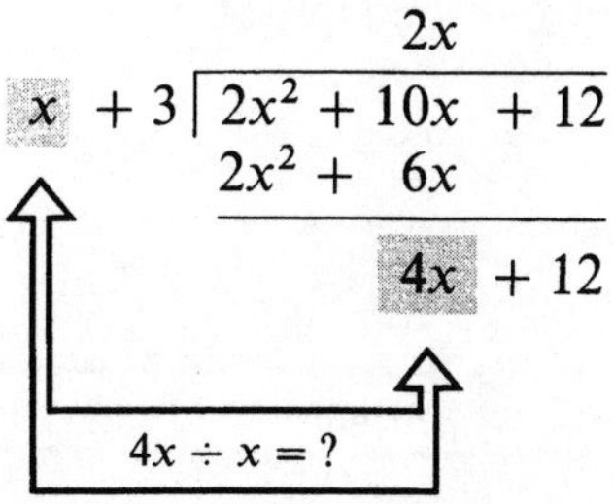

$$\begin{array}{r} 2x \\ x+3 \overline{) 2x^2 + 10x + 12} \\ \underline{2x^2 + 6x } \\ 4x + 12 \end{array}$$

$4x \div x = ?$

Step 6 Write the 4 as the next term in the answer.

$$\begin{array}{r} 2x + 4 \\ x+3 \overline{) 2x^2 + 10x + 12} \\ \underline{2x^2 + 6x } \\ 4x + 12 \end{array}$$

Step 7 Multiply this 4 by $x + 3$.

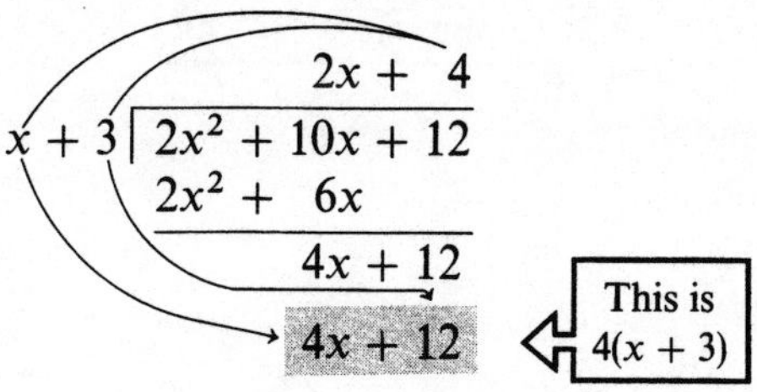

$$\begin{array}{r} 2x + 4 \\ x+3 \overline{) 2x^2 + 10x + 12} \\ \underline{2x^2 + 6x } \\ 4x + 12 \\ 4x + 12 \end{array}$$

This is $4(x + 3)$

Step 8 Subtract $4x + 12$ from $4x + 12$.

$$\begin{array}{r} 2x + 4 \\ x+3 \overline{) 2x^2 + 10x + 12} \\ \underline{2x^2 + 6x } \\ 4x + 12 \\ \underline{4x + 12} \\ 0 \end{array}$$

Our answer is $(2x^2 + 10x + 12) \div (x + 3) = 2x + 4$.

EXAMPLE 24 $(3n^2 - 14n + 8) \div (3n - 2)$

Part I

$$\begin{array}{r|l} & n \\ \hline 3n - 2 & 3n^2 - 14n + 8 \\ & 3n^2 - 2n \\ \hline & \underline{\qquad\quad} \end{array}$$

$3n^2 \div 3n = n$

$n(3n - 2)$

Subtract $3n^2 - 2n$ from $3n^2 - 14n + 8$

$-12n + 8$

Subtraction is defined as addition of the additive inverse. To avoid errors, it is better to *find the additive inverse and then add.*

The additive inverse of $3n^2 - 2n = -(3n^2 - 2n)$

$= -3n^2 + 2n$

That is, to subtract a polynomial, change the sign of every term and add.

$$\begin{array}{r|l} & \qquad\qquad n \\ \hline 3n - 2 & 3n^2 - 14n + 8 \\ & \overset{-}{}3n^2 \overset{+}{-} 2n \\ \hline & \underline{\qquad\quad} \end{array}$$

Change the signs

Add $3n^2 - 14n + 8$ and $-3n^2 + 2n$

$-12n + 8$

Part II

$$\begin{array}{r|l} & \qquad\qquad n - 4 \\ \hline 3n - 2 & 3n^2 - 14n + 8 \\ & \overset{-}{}3n^2 \overset{+}{-} 2n \\ \hline & \qquad - 12n + 8 \\ & \qquad \overset{+}{-} 12n \overset{-}{+} 8 \\ \hline & \qquad\quad \underline{\qquad} \end{array}$$

-4 is the result of $-12n \div 3n$

$-4(3n - 2) = -12n + 8$

Change signs and add

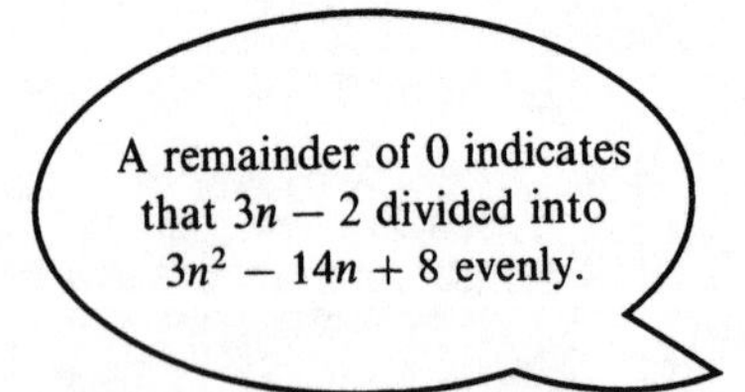

Therefore,

$$(3n^2 - 14n + 8) \div (3n - 2) = n - 4$$

EXAMPLE 25 Divide $(x^3 + 1) \div (x + 1)$.

In this case the x^2 and x terms are missing in the dividend $x^3 + 1$. Therefore, supply them and make their coefficients zero.

The complete dividend is $x^3 + 0x^2 + 0x + 1$.

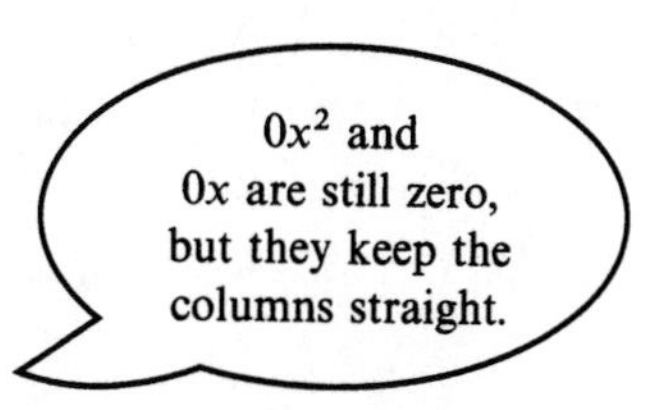

Part I

$$
\begin{array}{r|l l}
 & \quad x^2 & \Leftarrow \boxed{\text{Result of } x^3 \div x} \\
\cline{2-2}
x+1 & x^3 + 0x^2 + 0x + 1 & \\
 & \overset{-}{x^3} \overset{-}{+} x^2 & \text{Change signs} \\
\cline{2-2}
 & \quad - x^2 + 0x + 1 & \text{Add}
\end{array}
$$

Part II

$$
\begin{array}{r|l l}
 & \quad x^2 - x & \Leftarrow \boxed{\text{Result of } -x^2 \div x} \\
\cline{2-2}
x+1 & x^3 + 0x^2 + 0x + 1 & \\
 & \overset{-}{x^3} \overset{-}{+} x^2 & \\
\cline{2-2}
 & \quad - x^2 + 0x + 1 & \\
 & \quad \overset{+}{-} x^2 \overset{+}{-} x & \Leftarrow \boxed{-x(x+1)} \\
\cline{2-2}
 & \qquad\qquad x + 1 & \text{Change signs and add}
\end{array}
$$

Part III

$$
\begin{array}{r|l l}
 & \quad x^2 - x + 1 & \Leftarrow \boxed{\text{Result of } x \div x} \\
\cline{2-2}
x+1 & x^3 + 0x^2 + 0x + 1 & \\
 & \overset{-}{x^3} \overset{+}{+} x^2 & \\
\cline{2-2}
 & \quad - x^2 + 0x + 1 & \\
 & \quad \overset{+}{-} x^2 \overset{+}{-} x & \\
\cline{2-2}
 & \qquad\qquad x + 1 & \\
 & \qquad\qquad \overset{-}{x} \overset{-}{+} 1 & \Leftarrow \boxed{1(x+1)} \\
\cline{2-2}
 & \qquad\qquad\quad 0 & \text{Change signs and add}
\end{array}
$$

Our answer is $x^2 - x + 1$.

In the preceding examples, the remainders were zero. However, in many cases the remainder is not zero.

EXAMPLE 26 Divide $(7x + x^2 + 15) \div (x + 3)$.

$x^2 + 7x + 15$

First arrange the dividend in descending powers ______________.

$$
\begin{array}{r}
x + 4 \\
x + 3 \overline{\big)\; x^2 + 7x + 15} \\
\underline{-\; x^2 \mp 3x } \\
4x + 15 \\
\underline{-\; 4x \mp 12} \\
3
\end{array}
$$

3 ⇦ Remainder

In arithmetic, the remainder was frequently written as a fractional part of the divisor. The same procedure is used in algebra.

The answer is written $x + 4 + \dfrac{3}{x + 3}$.

EXAMPLE 27 Divide $(4x^3 - 6x^2 + 2x - 5) \div (2x + 1)$.

$$
\begin{array}{r}
2x^2 - 4x + 3 \\
2x + 1 \overline{\big)\; 4x^3 - 6x^2 + 2x - 5} \\
\underline{-\; 4x^3 \mp 2x^2 } \\
-8x^2 + 2x - 5 \\
\underline{\pm 8x^2 \pm 4x } \\
6x - 5 \\
\underline{-\; 6x \mp 3} \\
-8
\end{array}
$$

−8 ⇦ Remainder

The answer is written $2x^2 - 4x + 3 + \dfrac{-8}{2x + 1}$.

EXAMPLE 28 Divide $(9x^3 - 10x - 1) \div (3x - 2)$.

$9x^3 + 0x^2 - 10x - 1$

The complete dividend is ______________.

$$
\begin{array}{r}
3x^2 + 2x - 2 \\
3x - 2 \overline{\big)\; 9x^3 + 0x^2 - 10x - 1} \\
\underline{-\; 9x^3 \pm 6x^2 } \\
6x^2 - 10x - 1 \\
\underline{-\; 6x^2 \pm 4x } \\
-6x - 1 \\
\underline{\pm\; 6x \mp 4} \\
\underline{}
\end{array}
$$

⇦ Remainder

-5

$\dfrac{-5}{3x - 2}$

The complete answer is $3x^2 + 2x - 2 +$ ______________.

PROBLEM SET 9.4

Divide the following.

1. $(x^2 - 5x - 6) \div (x + 1)$
2. $(x^2 - 3x - 4) \div (x - 4)$
3. $(x^2 + 7x + 10) \div (x + 2)$
4. $(x^2 - 7x + 12) \div (x - 3)$
5. $(x^2 + 4x - 12) \div (x + 6)$
6. $(x^2 - 3x - 28) \div (x - 7)$
7. $(2x^2 - x - 6) \div (x - 2)$
8. $(3x^2 + 5x - 12) \div (x + 3)$
9. $(2x^2 - 3x - 20) \div (x - 4)$
10. $(4x^2 - 17x - 42) \div (x - 6)$
11. $(3x^2 + 9x - 30) \div (x + 5)$
12. $(5x^2 - 8x - 21) \div (x - 3)$
13. $(6x^2 + 23x + 25) \div (2x + 5)$
14. $(15x^2 + 2x - 18) \div (3x + 4)$
15. $(42x^2 + 17x - 19) \div (7x - 3)$
16. $(12x^2 - 13x - 41) \div (3x - 7)$
17. $(x^3 - 3x^2 + 5x - 3) \div (x - 1)$
18. $(x^3 + 6x^2 + 5x - 12) \div (x + 3)$
19. $(x^3 + 6x^2 + 3x - 10) \div (x + 2)$
20. $(x^3 - 3x^2 + 8x - 12) \div (x - 2)$
21. $(2x^3 + 9x^2 - x - 15) \div (2x + 3)$
22. $(3x^3 + 2x^2 - 23x + 20) \div (3x - 4)$
23. $(x^3 + 2x^2 + 32) \div (x + 4)$
24. $(x^3 - 12x - 16) \div (x - 4)$
25. $(-8x + 8x^3 + 12) \div (2x + 3)$
26. $(7x^2 + 6x^3 + 4) \div (3x - 1)$
27. $(10x^3 - 4x^2 + 5x) \div (5x - 2)$
28. $(12x^3 - 10x^2 + 5) \div (4x + 2)$
29. $(6x^4 - 4x^3 + 9x^2 - 4) \div (3x - 2)$
30. $(10x^4 + 9x^3 - 9x^2 - 5x) \div (5x - 3)$

Review Your Skills

Use the distributive property to find the following products.

31. $4(x + 2)$
32. $2x(x - 3)$
33. $(x + 2)(x - 4)$
34. $(2x-1)(3x-4)$

9 Review

NOW THAT YOU HAVE COMPLETED UNIT 9, YOU SHOULD BE ABLE TO APPLY THE FOLLOWING RULES:

To find the product of two polynomials, use the distributive property to multiply each term of one polynomial by each term of the other, and combine like terms.

To find the product of two binomials:

1. Find the product of the first terms of each binomial.
2. Add the product of the two outer terms plus the product of the two inner terms.
3. Add the product of the last terms of each binomial.

To find the square of a binomial:

1. Square the first term.
2. Add twice the product of the first and second terms.
3. Add the square of the second term.

$$(a + b)^2 = a^2 + 2ab + b^2$$

To find the product of a sum and difference of two terms, square the first term and subtract the square of the last term.

$$(a + b)(a - b) = a^2 - b^2$$

To divide a polynomial by a monomial, divide every term of the polynomial by the monomial using the laws of exponents.

To divide polynomials using long division:

1. Write them in order of decreasing powers.
2. Supply zero coefficients for any missing powers in the dividend.
3. Divide as you do in arithmetic.

Review Test 9

Find the following products. (9.1 and 9.2)

1. $4a(7a - 3) =$ ______
2. $(x + 5)(2x - 1) =$ ______
3. $(3x + 4)(5x + 3) =$ ______
4. $(a - 6)(a + 6) =$ ______
5. $(3x - 2)^2 =$ ______
6. $3x^2(x^2 + 4x + 7) =$ ______
7. $-2x^2(3x^2 - 4x - 5) =$ ______
8. $(3x - 2y)(3x + 2y) =$ ______
9. $(x + 3y)(x + 5y) =$ ______
10. $(a - 2b)(3a + b) =$ ______
11. $(5a - 4b)(5a + 4b) =$ ______
12. $(5a + 6b)^2 =$ ______
13. $(2x - 5y)(3x - 4y) =$ ______
14. $6ab^2(2a^2 - ab + b^2) =$ ______
15. $(5x^2 - 3x + 4)(-3x^2) =$ ______
16. $(4x - 2)(x^2 - 2x + 5) =$ ______
17. $(2x^2 + 3x - 4)(3x - 5) =$ ______
18. $(a^2 - 3a + 5)(3a^2 + 5a - 7) =$ ______

Find the following quotients. (9.3)

19. $\dfrac{10x^3 - 6x^2 + 2x}{2x} =$ ______
20. $\dfrac{16a^3b - 24a^2b^5}{8a^2} =$ ______
21. $\dfrac{6a^4b^2 - 12ab^3 - 18a^5b^3}{6a^2b^2} =$ ______

Use long division to find the following quotients. (9.4)

22. $(2x^2 + 7x + 3) \div (x + 3) =$ ______
23. $(6x^3 + 10x^2 - 19x + 3) \div (3x - 1) =$ ______
24. $(8x^3 - 1) \div (2x - 1) =$ ______
25. $(6x^3 - 5x^2 - 14x) \div (2x - 3) =$ ______

REVIEW YOUR SKILLS

26. Find the equation of the line with slope $\frac{-4}{3}$ and y-intercept -3.

27. Use the slope and y-intercept method to graph the line: $3x + 2y = 8$.

28. Use the addition method to find the simultaneous solution for the equations

$$2x - 3y = 3$$
$$-4x + 6y = -6$$

29. Use the substitution method to find the simultaneous solution for the equations

$$x + 4y = 5$$
$$-2x + 6y = -3$$

30. A retiree has 3 times as much money invested at 10% as at 8%. If her annual income from both investments is $4,560, how much is invested at each rate?

UNIT 10

FACTORING

OBJECTIVES

After you have successfully completed this unit, you will be able to:

1. Factor monomial factors from polynomials (10.1)
2. Factor trinomials into the product of two binomials (10.2 and 10.3)
3. Factor the difference of two squares (10.4)
4. Factor any polynomial completely (10.4)

10.1 MONOMIAL FACTORS

Factors

Algebraic expressions that are multiplied together are called *factors.*

$$3 \cdot 7 = 21$$

Factor — Factor — Product

Factoring

The process of finding which factors were multiplied to give a product is called *factoring.*

The first type of factoring we will do is an application of the distributive property. The *distributive property* changes a product into a sum.

Product — Sum

Distributive property

$$a \cdot (b + c) = a \cdot b + a \cdot c$$

In factoring, we change a sum into a product.

Sum — Product

$$a \cdot b + a \cdot c = a \cdot (b + c)$$

To factor a polynomial, first remove any monomial factors that are common to every term.

EXAMPLE 1 Factor $ax + ay$.

$$ax + ay = a(x + y)$$

a is common to both terms

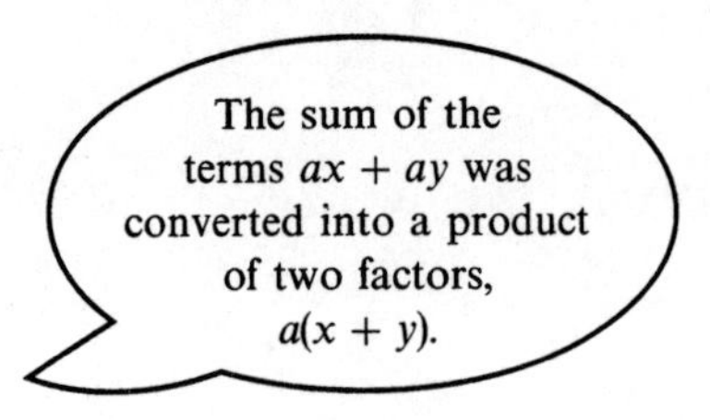

EXAMPLE 2 Factor $3x + 6$.

To work this problem, we need to see 6 as $3 \cdot 2$.

$$\begin{aligned} 3x + 6 &= 3x + 3 \cdot 2 \\ &= 3(x + 2) \end{aligned}$$

Frequently, an expression has more than one common factor in each term. In that case, we remove all common factors.

EXAMPLE 3 Factor $4x^2 + 8x$.

$$4x^2 + 8x = 4x(x + 2)$$

Both 4 and x are factors of each term.

EXAMPLE 4 Factor $x^2 + xy$.

$x + y$

$$x^2 + xy = x(\underline{\qquad\qquad})$$

EXAMPLE 5 Factor $5a^3 - 10a^2 + 5a$.

Each term contains a factor of $5a$. First write the common factor followed by an empty parentheses.

$$5a^3 - 10a^2 + 5a = 5a(\qquad)$$

To find what goes in the parentheses first ask, "What do I multiply $5a$ by to get $5a^3$?" Or, you can divide $5a^3$ by $5a$ to get the first term in the parentheses.

a^2

$$5a^3 - 10a^2 + 5a = 5a(\underline{\qquad}\qquad)$$

To find the second term divide $-10a^2$ by $5a$. This tells you what you need to multiply $5a$ by to get $-10a^2$.

$-2a$

$$5a^3 - 10a^2 + 5a = 5a(a^2\underline{\qquad}\qquad)$$

And finally, divide $5a$ by $5a$ to find the third term.

$+1$

$$5a^3 - 10a^2 + 5a = 5a(a^2 - 2a\underline{\qquad})$$

Check your work by multiplying the factors to obtain the original expression.

$$\begin{aligned} 5a(a^2 - 2a + 1) &= 5a \cdot a^2 - 5a \cdot 2a + 5a \cdot 1 \\ &= 5a^3 - 10a^2 + 5a \end{aligned}$$

EXAMPLE 6 Factor $40x^3y^2 + 20x^2y - 30x^3y^3$.

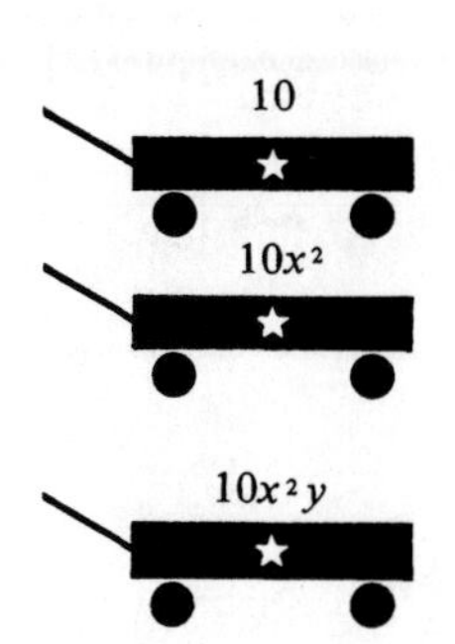

1. Look at the numerical coefficients and write the largest number that is a factor of all the coefficients. In this case, 10 is a factor of every term.
2. Look at the first variable in the first term. If it is contained in every other term, include it with your common factor. Raise it to the highest power used in *every* term. In this case x^2 is in every term.
3. Repeat Step 2 for all other variables. Only use variables that appear in every term.
4. Now that you have the common monomial, write it as a product followed by empty parentheses.

$10x^2y$ ( )

5. Fill in the parentheses with the factors that will produce the original expression if you multiply them.

$4xy, 2, 3xy^2$

$10x^2y(\_\_\_\_ + \_\_\_\_ - \_\_\_\_)$

PROBLEM SET 10.1

Use the distributive property to factor the following.

1. $2x + 4$
2. $3a + 3$
3. $4x - 10$
4. $5x - 15$
5. $6a - 10$
6. $8x - 12$
7. $5x^2 + 6x$
8. $4a^2 + 3a$
9. $4x^2 - 4x$
10. $3x^2 - 9x$
11. $a^3 - 4a^2$
12. $x^4 + 8x^2$
13. $y^3 - 2y^2 - 5y$
14. $a^3 + 3a^2 + a$
15. $8x^3 - 12x$
16. $15x^4 + 3x^3$
17. $4x^3 - 2x^2 + 10x$
18. $6a^4 + 9a^3 - 15a^2$
19. $7y^4 - 14y^2 - 21y$
20. $-5y^5 - 10y^3 + 15y$
21. $bx^2 + 3bx - 5b$
22. $ay^3 + 3ay^2 + 8ay$
23. $2ay^4 + 6ay^3 - 10ay^2$
24. $6ax^5 - 9ax^3 + 12ax$
25. $12a^3y^4 + 16a^2y^3 - 20a^2y^2$
26. $14b^4x^5 - 35b^3x^3 + 42b^2x$
27. $20a^5b^6 + 25a^4b^2 - 30a^3$
28. $-18a^4x^6 + 24a^3x^4 - 36ax^3$

Review Your Skills

Perform the indicated operations.

29. $-5 + 4$
30. $-6 + (-10)$
31. $-9 + 7 - 6$
32. $13 - 8 + (-7)$
33. $(-4)(-6)$
34. $(5)(-8)$
35. $(-6)(7)(-2)$
36. $(-2)(-8)(-4)$

NOTE TO TEACHERS: Sections 10.2 and 10.3 teach factoring by what is sometimes called the AC or key number method. An alternate method of factoring trinomials by inspection is used in the appendix, which precedes the answers to selected problems in the text. If you wish to use the alternate method, go to page 463 now.

10.2 FACTORING TRINOMIALS

In our earlier work, we multiplied two binomials. Now let's examine how the distributive law is used to find the product of two binomials.

EXAMPLE 7 Multiply $(x + 3)(x + 2)$.

Step 1

$(x + 3)(x + 2) = (x + 3)x + (x + 3)2$ — Distribute the $x + 3$

Step 2 $= x^2 + 3x + 2x + 6$ — First distribute the x, then the 2

Step 3 $= x^2 + 5x + 6$ — Collect like terms

EXAMPLE 8 Multiply $(x - 4)(x - 5)$.

Step 1

$(x - 4)(x - 5) = (x - 4)x + (x - 4)(-5)$ — Distribute $x - 4$

Step 2 $= x^2 - 4x - 5x + 20$

Step 3 $= x^2 - 9x + 20$

The process of determining which factors were multiplied to give the trinomial is called factoring. Factoring can be accomplished by following the process of multiplication backwards. First, we will practice undoing Step 1.

Sometimes a binomial can be removed as a common factor.

EXAMPLE 9

$$x(x + 4) + 3(x + 4) = (x + 3)(x + 4)$$

Common factor

$$a(a - 2) - 4(a - 2) = (a - 4)(a - 2)$$

Common factor

EXAMPLE 10 Remove the common binomial factor.

Answer	Problem
$(x + 4)(x - 3)$	a. $x(x - 3) + 4(x - 3) =$ ______________
$(x + 6)(x - 3)$	b. $x(x - 3) + 6(x - 3) =$ ______________
$(x + 6)(x + 5)$	c. $x(x + 5) + 6(x + 5) =$ ______________
$(x - 6)(x + 5)$	d. $x(x + 5) - 6(x + 5) =$ ______________
$(2x - 3)(x + 1)$	e. $2x(x + 1) - 3(x + 1) =$ ______________
$(3x + 4)(2x - 1)$	f. $3x(2x - 1) + 4(2x - 1) =$ ______________
$(4x - 1)(x - 2)$	g. $4x(x - 2) - (x - 2) =$ ______________
$(4x + 1)(x - 2)$	h. $4x(x - 2) + (x - 2) =$ ______________

Now we will practice undoing Step 2 in Example 7.

Recall $(x + 3)(x + 2)$

Step 1 $(x + 3)x + (x + 3)2$

Step 2 $x^2 + 3x + 2x + 6$

Step 3 $x^2 + 5x + 6$

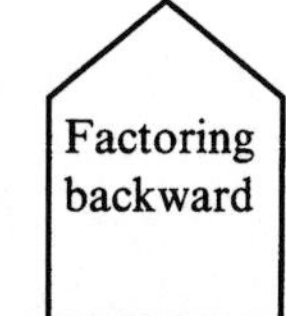

EXAMPLE 11 Factor $x^2 + 3x + 2x + 6$.

First, form two groups.

$$\underbrace{x^2 + 3x} \qquad \underbrace{+ 2x + 6}$$

Second, find a common factor in each group separately, then remove it.

$$x(x + 3) \qquad + 2(x + 3)$$

Third, remove the common binomial factor.

$$(x + 2)(x + 3)$$

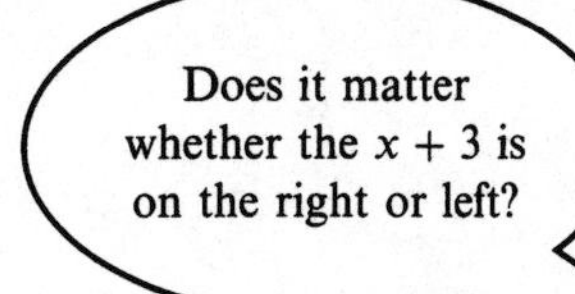

No. Because of the commutative property of multiplication, the final result could have been written $(x + 3)(x + 2)$.

EXAMPLE 12 Factor $x^2 + 3x - 2x - 6$.

First, form two groups.

$$\underbrace{x^2 + 3x} \qquad \underbrace{-2x - 6}$$

Second, find a common factor in each group separately, then remove it.

$$x\,(x + 3) \qquad -2\,(x + 3)$$

Third, remove the common binomial factor.

$$(x - 2)\,(x + 3)$$

EXAMPLE 13 Remove the binomial factor from each group of four terms by first grouping the terms in pairs and removing the common factor.

a. $x^2 - 3x + 2x - 6 = x^2 - 3x \quad +2x - 6$

$= x(x - 3) \quad +2(x - 3)$

$=$ __________

$(x + 2)(x - 3)$

b. $x^2 + 4x + 5x + 20 = x^2 + 4x \quad +5x + 20$

$= x(x + 4) \quad +5(x + 4)$

$=$ __________

$(x + 5)(x + 4)$

c. $x^2 + 4x - 3x - 12 = x^2 + 4x \quad -3x - 12$

$= x(x + 4) \quad -3(x + 4)$

$=$ __________

$(x - 3)(x + 4)$

d. $x^2 + 5x - 6x - 30 = x^2 + 5x \quad -6x - 30$

$= x(\_\_\_\_) \quad -6(\_\_\_\_)$

$x + 5, x + 5$

$=$ __________

$(x - 6)(x + 5)$

e. $x^2 - 5x - 3x + 15 = x(\_\_\_\_) \quad -3(\_\_\_\_)$

$(x - 5), (x - 5)$

$=$ __________

$(x - 3)(x - 5)$

f. $x^2 + 2x + x + 2 =$ __________

$x(x + 2) + (x + 2)$

$=$ __________

$(x + 1)(x + 2)$

g. $x^2 + 2x - x - 2 =$ __________

$x(x + 2) - (x + 2)$

$=$ __________

$(x - 1)(x + 2)$

Factoring requires the ability to select combinations of terms to yield a desired sum.

EXAMPLE 14 Find the proper combination of signs so that the desired sum is produced.

a. $\square 3x$, $\square 2x$, sum $-5x$

b. $\square 3x$, $\square 2x$, sum $-x$

c. $\square 3x$, $\square 2x$, sum $+x$

Answers: $- \; - \; +$; $- \; + \; -$

d. $\square 6x$, $\square 3x$, sum $+9x$

e. $\square 6x$, $\square 3x$, sum $-3x$

f. $\square 7x$, $\square 5x$, sum $+2x$

Answers: $+ \; - \; +$; $+ \; + \; -$

g. $\square 10x$, $\square 6x$, sum $-16x$

h. $\square 12x$, $\square 7x$, sum $+5x$

i. $\square 12x$, $\square 15x$, sum $-3x$

Answers: $- \; + \; +$; $- \; - \; -$

EXAMPLE 15 Find the missing sign so that the desired product is produced.

a. $(-3)(\square 4) = 12$ b. $(+3)(\square 4) = 12$ c. $(+3)(\square 4) = -12$

$- \; + \; -$

d. $(+4)(\square 2) = 8$ e. $(\square 4)(-2) = 8$ f. $(\square 4)(+3) = -12$

$+ \; - \; -$

EXAMPLE 16 Find two numbers whose product and sum are as follows.

	Product	Sum	Numbers
a.	12	7	3, 4
b.	−15	2	____, ____
c.	−8	−2	____, ____
d.	20	−9	____, ____

5, −3

−4, 2

−4, −5

PROBLEM SET 10.2A

Find two numbers whose product and sum are as follows.

1. Product 6, Sum 5
2. Product 6, Sum −5
3. Product −6, Sum 1
4. Product −6, Sum −1
5. Product 15, Sum −8
6. Product 10, Sum 7
7. Product −12, Sum 1
8. Product −15, Sum −2
9. Product −18, Sum −3
10. Product −36, Sum 5
11. Product −30, Sum −1
12. Product −27, Sum 6
13. Product 20, Sum −9
14. Product 18, Sum −9
15. Product −48, Sum 2
16. Product −48, Sum −8
17. Product −45, Sum −4
18. Product −28, Sum 3

Look at another example of multiplication of two binomials.

EXAMPLE 17 Multiply $(x + 5)(x + 4)$.

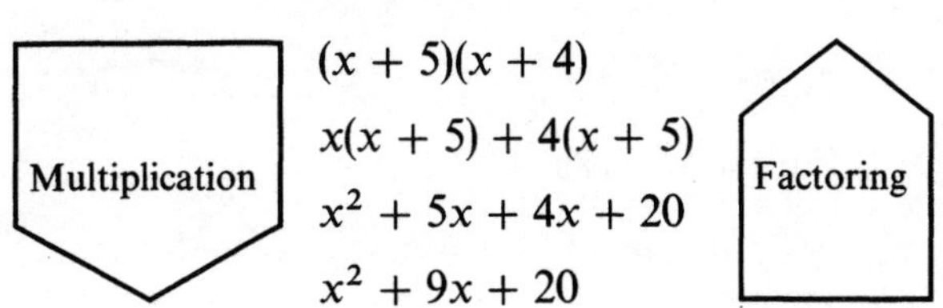

$(x + 5)(x + 4)$

$x(x + 5) + 4(x + 5)$

$x^2 + 5x + 4x + 20$

$x^2 + 9x + 20$

Now we will start with the trinomial and factor our way back to the product of two binomials.

$x^2 + 9x + 20$ ⇦ First look at the constant term.

⇧ Then note the numerical coefficient of x.

We need to find factors of the constant term whose sum is the coefficient of the x term. In this case, we need factors of 20 whose sum is $+9$.

Four and 5 are the right factors.

$4 \cdot 5 = 20$ and $4 + 5 = 9$

Rewrite the middle term of the trinomial as a sum using 4 and 5 as coefficients.

$$x^2 + 9x + 20 = x^2 + 4x + 5x + 20$$

Now proceed as we did in Examples 12 and 13.

$x^2 + 9x + 20 = x^2 + 4x + 5x + 20$

$= x(x + 4) + 5(x + 4)$ Remove common factors from each pair of terms

$= (x + 5)(x + 4)$ Factor out the binomial $(x + 4)$

Notice that the trinomial, which is a sum of three terms, has now been written as a product of two binomial factors. The ability to convert from a sum to a product is very useful in dealing with algebraic fractions and in solving equations that include trinomials.

EXAMPLE 18 Factor $x^2 - 7x + 10$.

Find two factors of $+10$ whose sum is -7. Since $+10$ is positive, we know the signs of the factors are either both positive or both negative. We are looking for a total of -7, therefore, we want two negative factors.

$(-2)(-5) = +10$ $(-2) + (-5) = -7$

$x^2 - 7x + 10 = x^2 - 2x - 5x + 10$

$= x(x - 2) - 5(x - 2)$

$= (x - 5)(x - 2)$

Why did you factor out a -5 instead of a $+5$?

Look at the first binomial. Here we have $x - 2$. To obtain $x - 2$ out of $-5x + 10$, we had to factor out a -5.

EXAMPLE 19 Factor $x^2 - 8x + 15$.

The factors whose product is $+15$ with a sum of -8 are ______ and ______.

$-3, -5$

$x^2 - 8x + 15 = x^2 - 3x - 5x + 15$

$= x(x - 3) - 5(x - 3)$

$=$ ________ $(x - 3)$

$x - 5$

EXAMPLE 20 Factor $x^2 + 8x + 15$.

The factors of 15 with a sum of $+8$ are ______ and ______.

$+3, +5$

$x^2 + 8x + 15 = x^2 + 3x +$ ______ $+ 15$

$5x$

$= x($________$) + 5($________$)$

$x + 3, x + 3$

$= ($________$)($________$)$

$x + 5, x + 3$

EXAMPLE 21 Factor $x^2 + 2x - 15$.

Because the product is a negative 15, one factor must be positive and the other negative. The magnitude of the sum of a positive and a negative number is the difference of their magnitudes. Ignoring signs for the minute, we will call this the numerical difference of the numbers.

To find factors of -15 with a sum of $+2$, we need factors with a numerical difference of 2. Three and 5 are such factors. Now we have only to arrange the signs so that the sum of the factors is $+2$.

5, -3

Factors of -15 whose sum is $+2$ are _____ and _____.

$$x^2 + 2x - 15 = x^2 + 5x - 3x - 15$$

$x + 5$

$$= x(x + 5) - 3(\_\_\_\_\_\_)$$

$x - 3$, $x + 5$

$$= (\_\_\_\_\_\_)(\_\_\_\_\_\_)$$

EXAMPLE 22 Factor $x^2 - 2x - 15$.

Factors of 15 whose numerical difference is 2 are 3 and 5.

-5, 3

Factors of -15 whose sum is -2 are _____ and _____.

$$x^2 - 2x - 15 = x^2 - 5x + 3x - 15$$

$x - 5$

$$= x(x - 5) + 3(\_\_\_\_\_\_)$$

$x + 3$, $x - 5$

$$= (\_\_\_\_\_\_)(\_\_\_\_\_\_)$$

We can check our answer by multiplication.

$$(x + 3)(x - 5) = x^2 - 5x + 3x - 15$$
$$= x^2 - 2x - 15$$

PROBLEM SET 10.2B

Factor the following trinomials. Check your answers by multiplication.

1. $x^2 + 3x + 2$	2. $x^2 - 3x + 2$	3. $x^2 + 7x + 6$	4. $x^2 - 7x + 6$
5. $x^2 + 6x + 5$	6. $x^2 + 6x + 8$	7. $x^2 - 6x + 8$	8. $x^2 + 9x + 8$
9. $x^2 - 9x + 8$	10. $x^2 + 11x + 10$	11. $x^2 - 11x + 10$	12. $x^2 - 7x + 10$
13. $x^2 + 7x + 10$	14. $x^2 + 3x - 10$	15. $x^2 - 3x - 10$	16. $x^2 + 9x - 10$
17. $x^2 - 9x - 10$	18. $x^2 - 3x - 4$	19. $x^2 - x - 2$	20. $a^2 - a - 6$
21. $x^2 + 3x - 4$	22. $y^2 + 2y - 3$	23. $x^2 + x - 20$	24. $x^2 - 11x + 30$
25. $x^2 + 8x + 7$	26. $x^2 + 8x + 15$	27. $a^2 - 7a - 18$	28. $x^2 - 12x + 35$
29. $x^2 + 2x - 8$	30. $y^2 + 3y - 18$	31. $x^2 + 11x + 28$	32. $x^2 - 11x + 24$
33. $x^2 + 2x - 24$	34. $b^2 - 14b + 24$	35. $y^2 - 13y + 30$	36. $x^2 - 12x + 32$
37. $x^2 - 8x + 12$	38. $x^2 + 13x + 40$	39. $r^2 - 3r - 18$	40. $a^2 - 7a + 12$
41. $x^2 - 13x - 30$	42. $x^2 + 8x - 48$	43. $x^2 + 12x + 20$	44. $m^2 + 12m + 27$
45. $x^2 - 2x - 24$	46. $t^2 - t - 42$	47. $x^2 + x - 72$	48. $x^2 + x - 56$
49. $x^2 - 11x - 42$	50. $x^2 + 10x - 56$	51. $b^2 - 8b - 65$	52. $x^2 - 6x - 72$
53. $z^2 + 4z - 77$	54. $x^2 - 18x + 65$	55. $y^2 - 17y + 42$	56. $x^2 - 17x + 66$
57. $x^2 - 6x - 55$	58. $x^2 + 15x + 44$	59. $a^2 - 15a + 26$	60. $x^2 - 18x + 45$

10.3 FACTORING MORE DIFFICULT TRINOMIALS

In the previous section, the trinomials you were asked to factor had an x^2 term whose coefficient was 1. This is not always the case.

In general, trinomials in one variable can be written in the form $Ax^2 + Bx + C$, where A, B, and C are constants. It is possible to factor trinomials like this if you can find factors of the product $A \cdot C$ whose sum is B.

EXAMPLE 23 Factor $3x^2 + 7x - 20$.

Coefficient of x^2 term is $+3$

Constant term is -20

First, find the product of the numerical coefficient of the x^2 term and the constant term. In this example, it is $(+3)(-20) = -60$.

We can call -60 the "key number." Because the key number is negative, look for factors of the key number whose numerical difference is the coefficient of the middle term.

Listing the factors of 60:

$1 \cdot 60$
$2 \cdot 30$
$3 \cdot 20$
$4 \cdot 15$
$5 \cdot 12$ ⇐ Numerical difference of this pair of factors is 7
$6 \cdot 10$

Thus, $+12$ and -5 are factors of -60 whose sum is $+7$.

Next, rewrite the middle term as a sum of two terms using -5 and 12 for coefficients.

$$3x^2 + 7x - 20 = 3x^2 - 5x + 12x - 20$$
$$= x(3x - 5) + 4(3x - 5)$$
$$= (x + 4)(3x - 5)$$

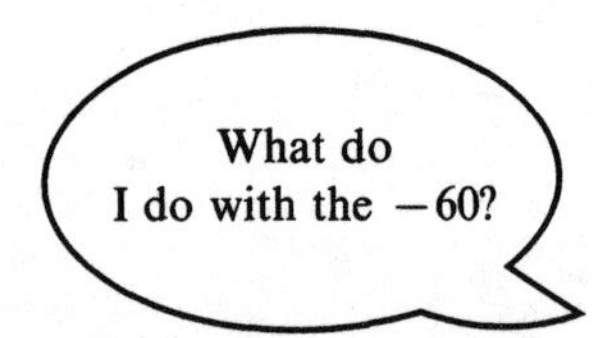

The -60 was only used to find two factors whose sum was 7.

Multiply to check the result.

$$(x + 4)(3x - 5) = 3x^2 + 7x - 20$$

EXAMPLE 24 Factor $3x^2 + 14x + 15$.

The product is $(3)(15) = 45$.

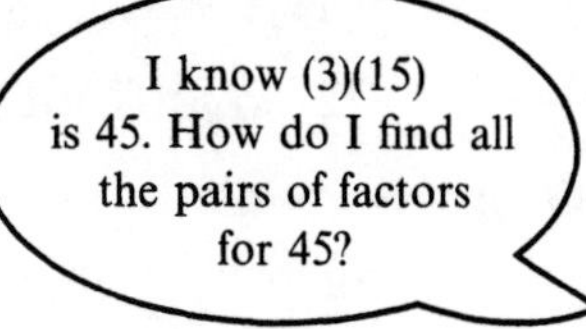

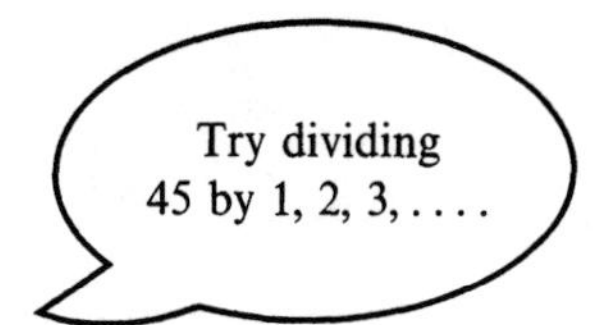

List the factors of 45.

$1 \cdot 45$
2 is not a factor of 45
$3 \cdot 15$
4 is not a factor of 45
$5 \cdot 9$
$6, 7, 8$ will not divide evenly into 45 and therefore are not possible factors
$9 \cdot 5$ ⇦ Notice, this is a repeat of a pair from above. Now you know you have all the possible cases.

Because the key number is positive, we look for factors of 45 whose numerical sum is 14.

$$5 \cdot 9 = 45 \quad \text{and} \quad 5 + 9 = 14$$

Rewrite the middle term.

$$\begin{aligned} 3x^2 + 14x + 15 &= 3x^2 + 5x + 9x + 15 \\ &= x(3x + 5) + 3(3x + 5) \\ &= (x + 3)(3x + 5) \end{aligned}$$

EXAMPLE 25 Factor $3x^2 - x - 24$.

The product is $(3)(-24) = -72$.

Because the key number is negative, we look for factors of 72 whose numerical difference is the absolute value of the middle term.

The coefficient of the middle term (-1) is numerically small. Hence, we will start with two factors that are nearly the same.

$$9 \cdot 8 = 72 \quad \text{and} \quad 9 - 8 = 1$$

Next, arrange the signs to produce a product of -72 and a sum of -1.

$$-9 \cdot 8 = -72$$

Rewrite the middle term using the coefficients we have found.

$$\begin{aligned} 3x^2 - x - 24 &= 3x^2 + 8x - 9x - 24 \\ &= x(3x + 8) - 3(3x + 8) \\ &= (\_\_\_\_\_\_)(\_\_\_\_\_\_) \end{aligned}$$

$x - 3, 3x + 8$

EXAMPLE 26 Factor $3x^2 - 34x - 24$.

The product is $(3)(-24) = -72$.

The key number is -72, so look for a difference of factors whose absolute value is 34.

The coefficient of the middle term (-34) is quite large. Hence we will start with one large factor and one small factor.

$(1)(72)$	Difference is 73
$(2)(-36)$	Difference is 34
$(2)(-36)$	The correct factors

Rewrite the middle term using the coefficients we have found.

$$3x^2 - 34x - 24 = 3x^2 + 2x - 36x - 24$$

$3x + 2$

$$= x(3x + 2) - 12(\_\_\_\_\_\_\_\_)$$

$(x - 12)(3x + 2)$

$$= \_\_\_\_\_\_\_\_\_\_\_\_$$

EXAMPLE 27 Factor $3x^2 - 73x + 24$.

The product is $(3)(24) = +72$.

73

We are now looking for factors of 72 whose numerical sum is _____.

$$(1)(72) = 72 \quad \text{and} \quad 1 + 72 = 73$$

Both signs must be negative to produce a middle term of $-73x$.

$(-1)(-72)$ are the factors we want.

Now we can rewrite the middle term using the coefficients we have found.

$$3x^2 - 73x + 24 = 3x^2 - 1x - 72x + 24$$

$3x - 1$

$$= x(\_\_\_\_\_\_\_\_) - 24(3x - 1)$$

$(x - 24)(3x - 1)$

$$= \_\_\_\_\_\_\_\_\_\_\_\_$$

Check using the FOIL method.

$$(x - 24)(3x - 1) = 3x^2 - 73x + 24$$

PROBLEM SET 10.3

Factor the following. Check your answers.

1. $2x^2 + x - 1$
2. $2x^2 - x - 1$
3. $2x^2 + 3x + 1$
4. $2x^2 - 3x + 1$
5. $3x^2 + 7x + 2$
6. $3x^2 - 7x + 2$
7. $3x^2 + 5x + 2$
8. $3x^2 - 5x + 2$
9. $3x^2 + x - 2$
10. $3x^2 - x - 2$
11. $3x^2 - 5x - 2$
12. $3x^2 + 5x - 2$
13. $8x^2 + 14x + 3$
14. $8x^2 + 10x - 3$
15. $8x^2 - 10x - 3$
16. $8x^2 - 14x + 3$
17. $8x^2 + 10x + 3$
18. $8x^2 - 10x + 3$
19. $8x^2 + 2x - 3$
20. $8x^2 - 2x - 3$
21. $6x^2 - 7x + 2$

22. $6x^2 + 7x + 2$

23. $6x^2 + x - 2$

24. $6x^2 - x - 2$

25. $7y^2 - 17y + 10$

26. $7y^2 + 17y + 10$

27. $7y^2 - 3y - 10$

28. $7y^2 + 3y - 10$

29. $5y^2 + 12y + 4$

30. $5y^2 - 12y + 4$

31. $5y^2 + 8y - 4$

32. $5y^2 - 8y - 4$

33. $7y^2 - 19y + 10$

34. $5x^2 + 41x + 8$

35. $3x^2 - 25x + 8$

36. $3a^2 + 23a + 14$

37. $3b^2 - 19b - 14$

38. $2x^2 + 13x + 20$

39. $3x^2 - 17x - 6$

40. $5x^2 + 47x - 30$

41. $3b^2 + 19b - 14$

42. $3x^2 - 14x + 15$

43. $6x^2 + 7x - 20$

44. $6x^2 - 5x - 6$

45. $11x^2 - 8x - 3$

46. $11x^2 - 14x + 3$

47. $12a^2 + 17a - 5$

48. $12a^2 - 4a - 5$

49. $24x^2 + 47x - 21$

50. $24x^2 + 10x - 21$

51. $24x^2 + 14x - 3$

52. $16x^2 + 32x + 15$

53. $20x^2 - 121x + 6$

54. $20x^2 - 9x - 18$

55. $5 + 49x - 10x^2$

56. $6x^2 + 29x + 28$

57. $20x^2 + 56x + 15$

58. $5 - 28b - 12b^2$

59. $5 - 12y - 9y^2$

60. $24a^2 - 26a - 15$

10.4 ADDITIONAL FACTORING TECHNIQUES

Factoring a Difference of Two Squares

Two binomials that are alike except for signs are called a sum and difference of two terms. For example, $(x + 4)$ and $(x - 4)$ are a sum and difference of two terms.

Multiplying the sum and difference above using the FOIL methods, we get

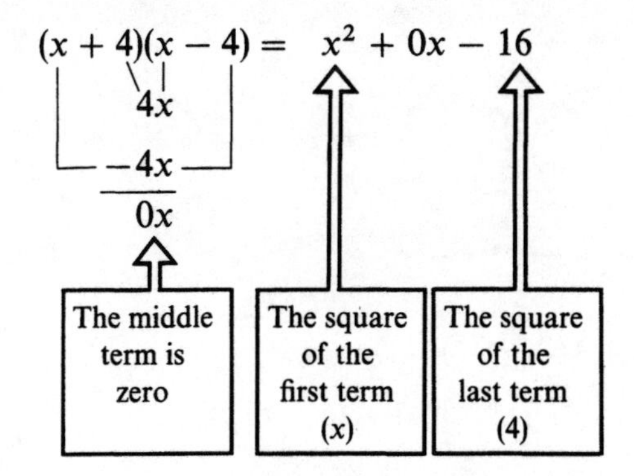

EXAMPLE 28 Multiply $(a + b)(a - b)$.

$$(a + b)(a - b) = a^2 - b^2$$

Square of a | Minus sign | Square of b

Product of sum and difference of two terms

> **10.4A PRODUCT OF SUM AND DIFFERENCE OF TWO TERMS**
>
> In the product of a sum and difference, the middle term is zero. The result is the difference of the squares of the two terms.

Therefore,

Factoring difference of two squares

> **10.4B FACTORING DIFFERENCE OF TWO SQUARES**
>
> The factors of the difference of two squares are the sum and difference of the numbers being squared.
>
> $$a^2 - b^2 = (a + b)(a - b)$$

EXAMPLE 29 Factor $x^2 - 9$.

$$x^2 - 9 = (x)^2 - (3)^2$$
$$= (x + 3)(x - 3)$$

Sum Difference

EXAMPLE 30 Factor $y^2 - 1$.

y, 1

$$y^2 - 1 = (\_\_\_\_)^2 - (\_\_\_\_)^2$$
$$= (y - 1)(y + 1)$$

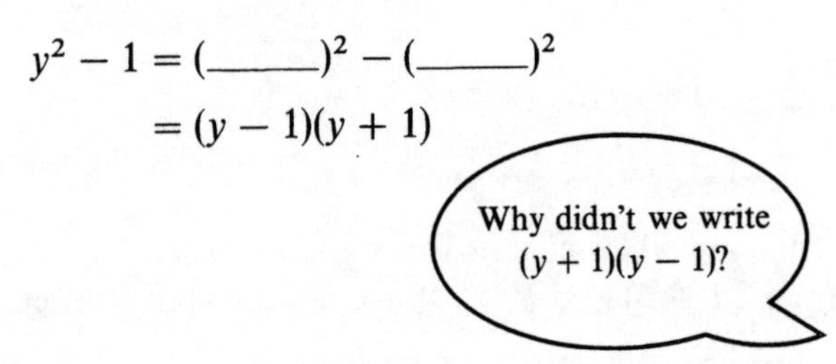

The commutative property of multiplication says it does not matter what order we multiply factors.

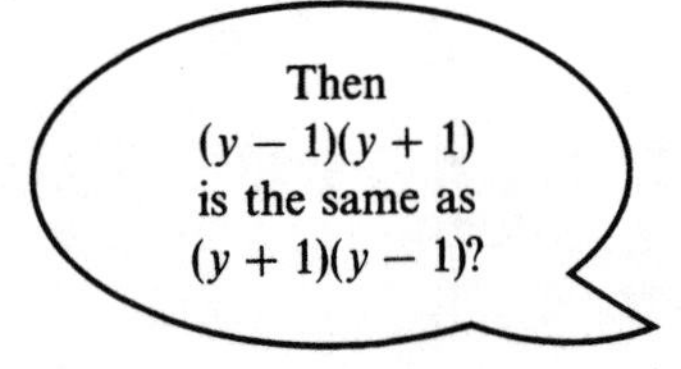

Yes.

EXAMPLE 31 Factor $4x^2 - 25y^2$.

2x, 5y

$$4x^2 - 25y^2 = (\_\_\_\_)^2 - (\_\_\_\_)^2$$
$$= (2x + 5y)(2x - 5y)$$

Can we factor the sum of two squares?

Look at Example 32.

EXAMPLE 32 Factor $x^2 + y^2$.

To get x^2, the first term of each binomial must be an x; to get y^2, the second term of each binomial must be a y.

$$(x \quad y)(x \quad y)$$

If we use opposite signs in the factors, we have the difference of two squares.

$$(x + y)(x - y) = x^2 - y^2$$

If both signs are positive or both are negative, we will get a middle term.

$$(x - y)(x - y) = x^2 - 2xy + y^2$$
$$(x + y)(x + y) = x^2 + 2xy + y^2$$

Since we do not have a middle term, $x^2 + y^2$ *cannot be factored.*

PROBLEM SET 10.4A

Factor the following expressions.

1. $x^2 - 4$	2. $x^2 - 25$	3. $x^2 - 36$	4. $y^2 - 49$
5. $y^2 - 64$	6. $4y^2 - 1$	7. $9x^2 - 4$	8. $9x^2 - 16$
9. $x^2 - y^2$	10. $9a^2 - b^2$	11. $x^2 - 16y^2$	12. $25a^2 - b^2$
13. $a^2 - 36b^2$	14. $4x^2 - 9y^2$	15. $25x^2 - 4y^2$	16. $49a^2 - 16x^2$

Factoring Completely

We opened this unit saying that the first step in factoring is to use the distributive property to remove common factors. We will also end this unit by pointing out that to factor an expression completely, use the distributive property to *remove any common factors first.*

EXAMPLE 33 Factor completely $10x^2 + 60x + 90$.

Each term has a common monomial factor of 10.

$$10x^2 + 60x + 90 = 10(x^2 + 6x + 9) \qquad \text{Remove common factor of 10}$$
$$= 10(x + 3)(x + 3) \qquad \text{Factor the trinomial}$$

EXAMPLE 34 Factor $3x^2 - 12$.

$$3x^2 - 12 = 3(x^2 - 4) \qquad \text{Remove common factor of 3}$$
$$= 3(\underline{\qquad\qquad})(\underline{\qquad\qquad}) \qquad \text{Factor the difference of two squares}$$

$x + 2, x - 2$

Factored completely

10.4C DEFINITION OF FACTORED COMPLETELY

An expression is *factored completely* if it is written as a product of polynomials, such that:

1. The coefficients in each polynomial factor are integers.
2. No polynomial factor can be factored further using integer coefficients.

EXAMPLE 35 Factor completely $25a^3 - 40a^2 - 20a$.

$5a$

Each term has a common factor of ______.

$5a^2 - 8a - 4$

$$25a^3 - 40a^2 - 20a = 5a(\_\_\_\_\_\_\_\_)$$ Remove common factor of $5a$

$$= 5a(5a + 2)(a - 2)$$ Factor the trinomial

EXAMPLE 36 Factor completely $4x^2y^2 - 4x^4y^2$.

$4x^2y^2$

Each term has a common monomial factor of ______.

$$4x^2y^2 - 4x^4y^2 = 4x^2y^2(1 - x^2)$$ Remove common factor $4x^2y^2$

$$= 4x^2y^2(1 + x)(1 - x)$$ Factor the difference of two squares

PROBLEM SET 10.4B

Factor the following completely.

1. $2x^2 - 8$
2. $3a^2 - 27$
3. $2x^2 - 4x + 2$
4. $3a^2 + 6a + 3$
5. $2b^2 - 18a^2$
6. $3a^2 - 3a - 9$
7. $2xy^2 + 9xy + 9x$
8. $9x^3y - 4xy^3$
9. $2x^2 + 12x + 18$
10. $ay^2 - 4ay + 4a$
11. $30 + 5x - 5x^2$
12. $36 - 6x - 6x^2$
13. $9x^3 - 21x^2 + 6x$
14. $3x^2 - 3x - 36$
15. $4ax^2 + 14ax - 8a$
16. $6by^2 - 3by - 9b$
17. $16xy^2 + 48xy + 36x$
18. $45ab^2 - 30ab + 5a$
19. $72bx^2 - 32b^3$
20. $9ab^3 - 144ab$
21. $18ab + 3ab^2 - ab^3$
22. $50x^2 - 10x - 10$
23. $128x^4 - 72x^2$
24. $4y^4 - 10y^3 - 50y^2$
25. $24xy^4 - 72xy^3 + 54xy^2$
26. $12x^2 - 2x^3 - 30x^4$
27. $6x^3 - 33x^4 - 63x^5$
28. $64x^2y^3 - 96x^2y^2 + 36x^2y$
29. $12a^4b^2 - 33a^3b^2 - 9a^2b^2$
30. $18ab^3 + 30a^2b^2 - 12a^3b$
31. $128ab^4 - 64a^2b^3 + 8a^3b^2$
32. $24a^3b^3 - 72a^2b^3 + 54ab^3$

33. $56ax^4 + 42ax^3 - 63ax^2$

34. $30xy^5 - 25xy^4 + 5xy^3$

35. $24b^4y - 150b^2y$

36. $36a^3b - 64ab^3$

37. $36a^3x^3 - 100a^3x$

38. $a^4b^2 - 6a^3b^3 + 9a^2b^4$

39. $3x^3a^3 - 11x^4a^2 - 20x^5a$

40. $8a^3b^2 - 24a^2b^3 + 18ab^4$

41. $3a^7b - 12a^5b^2 + 12a^3b^3$

42. $15by^3 + 80by^2 + 80by$

43. $2a^6b^3 - 2a^5b^3 + 2a^4b^3$

44. $3x^3y^2 - 15x^2y^2 + 21xy^2$

45. $3ab^4 - 18ab^3 + 27ab^2$

46. $12x^2 + 14ax - 10a^2$

47. $5x^3y - 50x^2y^2 + 45xy^3$

48. $45b^2x^3 - 60b^3x^2 + 20b^4x$

49. $81y^2 - 9y^4$

50. $64x^4 - 16x^6$

51. $12a^4y^2 + 4a^3y^2 - 56a^2y^2$

52. $12a^3y^3 + 45a^2y^3 - 12ay^3$

53. $125ax^3 - 20a^3x^3$

54. $5x^5y^3 - 22x^4y^3 + 8x^3y^3$

55. $18a^4 + 15a^3b - 18a^2b^2$

56. $80x^4y^2 - 45x^2y^4$

57. $63a^4b^3 - 28a^2b^7$

Review Your Skills

Multiply the following fractions. Write all answers in lowest terms. Use standard form for fractions.

58. $\frac{2}{7} \cdot \frac{3}{4}$

59. $\frac{-15}{12} \cdot \frac{8a}{-5}$

60. $-2ax \cdot \frac{-48ax^2}{-42}$

61. $\frac{27a^2x}{-16} \cdot \frac{-12ax^2}{9}$

62. $\frac{21ab}{-16} \cdot \frac{a^2}{14}$

10 REVIEW

NOW THAT YOU HAVE COMPLETED UNIT 10, YOU SHOULD BE ABLE TO APPLY THE FOLLOWING RULES:

To factor a monomial from a polynomial using the distributive property:

1. Look for the largest monomial factor that is common to all terms of the polynomial.
2. Write the common monomial as a product followed by parentheses.
3. Inside the parentheses, write the factors that will produce the original expression if you multiply them.

To factor a trinomial:

Multiply the coefficients of x and the constant term. Then find two factors of this product whose sum is the coefficient of the x terms.
Write the middle term as a sum using the two factors found in Step 1.
Form two groups and find the common factor in each group separately, then remove it.
Remove the common binomial factor.

To factor the differential of two squares:

$$a^2 - b^2 = (a + b)(a - b)$$

To factor an expression completely:

1. Use the distributive property to remove any common factors.
2. Apply the methods of factoring trinomials or the difference of two squares when possible.

REVIEW TEST 10

Factor the following completely. (10.1 through 10.4)

1. $3x^4 - 15x^2 + 3x =$ ____________

2. $2ay^4 - 6ay^3 + 10ay^2 =$ ____________

3. $x^2 - 2x - 24 =$ ____________

4. $x^2 - 5x + 6 =$ ____________

5. $x^2 + 4x - 21 =$ ____________

6. $3x^2 + 10x + 8 =$ ____________

7. $5x^2 - 29x + 20 =$ ____________

8. $6a^3 - 19a^2 + 3a =$ ____________

9. $2x^3 + 2x^2 - 24x =$ ____________

10. $6x^2 - x - 2 =$ ____________

11. $8x^4 - 2x^3 - 3x^2 =$ ____________

12. $24ay^3 - 8ay^2 - 10ay =$ ____________

13. $6x^2 - 27x + 12 =$ ____________

14. $4y^2 - 12y + 9 =$ ____________

15. $2x^2 - 5x - 12 =$ ____________

16. $6a^2 + 11a + 3 =$ ____________

17. $25x^2 - 16 =$ ____________

18. $4xy^3 - 36xy =$ ____________

19. $24y^3 - 54a^2y =$ ____________

20. $36bx^5 + 48bx^4 + 16bx^3 =$ ____________

REVIEW YOUR SKILLS

Use the addition method to find the simultaneous solution, if one exists.

21. $3x - 2y = -8$

    $6x + 3y = 5$

Use the substitution method to find the simultaneous solution, if one exists.

22. $2x + 4y = -5$

    $-x + 12y = 7$

23. A 64-inch string is cut into two pieces so that one piece is 3 times the length of the other. How long is each piece?

Simplify the following using the law of exponents. Write the results without negative exponents.

24. $-(-3x)^2$

25. $\left(\frac{-x^2}{y^3}\right)^2$

26. $(x^{-4}y^3)(x^2y^{-4})$

27. $\frac{12a^2b^3}{-18a^3b}$

Evaluate the following for $a = 2$, $b = -3$, $c = -1$.

28. $abc + b^3$

29. $(a + b - c)^3$

UNIT 11

Multiplication and Division of Algebraic Fractions

OBJECTIVES

After you have successfully completed this unit, you will be able to:

1. Reduce rational algebraic expressions to lowest terms (11.1)
2. Multiply rational algebraic expressions (11.2)
3. Divide rational algebraic expressions (11.3)

11.1 REDUCING FRACTIONS TO LOWEST TERMS

Algebraic fractions are multiplied the same way as arithmetic fractions are multiplied.

Multiplication of algebraic fractions

11.1A DEFINITION

To multiply two algebraic fractions, multiply their numerators to get the numerator of the product, then multiply the denominators to get the denominator of the product.

$$\frac{a}{b} \cdot \frac{c}{d} = \frac{a \cdot c}{b \cdot d}$$

EXAMPLE 1

$$\frac{3}{4} \cdot \frac{5}{4} = \frac{15}{16}$$

EXAMPLE 2

$$\frac{3}{4} \cdot \frac{a}{b} = \frac{3 \cdot a}{4 \cdot b} = \frac{3a}{4b}$$

These are two fractions with one multiplication sign.

This is one fraction with two multiplication signs.

EXAMPLE 3

$$\frac{a^2}{b} \cdot \frac{a^3}{b} = \frac{a^5}{b^2}$$

EXAMPLE 4

$$3a \cdot \frac{-2a^2}{5b} = \frac{3a}{1} \cdot \frac{-2a^2}{5b} = \frac{-6a^3}{5b}$$

PROBLEM SET 11.1A

Multiply the following.

1. $\frac{1}{2} \cdot \frac{3}{4}$
2. $\frac{4}{5} \cdot \frac{3}{5}$
3. $(-8)\left(\frac{2}{3}\right)$
4. $\frac{3}{4}(-5)$
5. $9 \cdot \frac{2}{7}$
6. $\frac{2}{3} \cdot 2$
7. $\frac{2}{3} \cdot \frac{x}{y}$
8. $\frac{a}{b} \cdot \frac{a}{c}$
9. $1 \cdot \frac{4x}{y}$
10. $x \cdot \frac{x}{yz}$
11. $x^2 \cdot \frac{xy}{z}$
12. $a^3 \cdot \frac{a^2b}{c}$
13. $\frac{x^2}{z} \cdot xy$
14. $\frac{ab}{c} \cdot \frac{ab}{c^2}$
15. $-1 \cdot \frac{r^2}{2s}$
16. $\frac{4x}{3y} \cdot \frac{-x}{3y^2}$

Reducing fractions to lowest terms or writing equivalent fractions rests on the ideas in the next two examples.

EXAMPLE 5

$$\frac{1}{1} \cdot \frac{x}{y} = \frac{1 \cdot x}{1 \cdot y} = \frac{x}{y}$$

This doesn't look like much but it's one of the best procedures we have: $\frac{1}{1} = 1$.

Recall that the identity element for multiplication is 1, and multiplying a number by 1 does not change the number.

EXAMPLE 6

$$\frac{4}{4} \cdot \frac{x}{y} = \frac{4 \cdot x}{4 \cdot y} = \frac{4x}{4y}$$

Hey! Isn't that the same routine as Example 5?

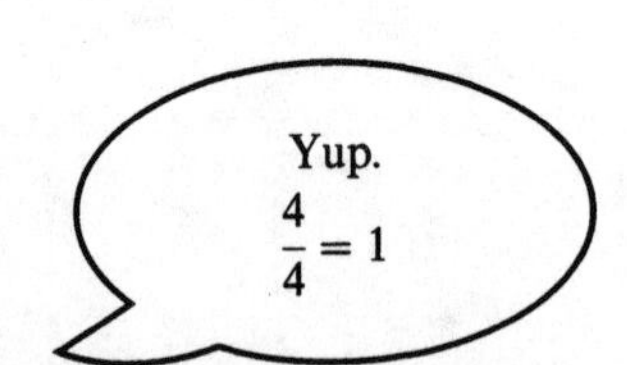

Lowest terms

11.1B DEFINITION

An algebraic fraction is reduced to *lowest terms* when the numerator and denominator contain no common factors.

Algebraic fractions are reduced in the same manner as arithmetic fractions. We use the old "multiply by 1" technique.

We know that $\frac{y}{y} = 1$. Let's use this to reduce the fraction $\frac{a \cdot y}{b \cdot y}$.

$$\frac{a \cdot y}{b \cdot y} = \frac{a}{b} \cdot \frac{y}{y}$$
$$= \frac{a}{b} \cdot 1$$
$$= \frac{a}{b}$$

This is just using the "multiply by 1" technique backwards.

Rather than write this all out, we usually just divide out like factors in the numerator and denominator.

$$\frac{a \cdot y}{b \cdot y} = \frac{a \cdot \overset{1}{\cancel{y}}}{b \cdot \underset{1}{\cancel{y}}} = \frac{a}{b}$$

Sometimes the word *cancel* is used to describe dividing out, but remember, we aren't eliminating anything. We are really just identifying multiplication by 1.

EXAMPLE 7 Reduce $\frac{a \cdot b}{b \cdot a}$.

$$\frac{a \cdot b}{b \cdot a} = \frac{a \cdot b}{a \cdot b}$$
$$= \frac{a}{a} \cdot \frac{b}{b}$$
$$= 1 \cdot 1$$
$$= \_\_\_\_$$

1

EXAMPLE 8 Reduce $\frac{x^4y^2z}{x^2yz^3}$.

First, we identify the multiplications by 1.

$$\frac{x^4y^2z}{x^2yz^3} = \frac{x^2}{x^2} \cdot \frac{x^2}{1} \cdot \frac{y}{y} \cdot \frac{y}{1} \cdot \frac{z}{z} \cdot \frac{1}{z^2}$$

$$= 1 \cdot \frac{x^2}{1} \cdot 1 \cdot \frac{y}{1} \cdot 1 \cdot \frac{1}{z^2}$$

$$= \frac{x^2y}{z^2}$$

Shortcut:

$$\frac{x^4y^2z}{x^2yz^3} = \frac{\overset{x^2}{\cancel{x^4}}\,\overset{y}{\cancel{y^2}}\,\overset{1}{\cancel{z}}}{\underset{1}{\cancel{x^2}}\,\underset{1}{\cancel{y}}\,\underset{z^2}{\cancel{z^3}}} = \frac{x^2y}{z^2}$$

x^2y is the product of the remaining factors in the numerator, and the z^2 is the product of the remaining factors in the denominator.

EXAMPLE 9 Reduce $\frac{18x^3y^2}{24x^4y}$.

$$\frac{18x^3y^2}{24x^4y} = \frac{6}{6} \cdot \frac{3}{4} \cdot \frac{x^3}{x^3} \cdot \frac{1}{x} \cdot \frac{y}{y} \cdot \frac{y}{1}$$

$$= \_\_\_\_ \cdot \frac{3}{4} \cdot \_\_\_\_ \cdot \frac{1}{x} \cdot \_\_\_\_ \cdot \frac{y}{1}$$

1, 1, 1

$$= \_\_\_\_$$

$\frac{3y}{4x}$

Shortcut:

$$\frac{18x^3y^2}{24x^4y} = \frac{\overset{3}{\cancel{18}}\,\overset{1}{\cancel{x^3}}\,\overset{y}{\cancel{y^2}}}{\underset{4}{\cancel{24}}\,\underset{x}{\cancel{x^4}}\,\underset{1}{\cancel{y}}} = \_\_\_\_$$

$\frac{3y}{4x}$

PROBLEM SET 11.1B

Reduce to lowest terms.

1. $\frac{2}{8}$ 2. $\frac{4}{10}$ 3. $\frac{6}{9}$ 4. $\frac{18}{24}$

5. $\frac{-12}{20}$ 6. $\frac{3x}{4x}$ 7. $\frac{ax}{bx}$ 8. $\frac{ax}{ay}$

9. $\frac{2x}{x^2}$ 10. $\frac{3y^3}{9y}$ 11. $\frac{6x^4}{9x^2}$ 12. $\frac{12a^2b}{15ab^2}$

13. $\frac{6a^4}{4a^6}$ 14. $\frac{8x^2y}{4xy}$ 15. $\frac{3a^4b^2}{a^3b^2}$ 16. $\frac{-24x^3}{16x^2y}$

17. $\frac{12ab^2c}{24b^4c}$ 18. $\frac{-36r^2s^2t^2}{6rt^2}$ 19. $\frac{3xy^2z^3}{8xyz^5}$ 20. $\frac{-18a^2bc^3}{12abc}$

Review Your Skills Factor the following.

21. $2x - 6$
22. $3ax^2 - 6ax$
23. $27ax^2 - 3ay^2$
24. $8a^2x^2 - 32a^2y^2$
25. $3x^2 - 15x + 18$
26. $4a^2x^2 - 24a^2x + 36a^2$

EXAMPLE 10 Reduce $\dfrac{3x + 6}{12}$.

The first step is to express this fraction so that both the numerator and denominator are products.

Sum of terms ⇨ $\dfrac{3x + 6}{12} = \dfrac{3(x + 2)}{12}$ ⇦ Product of factors

$$= \frac{3\,(x + 2)}{3 \cdot 4}$$

$$= \frac{3}{3} \cdot \frac{x + 2}{4}$$

$$= 1 \cdot \frac{x + 2}{4}$$

$$= \frac{x + 2}{4}$$

Why $3 \cdot 4$ instead of $2 \cdot 6$ in the denominator?

Because $\frac{3}{3} = 1$, and I saw it coming.

The short way:

Factors

$$\frac{3x + 6}{12} = \frac{3 \cdot (x + 2)}{12} \qquad \text{Factor}$$

$$= \frac{\overset{1}{\cancel{3}}(x + 2)}{\underset{4}{\cancel{12}}} \qquad \text{Divide out common factors}$$

$$= \frac{x + 2}{4}$$

If dividing out is so neat, why not divide the 2 into the 4 in $\dfrac{x + 2}{4}$?

Because the 2 in the numerator is a term (part of an addition) not a factor (part of a multiplication).

Let's demonstrate with numbers why you can't cancel 2.

Let $x = 5$ in $\dfrac{x + 2}{4}$. Then

$$\frac{5 + 2}{4} = \frac{7}{4}$$

If the 2 is divided out, then

$$\frac{5 + \overset{1}{\cancel{2}}}{\underset{2}{\cancel{4}}} \stackrel{?}{=} \frac{6}{2}$$

$\dfrac{6}{2}$ is not equal to $\dfrac{7}{4}$.

Therefore, be careful that you divide out only factors and not terms.

EXAMPLE 11 Reduce $\dfrac{x + 6}{2x + 12}$.

Before we can do anything to reduce this fraction, both the numerator and the denominator must be pure products. So, we factor a 2 from the denominator.

$$\frac{x + 6}{2x + 12} = \frac{(x + 6)}{2\ (x + 6)}$$

The bar in a fraction is a grouping symbol just like parentheses. $\dfrac{x + 6}{}$ means $(x + 6)$.

$$= \frac{1}{2} \cdot \frac{x + 6}{x + 6}$$

$$= \frac{1}{2} \cdot 1$$

$$= \frac{1}{2}$$

The short way:

$$\frac{x + 6}{2x + 12} = \frac{x + 6}{2(x + 6)} \qquad \text{Factor}$$

$$= \frac{\overset{1}{\cancel{(x + 6)}}}{2\underset{1}{\cancel{(x + 6)}}} \qquad \text{Divide out common factor}$$

$$= \frac{1}{2}$$

EXAMPLE 12 Reduce $\dfrac{x^2 + 4x - 5}{x^2 + 2x - 3}$.

Both the numerator and denominator must be factored before you can divide out.

$$\frac{x^2 + 4x - 5}{x^2 + 2x - 3} = \frac{(x - 1)(x + 5)}{(x - 1)(x + 3)} \qquad \text{Factor}$$

$$= \frac{\overset{1}{\cancel{(x - 1)}}(x + 5)}{\underset{1}{\cancel{(x - 1)}}(x + 3)} \qquad \text{Divide out}$$

$\dfrac{x + 5}{x + 3}$ $\qquad = \_\_\_\_\_\_\_\_$

EXAMPLE 13 Reduce $\dfrac{x - 3}{x + 6}$.

$$\frac{x - 3}{x + 6}$$

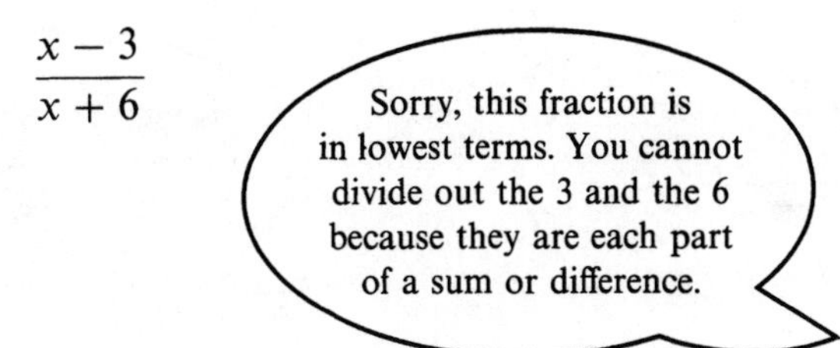

Now I get it!
We have terms not
factors. If it won't factor,
it can't be reduced.

EXAMPLE 14 Reduce $\dfrac{x^2 - y^2}{x + y}$.

$$\frac{x^2 - y^2}{x + y} = \frac{(x + y)(x - y)}{(x + y)} \qquad \text{Factor}$$

$$= \frac{\overset{1}{\cancel{(x + y)}}(x - y)}{\underset{1}{\cancel{(x + y)}}} \qquad \text{Divide out}$$

$$= \frac{x - y}{1}$$

$x - y$ $\qquad = \_\_\_\_\_\_\_\_$

Multiplying a polynomial by -1 changes all the signs in the polynomial. Notice the effect of multiplying the following difference by -1. Applying the distributive property, we obtain:

Sign changed

$$(-1)(\ a - b) = -a + b = b - a$$

Sign changed

If we apply the distributive property in reverse, that is, factor out a -1 from the expression $b - a$, we have

$$b - a = (-1)(-b + a) = (-1)(a - b)$$

EXAMPLE 15 Factor -1 from each of the following.

a. $(a - b) = -1(b - a)$

b. $(x - y) = -1(\underline{\qquad\qquad})$

c. $2a - b = -1(\underline{\qquad\qquad})$

d. $3p - 4q = -1(\underline{\qquad\qquad})$

e. $4 - x = -1(\underline{\qquad\qquad})$

$y - x$

$b - 2a$

$4q - 3p$

$x - 4$

EXAMPLE 16 Reduce $\dfrac{4 - x}{x^2 - 7x + 12}$.

$$\frac{4 - x}{x^2 - 7x + 12} = \frac{4 - x}{(x - 4)(x - 3)} \qquad \text{Factor}$$

$$= \frac{(4 - x)}{(x - 4)(x - 3)} \qquad \text{There are no common factors to divide out}$$

Isn't $4 - x$ the same as $x - 4$?

Nope!! The signs are different. We can write $4 - x$ as $(-1)(x - 4)$.

$$\frac{4 - x}{(x - 4)(x - 3)} = \frac{-1(x - 4)}{(x - 4)(x - 3)}$$

Now we have a common factor to divide out.

$$= \frac{-1\overset{1}{\cancel{(x - 4)}}}{\underset{1}{\cancel{(x - 4)}}(x - 3)}$$

$$= \frac{-1}{x - 3}$$

$$= \frac{-1}{x - 3} \cdot \frac{-1}{-1}$$

$$= \frac{1}{3 - x}$$

Why multiply by 1, written as $\dfrac{-1}{-1}$?

We do it to reduce the number of negative signs in the answer. However, $\dfrac{-1}{x - 3}$ and $\dfrac{1}{3 - x}$ are both correct answers.

PROBLEM SET 11.1C

Reduce the following to lowest terms.

1. $\frac{2x+4}{6}$ 2. $\frac{3y-6}{9}$ 3. $\frac{9}{6x-3}$ 4. $\frac{12}{3x-3}$

5. $\frac{5x+5y}{5x-5y}$ 6. $\frac{14a+7}{7a+7}$ 7. $\frac{5x-5y}{5y-5x}$ 8. $\frac{3x-9}{2x-6}$

9. $\frac{3x-3y}{4y-4x}$ 10. $\frac{5x-5}{2-2x}$ 11. $\frac{x^2+2x}{x}$ 12. $\frac{xy}{xy-x}$

13. $\frac{y^2-4y}{y^2}$ 14. $\frac{ax}{ax-ay}$ 15. $\frac{2x+2y}{2x-2y}$ 16. $\frac{2x-2y}{4x-4y}$

17. $\frac{ax+ay}{ax-ay}$ 18. $\frac{ax-ay}{bx-by}$ 19. $\frac{ax-ay}{by-bx}$ 20. $\frac{y^2+2y}{4+2y}$

21. $\frac{x^2+xy}{y^2+xy}$ 22. $\frac{x^2y-xy^2}{x^2y-xy^2}$ 23. $\frac{x^2-xy}{y^2-xy}$ 24. $\frac{x^3-x^2}{x^3+3x^2}$

25. $\frac{3x^2-x}{6x^3-2x^2}$ 26. $\frac{4a^2+6a}{6a^3+9a^2}$ 27. $\frac{2x-x^2}{x^3-2x^2}$ 28. $\frac{2y^3+2y^2}{2y^2-6y}$

29. $\frac{2a^3b-2a^2b^2}{2a^2b+2ab^2}$ 30. $\frac{x^4y^2+x^3y^2}{3x^2y^2+3xy^2}$ 31. $\frac{x+1}{x^2+3x+2}$ 32. $\frac{x^2-4x+4}{x-2}$

33. $\frac{5-x}{x^2-4x-5}$ 34. $\frac{x^2+2x-3}{x^2+5x+6}$ 35. $\frac{x^2+4x+4}{x^2-2x-8}$ 36. $\frac{x^2-6x+9}{x^2-9}$

37. $\frac{b^2-a^2}{a^2+2ab+b^2}$ 38. $\frac{x^2-16}{4-x}$ 39. $\frac{x^2-9}{x^2+5x+6}$ 40. $\frac{2x^2+12x+18}{4x^2+8x-12}$

41. $\frac{x^2+3x-10}{x^2-8x+12}$ 42. $\frac{x^2+9x+8}{x^2+5x-24}$ 43. $\frac{2a^2+5a-3}{2a^2-9a+4}$ 44. $\frac{6x^2+7x+2}{3x^2+5x+2}$

45. $\frac{2x^2-xy-3y^2}{3x^2+4xy+y^2}$ 46. $\frac{x^2-3x}{x^2+x-12}$ 47. $\frac{x^3-xy^2}{xy^2-x^2y}$ 48. $\frac{x^2y^2-x^3y}{x^3-xy^2}$

Review Your Skills

Multiply the following fractions. Write all answers in lowest terms. Use the standard form for fractions.

49. $\frac{3}{5}\cdot\frac{10a}{9}$ 50. $\frac{-8x}{-14}\cdot\frac{-21}{15x}$

51. $-6a\cdot\frac{-18a^2}{15a}$ 52. $\frac{-24a^3}{-15}\cdot(-25a)$

11.2 MULTIPLICATION OF ALGEBRAIC FRACTIONS

In Section 11.1, multiplication of algebraic fractions was defined: To multiply fractions, multiply the numerators to get the numerator of the product and multiply the denominators to get the denominator of the product.

To save work when multiplying algebraic fractions, divide out all possible factors first.

EXAMPLE 17 Multiply $\frac{a}{8b} \cdot \frac{4}{a}$.

$$\frac{a}{8b} \cdot \frac{4}{a} = \frac{\overset{1}{\cancel{a}}}{\underset{2}{\cancel{8}} \cdot b} \cdot \frac{\overset{1}{\cancel{4}}}{\underset{1}{\cancel{a}}}$$

$$= \frac{1}{2b}$$

Remember:
Dividing out can be done *only* when products are involved.

EXAMPLE 18 Multiply $\frac{2y^3}{x^2} \cdot \frac{3x}{4y}$.

$$\frac{2y^3}{x^2} \cdot \frac{3x}{4y} = \frac{\overset{1\ y^2}{\cancel{2}\cancel{y^3}}}{\underset{x}{\cancel{x^2}}} \cdot \frac{3\overset{1}{\cancel{x}}}{\underset{2\ 1}{\cancel{4}\cancel{y}}}$$ Divide out common factors

$= \underline{\qquad}$ Multiply the remaining factors

$\frac{3y^2}{2x}$

EXAMPLE 19 Multiply $\frac{a}{6a - 12} \cdot \frac{4a - 8}{a^3}$.

$$\frac{a}{6a - 12} \cdot \frac{4a - 8}{a^3} = \frac{a}{6(a - 2)} \cdot \frac{4(a - 2)}{a^3}$$ Factor completely

$$= \frac{\overset{1}{\cancel{a}}}{\underset{3}{\cancel{6}}\underset{1}{\cancel{(a - 2)}}} \cdot \frac{\overset{2}{\cancel{4}}\overset{1}{\cancel{(a - 2)}}}{\underset{a^2}{\cancel{a^3}}}$$ Divide out common factors

$= \underline{\qquad}$ Multiply

$\frac{2}{3a^2}$

EXAMPLE 20 Multiply $\frac{a^2-1}{a^2+a} \cdot \frac{a^3}{a^2+2a-3}$.

$$\frac{a^2-1}{a^2+a} \cdot \frac{a^3}{a^2+2a-3} = \frac{(a+1)(a-1)}{a(a+1)} \cdot \frac{a^3}{(a-1)(a+3)} \quad \text{Factor}$$

$$= \frac{\overset{1}{\cancel{(a+1)}}\overset{1}{\cancel{(a-1)}}}{\underset{1}{\cancel{a}}\underset{1}{\cancel{(a+1)}}} \cdot \frac{\overset{a^2}{\cancel{a^3}}}{\underset{1}{\cancel{(a-1)}}(a+3)} \quad \text{Divide out}$$

$\frac{a^2}{a+3}$

$$= \underline{\qquad\qquad} \quad \text{Multiply}$$

EXAMPLE 21 Multiply $(x^2+3x+2) \cdot \frac{x-1}{x^2+2x}$.

$$(x^2+3x+2) \cdot \frac{x-1}{x^2+2x} = \frac{(x+1)(x+2)}{1} \cdot \frac{x-1}{\underline{\qquad\qquad}} \quad \text{Factor}$$

$x(x+2)$

$$= \frac{(x+1)\overset{1}{\cancel{(x+2)}}}{1} \cdot \frac{(x-1)}{x\underset{1}{\cancel{(x+2)}}} \quad \text{Divide out}$$

$\frac{x^2-1}{x}$

$$= \underline{\qquad\qquad} \quad \text{Multiply}$$

EXAMPLE 22 Multiply $\frac{a^2+6a+9}{a^2-16} \cdot \frac{a^2-a-20}{a^2-2a-15}$.

$$\frac{a^2+6a+9}{a^2-16} \cdot \frac{a^2-a-20}{a^2-2a-15} = \frac{(a+3)(a+3)}{(a+4)(a-4)} \cdot \frac{(\underline{\qquad})(\underline{\qquad})}{(a-5)\quad(a+3)}$$

$a-5, a+4$

$$= \frac{(a+3)\overset{1}{\cancel{(a+3)}}}{\underset{1}{\cancel{(a+4)}}(a-4)} \cdot \frac{\overset{1}{\cancel{(a-5)}}\overset{1}{\cancel{(a+4)}}}{\underset{1}{\cancel{(a-5)}}\underset{1}{\cancel{(a+3)}}}$$

$\frac{a+3}{a-4}$

$$= \underline{\qquad\qquad}$$

POINTERS FOR FEWER ERRORS WHEN WORKING WITH FRACTIONS

Multiply $\frac{x^2 + 2x - 8}{x - 8} \cdot \frac{1 - x}{x^2 - 3x + 2}$.

First, recopy the problem exactly as it appears.

This equal sign says both sides are the same thing. Therefore, everything on the left side must also appear on the right.

$$\frac{x^2 + 2x - 8}{x - 8} \cdot \frac{1 - x}{x^2 - 3x + 2} = \frac{(x + 4)(x - 2)}{(x - 8)} \cdot \frac{(1 - x)}{(x - 1)(x - 2)}$$

This side has four distinct parts

$$\frac{A}{B} \cdot \frac{C}{D}$$

This side should have the same four parts, only written in factored form.

$$\frac{A}{B} \cdot \frac{C}{D}$$

Each of the four distinct parts should be equal to its corresponding part.

These match

$$\frac{x^2 + 2x - 8}{x - 8} \cdot \frac{C}{D} = \frac{(x + 4)(x - 2)}{x - 8} \cdot \frac{C}{D}$$

These match

The remaining elements also match

These match

$$\frac{x^2 + 2x - 8}{x - 8} \cdot \frac{1 - x}{x^2 - 3x + 2} = \frac{(x + 4)(x - 2) \cdot (1 - x)}{(x - 8) \cdot (x - 1)(x - 2)}$$

These match

Step 1 Recopied for easy reading:

$$\frac{x^2 + 2x - 8}{x - 8} \cdot \frac{1 - x}{x^2 - 3x + 2} = \frac{(x + 4)(x - 2) \cdot (1 - x)}{(x - 8) \cdot (x - 2)(x - 1)}$$

Continued on page 317

Step 2

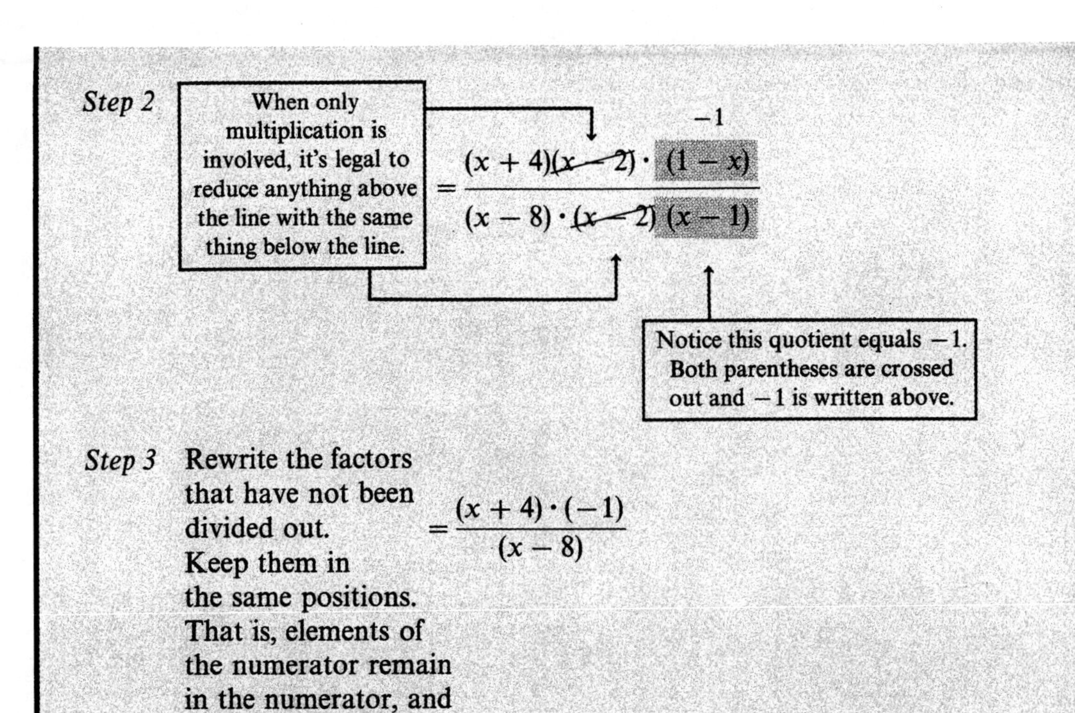

$$= \frac{(x+4)(x-2)\cdot(1-x)}{(x-8)\cdot(x-2)(x-1)}$$

Step 3 Rewrite the factors that have not been divided out. Keep them in the same positions. That is, elements of the numerator remain in the numerator, and elements of the denominator remain in the denominator. There is no way for part of a fraction to move from numerator to denominator.

$$= \frac{(x+4)\cdot(-1)}{(x-8)}$$

Step 4

$$= \frac{-(x+4)}{x-8}$$

There are several other ways to write this answer. The way you choose is no big deal, but here are the reasons we did not pick each of the following.

$\frac{-x-4}{x-8}$	Too many minus signs.
$\frac{-1(x+4)}{x-8}$	A good alternative, but most people find it simpler to omit the 1.
$\frac{x+4}{8-x}$	Another good alternative. It has the fewest minus signs but we prefer to write terms with the variable first.
$\frac{x+4}{-x+8}$	It's very easy to lose the negative sign in the denominator here. This can also lead to difficulties in finding common denominators.

PROBLEM SET 11.2

Multiply and reduce to lowest terms.

1. $\frac{5}{6}\cdot\frac{9}{10}$
2. $\frac{5}{6}\cdot\frac{3}{15}$
3. $\frac{4}{9}\cdot 3$
4. $\frac{3}{8}\cdot 4$
5. $\frac{4}{3}\cdot\frac{3}{4}$
6. $\frac{-1}{7}\cdot 7$
7. $\frac{a}{b}\cdot\frac{c}{d}$
8. $\frac{a}{b}\cdot ab$
9. $\frac{a}{b}\cdot\frac{1}{ab}$
10. $\frac{a}{b}\cdot\frac{b}{a}$
11. $ab\cdot\frac{1}{ab}$
12. $\frac{2x}{y}\cdot\frac{y}{6}$
13. $\frac{2r}{st}\cdot\frac{t}{4r}$
14. $\frac{x}{y^2}\cdot\frac{-y^4}{x^2y}$
15. $\frac{x^2}{2y^2}\cdot\frac{12y^2}{3x^3}$
16. $\frac{-4m}{9n}\cdot\frac{18n^3}{2m^4}$
17. $\frac{5r}{3s}\cdot\frac{6s^3t}{15r^4t^2}$
18. $\frac{x+2}{3}\cdot\frac{3}{2}$
19. $\frac{x+1}{4}\cdot\frac{4}{3}$
20. $\frac{a}{b+4}\cdot\frac{b}{a}$
21. $\frac{x}{y}\cdot\frac{y}{x+y}$
22. $(x+2)\cdot\frac{1}{x+2}$
23. $\frac{x+2}{3}\cdot\frac{2}{x+2}$
24. $\frac{x+2}{x+3}\cdot\frac{3}{x+2}$
25. $\frac{1}{x-y}\cdot(y-x)$
26. $\frac{2+2x}{2}\cdot\frac{3}{x+1}$
27. $\frac{a-2}{2}\cdot\frac{a}{a^2-2a}$
28. $\frac{x^2+2x}{x+2}\cdot\frac{x+1}{x}$
29. $\frac{3x-12}{5x+15}\cdot\frac{5x-20}{4x-16}$
30. $\frac{2c^2+2c}{c^2+2c}\cdot\frac{c+2}{2}$
31. $\frac{d^2+d}{3d}\cdot\frac{d+3}{3d+3}$
32. $\frac{y+1}{4y^2+4y}\cdot\frac{y^2+4y}{y-4}$
33. $\frac{2x-6}{x^2+3x}\cdot\frac{x^2}{12-3x}$
34. $\frac{x^3-2x^2}{x+2}\cdot\frac{x}{6x^4-3x^5}$
35. $\frac{3x-9}{2x^2-4x}\cdot\frac{4x}{3-x}$
36. $\frac{x^2}{x^3+x^2}\cdot(1+x)$
37. $\frac{y^3-ay^2}{y}\cdot\frac{a}{y-a}$
38. $\frac{x-3}{x^3-3x^2}\cdot\frac{3x^3}{3+x}$
39. $\frac{y^2-1}{y}\cdot\frac{1}{y+1}$
40. $\frac{m}{m-n}\cdot\frac{m^2-n^2}{mn}$
41. $\frac{x+9}{x^2+6x+9}\cdot\frac{x^2-9}{x-3}$
42. $\frac{x^2+6x+5}{x^2+6x+8}\cdot\frac{x+2}{x+1}$
43. $\frac{x^2+2x-8}{x-8}\cdot\frac{x-1}{x^2-3x+2}$
44. $\frac{3x-12}{12x^2}\cdot\frac{8x}{x^2-5x+4}$
45. $\frac{x^2}{4x^2+16x}\cdot\frac{x^2+x-12}{x^2-12x}$
46. $\frac{a^2-25a}{a^2-25}\cdot\frac{3a^2-13a-10}{3a^2+2a}$
47. $\frac{x^2-y^2}{x^2+y^2}\cdot\frac{x^2+2xy}{x^2+xy-2y^2}$
48. $\frac{a^2+4a}{a^2+4a+4}\cdot\frac{2a+4}{a+4}$
49. $\frac{x^2+8x+16}{x^2-16}\cdot\frac{4x+12}{x^2+7x+12}$
50. $\frac{x^2+2x}{x^2+3x+2}\cdot\frac{2x+2}{2x+1}$
51. $\frac{x-3}{x^2-6x+9}\cdot\frac{9-x^2}{3x^3+9x^2}$
52. $\frac{x^2-3x-10}{4x^3+8x^2}\cdot\frac{2x^2-9x-5}{x^2-10x+25}$

53. $\dfrac{2m - m^2}{2m + 6} \cdot \dfrac{m^2 + m - 6}{m^2 - 4m + 4}$

54. $\dfrac{y^2 - 3y - 4}{y^2 - y - 2} \cdot \dfrac{y^2 - 5y + 6}{y^2 - 2y - 3}$

55. $\dfrac{3x^2 - 4x - 4}{3x^2 + 11x + 6} \cdot \dfrac{2x^2 - 9x + 4}{x^2 - 6x + 8}$

56. $\dfrac{3x + 9y}{3x} \cdot \dfrac{x^2 - xy}{3x^2 - 3y^2}$

POINTERS FOR BETTER STYLE

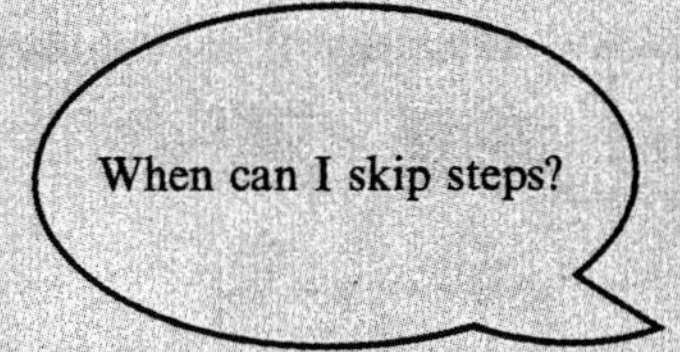

Usually, whenever it's comfortable. Read on.

Frequently when doing algebra, it's easy to see the result several steps ahead. For example, we know:

$$-1(a - b) = (b - a)$$

and

$$\frac{(a - b)}{(b - a)} = -1$$

In both these examples, several steps were skipped.

In general, you can skip steps whenever it's reasonable to expect the reader to be able to do the operations mentally.

Some examples

You can skip steps

- when a standard formula is used

  $(a + b)^2 = a^2 + 2ab + b^2$
- when the skipped steps are obvious
- when the skipped steps are simple arithmetic

You should not skip steps

- when you are doing a new process
- when it's possible to confuse your reader
- when it's possible to lose track of quantities or negative signs

If you try to skip too many steps at once you'll be in trouble.

I know, when I try to do too many things in my head, I make careless mistakes.

11.3 DIVISION OF ALGEBRAIC FRACTIONS

Inverses and Identity Elements

Remember that:

1. Zero is the identity element for addition.

$$a + 0 = a$$

Identity element

2. Two expressions are additive inverses of each other if their sum is zero.

$$2 + (-2) = 0 \quad \text{and} \quad -3x + 3x = 0$$

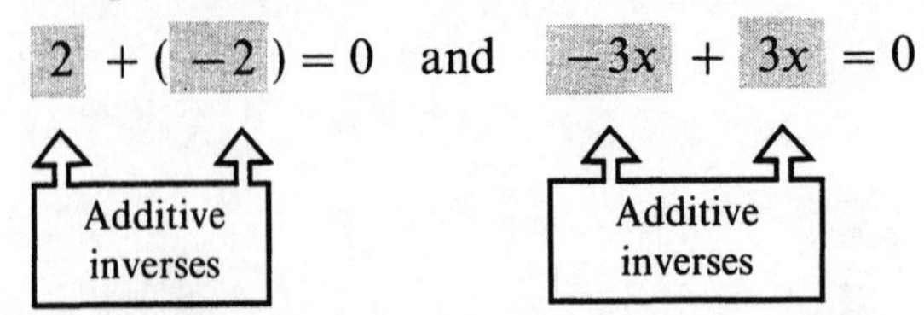

3. Subtraction of a number is defined as addition of the additive inverse of that number.

$$a - b = a + (-b)$$

We use similar properties for multiplication:

4. One is the identity element for multiplication.

$$a \cdot 1 = a$$

Identity element

5. Two expressions are multiplicative inverses of each other if their product is 1.

$$2 \cdot \frac{1}{2} = 1 \quad \text{and} \quad \frac{x}{3} \cdot \frac{3}{x} = 1$$

Multiplicative inverses   Multiplicative inverses

6. Division by a number is defined as multiplication by the multiplicative inverse of the number.

$$a \div b = a \cdot \frac{1}{b} \quad \text{where } b \neq 0$$

A comparison of addition and multiplication follows.

	Addition	Multiplication
Identity element	0	1
Identity law	$0 + a = a$	$1 \cdot a = a$
Inverses	$a + (-a) = 0$	$a \cdot \frac{1}{a} = 1$ where $a \neq 0$
How the inverse operation is defined	$a - b = a + (-b)$	$a \div b = a \cdot \frac{1}{b}$ where $b \neq 0$

EXAMPLE 23

Some pairs of multiplicative inverses.

Number	Multiplicative inverse	
4	$\frac{1}{4}$	
$\frac{1}{4}$	4	
$\frac{3}{4}$	$\frac{4}{3}$	
x	$\frac{1}{x}$	where $x \neq 0$
$\frac{x}{y}$	$\frac{y}{x}$	where $x, y \neq 0$
$\frac{-2a}{3b}$	$\frac{-3b}{2a}$	where $a, b \neq 0$
$x - 6$	$\frac{1}{x-6}$	where $x - 6 \neq 0$
$\frac{a+b}{a-b}$	$\frac{a-b}{a+b}$	where $a + b \neq 0$, $a - b \neq 0$

Try some of the products and see if they aren't equal to 1.

Why the restrictions?

You can't divide by zero.

Each pair of numbers in Example 23 are multiplicative inverses because their product is 1. Now, what can you multiply zero by to get 1 for an answer?

$$0 \cdot \square = 1$$

HOLD IT. ZERO TIMES ANYTHING IS ZERO, NOT 1.

The zero factor law says anything times zero is zero. Therefore, it is impossible to find a value to place in the box so that $0 \cdot \square = 1$. Zero is the only rational number without a multiplicative inverse.

Because division requires a multiplicative inverse, and zero has no multiplicative inverse,

DIVISION BY ZERO IS IMPOSSIBLE.

If you still don't believe division by zero is impossible consider this

$$6 \div 3 = 2 \quad \text{because} \quad 3 \cdot 2 = 6$$

Now try $6 \div 0$. Suppose there is an answer and we called it A. Then, if $6 \div 0 = A$, it would also be true that $0 \cdot A = 6$. There is no A such that $0 \cdot A = 6$. So we conclude that division by zero is impossible. This is why, in the table showing multiplicative inverses, values for variables that would attempt division by zero have been excluded. For the remainder of this book, we will assume that the values of all variables are restricted so that division by zero never occurs.

PROBLEM SET 11.3A

Give the multiplicative inverse of the following.

1. 3
2. $\frac{-1}{4}$
3. $\frac{3}{8}$
4. $\frac{-7}{9}$
5. $\frac{-4}{3}$
6. a
7. $\frac{1}{b}$
8. $\frac{a}{b}$
9. $\frac{4x}{3y}$
10. $\frac{-r}{5s}$
11. $\frac{6x^2}{7yz}$
12. $a + 2$
13. $3 - a$
14. $\frac{1}{5 - x}$
15. $\frac{1}{b - 3}$
16. $\frac{a + 2}{b - 3}$
17. $x + y$
18. $\frac{1}{x + y}$
19. $\frac{x + 2y}{3x - y}$
20. 1
21. 0
22. $1 - b$
23. -1
24. $b - 1$

Review Your Skills

Multiply the following fractions. Write all answers in lowest terms. Use the standard form for fractions.

25. $\frac{3}{5} \cdot \frac{2}{6}$
26. $-8 \cdot \frac{5}{16}$
27. $\frac{-15}{4} \cdot \frac{8b}{-5}$
28. $\frac{-6x}{35} \cdot 5$
29. $\frac{-33y^2}{15} \cdot \frac{25x}{-22}$
30. $\frac{81ay}{-32} \cdot \frac{-48y^2}{9}$

Division of Algebraic Fractions

Using the multiplicative inverse concept, we can define division for algebraic fractions.

Division of algebraic fractions

> **11.3A DEFINITION**
>
> To divide by an algebraic expression, multiply by its multiplicative inverse.

EXAMPLE 24 Divide $2x^2$ by $4x$.

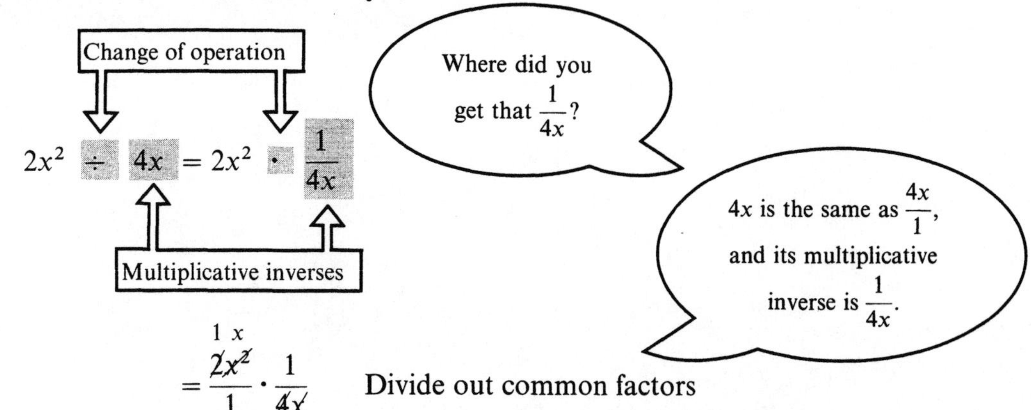

$= \frac{\overset{1\ x}{\cancel{2}\cancel{x^2}}}{1} \cdot \frac{1}{\underset{2\ 1}{\cancel{4}\cancel{x}}}$ Divide out common factors

$= \frac{x}{2}$ Multiply the remaining factors

EXAMPLE 25 Divide $\frac{3a}{4} \div \frac{15}{a^3}$.

$\frac{a^3}{15}$

$\frac{3a}{4} \div \frac{15}{a^3} = \frac{3a}{4} \cdot$ ______ Multiply by the multiplicative inverse

$= \frac{\overset{1}{\cancel{3}}a}{4} \cdot \frac{a^3}{\underset{5}{\cancel{15}}}$ Divide out

$\frac{a^4}{20}$

$=$ ______ Multiply

EXAMPLE 26 Divide $\frac{4p^2}{9q^3} \div \frac{16p^2q}{15}$.

$\frac{15}{16p^2q}$

$\frac{4p^2}{9q^3} \div \frac{16p^2q}{15} = \frac{4p^2}{9q^3} \cdot$ ________

$= \frac{\overset{1\ 1}{\cancel{4}\cancel{p^2}}}{\underset{3}{\cancel{9}}q^3} \cdot \frac{\overset{5}{\cancel{15}}}{\underset{4}{\cancel{16}}\,\underset{1}{\cancel{p^2}}q}$

$\frac{5}{12q^4}$

$=$ ________

PROBLEM SET 11.3B

Divide and reduce to lowest terms.

(4) Hint: $18 = \frac{18}{1}$

1. $\frac{4}{10} \div \frac{3}{5}$
2. $\frac{6}{9} \div \frac{2}{3}$
3. $\frac{4}{3} \div 12$
4. $\frac{6}{5} \div 18$
5. $9 \div \frac{3}{2}$
6. $\frac{6}{7} \div \frac{6}{7}$
7. $\frac{a}{b} \div \frac{c}{d}$
8. $\frac{a}{b} \div ab$
9. $\frac{a}{b} \div \frac{1}{ab}$
10. $\frac{a}{b} \div \frac{b}{a}$
11. $\frac{a}{b} \div \frac{a}{b}$
12. $\frac{x}{8} \div \frac{x}{2y}$
13. $x \div \frac{xy}{a}$
14. $ab \div \frac{a}{c}$
15. $xy \div \frac{x^2y}{z}$
16. $\frac{x^2}{y^2} \div x^2y$
17. $\frac{3r^2}{12s} \div \frac{r^2t}{8s^2}$
18. $\frac{-2ab^2}{9c^3} \div \frac{6a^3b^3}{3c^4}$

Division Involving Factoring

EXAMPLE 27 Divide $\frac{2x^2 + 7x + 6}{2x^2 - x - 15} \div \frac{6x + 9}{x^2 - 9}$.

$$\frac{2x^2 + 7x + 6}{2x^2 - x - 15} \div \frac{6x + 9}{x^2 - 9} = \frac{2x^2 + 7x + 6}{2x^2 - x - 15} \cdot \_\_\_\_\_\_ \quad \text{Multiply by the multiplicative inverse}$$

$$= \frac{(\_\_\_\_)(\_\_\_\_)}{(2x + 5)\quad(x - 3)} \cdot \frac{(x + 3)(x - 3)}{3(2x + 3)} \quad \text{Factor}$$

$$= \frac{\overset{1}{\cancel{(2x + 3)}}(x + 2)}{(2x + 5)\underset{1}{\cancel{(x - 3)}}} \cdot \frac{(x + 3)\overset{1}{\cancel{(x - 3)}}}{3\underset{1}{\cancel{(2x + 3)}}} \quad \text{Divide out common factors}$$

$$= \_\_\_\_\_\_ \quad \text{Multiply}$$

$\frac{x^2 - 9}{6x + 9}$

$2x + 3$, $x + 2$

$\frac{(x + 2)(x + 3)}{3(2x + 5)}$

We frequently leave answers like the one above in factored form. Not only is it less work, but it is often more useful.

EXAMPLE 28 Perform the indicated operations.

$$\frac{a^2 - b^2}{b^3 - ab^2 - 2a^2b} \div \frac{a - b}{2a^2 - 3ab + b^2} \cdot \frac{b^2}{a^2 - ab}$$

$$= \frac{a^2 - b^2}{b^3 - ab^2 - 2a^2b} \cdot \frac{2a^2 - 3ab + b^2}{a - b} \cdot \frac{b^2}{a^2 - ab}$$

$$= \frac{\overset{1}{\cancel{(a - b)}}\overset{1}{\cancel{(a + b)}}}{\underset{1}{\cancel{b}}(b - 2a)\underset{1}{\cancel{(b + a)}}} \cdot \frac{(2a - b)\overset{1}{\cancel{(a - b)}}}{\underset{1}{\cancel{a - b}}} \cdot \frac{\overset{b}{\cancel{b^2}}}{a\underset{1}{\cancel{(a - b)}}}$$

$\frac{b(2a - b)}{a(b - 2a)}$

$$= \underline{\qquad\qquad\qquad}$$

Can this be reduced further?

Yes. Read what follows.

Remember, we can factor out $a - 1$.

Therefore,

$$\frac{b\,(2a - b)}{a(b - 2a)} = \frac{b\,(-1)(-2a + b)}{a(b - 2a)}$$

$$= \frac{b(-1)(b - 2a)}{a(b - 2a)}$$

$$= \frac{b(-1)\overset{1}{\cancel{(b - 2a)}}}{a\underset{1}{\cancel{(b - 2a)}}}$$

$\frac{-b}{a}$

$$= \underline{\qquad}$$

PROBLEM SET 11.3C

Perform the indicated operation.

1. $\dfrac{x}{y} \div \dfrac{x}{x + y}$

2. $\dfrac{x + y}{x} \div \dfrac{x + y}{y}$

3. $\dfrac{a - b}{a} \div \dfrac{a - b}{b}$

4. $\dfrac{a^2 + b}{b} \div \dfrac{a^2 + b}{a}$

5. $(x - y) \div (x - y)$

6. $(x - y) \div (y - x)$

7. $\dfrac{a - 1}{a} \div (1 - a)$

8. $\dfrac{2h}{h - 2} \div \dfrac{h}{h - 2}$

9. $\dfrac{3x + 6}{6} \div \dfrac{x + 2}{x}$

10. $\dfrac{x + 3}{x^2} \div \dfrac{x - 3}{x^2 - 3x}$

11. $\dfrac{4x - 4}{4x + 4} \div \dfrac{3x - 3}{3x + 1}$

12. $\dfrac{5x - 5}{6x + 6} \div \dfrac{7x - 7}{5x + 1}$

13. $\dfrac{3x+3}{3x^4+3x^3} \div \dfrac{x+1}{x^5}$

14. $\dfrac{2a-4}{2a^2-4a} \div \dfrac{2}{a^2+2a}$

15. $\dfrac{x^2-xy}{x^2} \div \dfrac{y^2-xy}{y^2}$

16. $\dfrac{2s+2}{2s^3+2s^2} \div \dfrac{s+2}{s^3}$

17. $\dfrac{x^2-9}{x^2-3} \div (x-3)$

18. $\dfrac{a+b}{a-b} \div \dfrac{a^2+b^2}{a^2-b^2}$

19. $\dfrac{y^2+4y+4}{4y+4} \div \dfrac{y+2}{2y+2}$

20. $\dfrac{x^2+3x+2}{x^2-2x-3} \div \dfrac{3x+6}{x+3}$

21. $\dfrac{a^2+5a+6}{a^2+3} \div \dfrac{2a^2+5a+2}{2a^2+a}$

22. $\dfrac{4x^2-4x+1}{4x-2} \div \dfrac{4x^2-2x}{4x^2}$

23. $\dfrac{y^2-x^2}{x+y} \div \dfrac{2x^2-3xy+y^2}{x-2y}$

24. $(6x^2+12x) \div \dfrac{3x^2-3x-18}{4x^2-11x-3}$

25. $\dfrac{y^3-3y^2}{y^2+3y+2} \div \dfrac{y^3-9y}{3y^2-6y}$

26. $\dfrac{y^5-4y^4+4y^3}{4y^2-12y+8} \div \dfrac{y^3+2y^2-8y}{2y^2+6y-8}$

27. $\dfrac{x^2-4x}{2x^2-10x+8} \div \dfrac{2x^2-4x}{4x^2-4}$

28. $\dfrac{x^2}{6y} \div \dfrac{x^2}{4y^2} \cdot \dfrac{3}{4x^2}$

29. $\dfrac{x^2+1}{x^2} \div \dfrac{x^3}{x^2+x} \cdot \dfrac{x}{x+1}$

30. $\dfrac{2x+8}{x^2-3x} \cdot \dfrac{x^2}{ax+3a} \div \dfrac{x+4}{x^2-9}$

31. $\dfrac{x^2+5x+4}{x^2-2x-8} \div \dfrac{x^2-6x+9}{x^2-7x+12} \cdot \dfrac{x^2-2x-3}{x^2+4x}$

32. $\dfrac{x^2+2x+1}{x^2-x} \cdot \dfrac{x^2+2x-3}{x^2-3x-4} \div \dfrac{x^2+4x+3}{x^2+x-20}$

Review Your Skills Perform the indicated operation.

33. $\dfrac{13}{16} + \dfrac{-9}{16}$

34. $\dfrac{-8x}{15} + \dfrac{11x}{15}$

35. $\dfrac{-7}{18} + \dfrac{13}{18}$

36. $\dfrac{-13ax}{24} + \dfrac{11ax}{24}$

Review 11

NOW THAT YOU HAVE COMPLETED UNIT 11, YOU SHOULD BE ABLE TO APPLY THE FOLLOWING RULES:

To reduce an algebraic fraction to lowest terms, divide out common factors in the numerator and denominator.

$$\frac{a \cdot x}{b \cdot x} = \frac{a}{b}$$

To multiply algebraic fractions:

1. Factor completely.
2. Divide out common factors.
3. Multiply the remaining factors in the numerator to get the numerator of the product, then multiply the remaining factors in the denominator to get the denominator of the product.

To divide by an algebraic expression, multiply by its multiplicative inverse.

$$\frac{a}{b} \div \frac{c}{d} = \frac{a}{b} \cdot \frac{d}{c}$$

Review Test 11

Reduce to lowest terms. ***(11.1)***

1. $\dfrac{3a+9}{9} =$ ______

2. $\dfrac{a^2+ab}{b^2+ab} =$ ______

3. $\dfrac{x^2-5x+6}{3-x} =$ ______

4. $\dfrac{x^2+2x-8}{x^2-8x+12} =$ ______

Perform the indicated operations. Reduce all answers to lowest terms. ***(11.2 and 11.3)***

5. $\dfrac{x-3}{3} \cdot \dfrac{x}{x^2-3x} =$ ______

6. $\dfrac{a^2-b^2}{a^2+b^2} \cdot \dfrac{2a^2b-4ab^2}{a^2-ab-2b^2} =$ ______

7. $\dfrac{y^2-y-2}{y^2+2y} \cdot \dfrac{y^2+5y+6}{y^2+4y+3} =$ ______

8. $\dfrac{a}{b^3} \cdot \dfrac{-b^4}{a^2b} =$ ______

9. $\dfrac{a^2b}{a+b} \div \dfrac{a}{a+b} =$ ______

10. $\dfrac{3a-6}{2a^2-4a} \div \dfrac{3}{a^2+2a} =$ ______

11. $\dfrac{y^2-x^2}{x^2-3xy+2y^2} \div \dfrac{x^2+xy}{x-2y} =$ ______

12. $\dfrac{x^2-x-2}{x^2-3x-4} \div \dfrac{2x-4}{4x-x^2} =$ ______

13. $\dfrac{x^2+x-12}{x^2+6x+5} \cdot \dfrac{2x^2-8x}{x^2-2x-3} \div \dfrac{x^2-16}{x^2+2x+1} =$ ______

REVIEW YOUR SKILLS

Simplify. Write the results without negative exponents.

14. $(x^{-4}y^3)(x^{-4}y^{-5})$

15. $\left(\frac{-3x^2}{2yw}\right)^3$

Evaluate the following for $a = -2$, $b = 3$, $c = -1$.

16. $(a + b + c)^3 - a \cdot b \cdot c$

Find the following products.

17. $6a(5a - 4)$

18. $(b - 4)(b + 4)$

19. $(2x - 3y)(2x + 3y)$

20. $(3x - 4)(2x^2 + 3x - 6)$

Find the quotient.

21. $\frac{9a^2b - 15a^3b^3 - 18ab^4}{6a^2b^2}$

Use long division to find the quotient.

22. $(6x^3 - 7x^2 - 4x - 5) \div (3x - 2)$

UNIT 12

Addition and Subtraction of Fractions

OBJECTIVES

After you have successfully completed this unit, you will be able to:

1. Add and subtract algebraic fractions with common denominators (12.1)
2. Add and subtract algebraic fractions without common denominators (12.2)
3. Simplify complex fractions (12.3)
4. Solve equations involving fractions (12.4)
5. Solve word problems involving fractional equations (12.5)

12.1 ADDITION AND SUBTRACTION OF ALGEBRAIC FRACTIONS WITH COMMON DENOMINATORS

Recall from Unit 3 the definition of addition of fractions.

Addition of fractions

12.1A DEFINITION

To add two algebraic fractions with a common denominator, add the numerators and place the result over the common denominator.

$$\frac{a}{c} + \frac{b}{c} = \frac{a+b}{c}$$

EXAMPLE 1 Add $\frac{1}{x} + \frac{2}{x}$.

$$\frac{1}{x} + \frac{2}{x} = \frac{1+2}{x}$$

$$= \frac{3}{x}$$

This is what you do when you say $\frac{1}{4} + \frac{2}{4} = \frac{3}{4}$.

Yes, except now our denominators are both x instead of 4.

EXAMPLE 2 Add $\frac{a+4}{a}+\frac{a-1}{a}$.

$$\frac{a+4}{a}+\frac{a-1}{a}=\frac{a+4+a-1}{a}$$ Addition of fractions

$$=\frac{2a+3}{a}$$ Combine like terms

EXAMPLE 3 Add $\frac{x-3}{2x}+\frac{3-x}{2x}$.

$$\frac{x-3}{2x}+\frac{3-x}{2x}=\frac{x-3+3-x}{2x}$$ Add numerators

$$=\frac{0}{2x}$$ Combine like terms

0 $\quad = \_\_\_\_$

Subtraction of fractions with common denominators was also defined in Unit 3:

Subtraction of fractions

12.1B DEFINITION

To subtract a fraction, add its additive inverse.

$$\frac{a}{c}-\frac{b}{c}=\frac{a}{c}+\frac{-b}{c}$$

EXAMPLE 4 Subtract $\frac{5}{7}-\frac{3}{7}$.

$$\frac{5}{7}-\frac{3}{7}=\frac{5}{7}+\frac{-3}{7}$$ Rewrite as addition

$$=\frac{5+(-3)}{7}$$ Add numerators

$$=\frac{2}{7}$$

EXAMPLE 5 Subtract $\frac{3a}{4b}-\frac{5a}{4b}$.

When a fraction is negative, associate the negative sign with the numerator.

$$\frac{3a}{4b}-\frac{5a}{4b}=\frac{3a}{4b}+\frac{-5a}{4b}$$

$$=\frac{3a+(-5a)}{4b}$$

$$=\frac{-2a}{4b}$$ Reduces to

$\frac{-a}{2b}$ $\quad = \_\_\_\_$

EXAMPLE 6 Subtract $\dfrac{x}{x+1} - \dfrac{1}{x+1}$.

$$\frac{x}{x+1} - \frac{1}{x+1} = \frac{x}{x+1} + \frac{-1}{x+1} \quad \text{Rewrite as addition}$$

$$= \frac{x+(-1)}{x+1} \quad \text{Add numerators}$$

$$= \frac{x-1}{x+1} \quad \text{Simplify}$$

It looks like it would be easier to just subtract numerators.

That's true for simple cases, but with complex numerators it's better to use the definition of subtraction.

EXAMPLE 7 Subtract $\dfrac{6a+1}{a-b} - \dfrac{9a-3}{a-b}$.

Two expressions are additive inverses if their sum is zero.

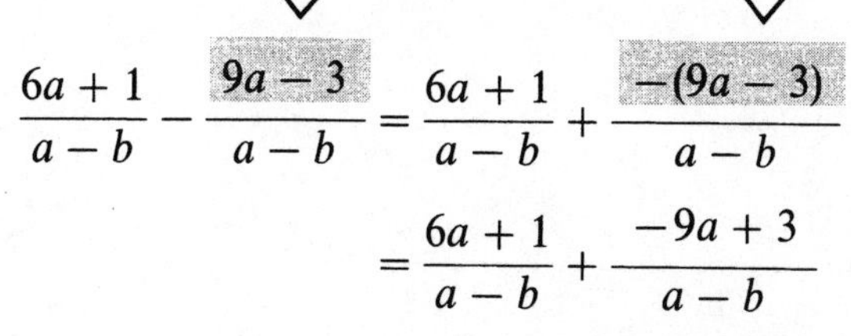

$$\frac{6a+1}{a-b} - \frac{9a-3}{a-b} = \frac{6a+1}{a-b} + \frac{-(9a-3)}{a-b} \quad \text{To subtract, add the additive inverse}$$

$$= \frac{6a+1}{a-b} + \frac{-9a+3}{a-b} \quad \text{Remove parentheses}$$

$$= \frac{6a+1+(-9a+3)}{a-b}$$

$$= \frac{-3a+4}{a-b} \quad \text{or}$$

$$= \frac{4-3a}{a-b}$$

It shows class to avoid negative signs in the first term and it uses one less symbol.

EXAMPLE 8 Subtract $\dfrac{3x-y}{x-y} - \dfrac{x+y}{x-y}$.

$$\frac{3x-y}{x-y} - \frac{x+y}{x-y} = \frac{3x-y}{x-y} + \frac{-(x+y)}{x-y} \quad \text{Rewrite as addition}$$

$-x-y$

$$= \frac{3x-y}{x-y} + \frac{\underline{\qquad}}{x-y} \quad \text{Remove parentheses}$$

$$= \frac{3x-y+(-x-y)}{x-y} \quad \text{Add numerators}$$

$2x-2y$

$$= \frac{\underline{\qquad}}{x-y} \quad \text{Combine like terms}$$

$$= \frac{2(x-y)}{x-y}$$

$$= 2$$

If we factor the numerator, we can reduce the answer.

Right.

PROBLEM SET 12.1

Add or subtract as indicated and reduce to lowest terms.

1. $\frac{1}{5}+\frac{3}{5}$

2. $\frac{3}{10}+\frac{5}{10}$

3. $\frac{5}{8}-\frac{3}{8}$

4. $\frac{7}{10}-\frac{3}{10}$

5. $\frac{6}{11}+\frac{5}{11}$

6. $\frac{4}{7}+\frac{3}{7}$

7. $\frac{x}{9}+\frac{y}{9}$

8. $\frac{a}{9}-\frac{3}{9}$

9. $\frac{x}{9}-\frac{y}{9}$

10. $\frac{x}{9}+\frac{x}{9}$

11. $\frac{x}{9}-\frac{x}{9}$

12. $\frac{2x}{9}+\frac{x}{9}$

13. $\frac{x^2}{9}+\frac{x}{9}$

14. $\frac{9}{x}+\frac{9}{x}$

15. $\frac{2}{x+2}+\frac{4}{x+2}$

16. $\frac{8}{x+2}-\frac{5}{x+2}$

17. $\frac{1}{a+5}+\frac{3}{a+5}$

18. $\frac{y}{y-3}+\frac{y+1}{y-3}$

19. $\frac{a}{y+1}-\frac{a+2}{y+1}$

20. $\frac{a-1}{y+2}-\frac{3}{y+2}$

21. $\frac{y}{y-3}-\frac{y+3}{y-3}$

22. $\frac{x+3}{x-1}+\frac{x-4}{x-1}$

23. $\frac{x+3}{x-1}-\frac{x-4}{x-1}$

24. $\frac{x+1}{2x+3}+\frac{x+2}{2x+3}$

25. $\frac{x-2}{x+1}+\frac{x+4}{x+1}$

26. $\frac{3y-3}{y-2}+\frac{2y-7}{y-2}$

27. $\frac{3y-3}{y-2}-\frac{y-1}{y-2}$

28. $\frac{x}{x^2-1}+\frac{1}{x^2-1}$

29. $\frac{m-6}{m^2-1}+\frac{m+8}{m^2-1}$

30. $\frac{m-6}{m^2-4}+\frac{m+2}{m^2-4}$

31. $\frac{4x+8}{x^2+x-12}+\frac{4-x}{x^2+x-12}$

32. $\frac{x^2+x}{x^2-9}-\frac{x^2+3}{x^2-9}$

33. $\frac{x^2+5x+6}{x^2+5x+6}+\frac{x^2+2x-3}{x^2+5x+6}$

34. $\frac{2x^2-3x-4}{x^2-4x+3}-\frac{x^2-x-1}{x^2-4x+3}$

35. $\frac{x^2+2x+1}{x^2+3x+2}-\frac{x^2+x-1}{x^2+3x+2}$

36. $\frac{3x^2+5x-1}{x^2+7x+12}-\frac{2x^2+3x+2}{x^2+7x+12}$

Review Your Skills Perform the indicated operations. Reduce all answers to lowest terms and write fractions in standard form.

37. $\frac{-9}{8}-\frac{-7a}{6}$

38. $\frac{-11xy}{18}-\frac{-7}{12}$

39. $\frac{-ay^2}{6}+\frac{9ay^2}{14}$

40. $\frac{-13bx^2}{18}-\frac{-5bx^2}{4}$

12.2 ADDITION AND SUBTRACTION OF ALGEBRAIC FRACTIONS WITHOUT COMMON DENOMINATORS

We can only add or subtract fractions if their denominators are identical.

In cases where two fractions do not have identical denominators, we make the denominators identical by using the "multiply by 1" routine.

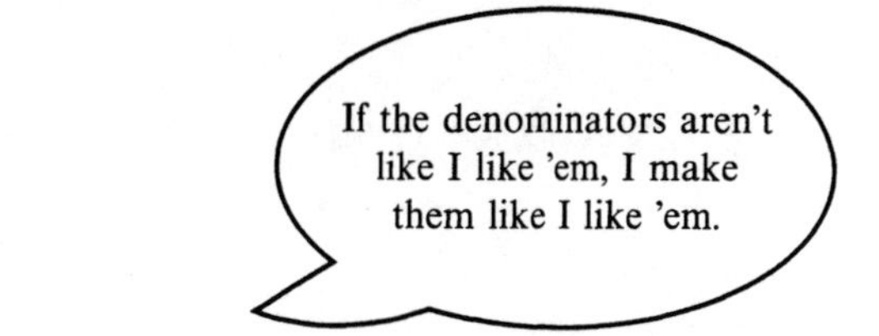

EXAMPLE 9 Add $\frac{3}{4} + \frac{5}{6}$.

We can find the least common denominator for two arithmetic fractions by using the method explained in Section 3.3.

For this example, the least common denominator is 12.

First rewrite the problem with more space, then use the "multiply by 1" technique to write both fractions as equivalent fractions with denominators of 12.

$$\frac{3}{4} + \frac{5}{6} = \frac{3}{4} \qquad + \frac{5}{6}$$

$$= \frac{3}{4} \cdot \left(\frac{3}{3}\right) + \frac{5}{6} \cdot \left(\frac{2}{2}\right)$$

The "multiply by 1" technique

$$= \frac{9}{12} + \frac{10}{12}$$

$$= \frac{19}{12}$$

EXAMPLE 10 Subtract $\frac{3}{8} - \frac{1}{6}$.

$$\frac{3}{8} - \frac{1}{6} = \frac{3}{8} \qquad + \frac{-1}{6}$$ Write as an addition
Leave space to multiply by 1

$$= \frac{3}{8}\left(\frac{3}{3}\right) + \frac{-1}{6}\left(\frac{4}{4}\right)$$ Multiply by 1 to get common demoninators

$$= \frac{9}{24} + \frac{-4}{24}$$

$$= \frac{9 + (-4)}{24}$$ Add numerators

$$= \underline{\qquad}$$

$\frac{5}{24}$

EXAMPLE 11 Add $\frac{3}{x} + \frac{4}{y}$.

For algebraic fractions, we build a common denominator that contains every factor used in any denominator of the fractions to be added. In this case, $x \cdot y$ has both x and y as factors so we'll use it for our common denominator.

$$\frac{3}{x} + \frac{4}{y} = \frac{3}{x}\left(\frac{y}{y}\right) + \frac{4}{y}\left(\frac{x}{x}\right)$$

The expressions you multiply each fraction by to get common denominators are always equal to 1.

$$= \frac{3y}{xy} + \frac{4x}{xy}$$

$$= \frac{3y + 4x}{xy}$$

Addition of fractions without common denominators

12.2 TO ADD FRACTIONS WITHOUT COMMON DENOMINATORS, FOLLOW THESE STEPS:

1. Find a common denominator.
   a. Factor each denominator.
   b. Write each different factor that appears in any denominator.
   c. Raise each factor to the highest power it has in any *single* denominator.

The result of this process is the least common denominator (LCD).

2. Convert all fractions to be added to equivalent fractions with the common denominator by supplying the missing factors as 1's.
3. Add the numerators of the equivalent fractions and place the result over the common denominator.
4. Reduce to lowest terms.

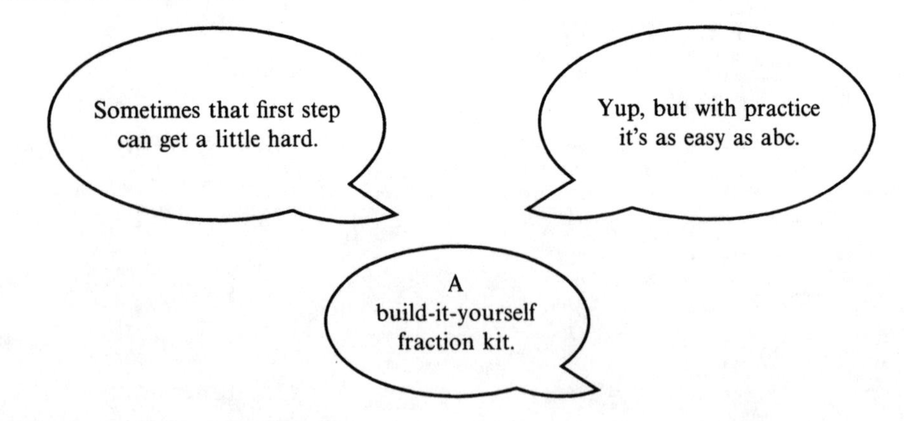

EXAMPLE 12 Add $\frac{5}{xy} + \frac{3}{y^2z}$.

Find the common denominator:

The denominators are already factored.

Write each factor.

$x \quad y \quad z$

Raise to highest powers.

$x \quad y^2 \quad z$

xy^2z is the least common denominator.

Convert to equivalent fractions with the common denominator xy^2z:

Supply the missing factors

$$\frac{5}{xy} + \frac{3}{y^2z} = \frac{5}{xy} \cdot \frac{yz}{yz} + \frac{3}{y^2z} \cdot \frac{x}{x}$$

$$= \frac{5yz}{xy^2z} + \frac{3x}{xy^2z}$$

Add:

$$= \frac{5yz + 3x}{xy^2z}$$

EXAMPLE 13 Subtract $\dfrac{5}{6pq^2} - \dfrac{7}{9p^3q}$.

To *subtract*, first convert to an addition problem and then add:

$$\frac{5}{6pq^2} - \frac{7}{9p^3q} = \frac{5}{6pq^2} + \frac{-7}{9p^3q}$$

Find the common denominator:

Factor each denominator.

$$6pq^2 = 2 \cdot 3 \cdot p \cdot q^2$$
$$9p^3q = 3^2 \cdot p^3 \cdot q$$

Write the different factors.

$$2 \cdot 3 \cdot p \cdot q$$

Raise each factor to its highest power in any one denominator.

$2 \cdot 3^2 \cdot p^3 \cdot q^2$

LCD = ______ · ______ · ______ · ______

Convert to fractions with common denominators:

$$\frac{5}{6pq^2} + \frac{-7}{9p^3q} = \frac{5}{2 \cdot 3 \cdot pq^2} + \frac{-7}{3 \cdot 3 \cdot p^3q}$$

$$= \frac{5}{2 \cdot 3 \cdot pq^2} \cdot \frac{3p^2}{3p^2} + \frac{-7}{3 \cdot 3 \cdot p^3q} \cdot \frac{2q}{2q}$$

If you leave the denominators in factored form, it's easier to see what is needed.

Missing factors

$$= \frac{15p^2}{18p^3q^2} + \frac{-14q}{18p^3q^2}$$

$\dfrac{15p^2 - 14q}{18p^3q^2}$

Add: = ______________

EXAMPLE 14 Subtract $\dfrac{a}{a+2} - \dfrac{5}{a}$.

Write the problem as an addition problem.

$\dfrac{a}{a+2} + \dfrac{-5}{a}$

$$\frac{a}{a+2} - \frac{5}{a} = \underline{\hspace{4cm}}$$

Find the common denominator:

The denominators are already factored.

The factors are $(a + 2)(a)$.

Since each factor is used only once, they are already raised to highest powers.

$(a + 2)(a)$

The LCD is ______.

Convert to fractions with a common denominator:

$$\frac{a}{a+2} + \frac{-5}{a} = \frac{a}{a+2} \cdot \left(\frac{a}{a}\right) + \frac{-5}{a} \cdot \left(\frac{a+2}{a+2}\right)$$

Supplying missing factors in the denominators

$$= \frac{a^2}{a(a+2)} + \frac{-5a-10}{a(a+2)}$$

$\dfrac{a^2 - 5a - 10}{a(a+2)}$

Add: $= \underline{\hspace{4cm}}$

We usually leave the denominator in factored form.

EXAMPLE 15 Add $\dfrac{8}{3a-3} + \dfrac{9}{a^2 - 2a + 1}$.

Find the common denominator:

Factor.

$3(a - 1)$

$3a - 3 =$ ______

$(a - 1)^2$

$a^2 - 2a + 1 =$ ______

List the factors.

$3 \cdot (a - 1)$

______ $\cdot$ ______

Raise to the highest power.

$3(a - 1)^2$

LCD = ______

Convert to common denominators:

$$\frac{8}{3a-3} + \frac{9}{a^2 - 2a + 1} = \frac{8}{3(a-1)} \cdot \qquad + \frac{9}{(a-1)^2}$$

$\dfrac{a-1}{a-1}, \dfrac{3}{3}$

$$= \frac{8}{3(a-1)} \cdot \frac{(\quad)}{(\quad)} + \frac{9}{(a-1)^2} \cdot \frac{(\quad)}{(\quad)}$$

Supply missing factors

$$= \frac{8a-8}{3(a-1)^2} + \frac{27}{3(a-1)^2}$$

Add: $$= \frac{8a+19}{3(a-1)^2}$$

EXAMPLE 16 Subtract $\dfrac{3x}{2x^2 - 8} - \dfrac{x + 2}{x^2 - x - 6}$.

Convert to an addition problem.

$-(x + 2)$

$$\frac{3x}{2x^2 - 8} - \frac{x + 2}{x^2 - x - 6} = \frac{3x}{2x^2 - 8} + \frac{\rule{2cm}{0.4pt}}{x^2 - x - 6}$$

Find the common denominator:

Factor.

$2(x - 2)(x + 2)$

$$2x^2 - 8 = \rule{4cm}{0.4pt}$$

$(x - 3)(x + 2)$

$$x^2 - x - 6 = \rule{4cm}{0.4pt}$$

List the factors.

$2(x - 2)(x + 2)(x - 3)$

$$\rule{5cm}{0.4pt}$$

Raise to powers.

$2(x - 2)(x + 2)(x - 3)$
Each factor is used only once

$$\text{LCD} = \rule{5cm}{0.4pt}$$

Supply missing factors to *convert* to common denominators:

$$\frac{3x}{2x^2 - 8} + \frac{-(x + 2)}{x^2 - x - 6} = \frac{3x}{2(x - 2)(x + 2)} + \frac{-(x + 2)}{(x + 2)(x - 3)}$$

$\dfrac{2(x - 2)}{2(x - 2)}$

$$= \frac{3x}{2(x - 2)(x + 2)} \cdot \frac{(x - 3)}{(x - 3)} + \frac{-(x + 2)}{(x + 2)(x - 3)} \cdot \frac{(\rule{1.5cm}{0.4pt})}{(\rule{1.5cm}{0.4pt})}$$

Keep the negative sign outside the bracket until you've done all the multiplication inside.

$$= \frac{3x(x - 3)}{2(x - 2)(x + 2)(x - 3)} + \frac{-[2(x + 2)(x - 2)]}{2(x - 2)(x + 2)(x - 3)}$$

$$= \frac{3x^2 - 9x}{2(x - 2)(x + 2)(x - 3)} + \frac{-[2x^2 - 8]}{2(x - 2)(x + 2)(x - 3)}$$

$$= \frac{3x^2 - 9x}{2(x - 2)(x + 2)(x - 3)} + \frac{-2x^2 + 8}{2(x - 2)(x + 2)(x - 3)}$$

Add:

$$= \frac{3x^2 - 9x - 2x^2 + 8}{2(x - 2)(x + 2)(x - 3)}$$

$$= \frac{x^2 - 9x + 8}{2(x - 2)(x + 2)(x - 3)}$$

EXAMPLE 17 Add $\frac{x}{x^2 - 49} + \frac{7}{7 - x}$.

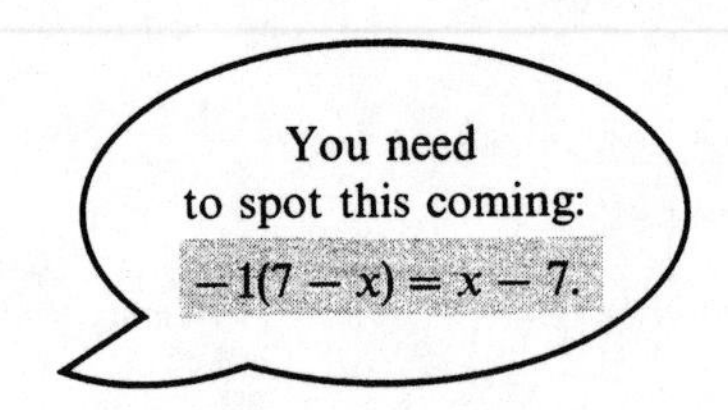

Multiplying the second fraction by $\frac{-1}{-1}$ will reduce the number of factors in the common denominator.

$$\frac{x}{x^2 - 49} + \frac{7}{(7 - x)} \cdot \frac{-1}{-1} = \frac{x}{x^2 - 49} + \frac{-7}{x - 7}$$

Find the common denominator:

Factor.

$$x^2 - 49 = (x - 7)(x + 7)$$
$$x - 7 = x - 7$$

List factors.

$$(x - 7)(x + 7)$$

Raise to powers.

$$\text{LCD} = (x - 7)(x + 7)$$

Convert to common denominators:

$\frac{x + 7}{x + 7}$

$$\frac{x}{x^2 - 49} + \frac{-7}{x - 7} = \frac{x}{(x - 7)(x + 7)} + \frac{-7}{x - 7} \cdot \frac{(\_\_\_\_\_\_)}{(\_\_\_\_\_\_)}$$

$$= \frac{x}{(x - 7)(x + 7)} + \frac{-7x - 49}{(x - 7)(x + 7)}$$

Add:

$$= \frac{x - 7x - 49}{(x - 7)(x + 7)}$$

$$= \frac{-6x - 49}{(x - 7)(x + 7)}$$

EXAMPLE 18 Subtract $\frac{5x + 2}{4x - 8} - \frac{2x + 5}{x^2 - x - 2}$.

If you leave the negative sign outside the parentheses, it will be easier to convert to equivalent fractions.

Convert to an addition problem.

$$\frac{5x + 2}{4x - 8} - \frac{2x + 5}{x^2 - x - 2} = \frac{5x + 2}{4x - 8} + \frac{-(2x + 5)}{x^2 - x - 2}$$

Find the common denominator:

Factor.

$4(x - 2)$

$(x - 2)(x + 1)$

$$4x - 8 = \_\_\_\_\_\_\_\_$$
$$x^2 - x - 2 = \_\_\_\_\_\_\_\_\_\_\_\_$$

List factors.

$4(x-2)(x+1)$ ______________

Raise to powers.

$4(x-2)(x+1)$ LCD = ______________

Convert to equivalent fractions with common denominators:

$$\frac{5x+2}{4x-8}+\frac{-(2x+5)}{x^2-x-2}=\frac{5x+2}{4(x-2)}+\frac{-(2x+5)}{(x-2)(x+1)}$$

$\frac{x+1}{x+1}, \frac{4}{4}$

$$=\frac{5x+2}{4(x-2)}\cdot\frac{(\_\_\_\_)}{(\_\_\_\_)}+\frac{-(2x+5)}{(x-2)(x+1)}\cdot\frac{(\_\_\_\_)}{(\_\_\_\_)}$$

$$=\frac{5x^2+7x+2}{4(x-2)(x+1)}+\frac{-(8x+20)}{4(x-2)(x+1)}$$

$$=\frac{5x^2+7x+2}{4(x-2)(x+1)}+\frac{-8x-20}{4(x-2)(x+1)}$$

Add:

$$=\frac{5x^2+7x+2-8x-20}{4(x-2)(x+1)}$$

$5x^2-x-18$

$$=\frac{\qquad\qquad}{4(x-2)(x+1)}$$

When the numerator is factorable, factor it. *Reduce* the fraction if possible.

$$=\frac{(5x+9)\overset{1}{\cancel{(x-2)}}}{4\underset{1}{\cancel{(x-2)}}(x+1)}$$

$$=\frac{5x+9}{4(x+1)}$$

PROBLEM SET 12.2

Add or subtract the following. Reduce your answers to lowest terms.

(15) Hint: $1=\frac{1}{1}=\frac{x}{x}$

1. $\frac{1}{2}+\frac{2}{3}$
2. $\frac{3}{4}-\frac{1}{12}$
3. $\frac{3}{5}-\frac{1}{10}$
4. $\frac{2}{3}+\frac{4}{5}$
5. $\frac{8x}{9}-\frac{5x}{6}$
6. $\frac{3a}{10}+\frac{b}{6}$
7. $\frac{1}{x}+\frac{1}{y}$
8. $\frac{1}{a}-\frac{1}{b}$
9. $\frac{1}{2x}+\frac{2}{3x}$
10. $\frac{5p}{12q}-\frac{p}{4q}$
11. $\frac{x+2}{4x}+\frac{1}{4}$
12. $\frac{5}{6}-\frac{x+2}{2x}$
13. $\frac{3}{2x}+\frac{1}{2y}$
14. $\frac{2}{5a}-\frac{7}{10b}$
15. $\frac{1}{x}+1$
16. $3-\frac{3}{4x}$
17. $\frac{1}{x^2}+\frac{2}{x}$
18. $\frac{3}{2c}-\frac{3}{c^2}$
19. $\frac{8}{x^2}-\frac{7}{xy}$
20. $\frac{3}{y^2}-\frac{5}{xy}$
21. $\frac{6}{x^2}-\frac{5}{xy}$
22. $\frac{5}{pq^2}+\frac{p}{3q}$
23. $\frac{1}{x+2}+\frac{1}{2}$
24. $\frac{3}{y+1}-\frac{2}{y}$

25. $\frac{3}{x-2}-\frac{1}{2x}$

26. $\frac{2}{x+2}+\frac{3}{x+3}$

27. $\frac{4}{x+4}-\frac{4}{x-4}$

28. $\frac{6}{r-5}-\frac{2}{r-1}$

29. $\frac{7}{x-4}+\frac{3}{4-x}$

30. $\frac{a}{a-b}+\frac{b}{b-a}$

31. $3-\frac{3x}{x+3}$

32. $\frac{2x}{2x-1}+3$

33. $7-\frac{7x}{x+6}$

34. $\frac{4x}{4x-1}+5$

35. $\frac{1}{2c+2}+\frac{1}{c+1}$

36. $\frac{3}{x-2}-\frac{2}{3x-6}$

37. $\frac{2}{x^2+2x}+\frac{1}{x+2}$

38. $\frac{2}{3y-9}-\frac{2}{y^2-3y}$

39. $\frac{4}{y-4}-\frac{3}{16-4y}$

40. $\frac{1}{4-2x}+\frac{1}{x^2-2x}$

41. $\frac{3}{x+3}+\frac{6}{x^2+4x+3}$

42. $\frac{4}{x^2-6x+8}+\frac{2}{x-2}$

43. $\frac{2}{x-1}+\frac{6x}{x^2-5x+4}$

44. $\frac{5}{x+3}-\frac{4x-13}{x^2+x-6}$

45. $\frac{5}{2x-10}-\frac{10}{x^2-6x+5}$

46. $\frac{4}{y^2+2y-3}-\frac{3}{y^2+3y}$

47. $\frac{1}{x-2}-\frac{4}{x^2-4}$

48. $\frac{9}{x^2-2xy-3y^2}-\frac{3}{x^2+xy}$

49. $\frac{3}{x^2+6x+9}-\frac{1}{x^2-9}$

50. $\frac{2}{a^2-9}+\frac{1}{3a+9}$

51. $\frac{8}{x^2-y^2}-\frac{4}{x^2+xy}$

52. $\frac{2}{m^2+2m+1}+\frac{2}{m^2-1}$

53. $\frac{3}{x^2-2x-8}-\frac{1}{x^2-6x+8}$

54. $\frac{5}{x^2+8x-9}-\frac{4}{x^2+10x+9}$

55. $\frac{1}{x^2+5x+6}+\frac{2}{x^2+8x+15}$

56. $\frac{4}{x^2-4}-\frac{2}{x^2+4x+4}$

57. $\frac{2}{x^2-6x}-\frac{1}{x^2-9x+18}$

58. $\frac{3}{x^2-6x+8}-\frac{3}{4-2x}$

59. $1-\frac{x^2+x}{x^2+3x+2}$

60. $\frac{x+1}{x^2-x}-2$

Review Your Skills Divide the following fractions. Reduce all answers to lowest terms and write all fractions in standard form.

61. $\frac{-4}{15}\div\frac{7}{9}$

62. $\frac{-24x}{25}\div\frac{-12}{15}$

63. $\frac{-24a}{-56}\div\frac{12}{-35}$

64. $\frac{-48x}{-56}\div\frac{-72}{42}$

Complex fraction

> **12.3 DEFINITION**
>
> A *complex fraction* is a fraction that contains fractions in its numerator or denominator or both.

For example,

$$\frac{\frac{1}{2}}{\frac{1}{3}}, \quad \frac{\frac{1}{x}}{2}, \quad \frac{1+\frac{1}{x}}{\frac{1}{2}}, \quad \frac{\frac{1}{2}+x}{\frac{1}{3}+\frac{x}{3}}$$

are complex fractions.

Now, the problem is to get these complex fractions simplified completely. That is, we want to write them as a single fraction. There are two ways to do this.

One way is to follow the directions inherent in the problem. The complex fractions above have a numerator that is to be divided by the denominator.

EXAMPLE 19 Simplify $\dfrac{\frac{1}{2}}{\frac{1}{3}}$. ⇐ Numerator ⇐ Denominator

Method I, Follow the Directions in the Problem

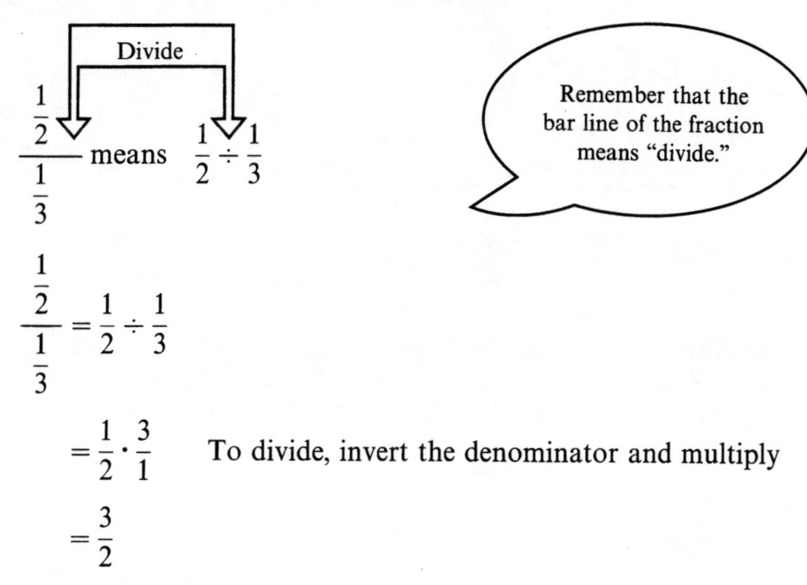

$$\frac{\frac{1}{2}}{\frac{1}{3}} \text{ means } \frac{1}{2} \div \frac{1}{3}$$

$$\frac{\frac{1}{2}}{\frac{1}{3}} = \frac{1}{2} \div \frac{1}{3}$$

$$= \frac{1}{2} \cdot \frac{3}{1} \qquad \text{To divide, invert the denominator and multiply}$$

$$= \frac{3}{2}$$

LCD

Method II, LCD Method

A second method to simplify complex fractions is called the LCD method.

Look at the two fractions that make up the numerator and the denominator of our complex fraction, that is, $\frac{1}{2}$ and $\frac{1}{3}$.

The LCD of $\frac{1}{2}$ and $\frac{1}{3}$ is 6.

What do you mean by LCD?

LCD is the abbreviation for *least common denominator.* Since 6 is the smallest number divisible by both 2 and 3, it is the LCD.

Multiply the numerator and the denominator by the LCD.

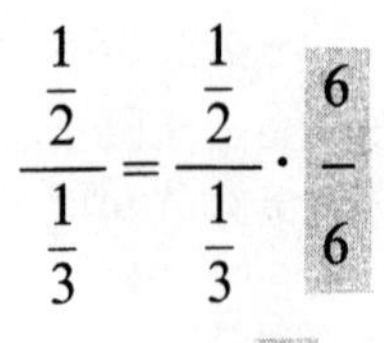

$$\frac{\frac{1}{2}}{\frac{1}{3}} = \frac{\frac{1}{2}}{\frac{1}{3}} \cdot \frac{6}{6}$$

$$= \frac{\frac{1}{2} \cdot \frac{6}{1}}{\frac{1}{3} \cdot \frac{6}{1}}$$

$$= \frac{\frac{6}{2}}{\frac{6}{3}}$$

$$= \frac{3}{2}$$

When we multiply by $\frac{6}{6}$, which equals 1, we don't change the value of the original problem.

Why multiply by the LCD?

So that the denominators of the two fractions will divide out and our answer is a single fraction.

EXAMPLE 20 Simplify $\frac{\frac{1}{x}}{2}$.

Method I, Follow the Directions in the Problem

$$\frac{\frac{1}{x}}{2} = \frac{1}{x} \div 2$$

$$= \frac{1}{x} \cdot \frac{1}{2}$$

$$= \underline{\qquad}$$

$\frac{1}{2x}$

The directions say divide the numerator $\frac{1}{x}$ by the denominator 2.

Method II, LCD Method

$$\frac{\frac{1}{x}}{2} = \frac{\frac{1}{x}}{\frac{2}{1}}$$

Write 2 as a fraction:

$$2 = \frac{2}{1}$$

The LCD of $\frac{1}{x}$ and $\frac{2}{1}$ is x. Multiply the numerator and the denominator by x.

$$= \frac{\frac{1}{x}}{\frac{2}{1}} \cdot \frac{x}{x}$$

$$= \frac{\frac{1}{x} \cdot \frac{x}{1}}{\frac{2}{1} \cdot \frac{x}{1}}$$

Rewrite x as a fraction: $\frac{x}{1}$.

$$= \frac{\frac{x}{x}}{\frac{2x}{1}}$$

$\frac{1}{2x}$ $\qquad = \underline{\qquad}$

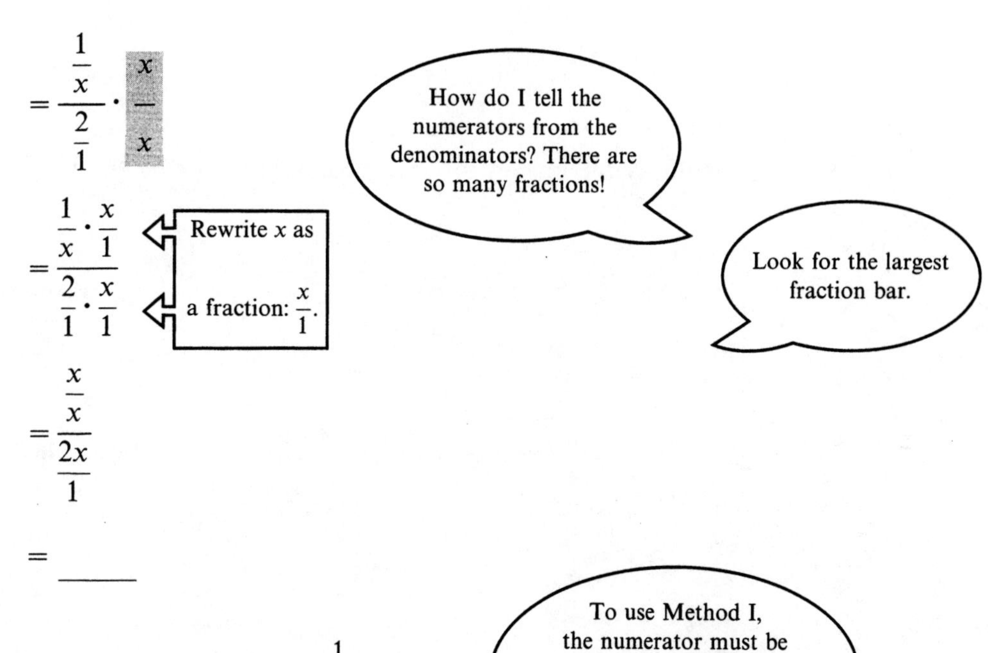

To use Method I, the numerator must be written as a single fraction, and the denominator must be a single fraction.

EXAMPLE 21 Simplify $\dfrac{1 + \frac{1}{x}}{\frac{1}{2}}$.

Method I, Follow the Directions in the Problem

Add in the numerator.

$$1 + \frac{1}{x} = \frac{x}{x} + \frac{1}{x}$$

$$= \frac{x + 1}{x}$$

Our original problem now looks like

$$\frac{1 + \frac{1}{x}}{\frac{1}{2}} = \frac{\frac{x+1}{x}}{\frac{1}{2}}$$

Divide.

$$= \frac{x+1}{x} \div \frac{1}{2}$$

$\frac{2}{1}$

$$= \frac{x+1}{x} \cdot \underline{\qquad}$$

$$= \frac{2x+2}{x}$$

Method II, LCD Method

Find the LCD of $1 = \frac{1}{1}$ and $\frac{1}{x}$ and $\frac{1}{2}$.

To find the LCD, be sure to include *every* factor of any denominator in the complex fraction.

$2x$

The LCD is ______.

Multiply the numerator and the denominator by the LCD.

$$\frac{1+\frac{1}{x}}{\frac{1}{2}} = \frac{\left(1+\frac{1}{x}\right)}{\frac{1}{2}} \cdot \frac{2x}{2x}$$

$$= \frac{1 \cdot 2x + \frac{1}{x} \cdot 2x}{\frac{1}{2} \cdot 2x}$$ Use the distributive law

$\frac{2x+2}{x}$

$$= \underline{\qquad\qquad}$$

EXAMPLE 22 Simplify $\dfrac{\frac{1}{2}+x}{\frac{1}{3}-\frac{x}{3}}$.

Method I The problem tells us what to do. First, add the terms in the numerator. Second, subtract the terms in the denominator. Third, divide the numerator by the denominator.

Add in the numerator.

$$\frac{1}{2} + x = \frac{1}{2} + \frac{x}{1} \cdot \frac{2}{2}$$

$\frac{1+2x}{2}$

$$= \underline{\qquad\qquad}$$

Now the numerator is a single fraction.

Subtract in the denominator.

$\frac{1-x}{3}$

$$\frac{1}{3} - \frac{x}{3} = \underline{\qquad\qquad}$$

And the denominator is a single fraction.

Divide the numerator by the denominator.

$$\frac{\frac{1}{2}+x}{\frac{1}{3}-\frac{x}{3}} = \frac{\frac{1+2x}{2}}{\frac{1-x}{3}}$$

$$= \frac{1+2x}{2} \div \frac{1-x}{3}$$

$$= \frac{1+2x}{2} \cdot \frac{3}{1-x}$$

$\frac{3+6x}{2-2x}$ $\qquad = \underline{\qquad\qquad}$

Method II, LCD Method

Find the LCD.

6 $\qquad \frac{\frac{1}{2}+x}{\frac{1}{3}-\frac{x}{3}} = \frac{\frac{1}{2}+\frac{x}{1}}{\frac{1}{3}-\frac{x}{3}}$ The LCD is ______

Multiply the numerator and the denominator by the LCD.

$$= \frac{\frac{1}{2}+\frac{x}{1}}{\frac{1}{3}-\frac{x}{3}} \cdot \frac{6}{6}$$

$$= \frac{\frac{1}{2}\cdot\left(\frac{6}{1}\right)+\frac{x}{1}\cdot\left(\frac{6}{1}\right)}{\frac{1}{3}\cdot\left(\frac{6}{1}\right)-\frac{x}{3}\cdot\left(\frac{6}{1}\right)}$$ Distribute

$\frac{3+6x}{2-2x}$ $\qquad = \underline{\qquad\qquad}$

PROBLEM SET 12.3

Use either method to simplify the following complex numbers. Reduce your answers to lowest terms.

1. $\frac{\frac{1}{2}}{\frac{3}{4}}$ 2. $\frac{\frac{3}{4}}{\frac{2}{3}}$ 3. $\frac{\frac{2}{3}}{\frac{5}{6}}$ 4. $\frac{\frac{2}{3}}{\frac{4}{5}}$ 5. $\frac{\frac{3}{4}}{\frac{4}{3}}$

6. $\frac{\frac{3}{4}}{\frac{1}{6}}$ 7. $\frac{1+\frac{1}{2}}{\frac{1}{4}}$ 8. $\frac{1-\frac{1}{4}}{\frac{1}{2}}$ 9. $\frac{\frac{2}{3}}{3-\frac{2}{3}}$ 10. $\frac{\frac{1}{3}}{2+\frac{1}{3}}$

11. $\frac{2-\frac{5}{3}}{1+\frac{2}{5}}$ 12. $\frac{3-\frac{1}{4}}{2+\frac{1}{4}}$ 13. $\frac{2-\frac{2}{3}}{\frac{1}{3}+\frac{1}{5}}$ 14. $\frac{2-\frac{3}{8}}{\frac{1}{2}+2}$ 15. $\frac{6}{1-\frac{2}{3}}$

16. $\dfrac{5 + \frac{5}{6}}{5}$ 17. $\dfrac{\frac{x}{3}}{\frac{2}{9}}$ 18. $\dfrac{\frac{x}{4}}{\frac{y}{5}}$ 19. $\dfrac{\frac{x}{10}}{\frac{x}{5}}$ 20. $\dfrac{\frac{2a}{3}}{\frac{a}{6}}$

21. $\dfrac{\frac{1}{x}}{\frac{1}{4}}$ 22. $\dfrac{\frac{2}{3}}{\frac{2}{x}}$ 23. $\dfrac{\frac{3}{x}}{\frac{5}{x}}$ 24. $\dfrac{12}{\frac{3}{x}}$ 25. $\dfrac{1 + \frac{1}{x}}{\frac{1}{x}}$

26. $\dfrac{1 - \frac{1}{x}}{\frac{1}{x}}$ 27. $\dfrac{1 - \frac{1}{a}}{a}$ 28. $\dfrac{1 + \frac{1}{a}}{a}$ 29. $\dfrac{x}{1 - \frac{3}{x}}$ 30. $\dfrac{\frac{3}{y}}{\frac{1}{y} - 2}$

31. $\dfrac{\frac{2}{a} - 3}{\frac{1}{a} + 2}$ 32. $\dfrac{\frac{4}{b} - 3}{5 + \frac{4}{b}}$ 33. $\dfrac{\frac{1}{2x} + 1}{\frac{1}{2}}$ 34. $\dfrac{\frac{1}{x} + \frac{1}{2}}{\frac{1}{2x}}$ 35. $\dfrac{1 - \frac{1}{2x}}{\frac{1}{x}}$

36. $\dfrac{2 + \frac{1}{x}}{\frac{1}{x} - \frac{1}{2}}$ 37. $\dfrac{\frac{1}{x} - \frac{1}{2x}}{\frac{1}{2} + \frac{1}{2x}}$ 38. $\dfrac{x + \frac{1}{x}}{\frac{1}{2}}$ 39. $\dfrac{y - \frac{1}{y}}{1 + \frac{1}{y}}$ 40. $\dfrac{x + \frac{1}{x}}{1 - \frac{1}{x}}$

41. $\dfrac{3 - \frac{1}{x}}{\frac{1}{x} + \frac{1}{3}}$ 42. $\dfrac{1 - \frac{1}{a^2}}{1 - \frac{1}{a}}$ 43. $\dfrac{1 + \frac{x}{y}}{1 + \frac{y}{x}}$ 44. $\dfrac{1 - \frac{b}{a}}{1 - \frac{a}{b}}$ 45. $\dfrac{\frac{1}{a} + \frac{1}{b}}{\frac{1}{a} - \frac{1}{b}}$

46. $\dfrac{\frac{1}{b} + a}{\frac{1}{a} + b}$ 47. $\dfrac{\frac{1}{x} - \frac{1}{y}}{\frac{1}{y} - \frac{1}{x}}$ 48. $\dfrac{\frac{1}{a} - b}{\frac{1}{b} - a}$

Review Your Skills Solve for the variable.

49. $\dfrac{2x - 3}{5} = 1$ 50. $\dfrac{2x}{5} - \dfrac{3}{5} = 1$ 51. $4x - \dfrac{2x}{3} = 5$ 52. $8 - \dfrac{3x}{7} = 5$

12.4 SOLVING EQUATIONS WITH FRACTIONS

In Unit 4, we defined the solution set of an equation as the set of replacements for the variable that makes the equation a true statement. We also defined equivalent equations as equations with the same solution sets. We use equivalent equations to solve equations with fractions.

The easiest way to solve equations involving fractions is to make them into equivalent equations that *do not involve* fractions. Now, as any good mathematician will tell you, you can't have a fraction without a denominator. Therefore, all we need to do is find something to divide out the denominators. Basically, we need an expression that has every denominator as one of its factors. The simplest expression with this property is the least common denominator of the fractions in the equation. So, to eliminate the fractions, we multiply both sides of the equation by the least common denominator.

EXAMPLE 23 Solve $\frac{x}{6} + \frac{x}{4} = 5$ for x.

Because the least common denominator of the fractions is 12, we multiply *both* sides by (12)

$$(12)\left(\frac{x}{6} + \frac{x}{4}\right) = 5 \cdot (12)$$

Use parentheses to ensure that the entire side is multiplied by 12.

$$(12) \cdot \frac{x}{6} + (12) \cdot \frac{x}{4} = 5 \cdot 12$$ Apply the distributive property

$$\frac{12x}{6} + \frac{12x}{4} = 60$$ Simplify and solve the resulting equation

$$2x + 3x = 60$$

$$5x = 60$$

$$x = 12$$

All solutions should be checked to detect errors and to ensure that the solution set satisfies the original equation.

Check.

$$\frac{x}{6} + \frac{x}{4} = 5$$

$$\frac{(12)}{6} + \frac{(12)}{4} \stackrel{?}{=} 5$$ Replace x with (12) in the original equation

$$2 + 3 \stackrel{?}{=} 5$$

$$5 = 5$$

EXAMPLE 24 Solve $\frac{2}{3x} + 3 = \frac{5}{6x}$ for x.

$6x$ The LCD is ______.

Multiply both sides by $6x$.

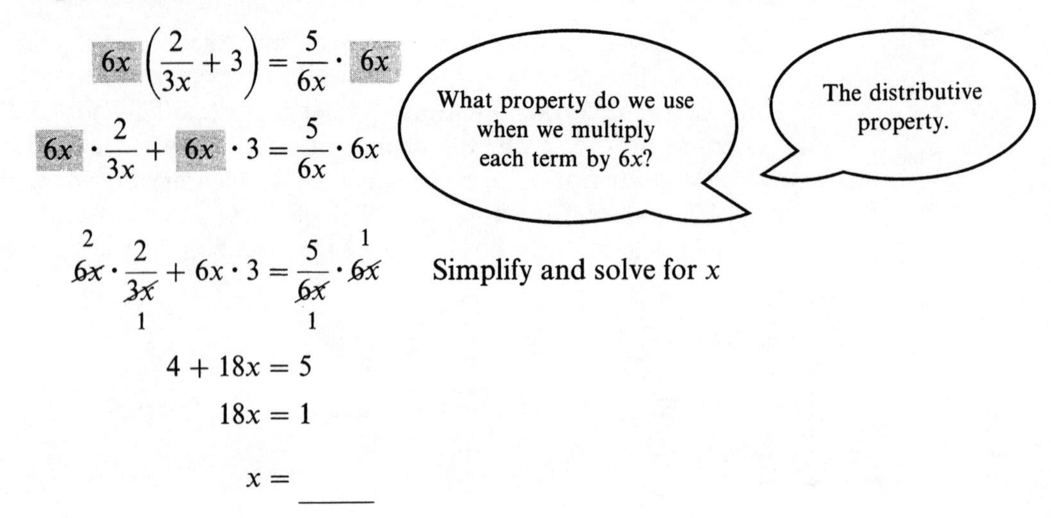

$$6x\left(\frac{2}{3x} + 3\right) = \frac{5}{6x} \cdot 6x$$

$$6x \cdot \frac{2}{3x} + 6x \cdot 3 = \frac{5}{6x} \cdot 6x$$

$$\overset{2}{\cancel{6x}} \cdot \frac{2}{\underset{1}{\cancel{3x}}} + 6x \cdot 3 = \frac{5}{\underset{1}{\cancel{6x}}} \cdot \overset{1}{\cancel{6x}} \qquad \text{Simplify and solve for } x$$

$$4 + 18x = 5$$

$$18x = 1$$

$\frac{1}{18}$ $$x = \underline{\qquad}$$

Check.

$$\frac{2}{3x} + 3 = \frac{5}{6x}$$

$$\frac{2}{3\left(\frac{1}{18}\right)} + 3 \stackrel{?}{=} \frac{5}{6\left(\frac{1}{18}\right)} \qquad \text{Replace } x \text{ with } \frac{1}{18}$$

$$\frac{2}{\frac{1}{6}} + 3 \stackrel{?}{=} \frac{5}{\frac{1}{3}} \qquad \text{Simplify each side separately}$$

$$2 \cdot \frac{6}{1} + 3 \stackrel{?}{=} 5 \cdot \frac{3}{1}$$

$$12 + 3 \stackrel{?}{=} 15$$

$$15 = 15$$

EXAMPLE 25 Solve $\frac{6x}{2x + 3} = 6$ for x.

The LCD is $2x + 3$.

Multiply both sides by the LCD.

$2x + 3,\ 2x + 3$ $$(\underline{\qquad}) \cdot \frac{6x}{2x + 3} = 6(\underline{\qquad})$$

$$\overset{1}{\cancel{(2x + 3)}} \cdot \frac{6x}{\underset{1}{\cancel{2x + 3}}} = 6(2x + 3) \qquad \text{Simplify and solve}$$

$12x + 18$ $$6x = \underline{\qquad} \qquad \text{Distributive property}$$

$$-6x = 18 \qquad \text{Subtract } 12x \text{ from both sides}$$

$$x = -3$$

Check.

$$\frac{6x}{2x+3} = 6$$

$$\frac{6(-3)}{2(-3)+3} \stackrel{?}{=} 6 \qquad \text{Replace } x \text{ with } (-3)$$

$$\frac{-18}{-6+3} \stackrel{?}{=} 6$$

$$\frac{-18}{-3} \stackrel{?}{=} 6$$

$$6 = 6$$

It checks!

Good.

EXAMPLE 26 Solve $\dfrac{4}{a+2} = \dfrac{1}{a-1}$ for a.

The LCD is $(a+2)(a-1)$.

Multiply both sides by the LCD.

$$(a+2)(a-1) \cdot \frac{4}{a+2} = \frac{1}{a-1} \cdot (a+2)(a-1)$$

$$\overset{1}{\cancel{(a+2)}}(a-1) \cdot \frac{4}{\underset{1}{\cancel{a+2}}} = \frac{1}{\underset{1}{\cancel{a-1}}} \cdot (a+2)\overset{1}{\cancel{(a-1)}}$$

$$4a - 4 = a + 2$$

$$3a = 6$$

$$a = 2$$

Check.

$$\frac{4}{a+2} = \frac{1}{a-1}$$

$$\frac{4}{(\_\_\_\_)+2} \stackrel{?}{=} \frac{1}{(\_\_\_\_)-1} \qquad \text{Replace } a \text{ with } (2)$$

2, 2

$$\frac{4}{4} \stackrel{?}{=} \frac{1}{1}$$

$$1 = 1$$

EXAMPLE 27 Solve $\dfrac{x+3}{x-3} + 3 = \dfrac{6}{x-3}$ for x.

The LCD is $x - 3$.

Multiply both sides by $x - 3$.

$$(x-3)\left[\frac{x+3}{x-3} + 3\right] = \frac{6}{x-3} \cdot (x-3)$$

$$\overset{1}{\cancel{(x-3)}} \cdot \frac{x+3}{\underset{1}{\cancel{x-3}}} + (x-3) \cdot 3 = \frac{6}{\underset{1}{\cancel{x-3}}} \cdot \overset{1}{\cancel{(x-3)}} \qquad \text{Distribute and simplify}$$

$$x + 3 + 3x - 9 = 6$$

$$4x - 6 = 6$$

$$4x = 12$$

$$x = 3$$

Check.

$$\frac{x+3}{x-3} + 3 = \frac{6}{x-3}$$

$$\frac{(3)+3}{(3)-3} + 3 \stackrel{?}{=} \frac{6}{(3)-3} \qquad \text{Replace } x \text{ with (3)}$$

$$\frac{6}{\_\_\_} + 3 \stackrel{?}{=} \frac{6}{\_\_\_}$$

Careful, we cannot divide by zero.

0, 0

Because *we can't divide by zero*, 3 cannot be considered a solution to the above equation.

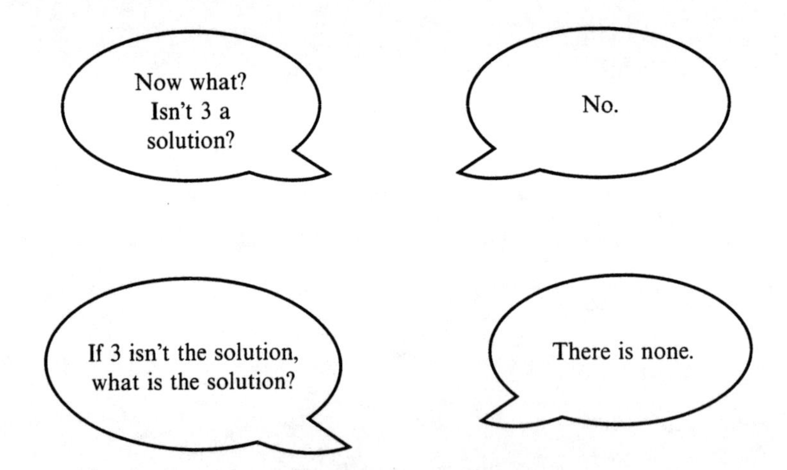

When the equations have variables in the denominator, it is particularly important to check your solution in the original equation. In Example 27, the solution $x = 3$ makes the original equation meaningless, because we can't divide by zero. Since $x = 3$ is not allowed, we say that the equation has no solution.

The last equation of Example 27, before the check, is not equivalent to the original equation because when we multiplied both sides by $(x - 3)$ in step one, we were multiplying by zero.

PROBLEM SET 12.4

Solve for the variable. Check your answers.

1. $\frac{x}{2} + 6 = x$
2. $p + \frac{p}{5} = 8$
3. $\frac{x}{2} + \frac{3x}{4} = \frac{5}{2}$
4. $\frac{y}{4} - 3 = \frac{9}{2}$
5. $\frac{4x}{7} - 6 = \frac{2x}{7}$
6. $\frac{5x}{6} - 6 = \frac{7}{3}$
7. $\frac{2x}{5} - 3 = \frac{3x}{5}$
8. $\frac{5}{a} + \frac{3}{a} = 2$
9. $\frac{8}{x} - \frac{2}{x} = 3$
10. $2 - \frac{3}{x} = 1$
11. $\frac{2}{3x} - \frac{1}{x} = \frac{2}{3}$
12. $\frac{3}{5x} - \frac{6}{x} = \frac{8}{5}$
13. $\frac{6}{t} + \frac{5}{t} = \frac{11}{4}$
14. $\frac{3}{x} + \frac{3}{4x} = 3$
15. $5 - \frac{5}{x} = \frac{5}{2}$
16. $\frac{1}{2x} + \frac{1}{3x} = \frac{1}{6}$
17. $\frac{4}{x} - \frac{3}{2x} = \frac{5}{6}$
18. $\frac{2}{x} + \frac{1}{2x} = \frac{5}{4}$
19. $\frac{3}{x} - \frac{5}{2x} = \frac{1}{6}$
20. $\frac{5}{x} - \frac{7}{3x} = \frac{-8}{9}$
21. $\frac{1}{2t} + \frac{1}{3t} + \frac{1}{6t} = 2$
22. $\frac{2}{x} - \frac{1}{3} = \frac{1}{6} - \frac{1}{x}$
23. $\frac{x+1}{2} - \frac{x-1}{3} = 4$
24. $\frac{x-1}{5} - \frac{2x-3}{3} = 1$
25. $\frac{t}{4} - \frac{t+1}{5} = 1$
26. $\frac{x+2}{3} - \frac{x-2}{6} = 2$
27. $\frac{x}{4} - 1 = \frac{x-4}{2}$
28. $\frac{4}{x+6} = 2$
29. $\frac{6}{x+4} = 2$
30. $\frac{2x}{5x-4} = 2$
31. $\frac{7x}{2x+5} = 6$
32. $\frac{6a}{2a+3} = 4$
33. $\frac{8}{x+5} = 1$
34. $\frac{3x}{x+5} = -2$
35. $\frac{3}{x-3} = \frac{x}{x-3}$
36. $\frac{3}{3-x} = \frac{x}{x-3}$
37. $\frac{9}{x-3} - \frac{5}{x-3} = 4$
38. $\frac{3}{a+2} - 2 = \frac{1}{a+2}$
39. $\frac{4x}{3-x} + \frac{5}{x-3} = 3$
40. $\frac{13}{y-1} - 3 = \frac{y}{y-1}$
41. $\frac{1}{x-2} = \frac{3}{x}$
42. $\frac{3}{x+2} - \frac{5}{x} = 0$
43. $\frac{3}{x+2} = \frac{5}{x}$
44. $\frac{2}{y-1} - \frac{4}{y} = 0$
45. $\frac{5}{x+3} = \frac{2}{x}$
46. $\frac{6}{x-4} = \frac{4}{2x}$
47. $\frac{x-6}{x-5} = \frac{4}{5}$
48. $\frac{x+1}{x-3} = \frac{5}{6}$
49. $\frac{3x-5}{x+4} = \frac{3}{2}$
50. $\frac{1}{x+1} = \frac{1}{2x}$
51. $\frac{6}{y-2} = \frac{3}{2y}$
52. $\frac{3}{2x+1} = \frac{3}{x+2}$
53. $\frac{3}{x+3} = \frac{3}{3x-1}$
54. $\frac{1}{2y+1} = \frac{1}{3y-1}$
55. $\frac{1}{3y-4} = \frac{2}{4y+5}$
56. $\frac{x+2}{2x-3} - 2 = \frac{2}{2x-3}$
57. $\frac{2x}{5x-7} - 2 = \frac{4}{5x-7}$
58. $\frac{4}{y+1} = \frac{2}{y+1} + \frac{1}{2}$
59. $\frac{x}{2x-1} = \frac{1}{2x-1} + \frac{1}{3}$

12.5 APPLICATIONS

Number Problems

To solve many number problems, we often have to write equations with algebraic fractions.

EXAMPLE 28 If the same number is added to the numerator and denominator of the fraction $\frac{3}{5}$, the result is equivalent to $\frac{4}{5}$. Find the number.

Make a model.

$$\frac{\boxed{\text{New numerator}}}{\boxed{\text{New denominator}}} = \frac{4}{5}$$

Express each component in algebraic terms.

Let $x =$ the number to be added.

$3 + x$ The original numerator was 3. The new numerator is ________.

5 The original denominator was ______.

$5 + x$ The new denominator is ________.

Translate the model into an equation.

$$\frac{3 + x}{5 + x} = \frac{4}{5}$$

$5(5 + x)$ The LCD is ________.

$$5(5 + x) \cdot \frac{3 + x}{5 + x} = \frac{4}{5} \cdot 5(5 + x)$$ Multiply by $5(5 + x)$ to clear the denominators

$$5\overset{1}{\cancel{(5 + x)}} \cdot \frac{3 + x}{\underset{1}{\cancel{5 + x}}} = \frac{4}{\underset{1}{\cancel{5}}} \cdot \overset{1}{\cancel{5}}(5 + x)$$ Divide out

$$5(3 + x) = 4(5 + x)$$ Simplify

$$15 + 5x = 20 + 4x$$ Distributive property

$$5x = 5 + 4x$$

$$x = 5$$

The number to be added is 5.

EXAMPLE 29 The denominator of a fraction exceeds the numerator by 3. If the numerator is decreased by 2 and the denominator is increased by 4, the value of the fraction is $\frac{1}{4}$. Find the value of the original fraction.

Make a model.

2

4

$$\frac{\boxed{\text{Original numerator} - \_\_\_\_}}{\boxed{\text{Original denominator} + \_\_\_\_}} = \frac{\boxed{\text{New numerator}}}{\boxed{\text{New denominator}}}$$

Express each component in algebraic terms.

Let n = the original numerator.

$n + 3$

Then the original denominator is ________.

Translate the model into an equation.

$$\frac{n-2}{(n+3)+4} = \frac{1}{4}$$

Solve.

$$\frac{n-2}{n+7} = \frac{1}{4}$$

$4(n + 7)$

The LCD is ____________.

$$4(n+7) \cdot \frac{n-2}{n+7} = \frac{1}{4} \cdot 4(n+7) \qquad \text{Multiply by the LCD}$$

$$\overset{1}{4\cancel{(n+7)}} \cdot \frac{n-2}{\underset{1}{\cancel{n+7}}} = \frac{1}{\underset{1}{\cancel{4}}} \cdot \overset{1}{\cancel{4}}(n+7) \qquad \text{Divide out}$$

$$4n - 8 = n + 7 \qquad \text{Simplify}$$

$$3n - 8 = 7$$

$$3n = 15$$

$$n = 5$$

We now know the original numerator was 5. To find the original fraction, we also need the value of the original denominator.

The original denominator was $n + 3 = 5 + 3 = 8$.

$\frac{5}{8}$

The original fraction was ______.

EXAMPLE 30 In a parking lot with many different colored cars, one-fifth of all the cars are red. One-fourth of all the cars in the lot are blue. Together, there are 180 red and blue cars in the lot. Find the total number of all the cars in the lot.

First make a model.

Number of red cars	+	Number of blue cars	=	180 cars

Now express each part of the model as an algebraic expression.

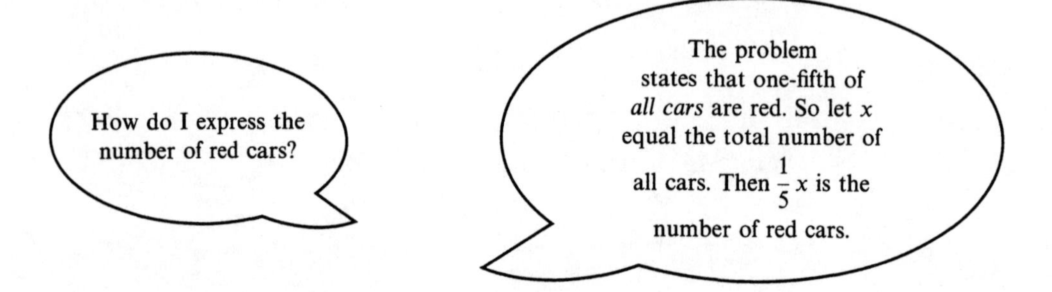

Let x = total number of all cars in the lot

$\frac{x}{5}$ = number of red cars

$\frac{x}{4}$

______ = number of blue cars

Translate the model into an equation.

$$\frac{x}{5} + \frac{x}{4} = 180$$

Solve.

$$20 \cdot \frac{x}{5} + 20 \cdot \frac{x}{4} = 180 \cdot 20$$

$$4x + 5x = 3600$$

$$9x = 3600$$

$$x = 400$$

Answer the question.

There are 400 cars in the lot.

EXAMPLE 31 Find two numbers so that one is four times the other and the sum of their reciprocals is $\frac{5}{12}$. (A reciprocal is a multiplicative inverse.)

Make a model for the situation.

Reciprocal of first number	+	Reciprocal of second number	=	$\frac{5}{12}$

Express each component in algebraic terms.

$\frac{1}{x}$

$4x$

$\frac{1}{4x}$

Let x represent the first number.

The reciprocal of the first number is ______.

The second number is four times the first. Therefore, the second number can be represented as ______.

The reciprocal of the second number can be expressed as ______.

Translate the model into an equation.

Reciprocal of first number	+	Reciprocal of second number	$= \frac{5}{12}$
$\frac{1}{x}$	+	$\frac{1}{4x}$	$= \frac{5}{12}$

Solve the equation.

$$12x\left[\frac{1}{x} + \frac{1}{4x}\right] = \frac{5}{12}(12x)$$

$$12 + 3 = 5x$$

$$15 = 5x$$

$$3 = x$$

Answer the question.

First number: $x = 3$

Second number: $4x = 4(3) = 12$

Work and Flow Problems

Some of the most common applications of algebraic fractions concern relative rates of doing some kind of a job.

EXAMPLE 32 Suppose that Grannie's old-fashioned rug factory is automated with a machine that can weave one rug per hour.

4 rugs

How many rugs can the machine weave in 4 hours? ______

12 rugs

How many rugs can the machine weave in 12 hours? ______

$\frac{1}{2}$ rug

How many rugs can the machine weave in $\frac{1}{2}$ hour? ______

t rugs

How many rugs can the machine weave in t hours? ______

$\frac{3}{2}t$ rugs

How many rugs can the machine weave in $\frac{3}{2}t$ hours? ______

EXAMPLE 33 Grannie recently met a salesperson who told her that a new machine could complete a rug in 40 minutes. A little skeptical, Grannie asked the following questions. If you can answer them, write your answers in the blanks. Otherwise refer to the salesperson's answers in the left-hand column.

How many rugs will the machine make in:

Answer	Question
$\frac{80}{40} = 2$ rugs	80 minutes? ________
$\frac{60}{40} = 1\frac{1}{2}$ rugs	60 minutes? ________
$\frac{17}{40}$ rugs	17 minutes? ________
$\frac{t}{40}$ rugs	t minutes? ________
$\frac{t+3}{40}$ rugs	$t + 3$ minutes? ________

The next day a used-machinery representative offered Grannie a deal she couldn't refuse on an older rug-weaving machine. The salesperson said that this machine could weave a rug in x hours. Before Grannie ventured to risk all of her capital, she asked a few questions of the smooth salesperson.

x hours

How long does it take the machine to weave 1 rug? ________

What part of a rug can the machine weave in:

Answer	Question
$\frac{1}{x}$ rugs	1 hour? ________
$\frac{2}{x}$ rugs	2 hours? ________
$\frac{15}{x}$ rugs	15 hours? ________
$\frac{x}{x} = 1$ rug	x hours? ________
$\frac{t}{x}$ rugs	t hours? ________

EXAMPLE 34 Grannie wanted a production of 10 rugs per day. As an enlightened employer, she also wanted to work her employees for a minimum time per day. (Grannie knew that machines were cheaper than people.) She asked her sales representative how long it would take to produce 10 rugs if she used 2 machines at the same time. She wanted to use one machine that could produce a rug every 2 hours and a second machine that could produce a rug every 3 hours.

Make a model for Grannie's problem.

Let t represent the number of hours Grannie will have to work the machines.

[Rugs by fast machine in t hours] + [Rugs by slow machine in t hours] = 10 rugs

Express each component of the model in algebraic terms.

How many rugs could the fast machine produce

$\frac{1}{2}$ rug

in 1 hour? ____________

$\frac{t}{2}$ rugs

in t hours? ____________

How many rugs could the slow machine produce

$\frac{1}{3}$ rug

in 1 hour? ____________

$\frac{t}{3}$ rugs

in t hours? ____________

Write an equation.

$$\frac{t}{2} + \frac{t}{3} = 10 \text{ rugs}$$

Solve.

$$6\left(\frac{t}{2} + \frac{t}{3}\right) = 10 \cdot 6$$

$$3t + 2t = 60$$

$$5t = 60$$

$$t = 12$$

Answer the question.

Grannie will have to work her machines 12 hours a day to produce 10 rugs a day.

EXAMPLE 35 Mario can mow his lawn with a hand mower in 4 hours, whereas his son, Alfredo, can mow the same lawn with a power mower in 2 hours. How long will it take Mario and Alfredo to mow the lawn if they work together?

Make a model

Part of lawn mowed by Mario	+	Part of lawn mowed by Alfredo	=	*One* completely mowed lawn

Express each component of the model.

Let t represent the time to complete the job working together.

The fractional part of the lawn Mario can mow in 1 hour is $\frac{1}{4}$.

$\frac{1}{2}$

The fractional part of the lawn Alfredo can mow in 1 hour is ______.

The fractional part of the lawn Mario will mow in t hours is $\frac{t}{4}$.

$\frac{t}{2}$

The fractional part of the lawn Alfredo will mow in t hours is ______.

Fractional part of lawn mowed by Mario in t hours	+	Fractional part of lawn mowed by Alfredo in t hours	= 1 (completed job)

Translate into an equation.

$$\frac{t}{4} + \frac{t}{2} = 1$$

Solve.

$$4\left(\frac{t}{4} + \frac{t}{2}\right) = 1 \cdot 4$$

$$t + 2t = 4$$

$$3t = 4$$

$$t = \frac{4}{3}$$

$$t = 1\frac{1}{3} \text{ hours}$$

Answer the question.

It will take Mario and Alfredo $1\frac{1}{3}$ hours, or 1 hour and 20 minutes, to mow the lawn together.

EXAMPLE 36 Joe and Mike want to earn some extra money for college. To do so, they agreed to do 15 similar lawns per week. If Joe can do one lawn in 3 hours and Mike can do a similar lawn in 4 hours, how long would it take Joe and Mike to do the 15 lawns together?

Make a table.

	Time to do one lawn	Part of lawn done in one hour	Number of lawns completed in t hours
Joe	3 hours	$\frac{1 \text{ lawn}}{3 \text{ hours}}$	$\frac{1 \text{ lawn}}{3 \cancel{\text{hours}}} \cdot t \cancel{\text{hours}} = \frac{t}{3}$ lawns
Mike	4 hours	$\frac{1 \text{ lawn}}{4 \text{ hours}}$	$\frac{1 \text{ lawn}}{4 \cancel{\text{hours}}} \cdot t \cancel{\text{hours}} = \frac{t}{4}$ lawns

Set up a model and express as an equation.

$$\boxed{\text{Number of lawns done by Joe in } t \text{ hours}} + \boxed{\text{Number of lawns done by Mike in } t \text{ hours}} = 15 \text{ lawns}$$

$$\frac{t}{3} \text{ lawns} + \frac{t}{4} \text{ lawns} = 15 \text{ lawns}$$

$$12 \cdot \frac{t}{3} + 12 \cdot \frac{t}{4} = 12 \cdot 15$$

$$4t + 3t = 180$$

$$7t = 180$$

$$t = 25\frac{5}{7} \text{ hours}$$

EXAMPLE 37 The maximum permissible level of particulate matter permitted by the clean air act is 75 micrograms per cubic meter of air. An electric generating plant can operate its new, low-pollution generator for 16 hours before the maximum permissible level of air pollutants is reached. The older generator, which emits more pullutants, can only be operated for 6 hours before the maximum level of pollutants is reached. During periods of peak load, it is sometimes necessary to operate both generators together.

If the low-pollution generator has been in operation for 5 hours before the high-pollution generator is brought on line, how long can both generators be operated before the maximum pollution level is reached?

Make a model.

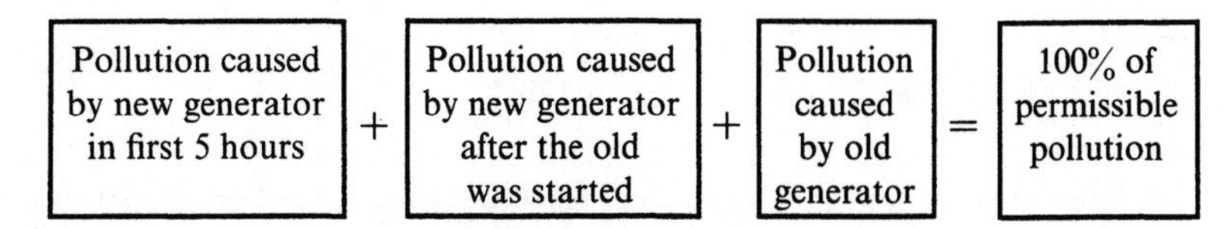

Express each component in algebraic form.

The new generator can operate for 16 hours before it builds the maximum pollution. What part of the maximum does it contribute in x hours? ______ $\frac{x}{16}$

In this problem, what part of the maximum pollution had the new generator contributed during the first 5 hours before the old generator was started? ______ $\frac{5}{16}$

In 1 hour, the old generator contributes $\frac{1}{6}$ of the maximum permissible pollution. What part of the maximum permissible pollution does the old generator contribute in x hours? ______ $\frac{x}{6}$

Translate the model into an equation.

$$\frac{5}{16}+\frac{x}{16}+\frac{x}{6}=1$$

Where did the 1 come from?

Look at the model above: 100% = 1.

Solve.

$$48\left(\frac{5}{16}+\frac{x}{16}+\frac{x}{6}\right)=1\cdot 48$$

Right, but 1 what?

1 whole polluted mess.

$$\overset{3}{\cancel{48}}\cdot\frac{5}{\underset{1}{\cancel{16}}}+\overset{3}{\cancel{48}}\cdot\frac{x}{\underset{1}{\cancel{16}}}+\overset{8}{\cancel{48}}\cdot\frac{x}{\underset{1}{\cancel{6}}}=48$$

$$15+3x+8x=48$$

$$15+11x=48$$

$$11x=33$$

$$x=3$$

Notice in these problems, the basic equation we write is that all the parts add up to one whole.

Answer the question.

Both generators can be operated for 3 additional hours before the maximum level of pollution is reached.

PROBLEM SET 12.5

1. When 36 is divided by a certain number, the result is $\frac{3}{5}$. Find the number.

2. When 45 is divided by a certain number, the result is $\frac{5}{9}$. Find the number.

3. The denominator of a fraction is 4 more than the numerator. Reduced to lowest terms, the fraction is $\frac{2}{3}$. Find the original fraction.

4. The denominator of a fraction is 5 more than the numerator. Reduced to lowest terms, the fraction is $\frac{7}{8}$. Find the original fraction.

5. If the same number is subtracted from both numerator and denominator of the fraction $\frac{17}{19}$, the result is equivalent to $\frac{5}{6}$. Find the number.

6. The denominator of a certain fraction is three times the numerator. If the numerator is multiplied by 2 and the denominator is increased by 4, the resulting fraction is $\frac{3}{5}$. What is the original fraction?

7. The denominator of a certain fraction is two times the numerator. If the numerator is increased by 5 and the denominator is decreased by 4, the value of the fraction is $\frac{3}{4}$. What is the fraction?

8. The denominator of a fraction exceeds the numerator by 4. If the numerator is increased by 3 and the denominator is decreased by 6, the value of the resulting fraction is $\frac{4}{3}$. What is the value of the original fraction?

9. Two-thirds of the students in an algebra class received a C grade. One-fourth of the class received B grades. Twenty-two students in all received a B or a C grade. How many students were in the class?

10. Find two numbers so that one number is three times the other and the sum of their reciprocals is $\frac{4}{15}$.

11. Joe, who is paid by the page, can type a 30-page manuscript in 10 hours. Debbie, who is paid by the hour, can type the 30-page manuscript in 15 hours. How long would it take Joe and Debbie to do the typing if they worked together?

12. Frank can grade a set of papers in 2 hours. Larry, who talks a lot, can grade the same set in 3 hours. How long would it take Frank and Larry to grade the papers if they worked together?

13. George and Edgar have summer jobs washing the elephants in a circus. George, who has a way with elephants, can wash the elephants in 4 hours. Edgar, who would rather be sailing, takes 6 hours to wash the elephants. How long would it take them to wash the elephants if they worked together?

14. Jack and his wife, Jill, planned a picnic for several friends. The noon menu included potato salad. For such a large group, a peck of potatoes was needed. Jack could peel the potatoes in 45 minutes, whereas Jill could peel them in 30 minutes. Jill started peeling them and had one-third of them peeled when Jack came along to help her. How long did it take them to finish the job working together?

15. A spacecraft can operate its television cameras for 1,000 hours on its batteries. It can run its ratio transmitter for 2,000 hours or it can use its heater for 500 hours. How long can the spacecraft function with all three units in operation simultaneously?

16. A swimming pool can be filled in 9 hours and it can be emptied in 12 hours. In error, the drain pipe was left open while attempting to fill the pool. How long did it take to fill the pool?

17. The cold water faucet can fill a sink in 20 seconds and the hot water faucet can fill it in 30 seconds. How long will it take to fill the sink if both faucets are left open?

18. Sally, who works at a root beer stand, observed that a keg could be filled by one pipe in 15 minutes and by another in 12 minutes. How long would it take to fill the keg three-fourths full using both pipes?

19. A cold water faucet can fill a tank in 45 minutes and a hot water faucet can fill it in 60 minutes. The drain can empty it in 75 minutes. If both faucets and drain are open, how long will it take to fill the tank?

20. The combined resistance of two resistors in parallel is given by the formula

$$\frac{1}{R_1} + \frac{1}{R_2} = \frac{1}{R_{\text{total}}}$$

Find the resistance of a 20-ohm resistor and a 40-ohm resistor in parallel.

21. The Free Flyer School of Hang Gliding sells hang gliders. In order to meet expenses, they must make a profit of at least one-third of the cost of each hang glider they sell. Down the street, the Hang-Loose Society offers a $99 special on hang gliders. What is the maximum amount that the Free Flyer School can pay for hang gliders if they are to compete with the Hang-Loose Society?

22. Julie, an experienced clerk in a mathematics laboratory, can sort and file a set of test papers in 9 minutes, whereas, Sam, a new employee, takes 15 minutes to sort and file a similar set of test papers. How long would it take Julie and Sam to sort and file a similar set of test papers if they worked together?

12 Review

NOW THAT YOU HAVE COMPLETED UNIT 12, YOU SHOULD BE ABLE TO:

Use the following rules and definitions.

To add two algebraic fractions with a common denominator, add the numerators and place the result over the common denominator.

$$\frac{a}{c}+\frac{b}{c}=\frac{a+b}{c}$$

To subtract a fraction, add its additive inverse.

$$\frac{a}{c}-\frac{b}{c}=\frac{a}{c}+\frac{-b}{c}$$

To add fractions without common denominators, follow these steps:

1. Find a common denominator.
   a. Factor each denominator.
   b. Write each different factor that appears in any denominator.
   c. Raise each factor to the highest power it has in any *single* denominator. The result of this process is the least common denominator (LCD).
2. Convert all fractions to be added to equivalent fractions with the common denominator by supplying the missing factors as 1's.
3. Add the numerators of the equivalent fractions and place the result over the common denominator.
4. Reduce to lowest terms.

To simplify complex fractions, use the following methods:

1. Division method
   a. Simplify and write both the numerator and denominator as single fractions by performing the indicated operations.
   b. To divide, multiply the numerator by the reciprocal of the denominator.
2. LCD method
   a. Find the LCD of all fractions in the numerator and denominator.
   b. Multiply the numerator and denominator of the given fraction by the LCD.
   c. Reduce to lowest terms.

To solve equations with fractions:

1. Multiply both sides of the equation by the LCD of all fractions.
2. Solve the resulting equation.
3. Check the solution in the *original* equation. (Division by zero is not possible.)

REVIEW TEST 12

Perform the indicated operations. Reduce all answers to lowest terms. (12.1 and 12.2)

1. $\frac{3x}{3x+1} - 1 =$ _____

2. $\frac{9}{x-2} + \frac{7}{2-x} =$ _____

3. $\frac{15}{x^2+x-6} - \frac{3}{x-2} =$ _____

4. $\frac{4}{x^2-4x} - \frac{7}{x^2-x-12} =$ _____

Simplify the following complex fractions. Reduce your answers to lowest terms. (12.3)

5. $\dfrac{2 + \frac{1}{2x}}{\frac{1}{x}} =$ _____

6. $\dfrac{1 + \frac{1}{3x}}{\frac{1}{x} - \frac{1}{3}} =$ _____

7. $\dfrac{\frac{1}{a^2} - 1}{1 - \frac{1}{a}} =$ _____

Solve for the variable. (12.4)

8. $\frac{3}{x} + \frac{1}{2x} = \frac{7}{4}$

9. $\frac{x-2}{5} - \frac{2x-7}{3} = 1$

10. $\frac{2}{2x-1} = \frac{4}{3x-5}$

Select a variable to represent the unknown quantity. Set up an equation and solve it. (12.5)

11. The denominator of a fraction exceeds the numerator by 6. If the numerator is increased by 4 and the denominator is decreased by 5, the value of the fraction is $\frac{5}{4}$. What is the value of the original fraction?

12. The cold water faucet can fill a sink in 12 seconds and the hot water faucet can fill it in 10 seconds. The drain can empty it in 15 seconds. How long will it take to fill the sink if all three are left open?

REVIEW YOUR SKILLS

Find the following products.

13. $(8x^2y - 4xy^2 + 6x^3y^4)(-4xy^5)$

14. $(5x - 3)(2x^2 - 5x + 6)$

Find the following quotient.

15. $\dfrac{8a^2b^4 - 12a^3b - 18ab^5}{6ab^2}$

Use long division to find the following quotient.

16. $(8x^3 + 10x^2 + 9) \div (4x - 3)$

Factor the following completely.

17. $4ax^4 + 2ax^3 - 30ax^2$

18. $4a^2x^4 - 6ax^3 + 8a^3x^2$

19. $16x^2 - 40xy + 25y^2$

20. $18a^2x^3 - 32a^2x$

Cumulative Review 1–12

Evaluate. ***(1.4)***

1. $-24 + 8 \div 2$
2. $[7 - (-5)] \div 6 - 3$
3. $[-2(-1) - 5]^2 - (-3)^2 + (-2)^3$
4. Find the area of a trapezoid with the following dimensions: height 6 meters, lower base 12 meters, and upper base 8 meters. Use the formula $A = \frac{1}{2}h(B + b)$. ***(1.4)***

Simplify the following.

5. $(-3a^0b^3)(-4a^0b^4c^0)$ ***(2.2)***
6. $-6ax^3(-8a^2x^4 + 9ax^3 - 6a^0x^0)$ ***(2.3)***
7. $-xy^2 - 4x^2y + 3xy - y + 7xy - 4xy^2 - 2x$ ***(2.4)***
8. $(3x^2 - y^2) - (15x^2 + 2xy - 3y^2) + (17xy - 4x^2 + 5y^2)$ ***(2.4)***

Multiply or divide the following fractions. Write all answers in lowest terms using standard form for fractions. ***(3.2)***

9. $-6a^2x \cdot \dfrac{-28ax^2}{-18}$
10. $\dfrac{-56by}{35} \div \dfrac{-40}{-75}$

Use the distributive property to simplify the following. Write the answer without parentheses. ***(3.2)***

11. $-\dfrac{3}{4}a\left(12a - \dfrac{18}{21}b\right)$

Perform the indicated operations. Reduce the answer to lowest terms and write the fraction in standard form. ***(3.3)***

12. $\dfrac{-5ax^2}{6} - \dfrac{-11ax^2}{14}$

Solve for the variable.

13. $6(x + 3) - 5(4 - x) = 3(2x - 2) + 3x$ ***(4.1)***
14. $2t - \dfrac{3t}{5} = 7$ ***(4.2)***
15. $A = \dfrac{h}{2}(B + b)$ for B ***(4.3)***

16. After shopping around for a new car, Yolanda found a dealer who offered her a car for 15% off the sticker price. If Yolanda paid \$10,625 plus taxes and license, what was the sticker price? *(4.5)*

Write the following English phrase as an algebraic expression.

17. If a number decreased by 5 is subtracted from three times the same number, the result is 35. *(5.1)*

Select a letter to represent the unknown quantity. Set up an equation and solve it.

18. Jamie is 3 years older than her sister Janiece. Two years ago, she was twice as old as her sister. How old are the two girls now? *(5.3)*

19. Chandra received a collection of coins for her birthday. The collection was made up of nickels, dimes, and quarters. If the entire collection was worth \$3.15 and there were 4 more dimes than quarters and 5 fewer nickels than dimes, how many of each coin was in the collection? *(5.4)*

20. John and Jonalee went from Los Angeles out to the desert to do some dirt bike riding. Due to heavy traffic, they could only average 40 mph. However, returning home, due to less traffic they could average 55 mph. If the total driving time was 4 hours and 45 minutes, how far did they travel going to the desert? *(5.5)*

21. Given $g(x) = \frac{-3}{4}x + 4$, find: *(6.1)*

    (a) $g(-4)$ (b) $g(0)$ (c) $g(5)$

22. Find the slope of the line through the given points. *(6.3)*

    $(-3, 4), (5, -2)$

23. Find the equation of the line passing through the given points. *(6.4)*

    $(2, 4), (-1, -5)$

24. Sketch the following graph using the slope and y-intercept method. *(6.5)*

    $$y = \frac{-2}{3}x + 4$$

25. Sketch the following graph using the intercepts method. *(6.5)*

    $$-3x + 4y = 12$$

26. Solve the following system using the addition method. *(7.2)*

    $$2x + 3y = -3$$
    $$-3x + 6y = 4$$

27. Use the substitution method to find the simultaneous solution for the following system of equations. *(7.3)*

    $$2x - y = 7$$
    $$3x + 4y = 5$$

28. Use any method to find the simultaneous solution for the following system of equations. *(7.2, 7.3)*

$$4x - y = 6$$
$$-6x + 2y = -11$$

29. Joe Dealer, an eager sales manager, had a particularly good year and was able to put aside $25,000 after a short period of time. He chose to invest part of the money in long-term certificates paying 14% interest and the remainder in a short-term certificate paying 8%. If his total income from the investments at the end of the year was $3,020, how much was invested at each rate? *(7.4)*

Simplify the following using the laws of exponents. Write the results without negative exponents.

30. $(-x^{-3}y^4)(3x^2y^{-4})$ *(8.1)*

31. $(-5a^3b^0c^5)^3$ *(8.2)*

32. $\dfrac{-15x^5y^4}{-18x^2y^7}$ *(8.1)*

33. $\left(\dfrac{-3m^5}{4n^3s^0}\right)^2$ *(8.2)*

34. Simplify using scientific notation. Write the answer in decimal notation. *(8.3)*

$$\frac{2400}{0.096}$$

Evaluate the following.

35. $a^2b - a^3bc$, if $a = -2$, $b = -1$, $c = 3$ *(2.1)*

36. $(5x^2 - 6x + 7) - (2x - 3)$, if $x = -2$ *(2.1)*

37. $\dfrac{2}{3}a^2 - \dfrac{1}{2}b^2 + \dfrac{1}{4}c^2$, if $a = 6$, $b = -2$, $c = -6$ *(3.4)*

38. $3(a^2 + b^2 - c^2) - \dfrac{9}{8}(abc - 9bc)$, if $a = 0$, $b = \dfrac{1}{3}$, $c = -\dfrac{1}{3}$ *(3.4)*

39. $a^3 + b^3 - c^3 + (a + b + c)^2$, if $a = -3$, $b = 2$, $c = -2$ *(8.2)*

40. $\left(\dfrac{x^2y^0}{ab^0}\right)^2$, if $a = 2$, $b = 3$, $x = -1$, $y = -2$ *(8.2)*

Find the following products.

41. $(3x - 5y^2)^2$ *(9.2)*

42. $(4a - 3b)(4a + 2b)$ *(9.2)*

43. $(4x^2 + 3x - 2)(3x - 4)$ *(9.1)*

Divide. Write with positive exponents only. *(9.3)*

44. $\dfrac{-6a^3b^2 + 9a^2b - 15a^4b^4}{-6a^2b^3}$

Divide.

45. $(6x^3 - 11x^2 + 13x - 12) \div (3x - 4)$ *(9.4)*

Factor the following completely.

46. $-12a^3x^5 + 20a^4x^3 - 32ax^4$ ***(10.1)***

47. $2x^2 - 5x - 12$ ***(10.2)***

48. $6a^2 + 13a - 15$ ***(10.3)***

49. $64x^4 - 9b^2$ ***(10.4)***

50. $-6a^7b + 24a^5b^2 - 24a^3b^3$ ***(10.4)***

Reduce to lowest terms.

51. $\dfrac{x^3y^2 - xy^4}{x^3y + 2x^2y^2 + xy^3}$ ***(11.1)***

Multiply and reduce to lowest terms.

52. $\dfrac{1}{a-b} \cdot (b-a)$ ***(11.2)***

53. $\dfrac{2a^2 + 6a}{a-4} \cdot \dfrac{a^2 - 16}{a^2 + 6a + 9}$ ***(11.2)***

Perform the indicated operations.

54. $\dfrac{a^2 - 16}{a^2 - 4} \div (a - 4)$ ***(11.3)***

55. $\dfrac{x^2 + 2x + 1}{x^2 - x} \cdot \dfrac{4x^2 - 1}{x^2 - 2x - 3} \div \dfrac{2x^2 + 3x + 1}{x^2 - 4x + 3}$ ***(11.3)***

Add or subtract as indicated and reduce to lowest terms.

56. $\dfrac{4x - 9}{x^2 - x - 12} - \dfrac{11 - x}{x^2 - x - 12}$ ***(12.1)***

57. $\dfrac{3x - 13}{x^2 - 8x + 15} - \dfrac{2}{x - 3}$ ***(12.2)***

Simplify the following complex number. Reduce your answer to lowest terms.

58. $\dfrac{\dfrac{1}{a^2} - 1}{\dfrac{1}{a} + 1}$ ***(12.3)***

Solve for the variable.

59. $\dfrac{x}{3} - 2 = \dfrac{x - 3}{2}$ ***(12.4)***

60. The cold water faucet can fill a sink in 15 seconds and the hot water faucet can fill it in 20 seconds. The drain can empty it in 30 seconds. If both faucets and the drain are open, how long will it take to fill the sink? ***(12.5)***

UNIT 13

Radical Expressions

OBJECTIVES

After you have successfully completed this unit, you will be able to:

1. Find the square root of perfect squares (13.1)
2. Simplify radical expressions that are not perfect squares (13.2)
3. Combine radical expressions using addition and subtraction (13.3)
4. Multiply and divide radical expressions (13.4)
5. Find the distance between two points (13.5)
6. Use the Pythagorean theorem to find parts of a right triangle (13.5)

13.1 RADICAL NOTATION

Square

A *square* is a product with the same factor used twice. Frequently, it is necessary to find the factor that was used to produce the given square.

For example,

Factor — Square

$6 \cdot 6 = 6^2 = 36$

We might ask, "What number was used as a factor twice to produce 36?"

$? \cdot ? = 36$

The above example shows that ? is 6.

A *factor* that is multiplied by itself to produce a given number is called the *square root* of that number. Hence, the square root of 36 is 6. In practice, we write this as $\sqrt{36} = 6$.

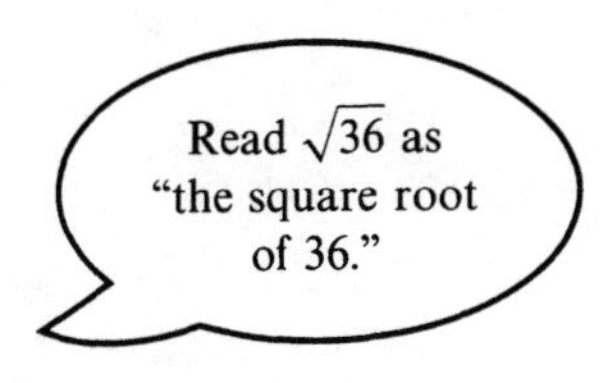

EXAMPLE 1 Find $\sqrt{16}$.

$\sqrt{16} = 4$ because $4 \cdot 4 = 16$

EXAMPLE 2 Find $\sqrt{7}$.

We are asking "What number multiplied by itself will produce 7?"

No

Can you find an integer? _____

To find the number, let us proceed as we did when we created a number to add to 7 to produce 0. We called it -7, the additive inverse of 7. When we wanted a number to multiply by 7 to give 1 we created $\frac{1}{7}$, the multiplicative inverse of 7.

Now, we will call the factor that is to be multiplied by itself to give 7 the *square root* of 7 and write it $\sqrt{7}$ so that $\sqrt{7} \cdot \sqrt{7} = 7$.

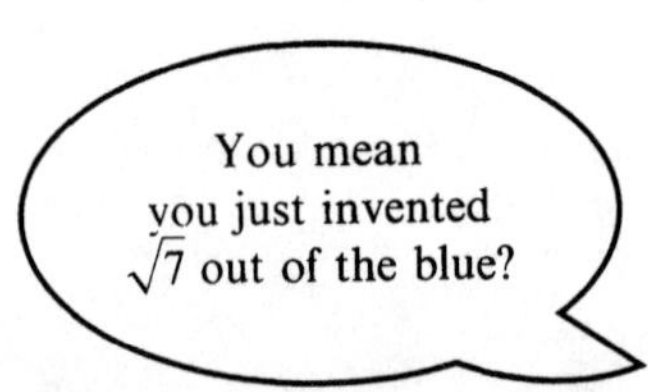

Almost. More accurately, mathematicians defined a symbol, $\sqrt{7}$, to represent a number with the desired properties.

Square root

> **13.1A DEFINITION: SQUARE ROOT**
>
> For all nonnegative numbers a and b,
>
> $$\sqrt{a} \cdot \sqrt{a} = a$$
>
> This is equivalent to saying
>
> $$\text{if } \sqrt{a} = b \quad \text{then} \quad a = b^2$$

Radical sign
Radicand

The symbol $\sqrt{}$ is called the *radical sign* and the number under the radical sign is called the *radicand.*

EXAMPLE 3

$\sqrt{13} \cdot \sqrt{13} = 13$ ⇦ Read: "The square root of 13 times the square root of 13 is 13."

19 — $\sqrt{19} \cdot \sqrt{19} =$ _____

16 — $\sqrt{16} \cdot \sqrt{16} =$ _____

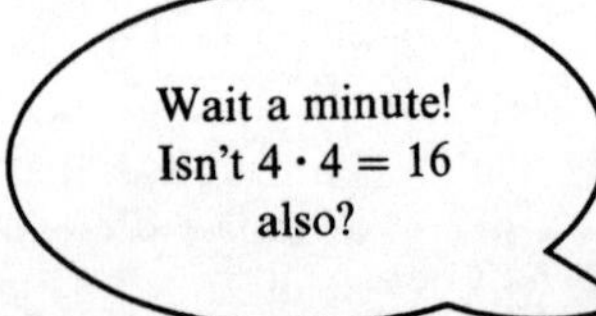

True. Therefore, we can say $\sqrt{16} = 4$.

We can give some square roots from our knowledge of multiplication.

EXAMPLE 4

$$\sqrt{16} \cdot \sqrt{16} = 16$$
$$\downarrow \quad \downarrow$$
$$4 \cdot 4 = 16$$

$$\sqrt{25} \cdot \sqrt{25} = 25$$
$$\downarrow \quad \downarrow$$

5, 5

$$\_\_\_\_ \cdot \_\_\_\_ = 25$$

$$\sqrt{64} \cdot \sqrt{64} = 64$$
$$\downarrow \quad \downarrow$$

8, 8

$$\_\_\_\_ \cdot \_\_\_\_ = 64$$

EXAMPLE 5

$\sqrt{9} = 3$ because $3 \cdot 3 = 9$

6, 6, 6, $\sqrt{36} = \_\_\_\_$ because $\_\_\_\_ \cdot \_\_\_\_ = 36$

7, 7, 7 $\sqrt{49} = \_\_\_\_$ because $\_\_\_\_ \cdot \_\_\_\_ = 49$

Principal square root

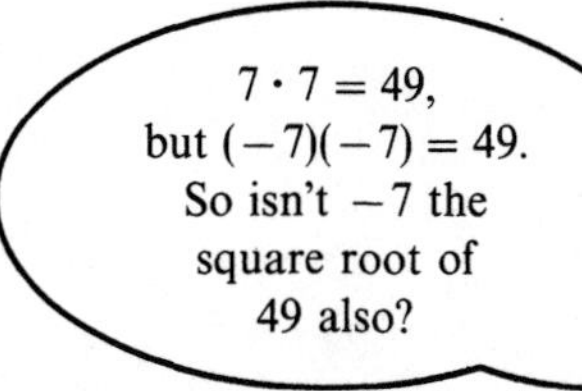

Yes, it is, but to avoid confusion, we want $\sqrt{49}$ to represent only one number. Therefore we choose the positive 7 and call it the *principal square root* of 49.

What if I want to use −7?

If you want to use −7, write $-\sqrt{49} = -7$.

EXAMPLE 6

$\sqrt{4} = 2$

$-\sqrt{9} = -3$

-5 $\quad -\sqrt{25} = \_\_\_\_$

8 $\quad \sqrt{64} = \_\_\_\_$

Are we limited to finding square roots of whole numbers only?

No. We can find square roots of fractions just as well.

EXAMPLE 7

$$\sqrt{\frac{4}{9}} = \frac{2}{3} \quad \text{because} \quad \frac{2}{3} \cdot \frac{2}{3} = \frac{4}{9}$$

$\frac{5}{6}, \frac{5}{6}, \frac{5}{6}$

$$\sqrt{\frac{25}{36}} = \_\_\_\_ \quad \text{since} \quad \_\_\_\_ \cdot \_\_\_\_ = \frac{25}{36}$$

The definition says for any nonnegative number. Can't we find the square root of a negative number?

With our present set of numbers, we can only find square roots of numbers that are greater than or equal to zero.

To see why we can't find the square root of a negative number, let's look for a square root of -4.

$$(2) \cdot (2) = 4$$
$$(-2) \cdot (-2) = 4$$
$$(?) \cdot (?) = -4$$

None

What number can you multiply by itself to give -4? ________

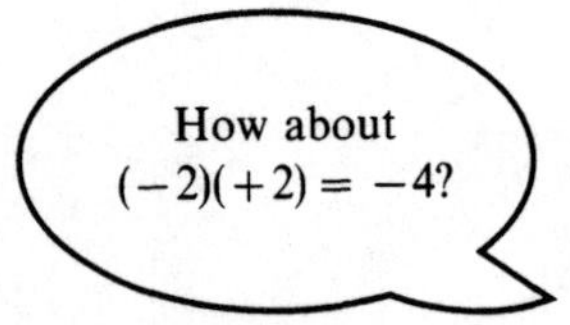

Nice try, but to be a square root, both factors must be identical.

EXAMPLE 8 Simplify the following.

9

$\sqrt{81} = \_\_\_\_$

-5

$-\sqrt{25} = \_\_\_\_$

$\frac{6}{7}$

$\sqrt{\frac{36}{49}} = \_\_\_\_$

Are all square roots rational numbers?

No. $\sqrt{2}$, for example, is not a rational number. It is a member of the set of *irrational numbers*.

Irrational numbers

> **13.1B DEFINITION: IRRATIONAL NUMBERS**
> All real numbers that cannot be written as the ratio of two integers are called *irrational numbers*.

Is that all the numbers there are?

No. The rational numbers combined with the irrational numbers make the *real number system*. Complex numbers, which include the real numbers, are beyond the scope of this book.

EXAMPLE 9 Simplify $\sqrt{25} + \sqrt{9}$.

5 $\quad \sqrt{25} + \sqrt{9} = \_\_\_\_ + 3$

8 $\quad = \_\_\_\_$

EXAMPLE 10 Simplify $\sqrt{\frac{36}{25}} \cdot \sqrt{\frac{9}{16}}$.

$\frac{6}{5}$ $\quad \sqrt{\frac{36}{25}} \cdot \sqrt{\frac{9}{16}} = \frac{\_\_\_\_}{\_\_\_\_} \cdot \frac{3}{4}$

$$= \frac{\overset{3}{\cancel{6}}}{5} \cdot \frac{3}{\underset{2}{\cancel{4}}}$$

$\frac{9}{10}$ $\quad = \frac{\_\_\_\_}{\_\_\_\_}$

EXAMPLE 11 Simplify $\frac{\sqrt{49} - \sqrt{16}}{\sqrt{9}}$.

3 $\quad \frac{\sqrt{49} - \sqrt{16}}{\sqrt{9}} = \frac{7 - 4}{\_\_\_\_}$

$$= \frac{3}{3}$$

1 $\quad = \_\_\_\_$

EXAMPLE 12 Simplify $\sqrt{\frac{100}{81}} \div \sqrt{\frac{64}{25}}$.

In this case, first simplify the radicals then divide.

$$\sqrt{\frac{100}{81}} \div \sqrt{\frac{64}{25}} = \frac{10}{9} \div \frac{8}{5}$$

$$= \frac{\overset{5}{\cancel{10}}}{9} \cdot \frac{5}{\underset{4}{\cancel{8}}}$$

$$= \frac{25}{36}$$

PROBLEM SET 13.1

Evaluate the following.

1. $\sqrt{25}$	2. $\sqrt{36}$	3. $-\sqrt{4}$
4. $-\sqrt{16}$	5. $\sqrt{49}$	6. $-\sqrt{9}$
7. $\sqrt{81}$	8. $\sqrt{121}$	9. $-\sqrt{100}$
10. $-\sqrt{64}$	11. $\sqrt{\frac{9}{4}}$	12. $\sqrt{\frac{16}{9}}$

13. $-\sqrt{\frac{25}{16}}$

14. $-\sqrt{\frac{36}{25}}$

15. $\sqrt{36} - \sqrt{4}$

16. $\sqrt{81} + \sqrt{25}$

17. $\sqrt{25} - \sqrt{16}$

18. $\sqrt{9} + \sqrt{16}$

19. $\sqrt{25} + \sqrt{16} - \sqrt{9}$

20. $\sqrt{81} - \sqrt{100} - \sqrt{36}$

21. $\sqrt{\frac{25}{16}} \cdot \sqrt{\frac{4}{9}}$

22. $\sqrt{\frac{81}{64}} \cdot \sqrt{\frac{36}{25}}$

23. $\sqrt{\frac{36}{16}} \cdot \sqrt{\frac{9}{25}} \cdot \sqrt{\frac{64}{81}}$

24. $\sqrt{\frac{49}{64}} \cdot \sqrt{\frac{36}{25}} \cdot \sqrt{\frac{121}{144}}$

25. $\sqrt{\frac{36}{49}} \div \sqrt{\frac{64}{81}}$

26. $\frac{\sqrt{25}}{\sqrt{16}} \div \frac{\sqrt{100}}{\sqrt{4}}$

27. $\frac{\sqrt{9}}{\sqrt{144}} \cdot \frac{\sqrt{81}}{\sqrt{36}}$

28. $\sqrt{\frac{16}{9}} \div \sqrt{\frac{64}{36}}$

29. $\sqrt{\frac{36}{25}} \div \sqrt{\frac{16}{49}}$

30. $\sqrt{\frac{81}{64}} \div \sqrt{\frac{144}{16}}$

31. $\frac{\sqrt{64} - \sqrt{49}}{\sqrt{9}}$

32. $\frac{\sqrt{144} - \sqrt{81}}{\sqrt{4}}$

33. $\frac{\sqrt{81} + \sqrt{25}}{\sqrt{49}}$

34. $\frac{\sqrt{64} + \sqrt{25} - \sqrt{9}}{\sqrt{25}}$

35. $\frac{\sqrt{16} - \sqrt{25} + \sqrt{64}}{\sqrt{49}}$

36. $\frac{\sqrt{36} - \sqrt{81} + \sqrt{9}}{\sqrt{144}}$

13.2 SIMPLIFYING RADICALS

It is not difficult to replace $\sqrt{9}$ with 3, but what about a number like $\sqrt{12}$? We may not be able to replace it with an integer, but we are often able to simplify it or change it to a desired form.

Consider the following example.

EXAMPLE 13

$$\sqrt{4} \cdot \sqrt{25} =$$
$$2 \cdot \quad 5 = 10$$

Also,

$$\sqrt{100} = 10$$

because two quantities equal to the same quantity are equal to each other.

$$\sqrt{4} \cdot \sqrt{25} = \sqrt{100}$$

Or factoring 100,

$$\sqrt{4} \cdot \sqrt{25} = \sqrt{4 \cdot 25}$$

This leads us to the multiplication law for radicals.

Multiplication law for radicals

13.2A MULTIPLICATION LAW FOR RADICALS

For all nonnegative numbers a and b,

$$\sqrt{a} \cdot \sqrt{b} = \sqrt{ab} \quad \text{or} \quad \sqrt{ab} = \sqrt{a} \cdot \sqrt{b}$$

This says that the square root of a product is equal to the product of the square roots of each of its factors.

EXAMPLE 14 Simplify $\sqrt{12}$.

$\sqrt{12} = \sqrt{4 \cdot 3}$

$= \sqrt{4} \cdot \sqrt{3}$ Multiplication law for radicals

$= 2 \cdot \sqrt{3}$

Could we have written $\sqrt{12} = \sqrt{2 \cdot 6}$?

Yes, you could but it wouldn't have done you any good. Try $\sqrt{2 \cdot 6} = \sqrt{2} \cdot \sqrt{6}$ What does $\sqrt{2} = ?$ What does $\sqrt{6} = ?$

How do we know what factors to use when writing a number as the product of two factors?

Just make sure that one of the factors is a *square* of a whole number.

EXAMPLE 15 Simplify $\sqrt{20}$.

$\sqrt{20} = \sqrt{4 \cdot 5}$

$= \sqrt{4} \cdot \sqrt{5}$

$= 2\sqrt{5}$

What does $\sqrt{5}$ equal?

$\sqrt{5}$ is an irrational number. Sorry I can't express its value exactly with decimals or fractions. However, $2 \cdot 2 = 4$ and $3 \cdot 3 = 9$. Therefore, $\sqrt{5}$ is between 2 and 3. Try $\sqrt{5} \approx 2.2$

$(2.2)(2.2) = 4.84$. I want it closer.

OK. Then guess a little bigger like $\sqrt{5} \approx 2.25$.

EXAMPLE 16 Simplify $\sqrt{48}$.

$\sqrt{48} = \sqrt{16 \cdot 3}$

$= \sqrt{16} \cdot \sqrt{3}$

$=$ ______ $\sqrt{3}$

4

EXAMPLE 17 Simplify $\sqrt{56}$.

$$\sqrt{56} = \sqrt{4 \cdot 14}$$

4, 14 $\quad = \sqrt{\_\_\_\_\_}\sqrt{\_\_\_\_\_}$

2 $\quad = \_\_\_\_\_ \sqrt{14}$

EXAMPLE 18 Is $\sqrt{9 + 16} = \sqrt{9} + \sqrt{16}$?

As the square root of a sum

$$\sqrt{9 + 16} = \sqrt{25}$$
$$= 5$$

However, as the sum of two square roots

$$\sqrt{9} + \sqrt{16} =$$
$$3 + \quad 4 \quad = 7$$

Notice

$$\sqrt{9 \cdot 16} = \sqrt{9} \cdot \sqrt{16}$$

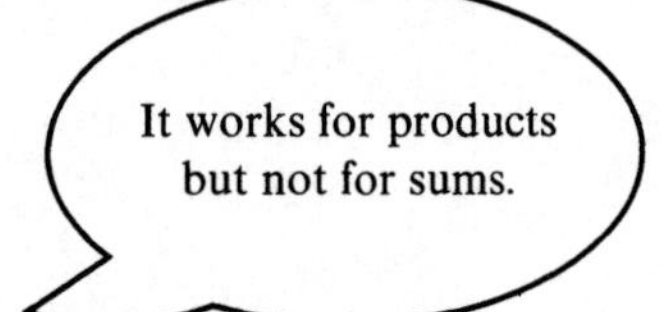

but $\sqrt{9 + 16}$ **is not equal to** $\sqrt{9} \cdot \sqrt{16}$.

Now let's look at the radical of a quotient.

EXAMPLE 19 $\sqrt{\dfrac{4}{9}} = \dfrac{2}{3}$.

Also

$$\frac{\sqrt{4}}{\sqrt{9}} = \frac{2}{3}$$

Therefore,

$$\sqrt{\frac{4}{9}} = \frac{\sqrt{4}}{\sqrt{9}}$$

What's the difference?

$\sqrt{\dfrac{4}{9}}$ is a fraction under a single radical. $\dfrac{\sqrt{4}}{\sqrt{9}}$ is a fraction made up of two separate radicals.

This leads us to the quotient law for radicals.

Quotient law for radicals

13.2B QUOTIENT LAW FOR RADICALS

For all nonnegative numbers a and b, where $b \neq 0$,

$$\sqrt{\frac{a}{b}} = \frac{\sqrt{a}}{\sqrt{b}} \quad \text{or} \quad \frac{\sqrt{a}}{\sqrt{b}} = \sqrt{\frac{a}{b}}$$

EXAMPLE 20 Simplify $\sqrt{\frac{3}{25}}$.

$$\sqrt{\frac{3}{25}} = \frac{\sqrt{3}}{\sqrt{25}}$$

$$= \frac{\sqrt{3}}{5}$$

EXAMPLE 21 Simplify $\frac{1}{\sqrt{3}}$.

In this case, *simplify* means to get an equivalent fraction without a radical in the denominator.

Recall that you can write an equivalent fraction by multiplying by 1.

We can use two facts to do this.

Fact 1

$$\sqrt{3} \cdot \sqrt{3} = 3$$

Fact 2

$$\frac{\sqrt{3}}{\sqrt{3}} = 1$$

Therefore,

$$\frac{1}{\sqrt{3}} = \frac{1}{\sqrt{3}} \cdot \frac{\sqrt{3}}{\sqrt{3}}$$

This is multiplication by 1.

$$= \frac{\sqrt{3}}{3}$$

You call $\frac{\sqrt{3}}{3}$ simpler than $\frac{1}{\sqrt{3}}$?

If you don't believe it, try evaluating $\frac{\sqrt{3}}{3}$ and $\frac{1}{\sqrt{3}}$ using $\sqrt{3} \approx 1.732$.

Simplifying radicals

13.2C RULE: A RADICAL IS SAID TO BE SIMPLIFIED IF:

1. The radicand does not contain a factor that can be written as the square of a whole number.
2. No fraction appears under the radical sign.
3. No radical appears in the denominator of a fraction.

EXAMPLE 22 Simplify $\sqrt{\frac{3}{2}}$.

$$\sqrt{\frac{3}{2}} = \frac{\sqrt{3}}{\sqrt{2}}$$

This is not completely simplified because there is a $\sqrt{2}$ in the denominator. We can remove this $\sqrt{2}$ from the denominator if we use the fact that $\sqrt{2} \cdot \sqrt{2} = 2$.

$$= \frac{\sqrt{3}}{\sqrt{2}} \cdot \frac{\sqrt{2}}{\sqrt{2}}$$ Multiplication by 1

$$= \frac{\sqrt{3 \cdot 2}}{2}$$ ← Multiplication law for radicals; ← Definition of square root

$\sqrt{6}$

$$= \frac{\quad}{2}$$

The 6 cannot be divided by 2 because the 6 is under the radical sign.

EXAMPLE 23 Simplify $\sqrt{\frac{9}{8}}$.

First, simplify the numerator and the denominator separately.

$$\sqrt{\frac{9}{8}} = \frac{\sqrt{9}}{\sqrt{8}}$$ ← 9 is a perfect square; ← $8 = 4 \cdot 2$

$$= \frac{3}{\sqrt{4}\sqrt{2}}$$

Because $\sqrt{4}$ is 2

$$= \frac{3}{2\sqrt{2}}$$

Now we will remove the $\sqrt{2}$ from the denominator.

$$= \frac{3}{2\sqrt{2}} \cdot \frac{\sqrt{2}}{\sqrt{2}}$$ Multiplication by 1

$$= \frac{3\sqrt{2}}{2\sqrt{2}\sqrt{2}}$$

$$= \frac{3\sqrt{2}}{2 \cdot 2}$$ ← $\sqrt{2} \cdot \sqrt{2} = 2$

$$= \frac{3\sqrt{2}}{4}$$

Eliminating the radical from the denominator is called rationalizing the denominator.

EXAMPLE 24 Simplify $\sqrt{\frac{10}{24}}$.

First, make any reduction possible under the radical.

$$\sqrt{\frac{10}{24}} = \sqrt{\frac{5}{12}}$$

Now, simplify the numerator and denominator as much as possible.

$$\frac{\sqrt{5}}{\sqrt{12}} = \frac{\sqrt{5}}{\sqrt{4} \cdot \sqrt{3}}$$

$$= \frac{\sqrt{5}}{2\sqrt{3}}$$

Now, get the $\sqrt{3}$ out of the denominator.

$$= \frac{\sqrt{5}}{2\sqrt{3}} \cdot \frac{\sqrt{3}}{\sqrt{3}} \quad \text{Multiplication by 1}$$

$\sqrt{15}$

$$= \frac{\quad}{2 \cdot 3}$$

$\sqrt{3}\sqrt{3} = 3$

$$= \frac{\sqrt{15}}{6}$$

That's called rationalizing the denominator.

PROBLEM SET 13.2A

Simplify the following radicals.

1. $\sqrt{8}$
2. $\sqrt{27}$
3. $\sqrt{24}$
4. $\sqrt{45}$
5. $\sqrt{20}$
6. $\sqrt{75}$
7. $\sqrt{125}$
8. $\sqrt{18}$
9. $\sqrt{32}$
10. $\sqrt{50}$
11. $\sqrt{72}$
12. $\sqrt{98}$
13. $\sqrt{147}$
14. $\sqrt{162}$
15. $\sqrt{48}$
16. $\sqrt{108}$
17. $\sqrt{243}$
18. $\sqrt{192}$
19. $\frac{1}{\sqrt{6}}$
20. $\frac{1}{\sqrt{11}}$
21. $\frac{1}{\sqrt{5}}$
22. $\frac{1}{\sqrt{7}}$
23. $\frac{1}{\sqrt{8}}$
24. $\frac{1}{\sqrt{12}}$
25. $\frac{3}{\sqrt{18}}$
26. $\frac{6}{\sqrt{27}}$
27. $\frac{4}{\sqrt{20}}$
28. $\frac{5}{\sqrt{45}}$
29. $\sqrt{\frac{9}{12}}$
30. $\sqrt{\frac{25}{27}}$
31. $\sqrt{\frac{7}{18}}$
32. $\sqrt{\frac{5}{32}}$
33. $\sqrt{\frac{36}{8}}$
34. $\sqrt{\frac{50}{125}}$
35. $\sqrt{\frac{36}{50}}$
36. $\sqrt{\frac{16}{72}}$
37. $\sqrt{\frac{25}{98}}$
38. $\sqrt{\frac{49}{128}}$
39. $\frac{\sqrt{9}}{\sqrt{8}}$
40. $\frac{\sqrt{36}}{\sqrt{75}}$
41. $\frac{\sqrt{36}}{\sqrt{108}}$
42. $\frac{\sqrt{25}}{\sqrt{20}}$
43. $\frac{\sqrt{45}}{\sqrt{81}}$
44. $\frac{\sqrt{75}}{\sqrt{125}}$
45. $\frac{\sqrt{36}}{\sqrt{45}}$
46. $\frac{\sqrt{27}}{\sqrt{80}}$
47. $\frac{\sqrt{50}}{\sqrt{125}}$
48. $\frac{\sqrt{72}}{\sqrt{180}}$
49. $\sqrt{\frac{36}{40}}$
50. $\sqrt{\frac{72}{90}}$

51. $\frac{\sqrt{36}}{\sqrt{160}}$ 52. $\frac{\sqrt{90}}{\sqrt{250}}$ 53. $\frac{\sqrt{4}\cdot\sqrt{9}}{\sqrt{4\cdot 9}}$ 54. $\frac{\sqrt{4}\cdot\sqrt{16}}{\sqrt{4\cdot 16}}$

55. $\frac{\sqrt{9}\cdot\sqrt{16}}{\sqrt{9\cdot 16}}$ 56. $\frac{\sqrt{9}+\sqrt{16}}{\sqrt{9+16}}$ 57. $\frac{\sqrt{25}-\sqrt{9}}{\sqrt{25-9}}$ 58. $\frac{\sqrt{25}\cdot\sqrt{9}}{\sqrt{25\cdot 9}}$

Simplifying Radicals with Variables

So far we have dealt only with square roots of constants. In this section, we will find square roots of algebraic expressions that also contain variables. We will assume that all the variables represent positive numbers.

EXAMPLE 25 Simplify the following radicals. Assume all variables represent positive numbers.

a. $\sqrt{x^2} = x$ because $x \cdot x = x^2$

b. $\sqrt{x^2 \cdot y^2} = \sqrt{x^2}\cdot\sqrt{y^2}$
$= x \cdot y$

c. $\sqrt{36x^2} = \sqrt{36}\cdot\sqrt{x^2}$
$= 6 \cdot x$

d. $\sqrt{25x^4} = \sqrt{25}\cdot\sqrt{x^4}$
$= 5x^2$

e. $\sqrt{\frac{x^2}{y^2}} = \frac{\sqrt{x^2}}{\sqrt{y^2}} = \frac{x}{y}$

f. $\sqrt{\frac{16x^2}{y^6}} = \frac{\sqrt{16x^2}}{\sqrt{y^6}} = \frac{4x}{y^3}$

$x^2 \cdot x^2 = x^4$.

All the exponents on the variables were even numbers.

That's because, if you multiply a number by itself, the resulting exponent is even. For example, $x^3 \cdot x^3 = x^6$. Therefore, $\sqrt{x^6} = x^3$.

It's simple to find the square root of a number with an even exponent. Just divide the exponent by 2.

Correct. However, if we have an odd-numbered exponent, we must treat it differently. Watch what happens in the next example.

EXAMPLE 26 Simplify $\sqrt{x^3}$.

$\sqrt{x^3} = \sqrt{x^2 \cdot x}$ Write as factors, making one factor an even power

$= \sqrt{x^2}\sqrt{x}$ Multiplication law for radicals

$= x \cdot \sqrt{x}$

When you have odd exponents under the radical, write the expression as two factors, one with an even exponent and the other to the first power.

EXAMPLE 27 Simplify $\sqrt{x^5y^3}$.

$\sqrt{x^5y^3} = \sqrt{x^4 \cdot x \cdot y^2 \cdot y}$ Write as factors

$= \sqrt{x^4 \cdot y^2 \cdot x \cdot y}$ Commutative property of multiplication

$= \sqrt{x^4}\sqrt{y^2}\sqrt{xy}$ Multiplication law

x^2y $= \_\_\_\_\_ \sqrt{xy}$

EXAMPLE 28 Simplify $\sqrt{\frac{x^3}{y^5}}$.

$\sqrt{\frac{x^3}{y^5}} = \frac{\sqrt{x^2 \cdot x}}{\sqrt{y^4 \cdot y}}$ Write as factors, one factor with even exponents

y^4 $= \frac{\sqrt{x^2} \cdot \sqrt{x}}{\sqrt{\_\_\_\_\_} \cdot \sqrt{y}}$ Multiplication law for radicals

$= \frac{x\sqrt{x}}{y^2\sqrt{y}}$ Take square roots

No Is this in simplest form? _____

Which form of 1 must we multiply by to write the denominator without a radical?

$\frac{\sqrt{y}}{\sqrt{y}}$ _____

$= \frac{x\sqrt{x}}{y^2\sqrt{y}} \cdot \left(\frac{\sqrt{y}}{\sqrt{y}}\right)$ Multiply by 1

$= \frac{x\sqrt{x}\sqrt{y}}{y^2\sqrt{y}\sqrt{y}}$

xy $= \frac{x\sqrt{\_\_\_\_\_}}{y^2 \cdot y}$

$\frac{x\sqrt{xy}}{y^3}$ $= \_\_\_\_\_$

$\sqrt{y} \cdot \sqrt{y} = y$
That's why we chose to multiply by $\frac{\sqrt{y}}{\sqrt{y}}$.

EXAMPLE 29 Simplify $\sqrt{24x^7y^3}$.

$$\sqrt{24x^7y^3} = \sqrt{4 \cdot 6 \cdot x^6 \cdot x \cdot y^2 \cdot y}$$
$$= \sqrt{4 \cdot x^6 \cdot y^2 \cdot 6 \cdot x \cdot y} \qquad \text{Rearranging}$$
$$= \_\_\_\_\_\_ \sqrt{6xy}$$

$2x^3y$

EXAMPLE 30 Simplify $\sqrt{\dfrac{27x^5y^2}{2w^7}}$.

$$\sqrt{\frac{27x^5y^2}{2w^7}} = \frac{\sqrt{9 \cdot 3 \cdot x^4 \cdot x \cdot y^2}}{\sqrt{2 \cdot w \cdot w^6}} \qquad \text{Separate into factors}$$
$$= \frac{\sqrt{9 \cdot x^4 \cdot y^2 \cdot 3 \cdot x}}{\sqrt{w^6 \cdot 2 \cdot w}} \qquad \text{Commutative property of multiplication}$$
$$= \frac{\sqrt{9}\sqrt{x^4}\sqrt{y^2}\sqrt{3x}}{\sqrt{w^6}\sqrt{2w}} \qquad \text{Multiplication law for radicals}$$
$$= \frac{\_\_\_\_\_\_ \sqrt{3x}}{w^3\sqrt{2w}}$$

$3x^2y$

Is this in simplest form? ______

No

By what must we multiply the fraction to eliminate the radical in the denominator?

______

$\dfrac{\sqrt{2w}}{\sqrt{2w}}$

$$\frac{3x^2y\sqrt{3x}}{w^3\sqrt{2w}} = \frac{3x^2y\sqrt{3x}}{w^3\sqrt{2w}} \cdot \frac{\sqrt{2w}}{\sqrt{2w}} \qquad \text{Multiplication by 1}$$
$$= \frac{3x^2y\sqrt{3x}\sqrt{2w}}{w^3\sqrt{2w}\sqrt{2w}}$$
$$= \frac{3x^2y \_\_\_\_\_\_}{w^3 \cdot 2w}$$

$\sqrt{6xw}$

$$= \_\_\_\_\_\_$$

$\dfrac{3x^2y\sqrt{6xw}}{2w^4}$

EXAMPLE 31 Simplify $\sqrt{\dfrac{75x^5a^3}{27x^2b^5}}$.

How about reducing the fraction under the radical sign before taking the square root?

If you do, it will cut down on your work. In the example, we will reduce it first. Try working it without reducing it first to see if you get the same result.

Reducing the fraction, we get

$$\frac{\overset{25x^3}{\cancel{75x^5}}a^3}{\underset{9}{\cancel{27x^2}}b^5} = \frac{25x^3a^3}{9b^5}$$

The radical now becomes

$$\sqrt{\frac{75x^5a^3}{27x^2b^5}} = \sqrt{\frac{25x^3a^3}{9b^5}}$$

$$= \frac{\sqrt{25x^3a^3}}{\sqrt{9b^5}} \qquad \text{Division of radicals}$$

$$= \frac{\sqrt{25 \cdot x^2 \cdot a^2 \cdot x \cdot a}}{\sqrt{9 \cdot b^4 \cdot b}} \qquad \text{Factored form}$$

$$= \frac{\sqrt{25}\sqrt{x^2}\sqrt{a^2}\sqrt{xa}}{\sqrt{9}\sqrt{b^4}\sqrt{b}} \qquad \text{Product rule}$$

$$= \frac{5 \cdot x \cdot a\sqrt{xa}}{\_\_\_\_\_\sqrt{b}} \qquad \text{Extracting roots}$$

$3b^2$

Is this in simplest form?

No. There still is a radical in the denominator. To take care of that, multiply the numerator and denominator by $\sqrt{b}$.

$$= \frac{5xa\sqrt{xa}}{3b^2\sqrt{b}} \cdot \frac{\sqrt{b}}{\sqrt{b}} \qquad \text{Multiply by 1}$$

$$= \frac{5xa\sqrt{xab}}{3b^2\,\_\_\_\_\_}$$

b

$$= \frac{5xa\sqrt{xab}}{3b^3}$$

PROBLEM SET 13.2B

Simplify the following.

1. $\sqrt{x^4}$
2. $\sqrt{y^6}$
3. $\sqrt{a^8}$
4. $\sqrt{a^{10}}$
5. $\sqrt{x^2y^4}$
6. $\sqrt{a^4b^4}$
7. $\sqrt{a^3}$
8. $\sqrt{b^5}$
9. $\sqrt{x^2y^3}$
10. $\sqrt{a^4b^5}$
11. $\sqrt{25x}$
12. $\sqrt{12y}$
13. $\sqrt{9a^3}$
14. $\sqrt{16b^5}$
15. $\sqrt{12x^2}$
16. $\sqrt{18y^4}$
17. $\sqrt{50x^3}$
18. $\sqrt{72a^5}$
19. $\sqrt{8a^5b^2}$
20. $\sqrt{27x^4y^5}$
21. $-\sqrt{45x^5a^4}$
22. $-\sqrt{48a^6b^7}$
23. $-\sqrt{50x^3a^5}$
24. $-\sqrt{32x^5y^3}$
25. $\sqrt{\frac{x^3}{y}}$
26. $\sqrt{\frac{x^2}{y^3}}$
27. $\sqrt{\frac{x^4}{y^3}}$
28. $\sqrt{\frac{a^5}{b^4}}$
29. $\sqrt{\frac{xy^4}{a^5}}$
30. $\sqrt{\frac{x^2a^3}{b^5}}$
31. $\sqrt{\frac{8a^3}{b^5}}$
32. $\sqrt{\frac{12a^5b}{b^7}}$
33. $\sqrt{\frac{27a^4x^5}{y^7}}$
34. $\sqrt{\frac{48a^5b^3}{x^5}}$
35. $\sqrt{\frac{45ab^3}{x^2y^3}}$
36. $\sqrt{\frac{72x^3y^2}{a^3b}}$

37. $\sqrt{\dfrac{98ax^5}{5b^3y^4}}$ 38. $\sqrt{\dfrac{125b^3x^5}{3ay^4}}$ 39. $\sqrt{\dfrac{10xy^3}{8a^3b^2}}$ 40. $\sqrt{\dfrac{27a^2b}{12x^4y}}$

41. $\sqrt{\dfrac{10ab^3}{45x^2y^3}}$ 42. $\sqrt{\dfrac{12x^3y}{27a^3b^2}}$ 43. $\sqrt{\dfrac{50x^7a}{72b^2y^3}}$ 44. $\sqrt{\dfrac{75a^5b}{48x^3y^5}}$

45. $\sqrt{\dfrac{60x^2b^3}{80a^4y^7}}$ 46. $\sqrt{\dfrac{90a^5b^6}{75a^4x^5}}$ 47. $\sqrt{\dfrac{90a^5y^3}{98b^4x^5}}$ 48. $\sqrt{\dfrac{150x^4y^3}{160a^5b^9}}$

13.3 ADDITION AND SUBTRACTION OF RADICAL EXPRESSIONS

Similar radicals

Square roots are *similar radicals* if they have the same radicand. Similar square roots can be combined the same way like terms were combined in Unit 2.

EXAMPLE 32

$\sqrt{a}$ and $2\sqrt{a}$ are similar radicals.
$\sqrt{a}$ and $\sqrt{2a}$ are *not* similar radicals.
$\sqrt{4x^3}$ and $-2\sqrt{4x^3}$ are similar radicals.
$\sqrt{5x^2}$ and $\sqrt{5x}$ are *not* similar radicals.

The important thing for similar radicals is that the quantities under the radical sign be identical.

EXAMPLE 33 Combine $3\sqrt{2} + 4\sqrt{2}$.

$3\sqrt{2} + 4\sqrt{2} = 7\sqrt{2}$

EXAMPLE 34 Combine $7\sqrt{3x} - 3\sqrt{3x} + 2\sqrt{3x}$.

6

$7\sqrt{3x} - 3\sqrt{3x} + 2\sqrt{3x} =$ _____ $\sqrt{3x}$ Combine similar radicals

EXAMPLE 35 Combine $4\sqrt{5} - 3\sqrt{2a} + 10\sqrt{5} + 8\sqrt{2a}$.

$4\sqrt{5} - 3\sqrt{2a} + 10\sqrt{5} + 8\sqrt{2a} = 4\sqrt{5} + 10\sqrt{5} - 3\sqrt{2a} + 8\sqrt{2a}$

Similar radicals

14, 5

$=$ _____ $\sqrt{5} +$ _____ $\sqrt{2a}$

EXAMPLE 36 Combine $\sqrt{8} + 6\sqrt{2}$.

No

Are $\sqrt{8}$ and $\sqrt{2}$ similar square roots? _____

Yes

Can $\sqrt{8}$ be simplified? _____

$\sqrt{8} = \sqrt{4}\sqrt{2}$

2

$=$ _____ $\sqrt{2}$

$\sqrt{8} + 6\sqrt{2} = 2\sqrt{2} + 6\sqrt{2}$ Rewrite $\sqrt{8}$

8

$=$ _____ $\sqrt{2}$

EXAMPLE 37 Combine $\sqrt{8a} + \sqrt{27a} - \sqrt{50a}$.

No — Are the square roots similar? _____

Yes — Can they be simplified? _____

Okay, then simplify each term separately.

Simplifying the first term:	Simplifying the second term:	Simplifying the third term:
$\sqrt{8a} = \sqrt{4 \cdot 2a}$	$\sqrt{27a} = \sqrt{9 \cdot 3a}$	$\sqrt{50a} = \sqrt{25 \cdot 2a}$
$= \sqrt{4}\sqrt{2a}$	$= \sqrt{9}\sqrt{3a}$	$= \sqrt{25}\sqrt{2a}$
$= 2\sqrt{2a}$	$= 3\sqrt{3a}$	$= 5\sqrt{2a}$

$$\sqrt{8a} + \sqrt{27a} - \sqrt{50a} = 2\sqrt{2a} + 3\sqrt{3a} - 5\sqrt{2a}$$

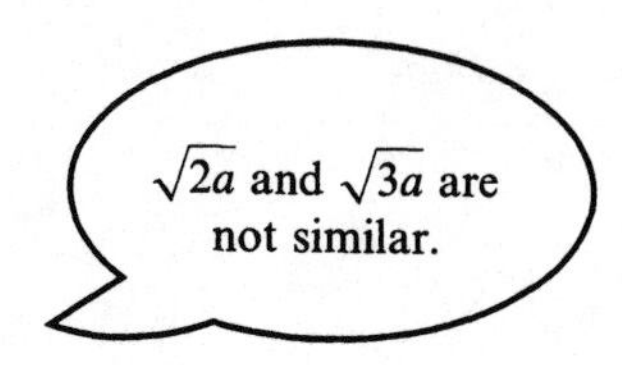

No — Are all radicals similar? _____

Yes — Can any be combined? _____

$$2\sqrt{2a} + 3\sqrt{3a} - 5\sqrt{2a} = 2\sqrt{2a} - 5\sqrt{2a} + 3\sqrt{3a}$$ Rearrange terms

-3 — $$= \_\_\_\_\_ \sqrt{2a} + 3\sqrt{3a}$$ Combine similar radicals

EXAMPLE 38 Simplify $\dfrac{11a\sqrt{2a} - \sqrt{8a^3}}{3}$.

Rewriting

$$\frac{11a\sqrt{2a} - \sqrt{8a^3}}{3} = \frac{11a\sqrt{2a} - \sqrt{4a^2 \cdot 2a}}{3}$$

$$= \frac{11a\sqrt{2a} - \sqrt{4a^2}\sqrt{2a}}{3}$$ Multiplication law of radicals

$2a$ — $$= \frac{11a\sqrt{2a} - \_\_\_\_\_\sqrt{2a}}{3}$$

$$= \frac{9a\sqrt{2a}}{3}$$ Combine similar radicals in the numerator

$3a\sqrt{2a}$ — $$= \_\_\_\_\_\_$$ Reduce

PROBLEM SET 13.3

Simplify the following.

1. $4\sqrt{2} + 3\sqrt{2}$
2. $8\sqrt{3} - 3\sqrt{3}$
3. $3\sqrt{5} - 8\sqrt{5}$
4. $6\sqrt{2} - 9\sqrt{2}$
5. $12\sqrt{5} - 2\sqrt{5} + 4\sqrt{5}$
6. $9\sqrt{7} + 3\sqrt{7} - 2\sqrt{7}$
7. $\sqrt{8} + \sqrt{18}$
8. $\sqrt{12} - \sqrt{27}$
9. $\sqrt{72} + \sqrt{50} - \sqrt{32}$
10. $\sqrt{48} + \sqrt{75}$
11. $\sqrt{45} - \sqrt{12} + \sqrt{75}$
12. $\sqrt{50} - \sqrt{12} + \sqrt{27}$
13. $\sqrt{8} + \sqrt{18} - \sqrt{45}$
14. $\sqrt{24} - \sqrt{54} + \sqrt{96}$
15. $3\sqrt{5} - 4\sqrt{2a} + 5\sqrt{5}$
16. $2\sqrt{3x} + 3\sqrt{12x}$
17. $5\sqrt{8} - 3\sqrt{32}$
18. $\sqrt{20x} + \sqrt{45x} - \sqrt{80x}$
19. $5x\sqrt{3x} + 4\sqrt{3x^3}$
20. $\dfrac{13\sqrt{2a} - \sqrt{18a}}{3}$
21. $\dfrac{15x\sqrt{3x} - \sqrt{27x^3}}{6}$
22. $\dfrac{\sqrt{28x^3} - 5x\sqrt{7x}}{9}$
23. $\dfrac{\sqrt{2x^3} - x\sqrt{32x}}{3}$
24. $\dfrac{8x^2\sqrt{3x} + \sqrt{16x^5}}{6}$
25. $\dfrac{5\sqrt{x^5} + x^2\sqrt{x}}{2}$
26. $\dfrac{9x\sqrt{x^3} - \sqrt{x^5}}{12}$
27. $\dfrac{2\sqrt{x^3} - x\sqrt{4x}}{2}$
28. $4\sqrt{5x^5} - 2x\sqrt{20x^3} - 3x^2\sqrt{45x}$
29. $\sqrt{28a^3} - 4a\sqrt{7a} + \sqrt{63a}$

Review Your Skills

Use the distributive property to change the following product into a sum or difference.

30. $2(x + 4)$
31. $6(a^2 - 2)$
32. $3a(a^2 + 9)$
33. $5a(a^3 - 3a)$
34. $8x^2(x^4 - 6x^2)$

13.4 MULTIPLICATION AND DIVISION OF RADICAL EXPRESSIONS

Products of certain radical expressions can be found by applying the distributive property directly or the FOIL method.

EXAMPLE 39 Find the product $\sqrt{2}(\sqrt{3} + \sqrt{5})$.

$$\sqrt{2}(\sqrt{3} + \sqrt{5}) = \sqrt{2}\sqrt{3} + \sqrt{2}\sqrt{5} \quad \text{Distributive property}$$
$$= \sqrt{6} + \sqrt{10} \quad \text{Product rule}$$

EXAMPLE 40 Multiply $\sqrt{3}(\sqrt{6} + \sqrt{12})$.

$$\sqrt{3}(\sqrt{6} + \sqrt{12}) = \sqrt{3}\sqrt{6} + \sqrt{3}\sqrt{12} \quad \text{Distributive property}$$
$$= \_\_\_\_ + \_\_\_\_ \quad \text{Product rule}$$
$$= \sqrt{9 \cdot 2} + 6 \quad \text{Simplify}$$
$$= \sqrt{9}\sqrt{2} + 6$$
$$= \_\_\_\_\sqrt{2} + 6$$

$\sqrt{18}$, $\sqrt{36}$

3

EXAMPLE 41 Multiply $\sqrt{3x}(\sqrt{2} + \sqrt{3x})$.

$$\sqrt{3x}(\sqrt{2} + \sqrt{3x}) = \sqrt{3x}\sqrt{2} + \sqrt{3x}\sqrt{3x} \qquad \text{Distributive property}$$
$$= \sqrt{6x} + \underline{\qquad} \qquad \text{Definition of square root}$$

$3x$

EXAMPLE 42 Multiply $(\sqrt{2} + 1)(\sqrt{2} + 5)$.

Use the FOIL method.

$$(\sqrt{2} + 1)(\sqrt{2} + 5) = \sqrt{2}\sqrt{2} + 5\sqrt{2} + 1\sqrt{2} + 1 \cdot 5$$
$$= \underline{\qquad} + 5\sqrt{2} + \sqrt{2} + 5$$
$$= 7 + \underline{\qquad}\sqrt{2} \qquad \text{Combine like terms}$$

2

6

Where did you get the 7?

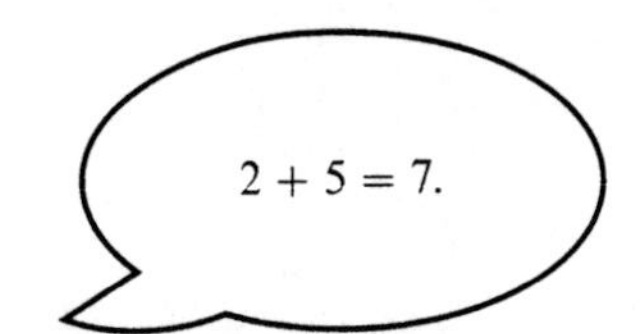

EXAMPLE 43 Multiply $(\sqrt{x} + 2)(\sqrt{x} - 3)$.

Use the FOIL method.

$$(\sqrt{x} + 2)(\sqrt{x} - 3) = \sqrt{x}\sqrt{x} - 3\sqrt{x} + 2\sqrt{x} + 2(-3)$$
$$= x - \sqrt{x} + \underline{\qquad}$$
$$= x - \sqrt{x} - 6$$

-6

EXAMPLE 44 Multiply $(\sqrt{2x} + 3)(\sqrt{2x} - 3)$.

This is the sum and difference of the same two terms.

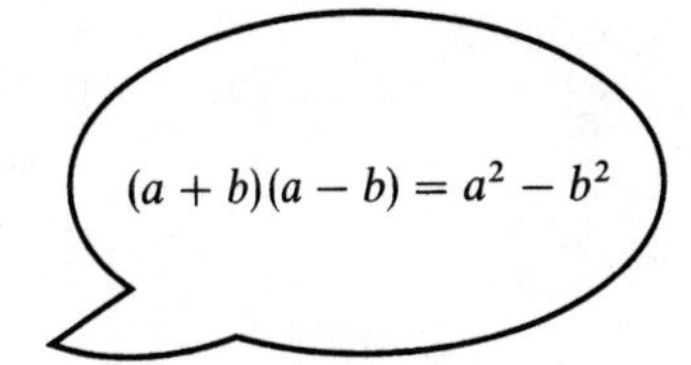

$$(\sqrt{2x} + 3)(\sqrt{2x} - 3) = (\sqrt{2x})^2 - 3^2$$

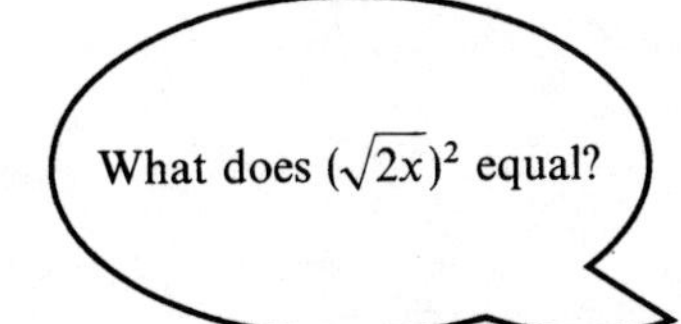

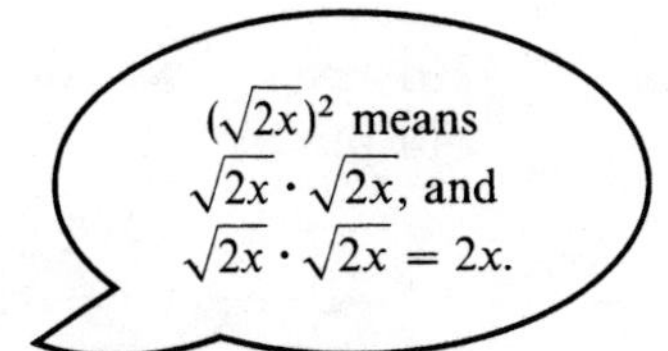

Therefore,

$$(\sqrt{2x} + 3)(\sqrt{2x} - 3) = 2x - 9$$

PROBLEM SET 13.4A

Change each product to a sum. Simplify each result.

1. $3(\sqrt{2} - 1)$
2. $5(\sqrt{3} - \sqrt{2})$
3. $3(2 - \sqrt{5})$
4. $5(\sqrt{5} - 2)$
5. $\sqrt{2}(2 + \sqrt{2})$
6. $\sqrt{3}(3 - \sqrt{3})$
7. $\sqrt{2}(3 - \sqrt{3})$
8. $\sqrt{3}(\sqrt{2} - \sqrt{3})$
9. $\sqrt{5}(\sqrt{20} - 2)$
10. $\sqrt{6}(\sqrt{3} - \sqrt{2})$
11. $\sqrt{2}(\sqrt{6} + 3)$
12. $\sqrt{5}(\sqrt{20} - \sqrt{5})$
13. $\sqrt{2x}(3 - \sqrt{2x})$
14. $\sqrt{5a}(\sqrt{10a} - 3)$
15. $\sqrt{3a}(\sqrt{3a} - \sqrt{2a})$
16. $\sqrt{7x}(\sqrt{14x} - \sqrt{7x})$
17. $\sqrt{3}(\sqrt{12} + \sqrt{3})$
18. $\sqrt{2}(\sqrt{18} - \sqrt{32})$
19. $\sqrt{3a}(\sqrt{3a} - \sqrt{6a})$
20. $\sqrt{6}(\sqrt{2} - \sqrt{3})$
21. $\sqrt{3a}(\sqrt{12a} - \sqrt{27a})$
22. $\sqrt{6x}(\sqrt{12x} - \sqrt{18x})$
23. $(1 - \sqrt{2})(4 + \sqrt{2})$
24. $(3 - \sqrt{3})(2 + \sqrt{3})$
25. $(\sqrt{6} - 1)(\sqrt{6} + 3)$
26. $(\sqrt{5} - 1)(\sqrt{5} + 4)$
27. $(5 - \sqrt{10})(3 + \sqrt{10})$
28. $(\sqrt{x} - 3)(\sqrt{x} + 3)$
29. $(4 - \sqrt{y})(4 + \sqrt{y})$
30. $(\sqrt{7} - 2)(\sqrt{7} + 3)$
31. $(\sqrt{2} + 3)^2$
32. $(\sqrt{3} - 4)^2$
33. $(\sqrt{3} - \sqrt{2})(\sqrt{3} + 2\sqrt{2})$
34. $(\sqrt{5} - \sqrt{3})(3\sqrt{5} + \sqrt{3})$
35. $(\sqrt{5} - \sqrt{2})(\sqrt{5} + \sqrt{2})$
36. $(\sqrt{5a} + \sqrt{3})(\sqrt{5a} - 2\sqrt{3})$

EXAMPLE 45 Simplify $\dfrac{\sqrt{5}}{\sqrt{5}+2}$.

In the previous examples, the radicals in the denominator were eliminated by multiplying the numerator and the denominator by the number in the original denominator. This is equivalent to multiplying the original fraction by 1.

Notice what happens if we choose to multiply this example by $\dfrac{\sqrt{5}+2}{\sqrt{5}+2}$.

$$\frac{\sqrt{5}}{\sqrt{5}+2}\cdot\frac{(\sqrt{5}+2)}{(\sqrt{5}+2)} = \frac{5+2\sqrt{5}}{5+4\sqrt{5}+4}$$

Yuck! The middle term of the denominator still has a radical.

However, if we remember that $(a + b)(a - b) = a^2 - b^2$ has no middle term, we can get out of this fix.

$$\frac{\sqrt{5}}{\sqrt{5}+2}\cdot\frac{(\sqrt{5}-2)}{(\sqrt{5}-2)} = \frac{5-2\sqrt{5}}{5-4} = \frac{5-2\sqrt{5}}{1} = 5-2\sqrt{5}$$

This is 1. | Look, no middle term. | That's what I call a simplified denominator.

EXAMPLE 46 Simplify $(\sqrt{7}+2)(\sqrt{7}-2)$.

$$(\sqrt{7}+2)(\sqrt{7}-2) = (\sqrt{7})^2 - 2^2 \qquad \text{Product of a sum and difference}$$
$$= 7 - 4$$
$$= \_\_\_\_\_$$

3

EXAMPLE 47 Simplify $(\sqrt{11}+\sqrt{6})(\sqrt{11}-\sqrt{6})$.

$$(\sqrt{11}+\sqrt{6})(\sqrt{11}-\sqrt{6}) = (\sqrt{11})^2 - (\sqrt{6})^2$$
$$= \_\_\_\_\_ - \_\_\_\_\_$$
$$= \_\_\_\_\_$$

11, 6

5

Notice in Examples 46 and 47 that the final result is a number that does not contain a radical. This is the result we want when simplifying fractions containing radicals in the denominator.

Conjugate

Two expressions are *conjugates* if one is a sum of two terms and the other is the difference of the same two terms.

$\sqrt{2}+1$ is the conjugate of $\sqrt{2}-1$

$\sqrt{3}-\sqrt{2}$ is the conjugate of $\sqrt{3}+\sqrt{2}$

$a+x$ is the conjugate of $a-x$

Rationalizing denominators

13.4 RULE

To *rationalize the denominator* of a fraction containing two terms, at least one of which is a radical, multiply the numerator and the denominator of the fraction by the conjugate of the denominator.

EXAMPLE 48 Rationalize the denominator and simplify $\dfrac{2}{\sqrt{3}-1}$.

The conjugate of $\sqrt{3}-1$ is ________

$\sqrt{3}+1$

$$\frac{2}{\sqrt{3}-1} = \frac{2}{\sqrt{3}-1}\cdot\frac{\sqrt{3}+1}{\sqrt{3}+1}$$

$\dfrac{\sqrt{3}+1}{\sqrt{3}+1}$ is equal to 1.

$$= \frac{2(\sqrt{3}+1)}{(\sqrt{3}-1)(\sqrt{3}+1)}$$
$$= \frac{2(\sqrt{3}+1)}{(\sqrt{3})^2 - 1^2} \qquad (a+b)(a-b) = a^2 - b^2$$
$$= \frac{2(\sqrt{3}+1)}{3-1}$$
$$= \frac{2(\sqrt{3}+1)}{2}$$
$$= \_\_\_\_\_\_\_\_ \qquad \text{Divide out the common factor}$$

$\sqrt{3}+1$

EXAMPLE 49 Rationalize the denominator and simplify $\frac{6\sqrt{2}}{\sqrt{5}+\sqrt{2}}$.

$\sqrt{5}-\sqrt{2}$

The conjugate of $\sqrt{5}+\sqrt{2}$ is ______

$$\frac{6\sqrt{2}}{\sqrt{5}+\sqrt{2}} = \frac{6\sqrt{2}}{\sqrt{5}+\sqrt{2}} \cdot \frac{\sqrt{5}-\sqrt{2}}{\sqrt{5}-\sqrt{2}}$$

$$= \frac{6\sqrt{2}\sqrt{5} - 6\sqrt{2}\sqrt{2}}{(\sqrt{5})^2 - (\sqrt{2})^2}$$

$\sqrt{10}$

$$= \frac{6 \cdot \_\_\_\_ - 6 \cdot 2}{5 - 2}$$

$$= \frac{6(\sqrt{10} - 2)}{3}$$ Factor out the 6 in the numerator and reduce

$2(\sqrt{10} - 2)$ or $2\sqrt{10} - 4$

$= \_\_\_\_\_\_\_\_$

EXAMPLE 50 Rationalize the denominator and simplify $\frac{\sqrt{3}-1}{\sqrt{5}-\sqrt{3}}$.

$\sqrt{5}+\sqrt{3}$

The conjugate of $\sqrt{5}-\sqrt{3}$ is ______

$$\frac{\sqrt{3}-1}{\sqrt{5}-\sqrt{3}} = \frac{\sqrt{3}-1}{\sqrt{5}-\sqrt{3}} \cdot \frac{\sqrt{5}+\sqrt{3}}{\sqrt{5}+\sqrt{3}}$$

$$= \frac{(\sqrt{3}-1)(\sqrt{5}+\sqrt{3})}{(\sqrt{5}-\sqrt{3})(\sqrt{5}+\sqrt{3})}$$

$$= \frac{\sqrt{3}\sqrt{5} + \sqrt{3}\sqrt{3} - 1\sqrt{5} - 1\sqrt{3}}{5 - 3}$$ FOIL

$\sqrt{15} + 3 - \sqrt{5} - \sqrt{3}$

$$= \frac{\_\_\_\_\_\_\_\_}{2}$$

PROBLEM SET 13.4B

Rationalize each denominator and simplify. Assume that no denominator equals zero.

1. $\frac{1}{3+\sqrt{2}}$
2. $\frac{2}{3-\sqrt{2}}$
3. $\frac{8}{\sqrt{5}-1}$
4. $\frac{4}{2-\sqrt{2}}$
5. $\frac{3}{\sqrt{3}-1}$
6. $\frac{5}{\sqrt{3}+2}$
7. $\frac{2}{\sqrt{x}+1}$
8. $\frac{3}{2-\sqrt{x}}$
9. $\frac{2}{\sqrt{2}-1}$
10. $\frac{\sqrt{3}}{1+\sqrt{3}}$
11. $\frac{2\sqrt{6}}{2-\sqrt{3}}$
12. $\frac{3\sqrt{5}}{2-\sqrt{5}}$
13. $\frac{6}{\sqrt{3}+\sqrt{2}}$
14. $\frac{\sqrt{10}}{\sqrt{5}+\sqrt{2}}$
15. $\frac{4\sqrt{6}}{\sqrt{3}-\sqrt{2}}$
16. $\frac{3\sqrt{5}}{\sqrt{5}+\sqrt{2}}$

17. $\dfrac{3\sqrt{15}}{\sqrt{3}+\sqrt{5}}$ 18. $\dfrac{2\sqrt{6}}{3\sqrt{3}-\sqrt{2}}$ 19. $\dfrac{\sqrt{5}-2}{\sqrt{5}+2}$ 20. $\dfrac{\sqrt{3}+1}{\sqrt{3}-1}$

21. $\dfrac{\sqrt{3}+2}{3+\sqrt{2}}$ 22. $\dfrac{\sqrt{2}+3}{3-\sqrt{2}}$ 23. $\dfrac{\sqrt{3}}{5+\sqrt{3}}$ 24. $\dfrac{2\sqrt{5}}{5+\sqrt{7}}$

25. $\dfrac{3\sqrt{5}+\sqrt{2}}{\sqrt{5}-\sqrt{2}}$ 26. $\dfrac{2\sqrt{7}-\sqrt{2}}{\sqrt{7}-\sqrt{2}}$ 27. $\dfrac{2\sqrt{3}-\sqrt{2}}{\sqrt{3}+\sqrt{2}}$ 28. $\dfrac{\sqrt{6}-\sqrt{2}}{\sqrt{6}+\sqrt{3}}$

29. $\dfrac{\sqrt{6}+\sqrt{3}}{\sqrt{2}+\sqrt{6}}$ 30. $\dfrac{2\sqrt{5}-\sqrt{3}}{\sqrt{5}+\sqrt{3}}$ 31. $\dfrac{6}{4+\sqrt{x}}$ 32. $\dfrac{7}{\sqrt{x}-3}$

13.5 APPLICATIONS

In plane geometry, a very useful formula relates the sides of a right triangle. It is called the *Pythagorean theorem.*

In the following right triangle, the side c, opposite the right angle, is called the *hypotenuse* of the right triangle.

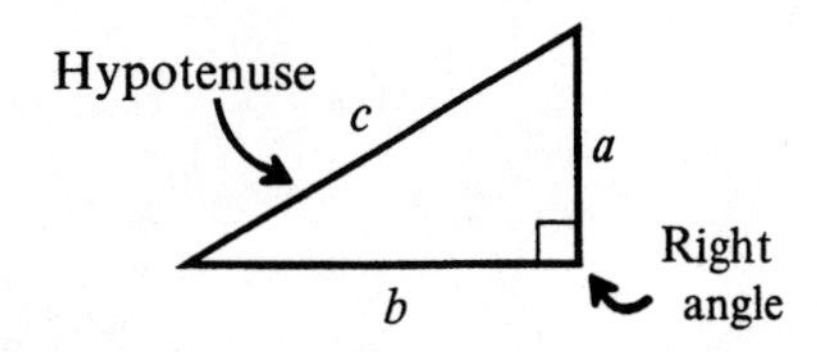

Pythagorean theorem

13.5A PYTHAGOREAN THEOREM

The *Pythagorean theorem* tells us that the square of the hypotenuse is equal to the sum of the squares of the other two sides. In symbols we have

$$c^2 = a^2 + b^2$$

If two sides of a right triangle are known, the third side can be found by using the Pythagorean theorem.

EXAMPLE 51 Given that $a = 3$ and $b = 4$, find side c in the right triangle below.

Substitute in the formula.

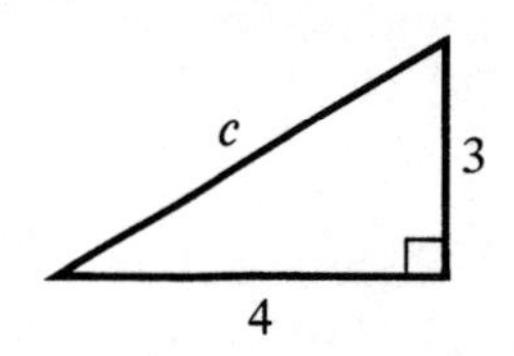

$$c^2 = a^2 + b^2$$
$$c^2 = (3)^2 + (4)^2$$
$$c^2 = 9 + 16$$
$$c^2 = 25$$
$$\sqrt{c^2} = \sqrt{25}$$

How can we find c?

Take the square root of both sides of the equation.

$$c = 5$$

EXAMPLE 52 Find b in the right triangle shown.

Substitute in the formula.

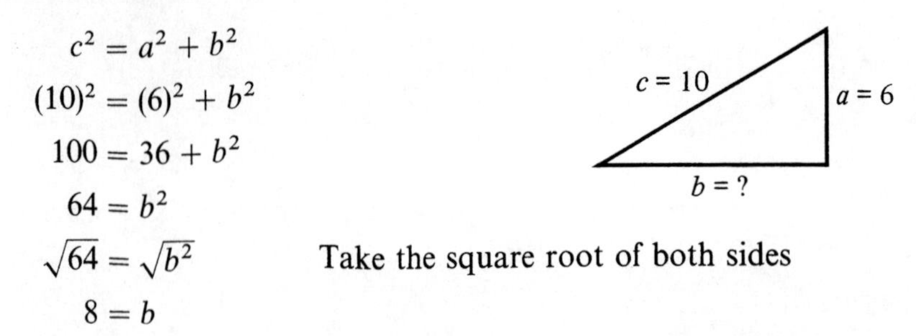

$$c^2 = a^2 + b^2$$
$$(10)^2 = (6)^2 + b^2$$
$$100 = 36 + b^2$$
$$64 = b^2$$
$$\sqrt{64} = \sqrt{b^2} \quad \text{Take the square root of both sides}$$
$$8 = b$$

Square roots are also useful in finding the distance between two points on a graph.

EXAMPLE 53 Find the distance between the points (2, 4) and (5, 8).

First, plot the two points on a graph and draw a right triangle as shown below.

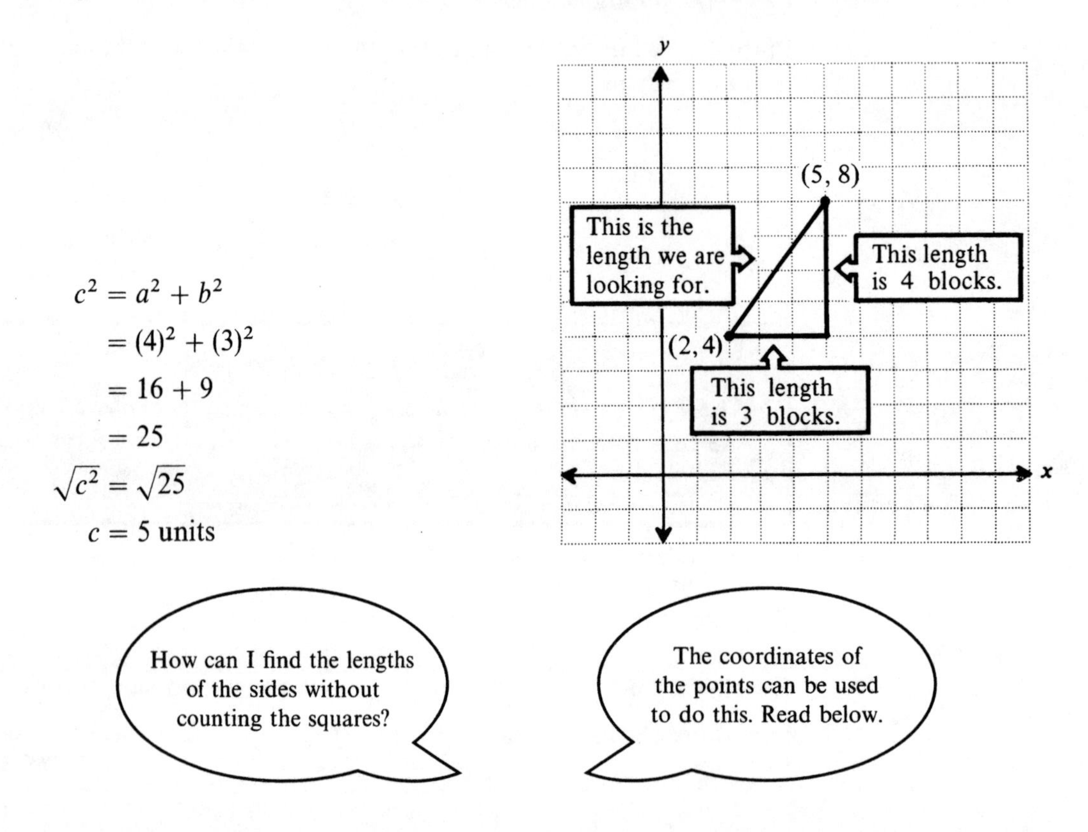

$$c^2 = a^2 + b^2$$
$$= (4)^2 + (3)^2$$
$$= 16 + 9$$
$$= 25$$
$$\sqrt{c^2} = \sqrt{25}$$
$$c = 5 \text{ units}$$

Notice in Example 53 that the 4-unit length of side a is equal to the difference of the y-coordinates of the two points: $(8 - 4) = 4$. The length of the base of the triangle is 3 units. That is, the base is the difference of the x-coordinates of the two points: $(5 - 2) = 3$.

In general,

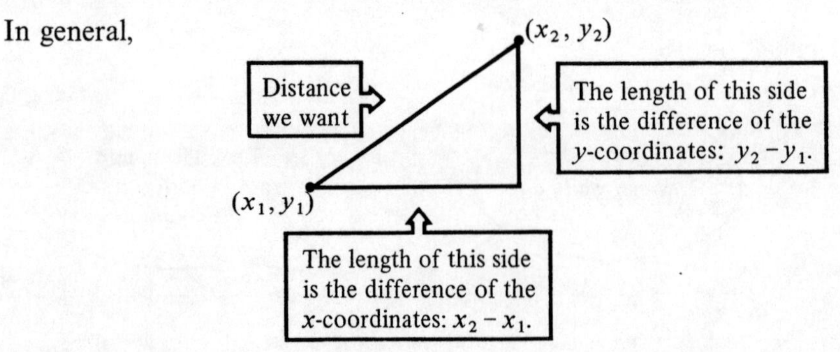

Distance between two points

13.5B DISTANCE FORMULA

The distance between any two points is given by the following formula.

$$d = \sqrt{(x_2 - x_1)^2 + (y_2 - y_1)^2}$$

where (x_1, y_1) and (x_2, y_2) represent the coordinates of any two points on the graph.

EXAMPLE 54 Find the distance between the points (2, 3) and (5, 7).

If we consider the point (2, 3) as (x_1, y_1) and (5, 7) as (x_2, y_2), we have

$$d = \sqrt{(x_2 - x_1)^2 + (y_2 - y_1)^2}$$
$$= \sqrt{(5 - 2)^2 + (7 - 3)^2}$$
$$= \sqrt{(\_\_\_\_)^2 + (4)^2} \quad 3$$
$$= \sqrt{9 + 16}$$
$$= \_\_\_\_ \quad \sqrt{25}$$
$$= 5$$

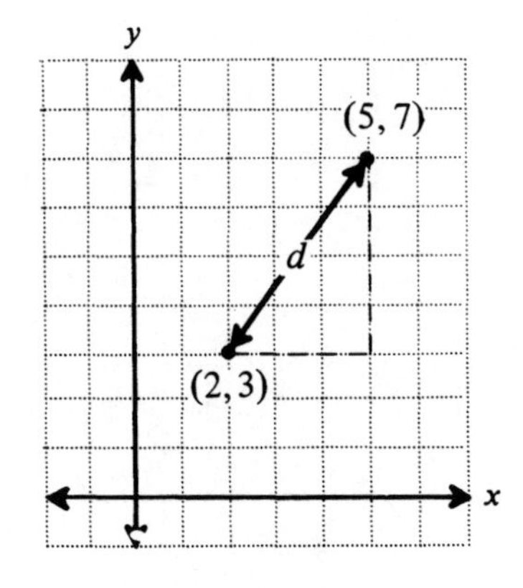

Does it make any difference which point is called (x_1, y_1) and which one is (x_2, y_2)?

Not really. Just be careful when subtracting. Squaring a negative number gives the same result as squaring a positive one. They are both positive.

EXAMPLE 55A Find the distance between the points (7, −3) and (1, 5).

Consider (7, −3) as (x_1, y_1) and (1, 5) as (x_2, y_2).

Substitute in the formula.

$$d = \sqrt{(x_2 - x_1)^2 + (y_2 - y_1)^2}$$
$$= \sqrt{(1 - 7)^2 + [5 - (-3)]^2}$$
$$= \sqrt{(-6)^2 + (\_\_\_\_)^2} \quad 8$$
$$= \sqrt{\_\_\_\_ + 64} \quad 36$$
$$= \sqrt{100}$$
$$= 10$$

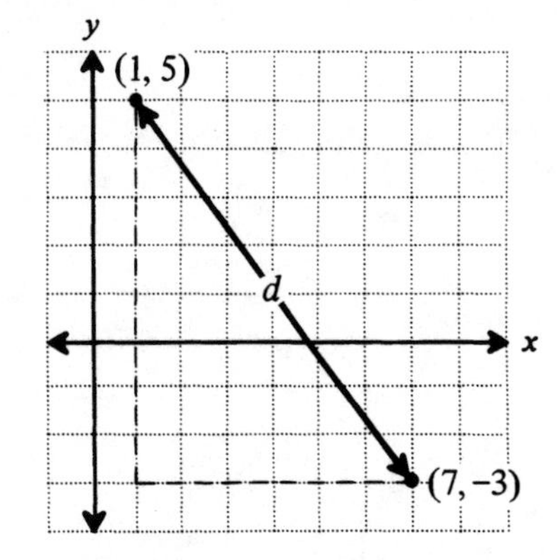

EXAMPLE 55B Now in Example 55A, consider (1, 5) as (x_1, y_1) and (7, −3) as (x_2, y_2).

Substitute in the formula.

$$d = \sqrt{(x_2 - x_1)^2 + (y_2 - y_1)^2}$$
$$= \sqrt{(7-1)^2 + (-3-5)^2}$$

−8 $$= \sqrt{6^2 + (\_\_\_\_)^2}$$

64 $$= \sqrt{36 + \_\_\_\_}$$

$$= \sqrt{100}$$
$$= 10$$

EXAMPLE 56 Find the distance between the points (−1, −2) and (2, 4).

Consider (−1, −2) as (x_1, y_1) and (2, 4) as (x_2, y_2).

Substitute in the formula.

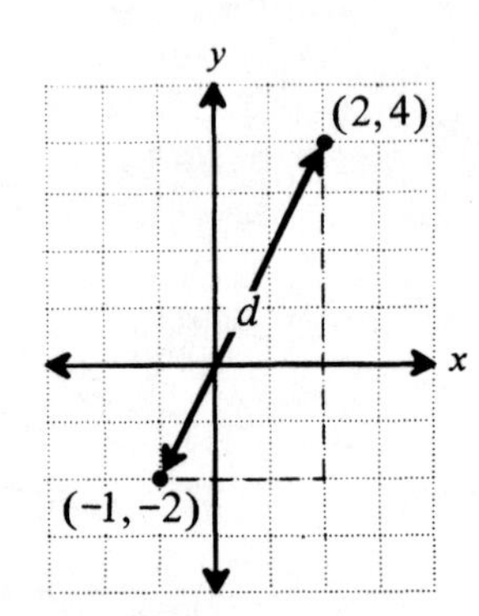

$$d = \sqrt{(x_2 - x_1)^2 + (y_2 - y_1)^2}$$
$$= \sqrt{[2-(-1)]^2 + [4-(-2)]^2}$$

3 $$= \sqrt{(\_\_\_\_)^2 + (6)^2}$$

36 $$= \sqrt{9 + \_\_\_\_}$$

$$= \sqrt{45}$$

No — Is $\sqrt{45}$ in simplest form? ____

9, 5 — $45 = \_\_\_\_ \cdot \_\_\_\_$

9 — $\sqrt{45} = \sqrt{\_\_\_\_ \cdot 5}$

$= \sqrt{9} \cdot \sqrt{5}$

$3\sqrt{5}$ — $= \_\_\_\_$

Therefore, $d = 3\sqrt{5}$.

EXAMPLE 57 Find the distance between the points (−3, 4) and (−6, −3).

Consider (−3, 4) as (x_1, y_1) and (−6, −3) as (x_2, y_2).

Substitute in the formula.

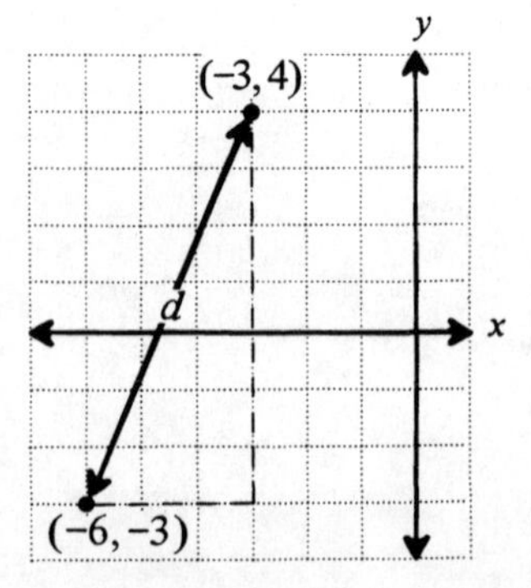

$$d = \sqrt{(x_2 - x_1)^2 + (y_2 - y_1)^2}$$
$$= \sqrt{[-6-(-3)]^2 + (-3-4)^2}$$

−3 $$= \sqrt{(\_\_\_\_)^2 + (-7)^2}$$

49 $$= \sqrt{9 + \_\_\_\_}$$

$$= \sqrt{58}$$

Yes — Is $\sqrt{58}$ in simplest form? ____

Assuming you want a decimal approximation of $\sqrt{58}$, you could do this:

7 The square root of 49 is ____

8 The square root of 64 is ____

Since the square root of 58 is a little more than halfway between 49 and 64, we guess that $\sqrt{58}$ is about 7.6. Of course, we could also use a calculator or a table to approximate $\sqrt{58}$.

EXAMPLE 58 Find the perimeter of a triangle whose vertices are $(-2, 2)$, $(2, 2)$, and $(2, 5)$.

If we sketch the triangle first, it's easier to visualize what we are doing.

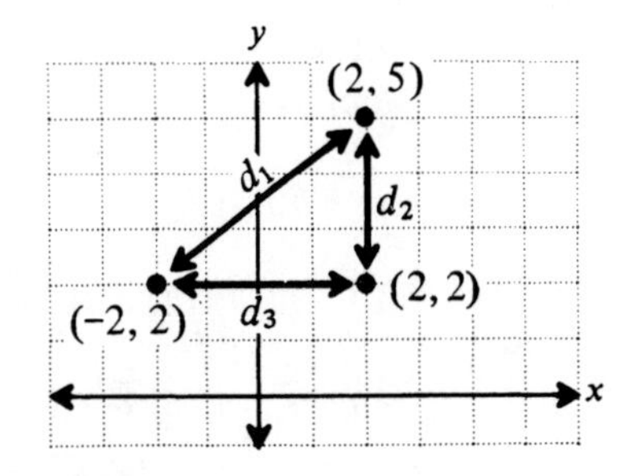

The *perimeter* is the distance around the outside of the triangle.

The distance from $(-2, 2)$ to $(2, 5)$ is

$$\begin{aligned} d_1 &= \sqrt{[2-(-2)]^2 + (5-2)^2} \\ &= \sqrt{4^2 + 3^2} \\ &= \sqrt{16 + 9} \\ &= \sqrt{25} \\ &= 5 \end{aligned}$$

The distance from $(2, 2)$ to $(2, 5)$ is

$$\begin{aligned} d_2 &= \sqrt{(2-2)^2 + (5-2)^2} \\ &= \sqrt{0^2 + 3^2} \\ &= \sqrt{9} \\ &= 3 \end{aligned}$$

The distance from $(2, 2)$ to $(-2, 2)$ is

$$\begin{aligned} d_3 &= \sqrt{(-2-2)^2 + (2-2)^2} \\ &= \sqrt{(-4)^2 + 0^2} \\ &= \sqrt{16} \\ &= 4 \end{aligned}$$

The perimeter is

$$\begin{aligned} P &= d_1 + d_2 + d_3 \\ &= 5 + 3 + 4 \\ &= 12 \end{aligned}$$

PROBLEM SET 13.5

Find the distance between the following pairs of points.

1. (4, 1) and (7, 5)
2. (2, 6) and (6, 3)
3. (9, 6) and (5, 3)
4. (8, 3) and (5, 7)
5. (5, 2) and (−7, 7)
6. (1, 2) and (−4, −10)
7. (−3, 2) and (4, −1)
8. (−2, 3) and (1, 5)
9. (−6, 2) and (−4, 5)
10. (−2, 4) and (3, −2)
11. (0, 0) and (6, 8)
12. (−4, 3) and (0, 0)
13. (2, −4) and (−2, 4)
14. (−5, −3) and (2, −1)

Find the perimeter of the following triangles having the given points as vertices. (Perimeter is the sum of the lengths of the sides.)

15. (0, 3), (4, 0), (0, 0)
16. (1, 3), (1, −1), (4, −1)
17. (2, 2), (2, −2), (−2, −2)
18. (−5, 12), (0, 0), (−5, 0)
19. (2, 4), (6, 4), (2, −1)
20. (3, 3), (3, −3), (−3, −3)

In the accompanying figures of right triangles, find the missing side x. Simplify all radicals in your answers.

21.

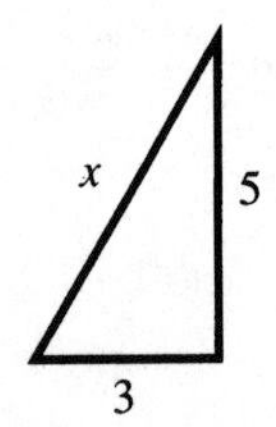

22.

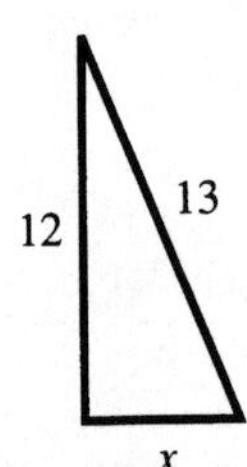

23.

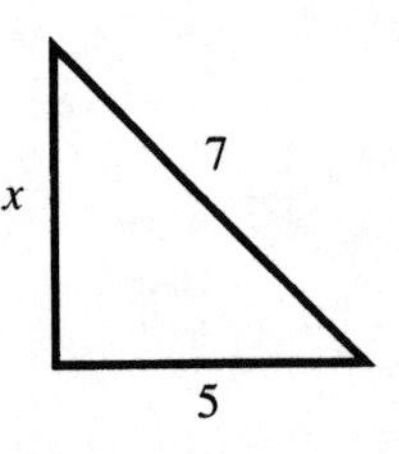

24.

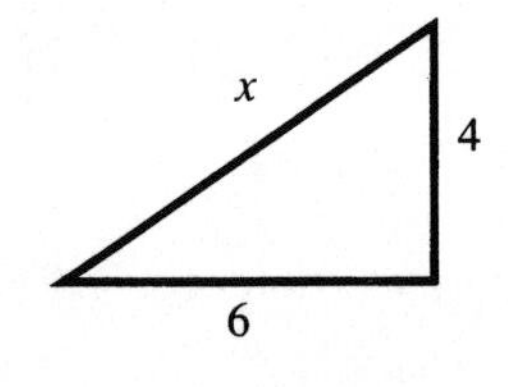

25. How far is the foot of a 25-foot ladder from the base of a wall if the top of the ladder reaches 20 feet up the wall?
26. A boy is flying a kite that is 100 meters above the ground and 200 meters away from him along the ground. What is the distance directly from his hand to the kite?
27. A boat is held by a 17-foot rope that is attached to the top of a dock. When the tide is out, the boat is 15 feet away from the dock. How high above the boat is the dock?
28. To check if the corner of a patio is square (a right angle) a mason measures along the two sides from the corner, 12 feet along one side to a point A, and 9 feet along the other side to a point B. What should the measurement be directly across the corner from point A to point B?
29. A rectangular desktop measures 6 feet by 3 feet. Find the length of a diagonal line drawn across the top of the desk.

Review Your Skills

Factor the following.

30. $x^2 - 3x - 10$
31. $4x^2 + 5x - 6$
32. $x^2 - 5x$
33. $4x^2 - 9$
34. $6x^2 - 5x - 6$
35. $x^2 - 8x + 16$

REVIEW 13

NOW THAT YOU HAVE COMPLETED UNIT 13, YOU SHOULD BE ABLE TO:

Use the following definitions in your working vocabulary.

Square root—A factor that is multiplied by itself to produce a given number

Radical sign—The symbol $\sqrt{\ }$

Radicand—The number under the radical sign

Similar radicals—Radicals with the same radicand

Conjugates—Two expressions are conjugates if one is a sum of two terms and the other is the difference of the same two terms

Rationalize the denominator—To rewrite an expression so that no radicals appear in the denominator

Apply the following rules and definitions:

Square Root

For all nonnegative numbers a and b,

$$\sqrt{a} \cdot \sqrt{a} = a$$

This is equivalent to saying

$$\text{if } \sqrt{a} = b \text{ then } a = b^2$$

Rules for Radicals

Product:

$$\sqrt{a} \cdot \sqrt{b} = \sqrt{a \cdot b}$$

Quotient:

$$\frac{\sqrt{a}}{\sqrt{b}} = \sqrt{\frac{a}{b}}$$

To write a radical in simplified form, use the following rules:

1. The radicand does not contain a factor that can be written as the square of a whole number.
2. No fraction appears under the radical sign.
3. No radical appears in the denominator of the fraction.

To add or subtract radicals:

1. Simplify each radical.
2. Use the distributive property to combine the radicals with like radicands.

To multiply expressions containing radicals, visualize the expressions containing radicals as variables. Follow the usual procedures for multiplication. Simplify the result by applying the laws for radicals.

To find the third side of a right triangle when two sides are known, use the *Pythagorean theorem.*

$$c^2 = a^2 + b^2$$

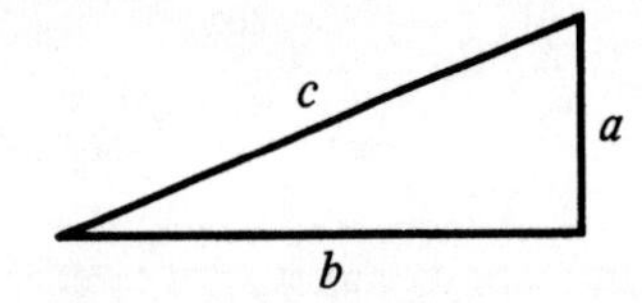

To find the distance between any two points, use the distance formula.

$$d = \sqrt{(x_2 - x_1)^2 + (y_2 - y_1)^2}$$

REVIEW TEST 13

Perform the indicated operations. Simplify all results. ***(13.1)***

1. $\dfrac{\sqrt{100}-\sqrt{4}}{\sqrt{16}} =$ ____________

2. $\sqrt{\dfrac{81}{36}} \div \sqrt{\dfrac{9}{64}} =$ ____________

Simplify the following. ***(13.2)***

3. $\sqrt{24} =$ ____________

4. $\dfrac{1}{\sqrt{7}} =$ ____________

5. $\sqrt{\dfrac{49}{18}} =$ ____________

6. $-\sqrt{\dfrac{72}{75}} =$ ____________

7. $\sqrt{20a^4b^3} =$ ____________

8. $\sqrt{\dfrac{12a^3x^5}{3y^5}} =$ ____________

Simplify the following additions and subtractions. ***(13.3)***

9. $3\sqrt{2}-\sqrt{72} =$ ____________

10. $7\sqrt{a}+3\sqrt{2a}-\sqrt{a} =$ ____________

11. $\sqrt{75}-\sqrt{12}+\sqrt{48} =$ ____________

12. $\dfrac{\sqrt{27a}+\sqrt{12a}}{5} =$ ____________

Simplify the following multiplications and divisions. ***(13.4)***

13. $\sqrt{3}(\sqrt{2}-3) =$ ____________

14. $(\sqrt{3a}+1)(\sqrt{3a}+4) =$ ____________

15. $(\sqrt{5}-\sqrt{2})(\sqrt{5}+6\sqrt{2}) =$ ____________

16. $(\sqrt{7}+3)(\sqrt{7}-3) =$ ____________

17. $\dfrac{3}{2+\sqrt{3}} =$ ____________

18. $\dfrac{3\sqrt{6}}{\sqrt{2}-\sqrt{3}} =$ ____________

19. $\dfrac{3}{\sqrt{x}-5} =$ ____________

20. $\dfrac{\sqrt{x}-6}{\sqrt{x}+3} =$ ____________

21. Find the distance between the points $(8, -3)$ and $(-2, 3)$. ***(13.5)***

22. Find the missing side x of the right triangle shown. Simplify your answer. ***(13.5)***

12

x

8

REVIEW YOUR SKILLS

Reduce to lowest terms.

23. $\dfrac{2x^2 - 7x - 4}{4 - x}$

Perform the indicated operations. Reduce all answers to lowest terms.

24. $\dfrac{x}{y^3} \cdot \dfrac{-y^5}{x^3 y}$

25. $\dfrac{b^2 - a^2}{a^2 - ab - 2b^2} \div \dfrac{a^2 - 2ab + b^2}{a^2 - 2ab}$

26. $\dfrac{2x}{(x - 3)(2x - 3)} - \dfrac{2}{(x - 3)}$

27. $\dfrac{1 - \dfrac{1}{x^2}}{\dfrac{1}{x} - 1}$

Solve for the variable.

28. $\dfrac{1}{2x} - \dfrac{2}{x} = \dfrac{1}{2}$

29. $\dfrac{x - 3}{4} + \dfrac{8 - x}{2} = 2$

UNIT 14

QUADRATIC EQUATIONS

OBJECTIVES

After you have successfully completed this unit, you will be able to:

1. Determine the constant term in a trinomial so that the result is a perfect square of a binomial (14.2)
2. Solve quadratic equations by factoring (14.1), extraction of roots (14.2), completing the square (14.2), and using the quadratic formula (14.3)
3. Solve word problems involving quadratic equations (14.4)
4. Graph quadratic equations (14.5)

14.1 SOLVING QUADRATIC EQUATIONS BY FACTORING

The word *quad* can mean *square* in Latin. Consequently, second-degree equations have long been called quadratic equations. One use for quadratic equations is in the description of the motion of a falling object.

Quadratic equation

14.1A DEFINITION

An equation of the form

$$ax^2 + bx + c = 0$$

is a *quadratic equation* in standard form where a, b, and c are constants.

Note: $a \neq 0$ or there is no second-degree term.

The solution of many quadratic equations depends on our ability to rewrite quadratic equations as a product of two factors and on the following theorem.

Product equal to zero

14.1B THEOREM

If the *product of two factors is zero*, one or both of the factors is zero.

If

$$a \cdot b = 0$$

then

$$a = 0 \quad \text{or} \quad b = 0$$

To solve equations that are products such as $(x + 4)(x - 3) = 0$, we use the "product equal to zero" theorem.

Since

$$(x + 4)(x - 3) = 0$$

then **either**

$$x + 4 = 0 \quad \textbf{or} \quad x - 3 = 0$$

$$x = -4 \quad \text{or} \quad x = 3$$

There are two solutions to the equation.

Check:

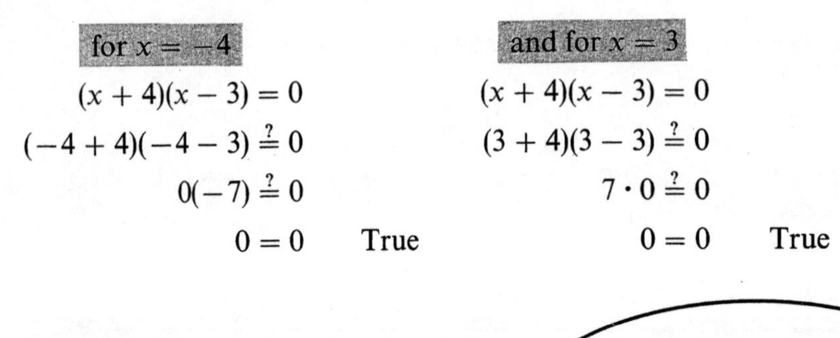

for $x = -4$

$$(x + 4)(x - 3) = 0$$
$$(-4 + 4)(-4 - 3) \stackrel{?}{=} 0$$
$$0(-7) \stackrel{?}{=} 0$$
$$0 = 0 \quad \text{True}$$

and for $x = 3$

$$(x + 4)(x - 3) = 0$$
$$(3 + 4)(3 - 3) \stackrel{?}{=} 0$$
$$7 \cdot 0 \stackrel{?}{=} 0$$
$$0 = 0 \quad \text{True}$$

Do quadratic equations always have two solutions?

Quadratic equations have at most two solutions. Sometimes they have only one solution. Some quadratics covered in advanced books have no real solutions.

EXAMPLE 1 Solve $x^2 + 2x - 3 = 0$.

Theorem 14.1B says that if a product is zero, one or both of its factors is zero.

But this quadratic equation is a sum, not a product!

True, however, factoring allows us to change a sum into a product.

Factoring,

$$x^2 + 2x - 3 = 0$$
$$(x + 3)(x - 1) = 0$$

Now apply the "product equal to zero" theorem.

$$x + 3 = 0 \quad \textbf{or} \quad x - 1 = 0$$

$$x = -3 \quad \text{or} \quad x = 1$$

Are you sure those answers are correct?

Let's check both answers in the original equation.

Check:

For $x = -3$

$$x^2 + 2x - 3 = 0$$
$$(-3)^2 + 2(-3) - 3 \stackrel{?}{=} 0$$
$$9 - 6 - 3 \stackrel{?}{=} 0$$
$$3 - 3 \stackrel{?}{=} 0$$
$$0 = 0$$

For $x = 1$

$$x^2 + 2x - 3 = 0$$
$$(1)^2 + 2(1) - 3 \stackrel{?}{=} 0$$
$$1 + 2 - 3 \stackrel{?}{=} 0$$
$$3 - 3 \stackrel{?}{=} 0$$
$$0 = 0$$

EXAMPLE 2 Solve $6x^2 + x - 2 = 0$.

We factor this trinomial to get a product we can set equal to zero.

$$6x^2 + x - 2 = 0$$
$$(3x + 2)(2x - 1) = 0$$

Then

$$3x + 2 = 0 \quad \text{or} \quad 2x - 1 = 0$$
$$3x = -2 \qquad 2x = 1$$
$$x = \frac{-2}{3} \qquad x = \frac{1}{2}$$

Checking the first solution,

$$6x^2 + x - 2 = 0$$
$$6\left(\frac{-2}{3}\right)^2 + \left(\frac{-2}{3}\right) - 2 \stackrel{?}{=} 0$$
$$6\left(\frac{4}{9}\right) - \frac{2}{3} - 2 \stackrel{?}{=} 0$$
$$\frac{8}{3} - \frac{2}{3} - 2 \stackrel{?}{=} 0$$
$$\frac{6}{3} - 2 \stackrel{?}{=} 0$$
$$2 - 2 \stackrel{?}{=} 0$$
$$0 = 0$$

Checking the second solution,

$$6x^2 + x - 2 = 0$$
$$6\left(\frac{1}{2}\right)^2 + \left(\frac{1}{2}\right) - 2 \stackrel{?}{=} 0$$
$$6\left(\frac{1}{4}\right) + \frac{1}{2} - 2 \stackrel{?}{=} 0$$
$$\frac{3}{2} + \frac{1}{2} - 2 \stackrel{?}{=} 0$$
$$\frac{4}{2} - 2 \stackrel{?}{=} 0$$
$$2 - 2 \stackrel{?}{=} 0$$
$$0 = 0$$

EXAMPLE 3 Solve $x^2 + 4x = 0$.

Use the distributive law to factor.

$$x^2 + 4x = 0$$

$$x(\_\_\_\_\_\_) = 0$$

$x + 4$

This quadratic equation looks incomplete.

Not really. In the equation $ax^2 + bx + c = 0$, the c is zero.

Then

$$x = 0 \quad \text{or} \quad x + 4 = 0$$

$$x = -4$$

The solutions are $x = 0$ or $x = -4$.

EXAMPLE 4 Solve $x^2 = 16$.

To solve this equation, first write it in standard form by getting all the terms on the left side.

$$x^2 = 16$$

$$x^2 - 16 = 0$$

In this quadratic equation, $b = 0$.

Yup. But if $a = 0$, it's not a quadratic equation.

$(x - 4)(x + 4) = 0$ By factoring

Then,

$0, 0$

$$x - 4 = \_\_\_\_ \quad \text{or} \quad x + 4 = \_\_\_\_$$

$$x = 4 \quad \text{or} \quad x = -4$$

EXAMPLE 5 Solve $\frac{2x}{15} + \frac{3}{5} = \frac{1}{3x}$.

First, remove the fractions by multiplying both sides of the equation by the least common denominator.

$15x$

The least common denominator is ____.

$$\frac{2x}{15} + \frac{3}{5} = \frac{1}{3x}$$

$$15x \cdot \left(\frac{2x}{15} + \frac{3}{5}\right) = \left(\frac{1}{3x}\right) \cdot 15x$$

$2x^2 + 9x = 5$ To use this method the equation must be in standard form; therefore, subtract 5 from both sides

$2x^2 + 9x - 5 = 0$ Now the equation is in standard form

$(2x - 1)(x + 5) = 0$ By factoring

$$2x - 1 = 0 \quad \text{or} \quad x + 5 = 0$$

$$2x = 1 \qquad x = -5$$

$$x = \frac{1}{2}$$

Check:

$$\frac{2x}{15}+\frac{3}{5}=\frac{1}{3x} \qquad \frac{2x}{15}+\frac{3}{5}=\frac{1}{3x}$$

$$\frac{2\left(\frac{1}{2}\right)}{15}+\frac{3}{5}\stackrel{?}{=}\frac{1}{3\left(\frac{1}{2}\right)} \qquad \frac{2(-5)}{15}+\frac{3}{5}\cdot\frac{3}{3}\stackrel{?}{=}\frac{1}{3(-5)}$$

$$\frac{1}{15}+\frac{3}{5}\cdot\frac{3}{3}\stackrel{?}{=}\frac{1}{\left(\frac{3}{2}\right)} \qquad \frac{-10}{15}+\frac{9}{15}\stackrel{?}{=}\frac{1}{-15}$$

$$\frac{1}{15}+\frac{9}{15}\stackrel{?}{=}\frac{2}{3} \qquad \frac{-1}{15}=\frac{-1}{15}$$

$$\frac{10}{15}\stackrel{?}{=}\frac{2}{3}$$

$$\frac{2}{3}=\frac{2}{3}$$

PROBLEM SET 14.1

Solve the following quadratic equations by factoring.

1. $x^2 - x - 6 = 0$
2. $x^2 + 5x - 14 = 0$
3. $x^2 - 7x - 18 = 0$
4. $x^2 - 6x + 8 = 0$
5. $x^2 - x - 12 = 0$
6. $x^2 - 6x + 5 = 0$
7. $x^2 - 5x - 6 = 0$
8. $x^2 - 8x + 15 = 0$
9. $x^2 + 7x - 8 = 0$
10. $x^2 - x - 20 = 0$
11. $x^2 + 4x - 12 = 0$
12. $x^2 - 12x + 20 = 0$
13. $2x^2 + 5x - 3 = 0$
14. $3x^2 - 11x - 4 = 0$
15. $3x^2 + 5x - 2 = 0$
16. $2x^2 - 9x - 5 = 0$
17. $3x^2 - 5x + 2 = 0$
18. $3x^2 - 7x - 6 = 0$
19. $3x^2 - x - 14 = 0$
20. $4x^2 + 4x - 15 = 0$
21. $4x^2 + x - 3 = 0$
22. $5x^2 - 8x + 3 = 0$
23. $3x^2 - 10x + 8 = 0$
24. $4x^2 - 11x + 6 = 0$
25. $x^2 - 9 = 0$
26. $x^2 - 16 = 0$
27. $4x^2 - 1 = 0$
28. $9x^2 - 1 = 0$
29. $x^2 - 4x = 0$
30. $x^2 + 3x = 0$
31. $16x^2 - 9 = 0$
32. $9x^2 - 16 = 0$
33. $2x^2 - 23x + 30 = 0$
34. $5x^2 + x - 4 = 0$
35. $x^2 = 5x$
36. $x^2 = 6x$
37. $25x^2 = 36$
38. $16x^2 = 25$
39. $6x^2 - 5x = 4$
40. $12x^2 - 4x = 5$
41. $8x^2 - 2x = 3$
42. $6x^2 + x - 5 = 0$
43. $3x^2 - 4x = 0$
44. $4x^2 - 25 = 0$
45. $8x^2 + 21x - 9 = 0$
46. $9x^2 - 2x - 7 = 0$
47. $12x^2 = 8x + 15$
48. $15x^2 = 17x - 4$

Solve the following fractional equations.

49. $\frac{3x}{8} + \frac{7}{4} + \frac{1}{x} = 0$

50. $\frac{3x}{2} - 3 = \frac{-4}{3x}$

51. $\frac{x}{10} + \frac{13}{30} = \frac{1}{3x}$

52. $\frac{5x}{8} + 2 = \frac{-3}{8x}$

53. $\frac{x}{3} + \frac{13}{12} = \frac{1}{x}$

54. $\frac{2x}{5} = \frac{1}{15} + \frac{1}{3x}$

55. $\frac{4x}{15} = \frac{1}{3} + \frac{7}{5x}$

56. $2x + \frac{1}{4} = \frac{3}{8x}$

57. $x + \frac{8}{9} = \frac{1}{9x}$

58. $\frac{2x}{7} + \frac{17}{21} = \frac{2}{3x}$

59. $x = \frac{16}{15} + \frac{1}{x}$

60. $x - \frac{1}{x} = \frac{5}{6}$

14.2 SOLVING QUADRATIC EQUATIONS BY COMPLETING THE SQUARE

In solving equations there is one basic rule: what you do to one side of the equation you must do to the other side. So far, there have only been four possible things to do to an equation. They are add, subtract, multiply, or divide. Now we introduce a fifth thing that can be done to both sides of an equation: taking a root.

Square roots of equations

14.2A PROPERTY OF SQUARE ROOTS OF EQUATIONS

If

$$a^2 = b$$

then either

$$a = \sqrt{b} \quad \text{or} \quad a = -\sqrt{b}$$

This property says that when the square root of both sides of an equation is taken, two new equations result. That is because

$$(\sqrt{b})^2 = b \quad \text{and also} \quad (-\sqrt{b})^2 = b$$

EXAMPLE 6 Solve $x^2 = 16$.

By Property 14.2A

$$x = \sqrt{16} \quad \text{or} \quad x = -\sqrt{16}$$

$$x = 4 \quad \text{or} \quad x = -4$$

Extraction of roots

This method is sometimes called *extraction of roots.*

EXAMPLE 7 Solve $x^2 = 7$.

$$x = \sqrt{7} \quad \text{or} \quad x = -\sqrt{7}$$

A shorthand way to say the same thing is

$$x = \pm\sqrt{7}$$

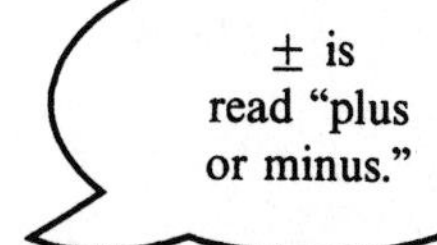

EXAMPLE 8 Solve $x^2 = 12$.

$$\begin{aligned} x &= \pm\sqrt{12} \\ &= \pm\sqrt{4}\sqrt{3} \\ &= \pm 2\sqrt{3} \end{aligned}$$

EXAMPLE 9 Solve $(x + 3)^2 = 25$.

First, take the square root of both sides.

$$\sqrt{(x+3)^2} = \sqrt{25} \quad \text{or} \quad \sqrt{(x+3)^2} = -\sqrt{25}$$

$$x + 3 = 5 \qquad\qquad x + 3 = -5$$

$$x = 2 \qquad\qquad x = -8$$

Check:

$$(x + 3)^2 = 25 \qquad\qquad (x + 3)^2 = 25$$

$$[(2) + 3]^2 \stackrel{?}{=} 25 \qquad\qquad [(-8) + 3]^2 \stackrel{?}{=} 25$$

$$5^2 \stackrel{?}{=} 25 \qquad\qquad (-5)^2 \stackrel{?}{=} 25$$

$$25 = 25 \qquad\qquad 25 = 25$$

EXAMPLE 10 Solve $(x - 5)^2 = 12$.

$$x - 5 = \sqrt{12} \quad \text{or} \quad x - 5 = -\sqrt{12}$$

$$x - 5 = 2\sqrt{3} \qquad\qquad x - 5 = -2\sqrt{3}$$

$$x = 5 + 2\sqrt{3} \qquad\qquad x = 5 - 2\sqrt{3}$$

These results can be written in a single statement:

$$x = 5 \pm 2\sqrt{3}$$

PROBLEM SET 14.2A

Solve each of the following equations by extraction of roots.

1. $x^2 = 36$
2. $x^2 = 25$
3. $x^2 = 23$
4. $x^2 = 13$
5. $x^2 = 17$
6. $x^2 = 20$
7. $x^2 = 18$
8. $x^2 = 100$
9. $(x + 2)^2 = 25$
10. $(x - 2)^2 = 25$
11. $(x + 15)^2 = 100$
12. $(x - 7)^2 = 49$
13. $(x + 7)^2 = 16$
14. $(x + 4)^2 = 16$
15. $(x + 4)^2 = 36$

(16) Hint: Write the equation with $(x + 4)^2$ alone on the left side.

16. $(x + 4)^2 - 25 = 0$
17. $(x - 6)^2 - 36 = 0$
18. $(x + 8)^2 - 49 = 0$
19. $(x + 3)^2 - 4 = 0$
20. $(x - 10)^2 - 9 = 0$
21. $(x + 5)^2 - 64 = 0$
22. $(x + 3)^2 = 7$
23. $(x - 5)^2 = 6$
24. $(x - 3)^2 = 20$
25. $(x + 7)^2 = 24$
26. $(x - 4)^2 = 28$
27. $(x + 3)^2 = 10$
28. $(x - 5)^2 = 15$

Completing the Square

An extension of the extraction of roots method, called completing the square, allows us to solve any quadratic equation.

First, we need to learn how to complete a square. Recall the rule for squaring a binomial.

Square of a binomial

14.2B SQUARING A BINOMIAL

The square of a binomial is the square of the first term plus twice the product of the terms plus the square of the second term.

$$(a + b)^2 = a^2 + 2ab + b^2$$

EXAMPLE 11 Square the following binomials.

$(x + 1)^2 = x^2 + 2x + 1$

4 — $(x + 2)^2 = x^2 + 4x +$ ______

Now let's study what happens when we square a binomial. We are looking for a relationship between the middle term of the trinomial and the constant term of the binomial.

EXAMPLE 12 Square the following binomial.

6, 9 — $(x + 3)^2 = x^2 +$ ______ $x +$ ______

3 — The constant term of the binomial that was squared is ______.

6 — The coefficient of the middle term of the trinomial is ______.

9 — The third term of the trinomial is ______.

Look at another.

EXAMPLE 13 Square the following binomial.

−8, 16 — $(x - 4)^2 = x^2 +$ ______ $x +$ ______

−4 — The constant term of the binomial is ______.

−8 — The coefficient of the middle term is ______.

The binomial constant is half the middle term of the trinomial. — What is the relationship between the constant of the binomial and the middle term of the trinomial? ______________________________

EXAMPLE 14 Square the following binomial.

$$(x + 6)^2 = x^2 + 12x + 36$$

$6 = \frac{1}{2}$ of 12

EXAMPLE 15 Square the following binomial.

−10, 25

$(x - 5)^2 = x^2 +$ ______ $x +$ ______

−10

The coefficient of the middle term of the trinomial is ______.

−5

Half of −10 is ______.

−5

The constant term of the binomial is ______.

25

The square of −5 is ______.

25

The third term of the trinomial is ______.

Perfect square trinomials

Trinomials that are perfect squares of a binomial have a constant term that is equal to the square of one-half the coefficient of the middle term. If you are given the first two terms of a trinomial in standard form, you can provide the third term required to make it a perfect square of a binomial.

EXAMPLE 16 Provide the missing term in each trinomial below so that the result is a perfect square of a binomial.

	Trinomial	Square of a binomial	
	$x^2 + 10x + 25$	$= (x + 5)^2$	25 is $\left[\frac{1}{2} \cdot 10\right]^2$
9, x + 3	$x^2 + 6x +$ ______	= (________)^2	
64, x − 8	$x^2 - 16x +$ ______	= (________)^2	
36, x + 6	$x^2 + 12x +$ ______	= (________)^2	
4, x − 2	$x^2 - 4x +$ ______	= (________)^2	

This only works if the coefficient of the x^2 term is 1.

The constant term of the binomial is always half the coefficient of the middle term of the trinomial.

The constant term of the binomial always has the same sign as the middle term of the trinomial.

Completing the square

This process of supplying the missing third term so that a trinomial is a perfect square of a binomial is called *completing the square.*

PROBLEM SET 14.2B

Provide the missing third term in each trinomial so that the result is a perfect square of a binomial.

1. $x^2 + 8x$
2. $x^2 + 4x$
3. $x^2 - 10x$
4. $x^2 + 14x$
5. $x^2 - 18x$
6. $x^2 - 22x$
7. $x^2 + 16x$
8. $x^2 + 24x$
9. $x^2 - 5x$
10. $x^2 + 7x$
11. $x^2 - 3x$
12. $x^2 + x$

Solving Quadratics by Completing the Square

The method of completing the square can be used to solve quadratic equations.

EXAMPLE 17 Solve $x^2 - 8x - 20 = 0$ using the method of completing the square. Write the equation with the constant term on the right side of the equation.

$$x^2 - 8x - 20 = 0$$
$$x^2 - 8x = 20$$

Determine what must be added to the left-hand side to make it a perfect square.

What is half the coefficient of $-8x$? ______ (−4)

What is $(-4)^2$? ______ (16)

What must be added to the left-hand side to make it a perfect square? ______ (16)

Since we must do the same thing to both sides of the equation, add a 16 to both sides.

$$x^2 - 8x + 16 = 20 + 16$$

Next, write the left member as a binomial squared.

$$(x - 4)^2 = 36$$

This binomial is always x + (half the coefficient of the middle term in the trinomial). Half of (-8) is -4.

Now take the square root of both sides.

$$x - 4 = \sqrt{36} \quad \text{or} \quad x - 4 = -\sqrt{36}$$
$$x - 4 = 6 \qquad x - 4 = -6$$
$$x = 10 \qquad x = -2$$

are solutions to the equation.

Check the solutions in the original equation.

$$x^2 - 8x - 20 = 0 \qquad x^2 - 8x - 20 = 0$$
$$(10)^2 - 8(10) - 20 \stackrel{?}{=} 0 \qquad (-2)^2 - 8(-2) - 20 \stackrel{?}{=} 0$$
$$100 - 80 - 20 \stackrel{?}{=} 0 \qquad 4 + 16 - 20 \stackrel{?}{=} 0$$
$$0 = 0 \qquad 0 = 0$$

Could the equation $x^2 - 8x - 20 = 0$ have been factored to find the solution?

Yes, it could. We just wanted to show the method. Some equations don't factor easily. After we have learned several methods, we usually pick the easiest method to solve a particular equation.

EXAMPLE 18 Solve $x^2 + 6x + 7 = 0$.

To use the method of completing the square, first write the equation with the constant term on the right-hand side.

$$x^2 + 6x + 7 = 0$$
$$x^2 + 6x \quad = -7$$

Leave a blank space here.

Now ask, "What must be added to the left-hand side so that it is a perfect square?" In this case, +9 would make the left side a perfect square because $\left[\frac{1}{2}(6)\right]^2 = 9$. Since we must do the same thing to both sides of the equation, add a 9 to both sides.

$$x^2 + 6x + 9 = -7 + 9$$
$$(x + 3)^2 = 2$$

Now take the square root of both sides.

$$x + 3 = +\sqrt{2} \quad \text{or} \quad x + 3 = -\sqrt{2}$$
$$x = -3 + \sqrt{2} \qquad x = -3 - \sqrt{2}$$

are the solutions to the equation.

I see what you mean when you say some equations don't factor easily. I don't know how I would have done this one by factoring.

EXAMPLE 19 Solve $x^2 + x - \frac{15}{4} = 0$ by completing the square.

First, write the equation with the constant on the right side.

$$x^2 + x - \frac{15}{4} = 0$$

$$x^2 + x \quad = \frac{15}{4}$$

The coefficient of the x term is ______.

One-half the coefficient of the x term is ______.

The square of half the coefficient of x is ______.

The quantity to add to both sides of the equation in order to complete the square is ______.

1

$\frac{1}{2}$

$\frac{1}{4}$

$\frac{1}{4}$

$$x^2 + x + \frac{1}{4} = \frac{15}{4} + \frac{1}{4}$$

$$\left(x + \frac{1}{2}\right)^2 = \frac{16}{4}$$

$$\left(x + \frac{1}{2}\right)^2 = 4$$

$$x + \frac{1}{2} = 2 \quad \text{or} \quad x + \frac{1}{2} = -2$$

$$x = \frac{3}{2} \qquad x = \frac{-5}{2}$$

"Half the coefficient of the middle term of the perfect square" is $\frac{1}{2}$:
$\left(x + \frac{1}{2}\right)^2 = x^2 + x + \frac{1}{4}$.

EXAMPLE 20 Solve $6x^2 - 5x - 6 = 0$ by completing the square.

First, get the constant term on the right.

$$6x^2 - 5x - 6 = 0$$
$$6x^2 - 5x \quad = 6$$

For the method of completing the square to work, the coefficient of the x^2 term must be 1. Therefore, divide both sides of the equation by 6.

$$\frac{6x^2}{6} - \frac{5x}{6} = \frac{6}{6}$$
$$x^2 - \frac{5}{6}x = 1$$

$\frac{-5}{6}$

The coefficient of the x term is ______.

$\frac{-5}{12}$

Half that coefficient of the x term is ______.

$\frac{25}{144}$

The square of half the coefficient of the x term is ______.

Complete the square by adding $\frac{25}{144}$ to both sides.

$$x^2 - \frac{5}{6}x + \frac{25}{144} = 1 + \frac{25}{144}$$
$$x^2 - \frac{5}{6}x + \frac{25}{144} = \frac{144}{144} + \frac{25}{144} \qquad \text{Common denominator}$$
$$\left(x - \frac{5}{12}\right)^2 = \frac{169}{144}$$

Take the square root of both sides.

$$x - \frac{5}{12} = \sqrt{\frac{169}{144}} \quad \text{or} \quad x - \frac{5}{12} = -\sqrt{\frac{169}{144}}$$
$$x - \frac{5}{12} = \frac{13}{12} \quad \text{or} \quad x - \frac{5}{12} = \frac{-13}{12}$$
$$x = \frac{18}{12} \qquad x = \frac{-8}{12}$$
$$x = \frac{3}{2} \qquad x = \frac{-2}{3}$$

Check both solutions in the original equation.

$$6x^2 - 5x - 6 = 0 \qquad\qquad 6x^2 - 5x - 6 = 0$$

$$6\left(\frac{3}{2}\right)^2 - 5\left(\frac{3}{2}\right) - 6 \stackrel{?}{=} 0 \qquad 6\left(\frac{-2}{3}\right)^2 - 5\left(\frac{-2}{3}\right) - 6 \stackrel{?}{=} 0$$

$$6\left(\frac{9}{4}\right) - 5\left(\frac{3}{2}\right) - 6 \stackrel{?}{=} 0 \qquad 6\left(\frac{4}{9}\right) - 5\left(\frac{-2}{3}\right) - 6 \stackrel{?}{=} 0$$

$$\frac{27}{2} - \frac{15}{2} - 6 \stackrel{?}{=} 0 \qquad \frac{8}{3} + \frac{10}{3} - 6 \stackrel{?}{=} 0$$

$$\frac{12}{2} - 6 \stackrel{?}{=} 0 \qquad \frac{18}{3} - 6 \stackrel{?}{=} 0$$

$$6 - 6 \stackrel{?}{=} 0 \qquad 6 - 6 \stackrel{?}{=} 0$$

$$0 = 0 \qquad 0 = 0$$

EXAMPLE 21 Solve $2x^2 + 7x - 4 = 0$ for x by completing the square.

First, get the constant term on the right side.

$$2x^2 + 7x - 4 = 0$$

$$2x^2 + 7x = 4$$

The equal sign says that we have equivalent quantities on both sides of the equation. It is like a balance.

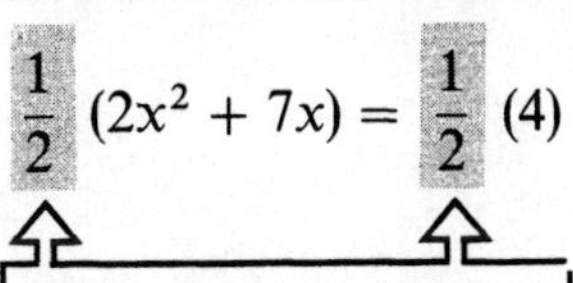

$$\frac{1}{2}(2x^2 + 7x) = \frac{1}{2}(4)$$

To keep our balance, we must do the same thing to both sides. Multiply both sides by $\frac{1}{2}$.

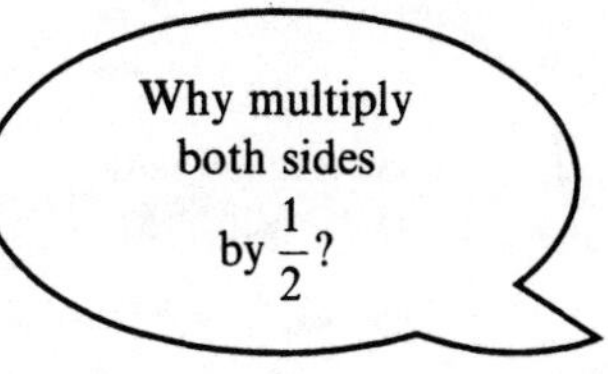

To complete the square, the coefficient of x^2 must be one.

$$x^2 + \frac{7}{2}x = 2$$

$\frac{7}{2}$

The coefficient of the x term is ______.

Half that coefficient is $\frac{1}{2}\left(\frac{7}{2}\right) = \frac{7}{4}$.

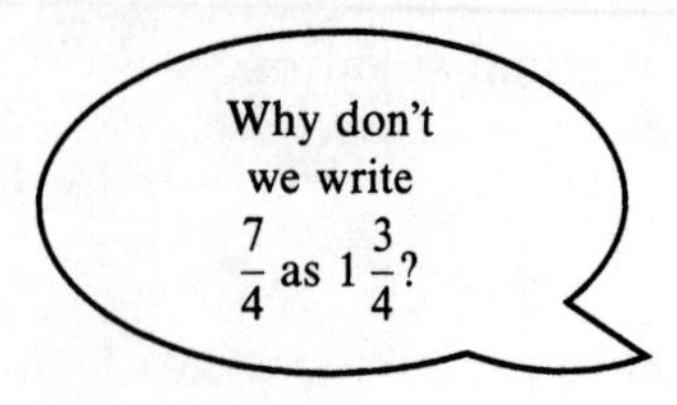

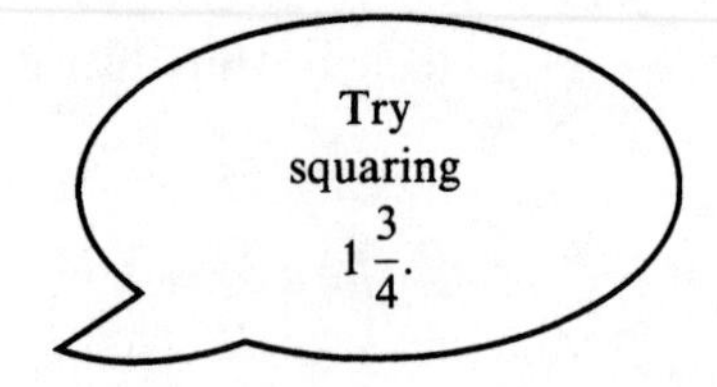

$\frac{49}{16}$

The square of half the coefficient of the x term is ______.

Complete the square by adding $\frac{49}{16}$ to both sides.

$$x^2 + \frac{7}{2}x + \frac{49}{16} = 2 + \frac{49}{16}$$

To keep our balance, we must add the same amount to both sides.

$$x^2 + \frac{7}{2}x + \frac{49}{16} = 2\left(\frac{16}{16}\right) + \frac{49}{16} \qquad \text{Find a common denominator.}$$

$$x^2 + \frac{7}{2}x + \frac{49}{16} = \frac{32}{16} + \frac{49}{16} \qquad \text{Common denominator.}$$

$$x^2 + \frac{7}{2}x + \frac{49}{16} = \frac{81}{16}$$

Write the left side in factored form.

$$\left(x + \frac{7}{4}\right)^2 = \frac{81}{16}$$

How do I know the left side can be factored?

When you added $\frac{49}{16}$ you made a trinomial that would factor.

Take the square root of both sides.

$$\sqrt{\left(x + \frac{7}{4}\right)^2} = \sqrt{\frac{81}{16}} \quad \text{or} \quad \sqrt{\left(x + \frac{7}{4}\right)^2} = -\sqrt{\frac{81}{16}}$$

$$x + \frac{7}{4} = \frac{9}{4} \qquad\qquad x + \frac{7}{4} = -\frac{9}{4}$$

Solve both equations for x.

$$x + \frac{7}{4} = \frac{9}{4} \quad \text{or} \quad x + \frac{7}{4} = -\frac{9}{4}$$

$$x = \frac{9}{4} - \frac{7}{4} \qquad\qquad x = -\frac{9}{4} - \frac{7}{4}$$

$$x = \frac{1}{2} \qquad\qquad x = -4$$

EXAMPLE 22 Solve $2x^2 + 10x + 11 = 0$ by completing the square.

Write the equation with the constant term on the right side.

$$2x^2 + 10x + 11 = 0$$
$$2x^2 + 10x = -11$$

Make the coefficient of the x^2 term a "+1" by dividing both sides of the equation by the coefficient of x^2, which is 2.

$$\frac{2x^2}{2} + \frac{10x}{2} = \frac{-11}{2}$$
$$x^2 + 5x = \frac{-11}{2}$$

Complete the square by adding the square of half the coefficient of the x term to both sides.

$$x^2 + 5x + \left(\frac{5}{2}\right)^2 = \frac{-11}{2} + \left(\frac{5}{2}\right)^2$$
$$\left(x + \frac{5}{2}\right)^2 = \frac{-11}{2} + \frac{25}{4}$$
$$\left(x + \frac{5}{2}\right)^2 = \frac{-22}{4} + \frac{25}{4}$$
$$\left(x + \frac{5}{2}\right)^2 = \frac{3}{4}$$

Take the square root of both sides.

$$x + \frac{5}{2} = \frac{\sqrt{3}}{2} \quad \text{or} \quad x + \frac{5}{2} = \frac{-\sqrt{3}}{2}$$
$$x = \frac{-5}{2} + \frac{\sqrt{3}}{2} \qquad x = \frac{-5}{2} + \frac{-\sqrt{3}}{2}$$
$$x = \frac{-5 + \sqrt{3}}{2} \qquad x = \frac{-5 - \sqrt{3}}{2}$$

Check these values in the original equation.

$$2x^2 + 10x + 11 = 0$$
$$2\left(\frac{-5 + \sqrt{3}}{2}\right)^2 + 10\left(\frac{-5 + \sqrt{3}}{2}\right) + 11 \stackrel{?}{=} 0$$
$$2\left(\frac{25 - 10\sqrt{3} + 3}{4}\right) + 5(-5 + \sqrt{3}) + 11 \stackrel{?}{=} 0$$
$$\left(\frac{28 - 10\sqrt{3}}{2}\right) - 25 + 5\sqrt{3} + 11 \stackrel{?}{=} 0$$
$$14 - 5\sqrt{3} + 5\sqrt{3} - 14 \stackrel{?}{=} 0$$
$$0 = 0$$

$$2x^2 + 10x + 11 = 0$$
$$2\left(\frac{-5 - \sqrt{3}}{2}\right)^2 + 10\left(\frac{-5 - \sqrt{3}}{2}\right) + 11 \stackrel{?}{=} 0$$
$$2\left(\frac{25 + 10\sqrt{3} + 3}{4}\right) + 5(-5 - \sqrt{3}) + 11 \stackrel{?}{=} 0$$
$$\left(\frac{28 + 10\sqrt{3}}{2}\right) - 25 - 5\sqrt{3} + 11 \stackrel{?}{=} 0$$
$$14 + 5\sqrt{3} - 5\sqrt{3} - 14 \stackrel{?}{=} 0$$
$$0 = 0$$

PROBLEM SET 14.2C

Solve the following quadratic equations by completing the square.

1. $x^2 + 2x - 8 = 0$	2. $x^2 - 2x - 15 = 0$	3. $x^2 - 6x - 16 = 0$
4. $x^2 + 8x + 12 = 0$	5. $x^2 - 8x + 15 = 0$	6. $x^2 + 6x - 7 = 0$
7. $x^2 + 10x + 21 = 0$	8. $x^2 + 6x - 27 = 0$	9. $x^2 - 10x = 0$
10. $x^2 + 12x = 0$	11. $x^2 + 2x - 35 = 0$	12. $x^2 + 4x - 45 = 0$
13. $x^2 - 10x - 11 = 0$	14. $x^2 + 8x - 33 = 0$	15. $3x^2 - 14x - 5 = 0$
16. $5x^2 - 12x - 9 = 0$	17. $2x^2 - 7x - 4 = 0$	18. $4x^2 - 5x - 6 = 0$
19. $2x^2 + 9x - 18 = 0$	20. $3x^2 + 5x - 12 = 0$	21. $2x^2 - 32 = 0$
22. $2x^2 - 18 = 0$	23. $3x^2 - 12x = 0$	24. $3x^2 - 6x = 0$
25. $4x^2 - 8x + 3 = 0$	26. $6x^2 - 24 = 0$	27. $6x^2 + 5x - 3 = 0$
28. $2x^2 + 7x + 6 = 0$	29. $5x^2 - 4x - 2 = 0$	30. $3x^2 - 8x + 2 = 0$
31. $5x^2 - 20x = 0$	32. $4x^2 - 10x + 5 = 0$	33. $2x^2 + 12x - 7 = 0$
34. $3x^2 + 4x - 1 = 0$	35. $3x^2 + 2x - 3 = 0$	36. $5x^2 - 8x + 1 = 0$

14.3 THE QUADRATIC FORMULA

Recall from Section 14.1 that the standard form of a quadratic equation is

$$ax^2 + bx + c = 0$$

where a, b, and c are constants and $a \neq 0$.

In the equation $2x^2 + 6x - 7 = 0$, $a = 2$, $b = 6$, and $c = -7$.

EXAMPLE 23 Give the value of a, b, and c for each of the following quadratic equations. If the equation is not in standard form, put it in standard form before identifying a, b, and c.

Answers	Equation			
5, 8, 1	$5x^2 + 8x + 1 = 0$	$a =$ ____	$b =$ ____	$c =$ ____
3, −7, −3	$3x^2 - 7x - 3 = 0$	$a =$ ____	$b =$ ____	$c =$ ____
1, −14, 4	$x^2 - 14x + 4 = 0$	$a =$ ____	$b =$ ____	$c =$ ____
−4, 19, −3	$19x - 4x^2 = 3$	$a =$ ____	$b =$ ____	$c =$ ____
1, 7, 0	$x^2 + 7x = 0$	$a =$ ____	$b =$ ____	$c =$ ____
7, −12, 3	$7x^2 = 12x - 3$	$a =$ ____	$b =$ ____	$c =$ ____

Any quadratic equation can be written in standard form. Next, we will derive a formula that will give the solutions of any quadratic equation in terms of its coefficients a, b, and c.

To derive the quadratic formula, it is necessary to solve for x in the general form of a quadratic equation by completing the square.

$ax^2 + bx + c = 0$ Subtract c from both sides of the equation

$ax^2 + bx = -c$

$\frac{ax^2}{a} + \frac{bx}{a} = \frac{-c}{d}$ To make the coefficient of x^2 equal to 1, divide both sides by a

$x^2 + \frac{b}{a}x = \frac{-c}{a}$

The coefficient of the x term is $\frac{b}{a}$.

One-half the coefficient of the x term is $\frac{1}{2} \cdot \frac{b}{a} = \frac{b}{2a}$.

The square of $\frac{b}{2a}$ is $\frac{b^2}{4a^2}$.

Adding $\frac{b^2}{4a^2}$ to both sides completes the square of the left-hand side.

$x^2 + \frac{b}{a}x + \frac{b^2}{4a^2} = \frac{b^2}{4a^2} + \frac{-c}{a}$

$\left(x + \frac{b}{2a}\right)^2 = \frac{b^2}{4a^2} + \frac{-c}{a}$ Write the left side as a binomial squared

$\left(x + \frac{b}{2a}\right)^2 = \frac{b^2}{4a^2} + \frac{-c}{a}\left(\frac{4a}{4a}\right)$ Make the denominators alike on the right side

$\left(x + \frac{b}{2a}\right)^2 = \frac{b^2 - 4ac}{4a^2}$ Add the fractions on the right

$\sqrt{\left(x + \frac{b}{2a}\right)^2} = \pm\sqrt{\frac{b^2 - 4ac}{4a^2}}$ Take the square root of both sides

$x + \frac{b}{2a} = \pm\frac{\sqrt{b^2 - 4ac}}{2a}$

$x = \frac{-b}{2a} \pm \frac{\sqrt{b^2 - 4ac}}{2a}$ Subtract $\frac{b}{2a}$ from both sides

$x = \frac{-b \pm \sqrt{b^2 - 4ac}}{2a}$ Write the result over the common denominator

The solutions to the quadratic equation are

$$x = \frac{-b + \sqrt{b^2 - 4ac}}{2a} \quad \text{and} \quad x = \frac{-b - \sqrt{b^2 - 4ac}}{2a}$$

Quadratic formula

14.3 QUADRATIC FORMULA

The roots of any quadratic equation in the form $ax^2 + bx + c = 0$ are

$$x = \frac{-b \pm \sqrt{b^2 - 4ac}}{2a}$$

where $a \neq 0$.

EXAMPLE 24 Solve $2x^2 - 11x + 5 = 0$ using the quadratic formula.

In $2x^2 - 11x + 5 = 0$, $a = 2$, $b = -11$, and $c = 5$.

Write the quadratic formula

$$x = \frac{-b \pm \sqrt{b^2 - 4ac}}{2a}$$

Substitute the values of a, b, and c.

$$x = \frac{-(-11) \pm \sqrt{(-11)^2 - 4(2)(5)}}{2(2)}$$

$$x = \frac{11 \pm \sqrt{121 - 40}}{4}$$

$$x = \frac{11 \pm \sqrt{81}}{4}$$

$$x = \frac{11 \pm 9}{4}$$

$$x = \frac{11 + 9}{4} = \frac{20}{4} = 5$$

or

$$x = \frac{11 - 9}{4} = \frac{2}{4} = \frac{1}{2}$$

I could have solved that by factoring.

Next is one you can't factor. The formula works in all cases.

EXAMPLE 25 Solve $4x^2 + 7x + 2 = 0$.

4, 7, 2

In $4x^2 + 7x + 2 = 0$, $a =$ ____, $b =$ ____, $c =$ ____.

Write the formula

$$x = \frac{-b \pm \sqrt{b^2 - 4ac}}{2a}$$

Substitute

$$x = \frac{-(7) \pm \sqrt{(7)^2 - 4(4)(2)}}{2(4)}$$

$$x = \frac{-7 \pm \sqrt{49 - 32}}{8}$$

$$x = \frac{-7 \pm \sqrt{17}}{8}$$

You can leave the answer in this form.

What if I don't believe that's the solution?

Check it by substituting for x in the original equation.

Check:

Substituting $x = \frac{-7 + \sqrt{17}}{8}$ in $4x^2 + 7x + 2 = 0$

$$4\left(\frac{-7 + \sqrt{17}}{8}\right)^2 + 7\left(\frac{-7 + \sqrt{17}}{8}\right) + 2 \stackrel{?}{=} 0$$

$$4\left(\frac{49 - 14\sqrt{17} + 17}{64}\right) - \frac{49}{8} + \frac{7\sqrt{17}}{8} + 2 \stackrel{?}{=} 0$$

$$\overset{1}{\cancel{4}}\left(\frac{66 - 14\sqrt{17}}{\underset{16}{\cancel{64}}}\right) - \frac{49}{8} + \frac{7\sqrt{17}}{8} + 2 \stackrel{?}{=} 0$$

$$\frac{66}{16} - \frac{14\sqrt{17}}{16} - \frac{49}{8} + \frac{7\sqrt{17}}{8} + 2 \stackrel{?}{=} 0$$

$$\frac{33}{8} - \frac{7\sqrt{17}}{8} - \frac{49}{8} + \frac{7\sqrt{17}}{8} + \frac{16}{8} \stackrel{?}{=} 0$$

$$\frac{33 - 49 + 16}{8} \stackrel{?}{=} 0$$

$$\frac{0}{8} \stackrel{?}{=} 0$$

$$0 = 0$$

Now, do you want to see me prove $x = \frac{-7 - \sqrt{17}}{8}$?

That's okay. I believe you.

EXAMPLE 26 Solve $2x^2 = 4x + 9$.

First, write the equation in standard form.

2, −4, −9

In $2x^2 - 4x - 9 = 0$, $a =$ ____, $b =$ ____, $c =$ ____.

Write the formula

$$x = \frac{-b \pm \sqrt{b^2 - 4ac}}{2a}$$

Substitute

$$x = \frac{-(-4) \pm \sqrt{(-4)^2 - 4(2)(-9)}}{2(2)}$$

$$x = \frac{4 \pm \sqrt{16 + 72}}{4}$$

$$x = \frac{4 \pm \sqrt{88}}{4}$$

$$x = \frac{4 \pm 2\sqrt{22}}{4}$$

$$x = \frac{2(2 \pm \sqrt{22})}{4}$$

$$x = \frac{\overset{1}{\cancel{2}}(2 \pm \sqrt{22})}{\underset{2}{\cancel{4}}}$$

$$x = \frac{2 \pm \sqrt{22}}{2}$$

EXAMPLE 27 Solve $3x^2 + 5x = 0$ using the quadratic formula.

In $3x^2 + 5x = 0$, $a = 3$, $b = 5$, $c = 0$.

Write the formula

$$x = \frac{-b \pm \sqrt{b^2 - 4ac}}{2a}$$

Substituting $a = 3$, $b = 5$, and $c = 0$, we get

$$x = \frac{-(5) \pm \sqrt{(5)^2 - 4(3)(0)}}{2(3)}$$

$$x = \frac{-5 \pm \sqrt{25}}{6}$$

$$x = \frac{-5 \pm 5}{6}$$

$$x = \frac{-5 + 5}{6} = \frac{0}{6} = 0 \quad \text{or} \quad x = \frac{-5 - 5}{6} = \frac{-10}{6} = \frac{-5}{3}$$

POINTERS ABOUT THE QUADRATIC FORMULA

$x = \dfrac{-b \pm \sqrt{b^2 - 4ac}}{2a}$ is really two sentences that give the roots of a quadratic equation.

However, for the formula to make sense, a must not equal zero or we will have division by zero. This is not a problem because if $a = 0$, $ax^2 + bx + c = 0$ is not a quadratic equation.

Notice the quantity under the radical sign. Since we can't find the square root of a negative number with the tools we have, the quantity $b^2 - 4ac$ must be greater than or equal to zero. The solutions of quadratic equations with $b^2 - 4ac$ less than zero are complex numbers, which are beyond the scope of this book.

Another pointer deals with the division bar. Be sure the division bar includes the $-b$ portion of the numerator.

PROBLEM SET 14.3

Solve the following quadratic equations using the quadratic formula.

1. $x^2 + 3x - 10 = 0$
2. $x^2 + 7x + 12 = 0$
3. $x^2 - 11x + 30 = 0$
4. $x^2 + 6x + 5 = 0$
5. $2x^2 - 7x - 15 = 0$
6. $3x^2 - 8x + 4 = 0$
7. $4x^2 - 8x - 10 = 0$
8. $5x^2 + 11x - 12 = 0$
9. $4x^2 + 9x = 0$
10. $5x^2 + 8x = 0$
11. $6x^2 + x - 5 = 0$
12. $5x^2 - 4x - 12 = 0$
13. $6x^2 + 5x - 6 = 0$
14. $6x^2 - x - 12 = 0$
15. $12x^2 - 4x - 5 = 0$
16. $6x^2 - 7x - 3 = 0$
17. $10x^2 - 4x - 1 = 0$
18. $8x^2 + 3x - 2 = 0$
19. $2x^2 - 16 = 0$
20. $3x^2 - 24 = 0$
21. $4x^2 - 9x = 0$
22. $3x^2 - 7x + 2 = 0$
23. $4x^2 + 5x - 2 = 0$
24. $6x^2 - 7x = 0$
25. $2x^2 + 7x = 3$
26. $5x^2 = 8x - 1$
27. $3x^2 = 6x + 2$
28. $2x^2 + 5 = 8x$
29. $3x^2 - 12 = 0$
30. $4x^2 - 5x = 1$
31. $5x^2 + 9x = -2$
32. $5x^2 + 11x + 2 = 0$
33. $4x^2 + 9x - 9 = 0$
34. $5x^2 - 45 = 0$
35. $3x^2 = 2x + 3$
36. $2x^2 + 5x = 1$

14.4 APPLICATIONS

The ability to solve quadratic equations permits us to solve word problems that include the square of a variable.

EXAMPLE 28 A number plus its square is equal to 12. Find the number.

Make a model.

[Number] + [Number squared] = 12

Let n be the number; express as an equation.

A number plus its square is equal to 12

$$n + n^2 = 12$$

Write the equation in standard form.

$$n^2 + n - 12 = 0$$

$$(n - 3)(n + 4) = 0 \quad \text{Factor}$$

$$n - 3 = 0 \quad \text{or} \quad n + 4 = 0$$

$$n = 3 \qquad n = -4$$

Check:

$$n + n^2 = 12 \qquad n + n^2 = 12$$

$$(3) + (3)^2 \stackrel{?}{=} 12 \qquad (-4) + (-4)^2 \stackrel{?}{=} 12$$

$$12 = 12 \qquad 12 = 12$$

This problem has two answers, $n = 3$ and $n = -4$.

EXAMPLE 29 A patio is 3 yards longer than it is wide. If the total area of the patio is 28 square yards, find the dimensions of the patio.

$w + 3$

Let width = w. Then length = ________

$l = w + 3$

w Area = $l \cdot w$

$$A = l \cdot w$$

$$28 = (w + 3)w$$

$$28 = w^2 + 3w$$

$$0 = w^2 + 3w - 28$$

$$0 = (w + 7)(w - 4) \quad \text{Factor}$$

$$w + 7 = 0 \quad \text{or} \quad w - 4 = 0$$

$$w = -7 \qquad w = 4$$

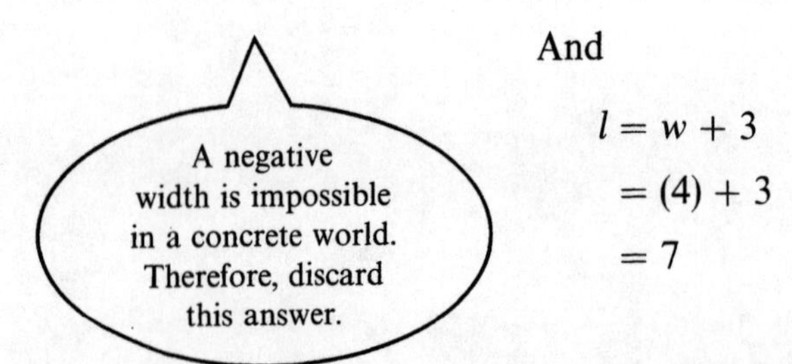

And

$$l = w + 3$$

$$= (4) + 3$$

$$= 7$$

The dimensions of the patio are 4 yards by 7 yards.

The kinetic energy of an object in motion is given by the formula

$$K = \frac{1}{2}mv^2$$

where K = kinetic energy in joules

m = mass of object in kilograms

v = velocity in meters/second

EXAMPLE 30 Determine the kinetic energy of a car with a mass of 1,000 kilograms traveling at 20 meters/second (approximately 45 miles per hour).

$$\begin{aligned} K &= \frac{1}{2}mv^2 \\ &= \frac{1}{2}(1{,}000)(20)^2 \\ &= (500)(400) \\ &= 200{,}000 \text{ joules} \end{aligned}$$

1000 kilograms is a very light car.

A joule is a metric unit of energy. To give you some idea of the energy involved, this is enough energy to lift the car approximately 67 feet straight up in the air.

EXAMPLE 31 Find the speed of a 1,000-kilogram car if its kinetic energy is 50,000 joules.

$$\begin{aligned} K &= \frac{1}{2}mv^2 \\ 50{,}000 &= \frac{1}{2}(1{,}000)v^2 \\ 50{,}000 &= 500v^2 \\ 100 &= v^2 \\ 10 &= v \qquad \text{Take the square root of both sides} \\ v &= 10 \text{ meters per second} \end{aligned}$$

A 55-mph speed limit is approximately 90 kilometers per hour or 25 meters per second.

Notice that when a car's speed is doubled, its kinetic energy is increased four times.

EXAMPLE 32 The height attained by an object shot upward by a gun is given by the following formula.

$$h = vt - 4.9t^2$$

where v = vertical velocity in meters per second

t = time in seconds

h = distance above starting point in meters

On level ground when the object returns to earth, its height h will be zero. This fact allows us to calculate the time an object will remain in the air if we know its initial vertical velocity.

How long will a bullet shot upward with a vertical velocity of 49 meters per second remain in the air?

$$h = vt - 4.9t^2$$

At the time the bullet strikes ground, $h = 0$.

$$0 = (49)t - 4.9t^2$$

$49 - 4.9t$

$$0 = t(\underline{\qquad\qquad}) \qquad \text{Factor}$$

$$t = 0 \quad \text{or} \quad 49 - 4.9t = 0$$

This is the time when the bullet left the gun on the way up.

$$49 = 4.9t$$

$$\frac{49}{4.9} = t$$

$$10 = t$$

Time when the bullet strikes the ground.

The bullet remains in the air 10 seconds.

Now that we know the time the bullet was in the air, and if we assume it took half the time going up and half the time coming down, we can calculate the maximum height attained. In this case, maximum height is attained in $t = 5$ seconds.

$$\begin{aligned} h &= vt - 4.9t^2 \\ &= 49(5) - 4.9(5)^2 \\ &= 245 - 4.9(25) \\ &= 245 - 122.5 \\ &= 122.5 \text{ meters} \end{aligned}$$

EXAMPLE 33 Water is shot upward from a fire hose at a velocity of 19.6 meters per second. (a) How long will the water remain in the air? (b) What is the maximum height water can reach at this velocity?

$h = vt - 4.9t^2$

$0 = 19.6t - 4.9t^2$

$0 = t(19.6 - 4.9t)$

$19.6 - 4.9t = 0 \quad \text{or} \quad t = 0$

$t = \dfrac{-19.6}{-4.9}$

$t = 4$ seconds in air

2

The water reached its maximum height in $t =$ _____ seconds. If we substitute $t = 2$ in the equation, we can solve for the maximum height.

$$\begin{aligned} h &= 19.6t - 4.9t^2 \\ &= 19.6(2) - 4.9(2)^2 \\ &= 19.6(2) - 4.9(4) \\ &= 39.2 - 19.6 \\ &= 19.6 \text{ meters} \end{aligned}$$

PROBLEM SET 14.4

Work the following problems using these steps:

a. Indicate what the variable represents.
b. Set up an equation.
c. Solve for the variable.

1. A number plus its square is 20. Find the number.
2. Two positive integers differ by 3. If the sum of their squares is 65, find the integers.
3. The product of two positive, consecutive, even integers is 48. Find the integers.
4. The product of two positive, consecutive, odd integers is 35. Find the integers.
5. The product of two positive, consecutive integers is 42. Find the integers.
6. A positive integer subtracted from its square is 20. Find the integer.
7. The square of a positive integer diminished by four times that integer is 12. Find the integer.
8. The square of a positive integer increased by twice the integer is 63. Find the integer.
9. Three times the square of a positive integer diminished by four times the integer is 32. Find the integer.
10. Two times the square of a positive integer increased by three times the integer is 65. Find the integer.
11. A swimming pool is 8 meters longer than it is wide. If the area is 65 square meters, find the dimensions of the swimming pool.

12. A garden has an area of 150 square meters. If the length of the garden is 5 meters more than its width, find the dimensions of the garden.

13. A patio is twice as long as it is wide. If the total area of the patio is 72 square meters, find the dimensions of the patio.

14. The width of a swimming pool is two-thirds its length. If the total area of the swimming pool is 96 square meters, find the dimensions of the swimming pool.

15. The width of a warehouse is three-fourths its length. If the total area of the warehouse floor is 768 square meters, find the dimensions of the warehouse floor.

16. A classroom is 4 feet longer than it is wide. If the area is 1,440 square feet, find the dimensions of the classroom.

17. Use the kinetic energy formula $K = \frac{1}{2}mv^2$ to find the speed of a 1,200-kilogram car if its kinetic energy is 240,000 joules.

18. Use the kinetic energy formula $K = \frac{1}{2}mv^2$ to find the speed of a 1,500-kilogram car if its kinetic energy is 160,000 joules.

19. How long will it take for a baseball thrown into the air vertically to fall to the ground if thrown with a vertical velocity of 19.6 meters per second? Use $h = vt - 4.9t^2$.

20. A stone is thrown vertically upward with a velocity of 24.5 meters per second. How long will it take the stone to strike the ground?

21. The area of a triangle is given by the formula $A = \frac{1}{2}bh$, where A is the area, b is the length of the base, and h is the length of the height. Find the length of the base of a triangle if the height is 4 centimeters more than the base and the area is 96 square centimeters.

22. The formula $s = 4.9t^2$ is used to find the distance a body falls in t seconds. How long will it take for a ball to fall 176.4 meters?

23. Use the method of Example 32 and Example 33 to calculate how far into the air the car in Example 31 would fly if its forward velocity of 25 meters per second were suddenly transformed into upward velocity.

Review Your Skills

Given $f(x) = 3x - 4$, find the following.

24. $f(0)$ 25. $f(3)$ 26. $f(-4)$

Given $g(x) = -\frac{3}{4}x + 5$, find the following.

27. $g(0)$ 28. $g(-4)$ 29. $g(8)$

14.5 GRAPHING QUADRATIC FUNCTIONS

Parabola

An equation of the form $y = ax^2 + bx + c$ is a quadratic function. Its graph is called a *parabola*. A parabola can be graphed by evaluating the quadratic function for several values of the variable.

EXAMPLE 34 Graph $y = x^2 + 2x + 1$.

To draw a graph, we need to evaluate $y = f(x)$ for several values of x.

x	$f(x)$	Ordered pairs
x	$x^2 + 2x + 1$	(x, y)
0	$(0)^2 + 2(0) + 1 = 1$	$(0, 1)$
1	$(1)^2 + 2(1) + 1 = 4$	$(1, 4)$
-1	$(-1)^2 + 2(-1) + 1 = 0$	$(-1, 0)$
2	$(2)^2 + 2(2) + 1 = 9$	$(2, 9)$
-2	$(-2)^2 + 2(-2) + 1 = 1$	$(-2, 1)$
-3	$(-3)^2 + 2(-3) + 1 = 4$	$(-3, 4)$
-4	$(-4)^2 + 2(-4) + 1 = 9$	$(-4, 9)$

Then we plot the points that correspond to the ordered pairs.

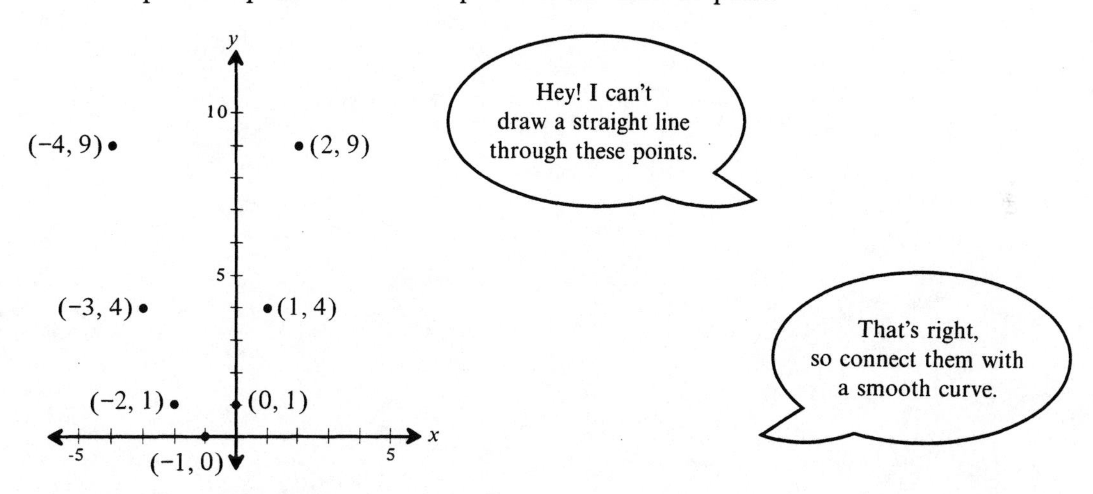

We draw a curve through the points, estimating its path between the points we know.

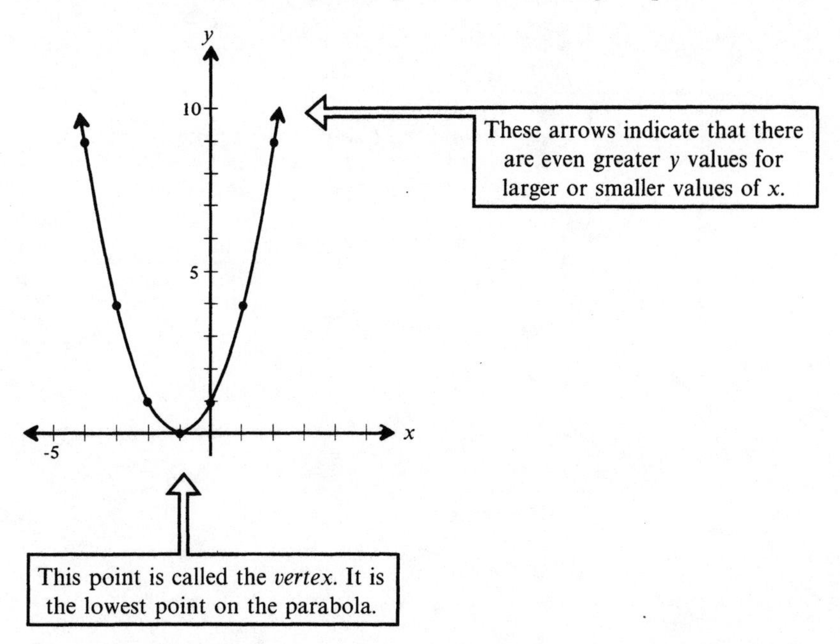

Vertex

If you are uncertain of where to draw the curve, pick additional x values to help fill in the blank spots.

EXAMPLE 35 Graph $y = +2x^2 - 7$.

First, make a table of values.

x	$f(x)$	Ordered pairs
x	$+2x^2 - 7$	(x, y)
0	$+2(0)^2 - 7 = -7$	$(0, -7)$
1	$+2(1)^2 - 7 = -5$	$(1, -5)$
-2	$+2(-2)^2 - 7 = 1$	$(-2, 1)$
-3	$+2(-3)^2 - 7 = 11$	$(-3, 11)$
2	$+2(2)^2 - 7 = 1$	$(2, 1)$
3	$+2(3)^2 - 7 = 11$	$(3, 11)$

How do you know which points to pick?

Start with the easy ones like 0, 1, -1, 2. Then watch the resulting values for $f(x)$. If the magnitude is too great, change the choice for x.

Plot the points and sketch the curve.

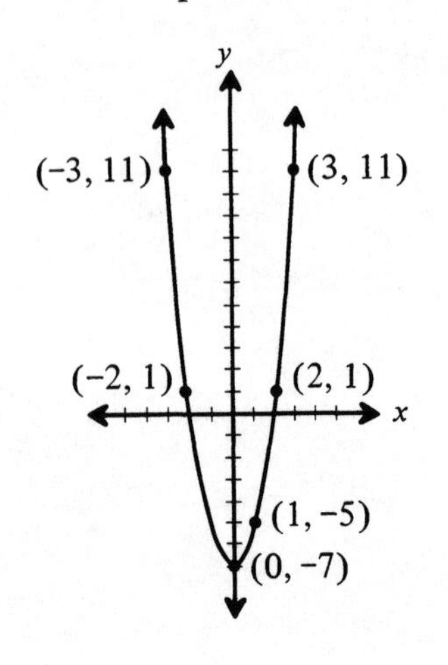

How can I tell where the vertex is?

Try x values slightly different from what you believe to be the x value of the vertex and see what happens to y.

To confirm that $(0, -7)$ is the vertex, evaluate $f(x)$ for $x = \frac{1}{2}$ and $x = \frac{-1}{2}$.

x	$f(x)$
$\frac{1}{2}$	$2\left(\frac{1}{2}\right)^2 - 7 = -6\frac{1}{2}$
$\frac{-1}{2}$	$2\left(\frac{-1}{2}\right)^2 - 7 = -6\frac{1}{2}$

Both of these values are equal and larger than -7. Therefore, we are fairly sure that the vertex is $(0, -7)$.

EXAMPLE 36 Graph $y = -2x^2 + 9$.

First, make a table of values.

x	$f(x)$	Coordinates
x	$-2(x)^2 + 9$	(x, y)
0	$-2(0)^2 + 9 = 9$	(0, 9)
1	$-2(1)^2 + 9 = 7$	(1, 7)
-1	$-2(-1)^2 + 9 = 7$	$(-1, 7)$
2	$-2(2)^2 + 9 = 1$	(2, 1)
-2	$-2(-2)^2 + 9 =$ ____	(____)
3	$-2(3)^2 + 9 =$ ____	(____)
-3	$-2(-3)^2 + 9 =$ ____	(____)

1 $(-2, 1)$

-9 $(3, -9)$

-9 $(-3, -9)$

Plot the points and sketch the curve.

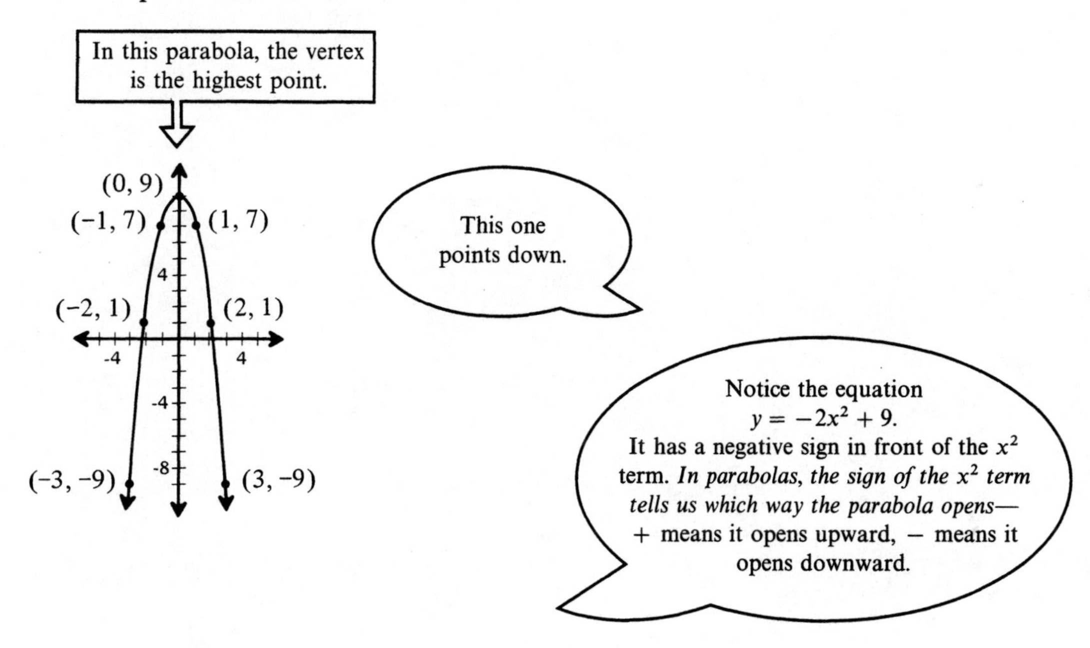

PROBLEM SET 14.5

Graph each of the following.

1. $y = 2x^2 + 4x - 5$
2. $y = x^2 + 4x - 5$
3. $y = 2x^2 - 5$
4. $y = x^2 - 5$
5. $y = 6 - x^2$
6. $y = 3 - 2x^2$

On the same set of axes, graph the following.

7. $y = x^2$
8. $y = 2x^2$
9. $y = \frac{1}{2}x^2$
10. $y = 4x^2$
11. $y = -x^2$
12. $y = -2x^2$
13. $y = \frac{-1}{2}x^2$
14. $y = -4x^2$

On the same set of axes, graph each of the following.

15. $y = x^2$
16. $y = x^2 + 2$
17. $y = x^2 + 4$
18. $y = x^2 + 6$
19. $y = x^2 - 2$
20. $y = x^2 - 4$

Graph each of the following.

21. $y = 3x^2 + 6x - 10$
22. $y = 3x^2 - 6x - 10$
23. $y = 2x^2 + 8x + 8$
24. $y = -2x^2 - 8x - 8$
25. $y = -2x^2 - 4x + 3$
26. $y = -3x^2 + 6x + 2$
27. $y = 3x^2 + 7x - 6$
28. $y = 2x^2 + 3x - 2$

Review 14

NOW THAT YOU HAVE COMPLETED UNIT 14, YOU SHOULD BE ABLE TO:

Use the following definitions in your working vocabulary:

Quadratic equation—An equation of the form $ax^2 + bx + c = 0$, where a, b, and c are constants and $a \neq 0$

Parabola—The graph of a quadratic equation

Vertex—The highest or lowest point on a parabola

Apply the following rules and definitions:

If the product of two factors is zero, one or both of the factors is zero.

To solve a quadratic equation by factoring, set each factor equal to zero and solve for the variable.

To solve equations of the type $a^2 = b$, *extract roots* by finding the square root of both sides of the equation, obtaining the results

$$a = \sqrt{b} \quad \text{or} \quad a = -\sqrt{b}$$

To provide the third term of a trinomial so that the result is a *perfect square* of a binomial, square one-half the coefficient of the term with an exponent of 1.

To solve a quadratic equation using the method of completing the square, follow these steps:

1. Transform the equation so that the left member contains only the two terms containing the variable.
2. If the coefficient of the squared term is not 1, divide the entire equation by the coefficient of the squared term.
3. For the left side of the equation, find the third term of the trinomial to make the result a perfect square.

4. Add the missing quantity to both sides of the equation.
5. Write the left side of the equation as a binomial squared and combine the terms on the right side of the equation.
6. Extract the square root of both sides of the equation.
7. Solve for the variable.

To solve a quadratic equation using the quadratic formula:

1. Write the equation in standard form.

$$ax^2 + bx + c = 0$$

2. Substitute the values for a, b, and c in the quadratic formula.

$$x = \frac{-b \pm \sqrt{b^2 - 4ac}}{2a}$$

3. Simplify the fraction and write the solution as two separate numbers.

To solve word problems involving quadratic equations, use the methods for setting up equations discussed in Unit 5. Solve the resulting quadratic equation using one of the three methods:

1. Factoring
2. Completing the square
3. The quadratic formula

To sketch the graph of a quadratic equation:

1. Make a table of values, choosing a sufficient number of values to complete a parabola.
2. Plot the points on graph paper.
3. Draw a smooth curve through the points. (Remember: Successive points are not connected by straight lines.)

REVIEW TEST 14

Solve the following quadratic equations by factoring. (14.1)

1. $x^2 + 3x - 4 = 0$
2. $x^2 + 5x + 6 = 0$
3. $x^2 - 6x - 7 = 0$

Provide the missing third term in each trinomial so that the result is a perfect square of a binomial. (14.2)

4. $x^2 - 12x +$ _____
5. $x^2 + 20x +$ _____
6. $x^2 + 7x +$ _____

Solve the following quadratic equations by completing the square. (14.2)

7. $x^2 + 6x + 7 = 0$
8. $3x^2 - 4x - 2 = 0$
9. Write the quadratic formula. *(14.3)* ____________________

Solve the following quadratic equations using the quadratic formula. (14.3)

10. $x^2 - 5x + 3 = 0$
11. $5x^2 + 8x + 2 = 0$

Solve the following equations. (14.1 through 14.3)

12. $12x^2 - 16x - 3 = 0$
13. $x^2 + 8x + 5 = 0$
14. $\frac{x}{2} + 1 = \frac{4}{x}$
15. $x^2 = 50$

Solve the following. (14.4)

16. Three times the square of an integer diminished by ten times the integer is 48. Find the integer.
17. The length of a single-story science building is 75 meters more than its width. If the area of the floor space is 2,500 square meters, what are the dimensions of the building?

Graph the following equations. *(14.5)*

18. $y = -x^2 + 4$

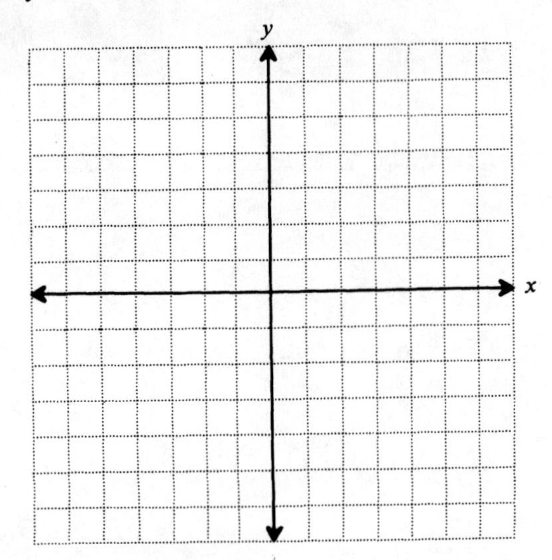

19. $y = 2x^2 + 4x - 5$

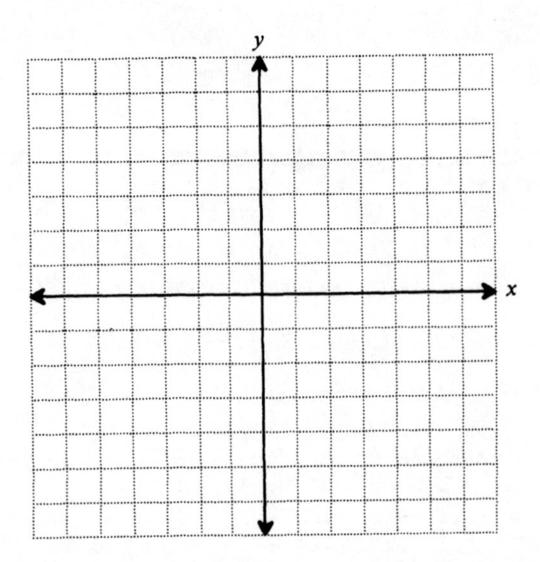

REVIEW YOUR SKILLS

Perform the indicated operations.

20. $\dfrac{2-x}{x^2} \cdot \dfrac{-x^4}{x^3-4x}$

21. $\dfrac{x^2+4x+4}{x^2} \div \dfrac{x^2+x-2}{x^4+x^3}$

22. $\dfrac{9}{(x-2)(x+1)} + \dfrac{3}{x+1}$

23. $\dfrac{1-\dfrac{1}{2x}}{\dfrac{1}{x}-\dfrac{1}{2}}$

Solve for the variable.

24. $\dfrac{3}{3x-1} = \dfrac{6}{4x-7}$

UNIT 15

Inequalities and Absolute Value

OBJECTIVES

After you have successfully completed this unit, you will be able to:

1. Graph inequalities on a number line (15.1)
2. Solve absolute value equations (15.2)
3. Graph absolute value equations (15.2)
4. Solve inequalities (15.3)
5. Graph linear inequalities (15.4)

15.1 INEQUALITIES

Inequalities

Most of our attention in this book has been devoted to equations that are statements about two equal quantities. It is also possible to express relationships between quantities that are unequal. These relationships are called *inequalities.*

$3 < 5$ is an example of an inequality. It is read "3 is less than 5."

Inequalities using numerals are either true or false. When variables are used, the truth of the inequality depends on the value of the variable.

$x < 5$ is true for all values of x less than 5. Some values of x that make this inequality true are 4, 4.99, 0, -2000. There are obviously many more replacements for x that make the inequality true. One of the values that make $x < 5$ false is 5, because 5 is not less than 5.

One way to represent the replacements of x that make an inequality true is with a number line.

EXAMPLE 1

$x < 5$

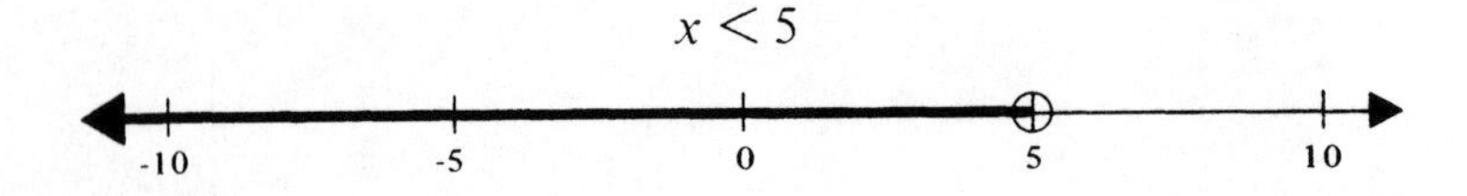

Things to notice about this graph are:

1. The line itself has arrows on both ends. These arrows mean the line extends forever in both directions.
2. The heavy portion of the line represents the values of x that make the inequality true. It has an arrow only on the left side, which means that the shaded portion and all numbers to the left make the inequality true. Therefore, the graph includes 4, -8, and -101.
3. The point 5 has an open circle around it. This indicates that 5 is not included. If 5 was included in the replacement set, then the circle would be solid.
4. It is possible to test if the graph is shaded correctly by substituting values from the shaded region in the inequality. These values should make the inequality true. Values from the unshaded region should make the inequality false.

$x < 5$

Value	Region	Substitution	Truth
0	shaded	$0 < 5$	true
-6	shaded	$-6 < 5$	________
6	unshaded	$6 < 5$	false
5	unshaded	$5 < 5$	false
7	unshaded	$7 < 5$	________

true

false

Notice that the larger number is always on the right on a number line.

Just like reading from left to right.

EXAMPLE 2

$5 > 3$ is an inequality. It is read "5 is greater than 3."

There are only three possible relationships that can exist between two numbers. The trichotomy axiom states these relationships.

Trichotomy

15.1A TRICHOTOMY AXIOM

For any real numbers, a and b, exactly one of the following is true:

$a < b$, $\quad a = b$, $\quad a > b$

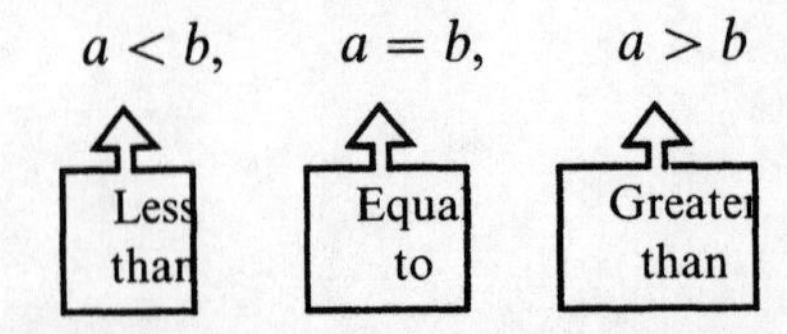

Axiom

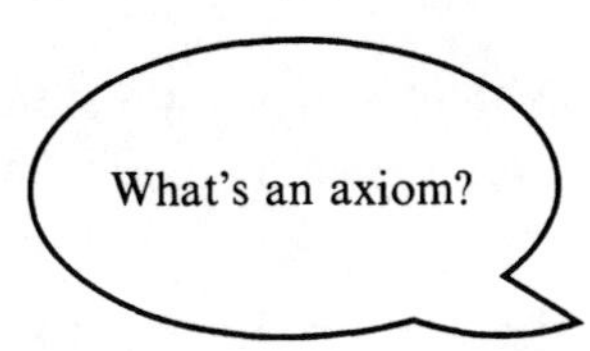

An *axiom* is a statement that we accept as true without proof. While trichotomy looks very reasonable, it's impossible to prove. Therefore, trichotomy is an axiom.

A second important property of real numbers is the transitive property of inequality.

Transitive property of inequality

15.1B TRANSITIVE PROPERTY OF INEQUALITY

For any real numbers a, b, and c,

if $a < b$ and $b < c$, then $a < c$.

EXAMPLE 3 $5 < 7$ and $7 < 10$, therefore $5 < 10$ is an example of the transitive property of inequality.

Here is a summary of the symbols we can use to compare two numbers.

Symbol	How read
$a = b$	a is equal to b
$a > b$	a is greater than b
$a < b$	a is less than b
$a \leqslant b$	a is less than or equal to b
$a \geqslant b$	a is greater than or equal to b
$a \neq b$	a is not equal to b

EXAMPLE 4 Here are some typical inequalities and their graphs.

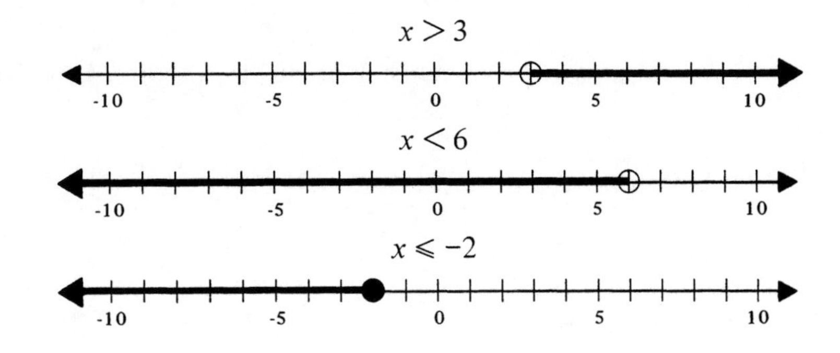

$x \leqslant -2$ is really two sentences, $x < -2$ or $x = -2$.

It is also possible to combine two statements of inequality.

EXAMPLE 5 $3 \leq x < 7$ is equivalent to the two statements $3 \leq x$ and $x < 7$. It is read "3 is less than or equal to x, and x is less than 7." We shade the values of x that make both statements true.

We are looking for the points on the number line that satisfy both inequalities.

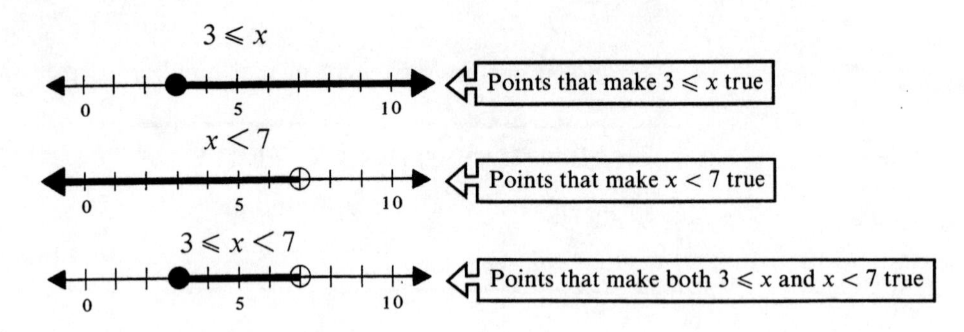

Some values of x that make the inequality $3 \leq x < 7$ true are 3, 3.001, 5.7, and 6.9999. The number 7 is greater than 3 but is not less than 7, therefore, it is not included in the set of values for x that make this inequality true.

EXAMPLE 6 Here are some other double inequality statements and their graphs.

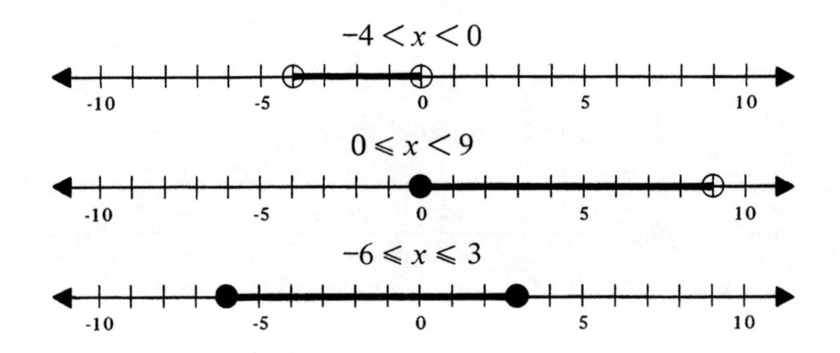

Double inequalities make sense only if they satisfy the transitive law of inequality. $7 < x < 3$ is meaningless because $7 < 3$ is false.

PROBLEM SET 15.1

Graph the following on a number line.

1. $x > 2$
2. $x > 3$
3. $x < 5$
4. $x < 4$
5. $x \geq -2$
6. $x \geq 3$
7. $4 \leq x$
8. $-3 \leq x$
9. $-6 \geq x$

10. $5 \geqslant x$

11. $x > 1$

12. $x \geqslant 1$

13. $1 < x$

14. $1 \leqslant x$

15. $x > 0$

16. $x < 0$

17. $x \geqslant -4$

18. $x \leqslant -4$

19. $4 < x < 8$

20. $2 < x < 6$

21. $-3 < x < 4$

22. $-2 < x < 3$

23. $-2 \leqslant x \leqslant 3$

24. $1 \leqslant x \leqslant 6$

25. $-2 \leqslant x < 5$

26. $-5 \leqslant x \leqslant 2$

27. $-7 \leqslant x < -4$

28. $-8 < x \leqslant -2$

29. $3 < x < 8$

30. $2 < x \leqslant 10$

31. $-4 < x \leqslant 2$

32. $-3 \leqslant x \leqslant 0$

33. $-7 < x \leqslant -1$

34. $-10 \leqslant x \leqslant -5$

35. $-6 < x < -2$

36. $-1 \leqslant x < 1$

15.2 ABSOLUTE VALUE

There are times when how much or how far is more important than direction. For example, when we say a car can travel 400 miles on a tank of gas, we don't care if the car is going north, east, south, or west. We only care how far. For those times in mathematics when we only care *how far a number is from zero*, we use the term *absolute value. The absolute value of a number refers to its magnitude. It does not tell us the direction from zero.*

In symbols	In English
$\|3\| = 3$	The absolute value of 3 is 3.
$\|-3\| = 3$	The absolute value of negative 3 is 3.

The absolute value of a number can be viewed as the number without its sign. However, in algebra there is no such operation as "take away the sign." It is also impossible to program a computer to remove the sign. What we could tell a computer is this:

To find the absolute value of a number:

a. If the number is zero or greater, its absolute value is the number itself.

$$|3| = 3 \qquad |0| = 0 \qquad |212| = 212$$

b. If the number is negative, its absolute value is the negative of the number.

$$|-3| = -(-3) = 3 \qquad |-212| = -(-212) = 212$$

Absolute value

15.2 DEFINITION OF ABSOLUTE VALUE

For any number a

if $a \geqslant 0$ then $|a| = a$

if $a < 0$ then $|a| = -a$

EXAMPLE 7 Graph $|x| = 4$.

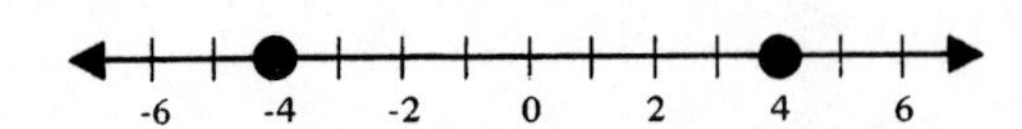

Because $|+4| = 4$ and $|-4| = 4$, x can be either $+4$ or -4, and both $+4$ and -4 are graphed.

EXAMPLE 8 Solve and graph $|x| + 3 = 5$.

$|x| + 3 = 5$ Subtract 3 from both sides

$|x| = 2$

Therefore,

2, −2

$x =$ _____ or $x =$ _____

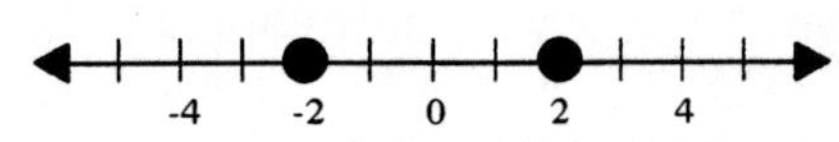

EXAMPLE 9 Graph $|x| < 7$.

The solution is all numbers less than 7 units from zero.

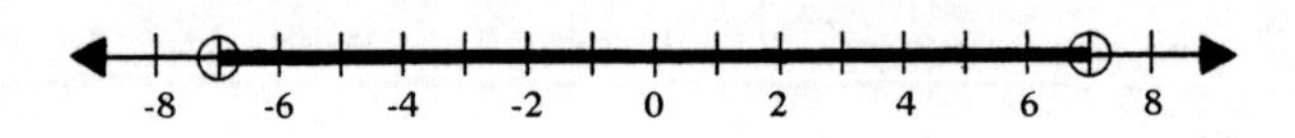

How do I know that's right?

Test some values from the darker portion of the line in the inequality: $|0| < 7$, $|-6| < 7$, $|+4| < 7$.

EXAMPLE 10 Graph $|x| \geqslant 3$.

The solution is all numbers that are 3 or more units from zero.

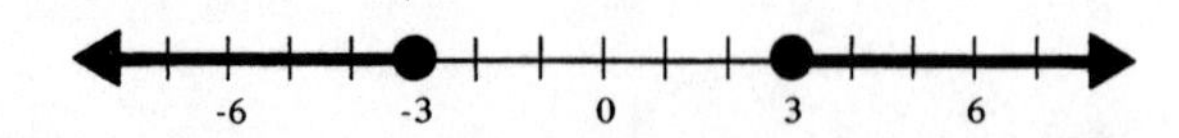

EXAMPLE 11 Graph $y = |x|$.

Before we graph $y = |x|$, build a table of values.

x	$y = \|x\|$
0	0
1	+1
2	+2
3	+3
−1	+1
−2	+2
−3	+3

Plot the points of the table on a rectangular graph.

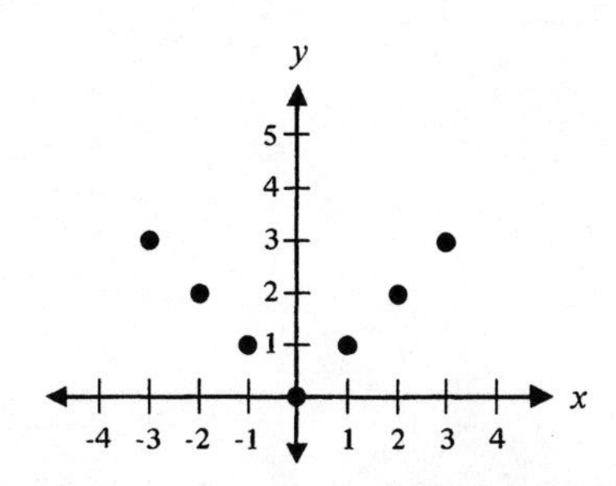

Since absolute value is defined for all real numbers, we may connect the dots.

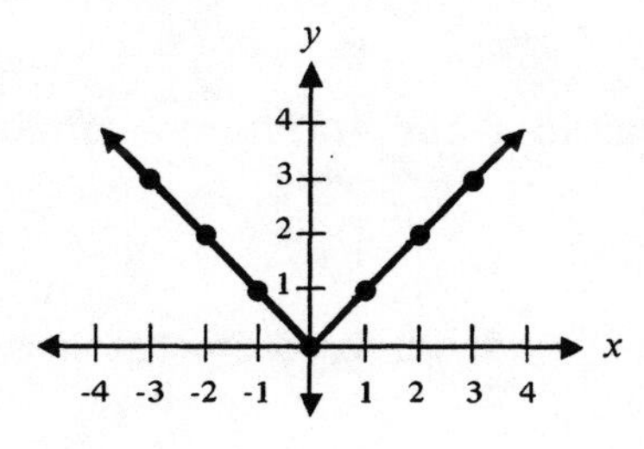

Do all absolute value graphs look like this one?

For first-degree variables, they have a V shape. The slope of the lines and the y-intercept may change.

EXAMPLE 12 Graph $y = |2x|$.

Make a table.

x	$y = \|2x\|$
0	0
1	2
2	4
−2	4
−1	2

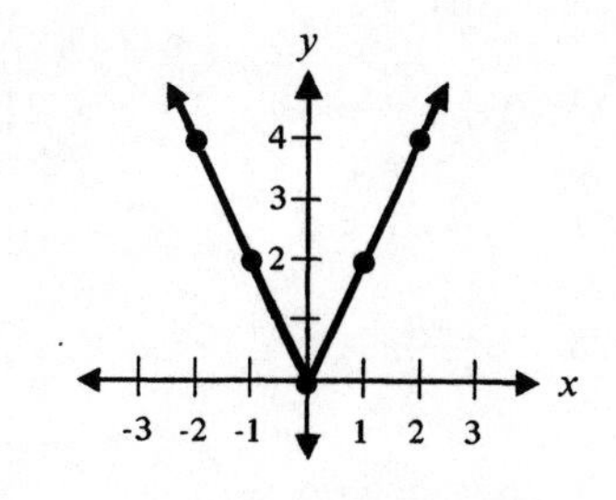

EXAMPLE 13 Graph $y = |x| - 3$.

Make a table of values.

x	$y = \|x\| - 3$
0	-3
2	-1
3	0
-3	0
-2	-1

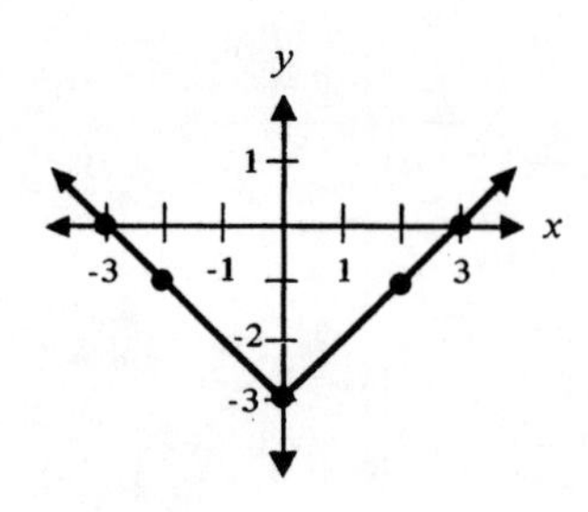

EXAMPLE 14 Graph $y = |2x| - 3$.

Make a table.

x	$y = \|2x\| - 3$
0	-3
1	-1
2	1
-2	1
-1	-1

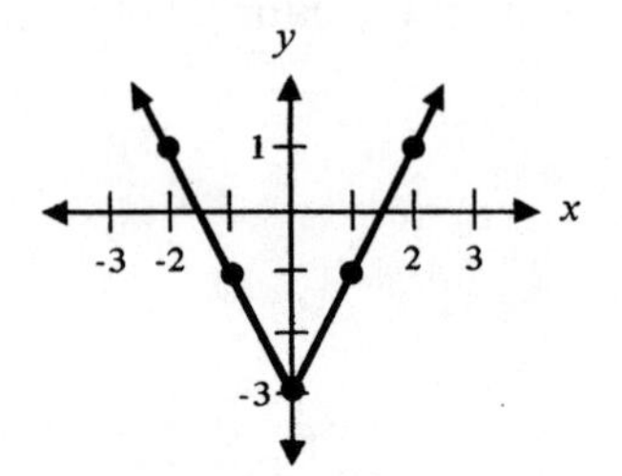

Notice from Examples 13 and 14 that changing the coefficient of x in an absolute value graph changes the slope of the characteristic V. A constant term added outside the absolute value term raises or lowers the V on the graph, as illustrated by Examples 12 and 14.

PROBLEM SET 15.2

Solve the following absolute value equations and graph on a number line.

1. $|x| = 1$
2. $|x| = 5$
3. $|x| = 2$
4. $|x| = 7$
5. $|x| + 2 = 3$
6. $|x| + 1 = 4$
7. $|x| - 1 = 1$
8. $|x| - 3 = 6$
9. $|x| - 4 = 1$
10. $|x| - 2 = 6$
11. $|x| + 3 = 4$
12. $|x| + 4 = 4$
13. $|x| - 1 = 5$
14. $|x| - 3 = 3$

Graph the following absolute value inequalities on a number line.

15. $|x| > 6$
16. $|x| > 2$
17. $|x| \geqslant 1$
18. $|x| \geqslant 5$
19. $|x| < 4$
20. $|x| > 4$
21. $|x| \leqslant 3$
22. $|x| \leqslant 8$
23. $|x| + 2 < 4$
24. $|x| + 1 < 5$
25. $|x| - 3 < 2$
26. $|x| - 2 < 3$
27. $|x| + 3 > 5$
28. $|x| + 6 > 8$
29. $|x| + 3 < 6$
30. $|x| + 4 \leqslant 8$
31. $|x| - 1 \leqslant 5$
32. $|x| - 4 \leqslant 3$
33. $|x| + 1 \geqslant 4$
34. $|x| + 2 \geqslant 5$
35. $|x| - 3 \geqslant 5$
36. $|x| - 5 \geqslant 3$

Graph the following absolute value equations.

37. $y = |x| + 1$ 38. $y = |x| + 3$ 39. $y = |x| - 4$ 40. $y = |x| - 3$

41. $y = |3x|$ 42. $y = |2x|$ 43. $y = \left|\frac{1}{2}x\right|$ 44. $y = \left|\frac{1}{3}x\right|$

45. $y = |2x| - 1$ 46. $y = |3x| - 5$ 47. $y = \left|\frac{1}{3}x\right| + 2$ 48. $y = \left|\frac{1}{2}x\right| + 3$

15.3 SOLVING FIRST-DEGREE INEQUALITIES IN ONE VARIABLE

The rules for solving inequalities are very similar to the rules for solving equations. The basic rule remains: *What you do to one side of the inequality you must do to the other.*

To develop the rules for the solution of inequalities, we will use three true inequalities as examples.

$3 < 5 \qquad -2 < 3 \qquad -6 < -4$

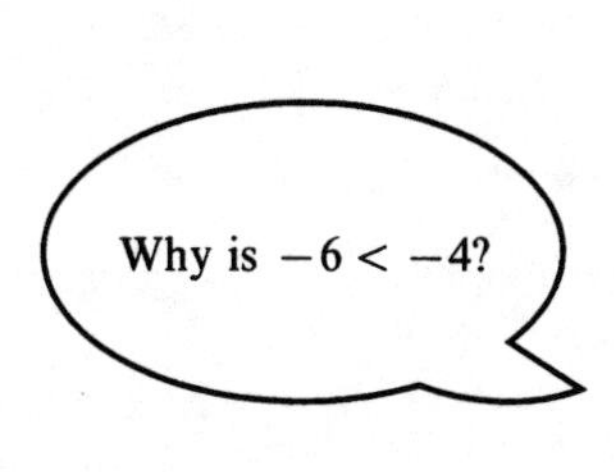

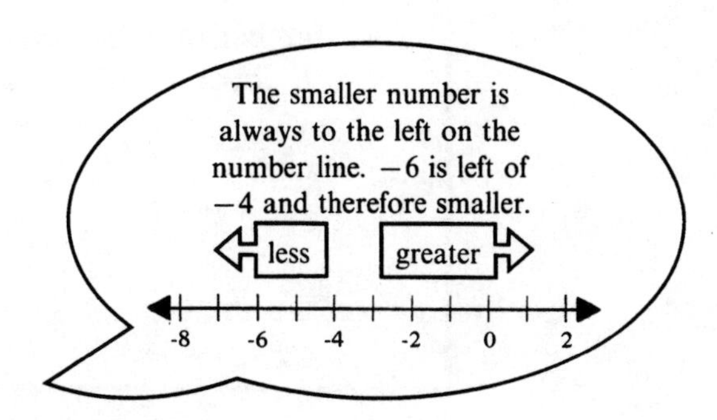

A positive number may be added to both sides of an inequality.

Next we have added +2 to both sides of all three inequalities.

EXAMPLE 15

$$\begin{array}{ccc} 3 < 5 & -2 < 3 & -6 < -4 \\ 3 + 2 < 5 + 2 & -2 + 2 < 3 + 2 & -6 + 2 < -4 + 2 \\ 5 < 7 & 0 < 5 & -4 < -2 \end{array}$$

A negative number may also be added to both sides of an inequality.

Next we have added −2 to both sides of all three inequalities.

EXAMPLE 16

$$\begin{array}{ccc} 3 < 5 & -2 < 3 & -6 < -4 \\ 3 + (-2) < 5 + (-2) & -2 + (-2) < 3 + (-2) & -6 + (-2) < -4 + (-2) \\ 1 < 3 & -4 < 1 & -8 < -6 \end{array}$$

A positive or negative number may be subtracted from both sides of an inequality.

EXAMPLE 17

Subtracting $+2$ from each inequality,

$3 < 5$	$-2 < 3$	$-6 < -4$
$3 - 2 < 5 - 2$	$-2 - 2 < 3 - 2$	$-6 - 2 < -4 - 2$
$1 < 3$	$-4 < 1$	$-8 < -6$

Subtracting -2 from each inequality,

$3 < 5$	$-2 < 3$	$-6 < -4$
$3 - (-2) < 5 - (-2)$	$-2 - (-2) < 3 - (-2)$	$-6 - (-2) < -4 - (-2)$
$5 < 7$	$0 < 5$	$-4 < -2$

Addition and subtraction law for inequalities

15.3A ADDITION AND SUBTRACTION LAW FOR INEQUALITIES

Any number may be added to or subtracted from both sides of an inequality and the direction of the inequality will remain the same.

If $a < b$, then

$$a + c < b + c$$

and

$$a - c < b - c$$

EXAMPLE 18 Solve $x + 3 < 2$ for x.

$$x + 3 < 2$$

$$x + 3 - 3 < 2 - 3 \quad \text{Subtract 3 from both sides}$$

$$x < -1$$

EXAMPLE 19 Solve $2x - 7 \leqslant x + 5$ for x.

$$2x - 7 \leqslant x + 5$$

$$2x \leqslant x + 12 \quad \text{Add 7 to both sides}$$

$$x \leqslant 12 \quad \text{Subtract } x \text{ from both sides}$$

EXAMPLE 20 Solve $9 - 5x > 3 - 6x$ for x.

$$9 - 5x > 3 - 6x$$

$$9 + x > 3 \quad \text{Add } 6x \text{ to both sides}$$

$$x > -6 \quad \text{Subtract 9 from both sides}$$

Multiplication and division are almost as easy.

EXAMPLE 21 Consider the true inequality $3 < 5$.

Multiply both sides by $+2$	Multiply both sides by (-2)
$3 < 5$	$3 < 5$
$3(2) \; ? \; 5(2)$	$3(-2) \; ? \; 5(-2)$
$6 < 10$	$-6 > -10$
Direction of inequality sign unchanged	Direction of inequality sign changed

EXAMPLE 22 Consider the true inequality $-2 < 3$.

Multiply both sides by $+2$	Multiply both sides by (-2)
$-2 < 3$	$-2 < 3$
$-2(2) \; ? \; 3(2)$	$-2(-2) \; ? \; 3(-2)$
$-4 < 6$	$4 > -6$

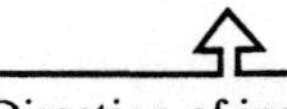

Direction of inequality sign unchanged	Direction of inequality sign changed

EXAMPLE 23 Consider $10 > 4$.

Divide both sides by $+2$	Divide both sides by (-2)
$10 > 4$	$10 > 4$
$\frac{10}{2} \; ? \; \frac{4}{2}$	$\frac{10}{-2} \; ? \; \frac{4}{-2}$
$5 > 2$	$-5 < -2$

Direction of inequality sign unchanged	Direction of inequality sign changed

Notice whenever an inequality is multiplied or divided by a *positive* number, the direction of the inequality remains the *same*. However, multiplying or dividing an inequality by a *negative* number *reverses* the direction of the inequality symbol.

Multiplication and division law for inequalities

15.3B MULTIPLICATION AND DIVISION LAW FOR INEQUALITIES

1. If both sides of a true inequality are multiplied or divided by the same positive number, n, another true inequality in the same direction is produced.

   If $a < b$, then

   $$a \cdot (n) < b \cdot (n)$$

   and

   $$\frac{a}{n} < \frac{b}{n} \quad \text{where } n \text{ is a positive number}$$

2. If both sides of a true inequality are multiplied or divided by the same negative number, $-n$, another true inequality in the *opposite* direction is produced.

   If $a < b$, then

   $$a(-n) > b(-n)$$

   and

   $$\frac{a}{(-n)} > \frac{b}{(-n)} \quad \text{where } (-n) \text{ is a negative number}$$

EXAMPLE 24 Solve $2x < 4$.

$$2x < 4$$

$$\frac{2x}{2} < \frac{4}{2} \quad \text{Divide both sides by } +2$$

$$x < 2$$

EXAMPLE 25 Solve $-2x < 4$.

$$-2x < 4$$

$$\frac{-2x}{-2} > \frac{4}{-2} \quad \text{Divide both sides by } -2\text{, change the direction of the inequality sign}$$

$x >$ ______

-2

Notice the direction of the inequality sign changes as soon as both sides are divided by a negative number.

EXAMPLE 26 Solve $\frac{x}{3} > 4$.

$$\frac{x}{3} > 4$$

$$(3)\frac{x}{3} > 4(3) \qquad \text{Multiply both sides by } +3$$

$$x > 12$$

EXAMPLE 27 Solve $\frac{x}{-3} > 4$.

$$\frac{x}{-3} > 4$$

$$(-3)\frac{x}{-3} \;\_\_\_\_\_\; 4(-3) \qquad \text{Multiply by } -3\text{, change direction of inequality}$$

$$x < -12$$

EXAMPLE 28 Solve $2x + 3 \geqslant 7$.

$2x + 3 \geqslant 7$	
$2x \geqslant 4$	Subtract 3
$x \geqslant 2$	Divide by 2

EXAMPLE 29 Solve $6 - 5x > 3x + 14$.

$6 - 5x > 3x + 14$	
$6 > 8x + 14$	Add $5x$
$-8 > 8x$	Subtract 14
$\frac{-8}{8} > \frac{8x}{8}$	Divide by $+8$
$-1 > x$	

What would happen if I subtracted $3x$ first?

$6 - 5x > 3x + 14$	
$6 - 8x > 14$	Subtract $3x$
$-8x > 8$	Subtract 6
$\frac{-8x}{-8} < \frac{8}{-8}$	Divide by -8
$x < -1$	

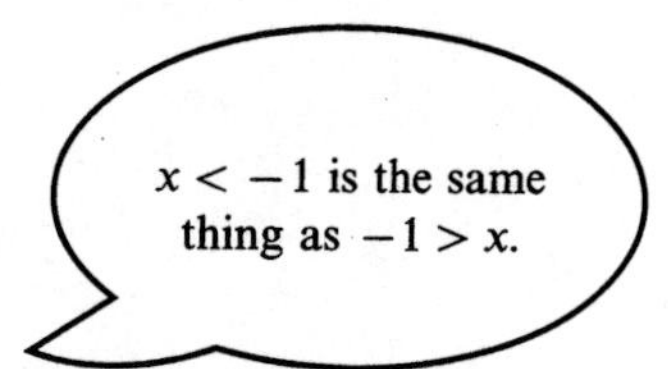

PROBLEM SET 15.3

Solve the following inequalities.

1. $x - 3 < 1$
2. $x - 5 < 6$
3. $x - 2 > 1$
4. $x - 5 > 6$
5. $x + 2 < 4$
6. $x + 3 < 6$
7. $x + 4 > 5$
8. $x + 6 > 5$
9. $x - 4 \leqslant 8$
10. $x - 5 \leqslant 10$
11. $x + 3 \geqslant 7$
12. $x + 4 \geqslant 6$
13. $2x - 5 < 7$
14. $2x + 3 < 9$
15. $3x + 4 \geqslant 10$
16. $3x - 2 \geqslant 7$
17. $2x + 7 < x - 6$
18. $3x - 4 < 2x - 1$
19. $4x - 3 \leqslant 8 + 3x$
20. $8x - 4 \leqslant 6 + 7x$
21. $6x - 3 > 5 + 5x$
22. $4x - 12 > 6 + 3x$
23. $4x - 3 \geqslant 2 + 3x$
24. $7x - 1 \geqslant 6x + 4$
25. $8x - 7 < 6x + 5$
26. $9x - 6 < 6x + 9$
27. $-6x + 3 > -4x + 7$
28. $-8x - 4 > -10x + 6$
29. $x - 7 \leqslant 4x + 2$
30. $-2x - 9 \geqslant 3x + 1$
31. $\frac{x}{4} > 9$
32. $\frac{x}{3} \leqslant 12$
33. $\frac{-x}{2} \geqslant 11$
34. $\frac{-x}{4} \leqslant 8$
35. $\frac{x}{4} - 3 < 1$
36. $\frac{x}{5} + 4 > 6$
37. $\frac{x}{4} - 2 \leqslant \frac{3}{4}x + 5$
38. $\frac{5x}{6} - 5 \geqslant \frac{x}{6} - 9$
39. $3 - 4x \leqslant x + 5$
40. $8 - 5x < 3x - 6$
41. $\frac{-x}{3} + 4 \geqslant 3$
42. $\frac{-x}{4} - 5 \leqslant 6$
43. $4 < 3 + \frac{x}{3}$
44. $6 \geqslant -1 - \frac{x}{5}$
45. $5 > 5 - \frac{x}{2}$
46. $8 \leqslant -3 + \frac{x}{6}$
47. $-2x + 7 \leqslant 5x - 3$
48. $-3x - 9 \geqslant -x + 7$
49. $4x + 3 \leqslant 9x - 7$

15.4 GRAPHING LINEAR INEQUALITIES

Just as equations in two variables have an infinite number of ordered pairs that make them true, so do inequalities in two variables.

An inequality in two variables divides the graph into two regions. All points whose coordinates make the inequality true are in the shaded region, and all points that make the inequality false are in the unshaded region.

EXAMPLE 30 Graph $y \geqslant 3x - 4$.

First, graph the line $y = 3x - 4$.

Every point on the line satisfies the inequality because the inequality includes the case $y = 3x - 4$.

Every point above the line $y = 3x - 4$ satisfies the inequality because these are the points where $y > 3x - 4$.

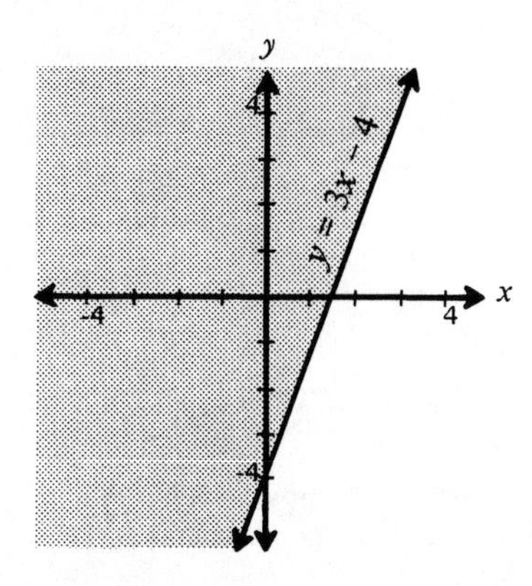

You can think of an equation as a boundary or fence between two portions of the graph. The only question is who owns the fence. If the inequality includes an equal sign, the line of the equation is included in the graph. If the inequality does not contain an equal sign, represent the line with a dashed line to show it is not included in the shaded area.

How can you test if the graph is shaded correctly?

Try a few points on either side of the line.

Value	Region	Substitution	Inequality	Truth
		$y \geqslant 3x - 4$		
$(-4, 0)$	shaded	$0 \geqslant 3(-4) - 4$	$0 \geqslant -16$	True
$(-2, 5)$	shaded	$5 \geqslant 3(-2) - 4$	$5 \geqslant -10$	True
$(0, 0)$	shaded	$0 \geqslant 3(0) - 4$	$0 \geqslant -4$	________
$(2, 2)$	on line	$2 \geqslant 3(2) - 4$	$2 \geqslant 2$	True
$(4, 0)$	unshaded	$0 \geqslant 3(4) - 4$	$0 \geqslant 8$	False
$(5, -6)$	unshaded	$-6 \geqslant 3(5) - 4$	$-6 \geqslant 11$	________

True

False

The shaded portion of the graph represents the points that make the inequality true.

EXAMPLE 31 Graph $y < -2x + 3$.

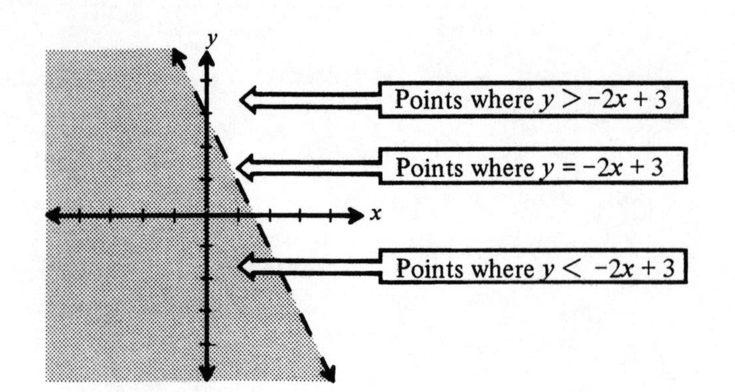

How do I know which side of the line to shade?

First draw the line, then pick any point not on the line. (0, 0) is the easiest. If it satisfies the inequality, shade that side. If it doesn't, shade the other side.

EXAMPLE 32 Graph $y > \frac{1}{2}x - 2$.

Step 1

Draw the line $y = \frac{1}{2}x - 2$.

Since the inequality does not contain an equal sign, use a dashed line for the graph of the line.

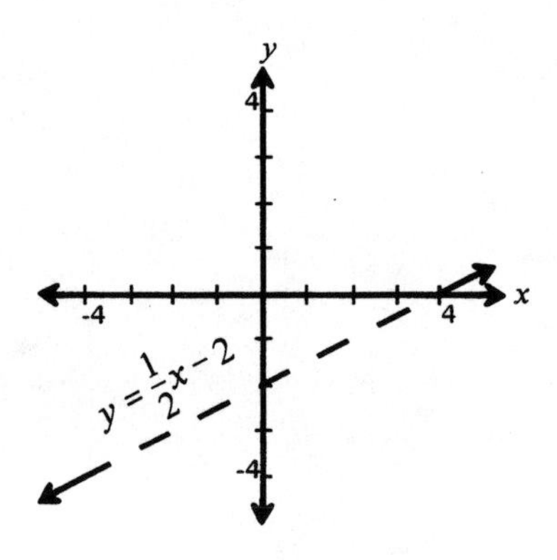

Step 2

Try (0, 0) in the inequality.

$$y > \frac{1}{2}x - 2$$

$$0 > \frac{1}{2}(0) - 2$$

$$0 > -2 \qquad \textit{True!}$$

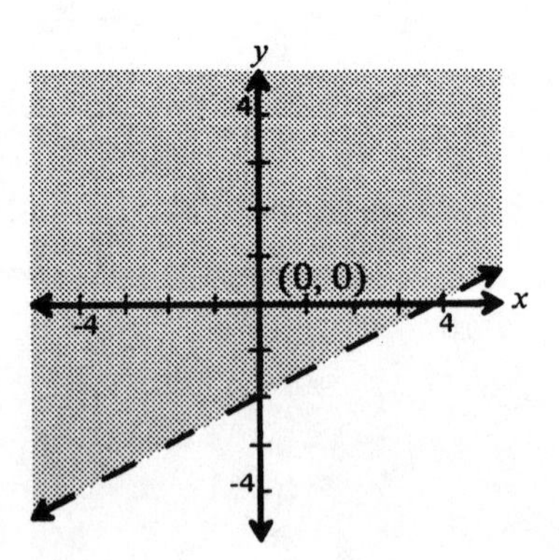

OK, then shade the side that includes (0, 0).

EXAMPLE 33 Graph $x \leqslant 4$.

Step 1

Graph $x = 4$.

Since the inequality includes an equal sign, draw a solid line.

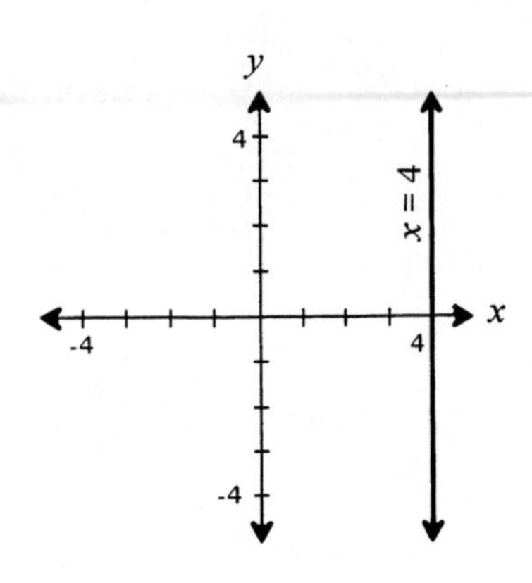

Step 2

Test (0, 0) in the inequality.

$x \leqslant 4$

$0 \leqslant 4$ is *true*

Therefore, shade the side that contains (0, 0).

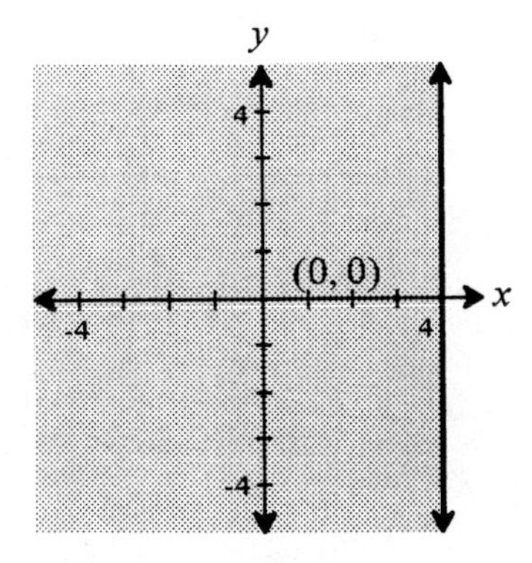

EXAMPLE 34 Graph $y < -2$.

Step 1

Graph $y = -2$.

Since no equal sign is included in the inequality, draw a dashed line.

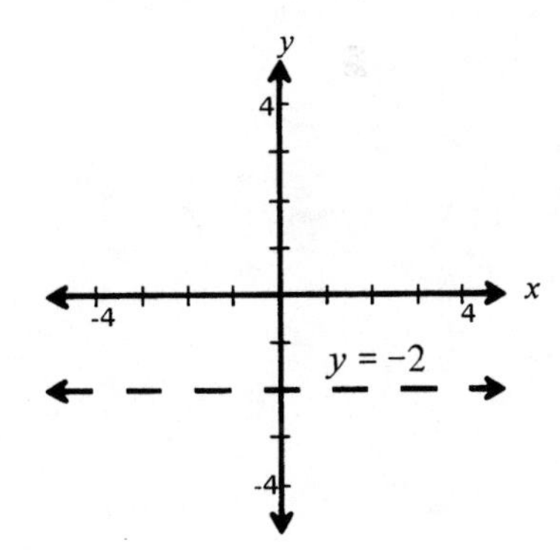

Step 2

Try (0, 0) in the inequality.

$y < -2$

$0 < -2$ is *false*

Therefore, shade the side that does not contain (0, 0).

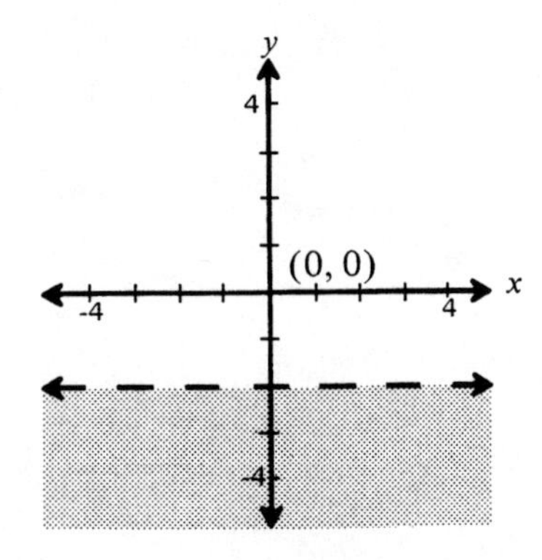

What if (0, 0) falls on the line?

No big thing. Just pick another point and use it to test.

PROBLEM SET 15.4

Graph the following linear inequalities.

1. $y > x$
2. $y > x + 3$
3. $y < x - 3$
4. $y < x - 2$
5. $y \geqslant x - 2$
6. $y \geqslant x + 1$
7. $y < x + 1$
8. $y > x + 2$
9. $y \leqslant x + 2$
10. $y \leqslant x$
11. $y > 0$
12. $y > -2$
13. $y \leqslant 3$
14. $x \geqslant 0$
15. $x \leqslant -1$
16. $y \geqslant -3$
17. $y > 2x + 2$
18. $y \geqslant 3x + 2$
19. $y \leqslant 2x - 4$
20. $y < 3x - 3$
21. $y \geqslant -3x + 4$
22. $y > -2x + 3$
23. $y < -4x - 2$
24. $y \leqslant -3x + 4$
25. $y \leqslant \frac{1}{2}x$
26. $y < \frac{1}{2}x + 1$
27. $y > \frac{1}{3}x - 1$
28. $y \geqslant \frac{1}{3}x + 2$
29. $y < \frac{-1}{2}x + 1$
30. $y \leqslant \frac{-1}{2}x$
31. $y \geqslant \frac{-2}{3}x - 4$
32. $y > \frac{-2}{3}x + 3$
33. $y > -3x + 5$
34. $y \geqslant -4x - 1$
35. $x \geqslant -2$
36. $x < 3$
37. $y \leqslant \frac{-3}{4}x + 2$
38. $y < \frac{-2}{3}x - 1$
39. $y > 4 - \frac{2}{3}x$
40. $y \geqslant 3 - \frac{3}{4}x$

Review 15

NOW THAT YOU HAVE COMPLETED UNIT 15, YOU SHOULD BE ABLE TO:

Use the following definitions in your working vocabulary:

Inequality—A relationship between quantities that are not equal

Trichotomy axiom—For any real numbers a, b, and c, exactly one of the following is true:

$$a < b \qquad a = b \qquad a > b$$

Transitive property of inequality—For any real numbers a, b, and c,

if $a < b$ and $b < c$, then $a < c$

Absolute value—For every number a,

if $a \geqslant 0$ then $|a| = a$
if $a < 0$ then $|a| = -a$

Apply the following rules, definitions, and procedures:

To graph an inequality in one variable on a number line, determine the points on the line that make the inequality true, then shade that part of the number line.

To graph an absolute value or absolute value inequality on a number line, determine the points on the line that make the absolute value or absolute value inequality true, then shade that part of the line.

To graph an absolute value equation in two variables:

1. Make a table of values.
2. Plot the points.
3. Draw a V-shaped line through the points.

To solve a first-degree inequality in one variable, use the following laws:

1. Any number may be added or subtracted from both sides of an inequality and the direction of the inequality will remain the same.

   If $a < b$, then

   $$a + c < b + c$$

   and

   $$a - c < b - c$$

2. If both sides of a true inequality are multiplied or divided by the same positive number, n, another true inequality in the same direction is produced.

   If $a < b$, then

   $$a \cdot (n) < b \cdot (n)$$

   and

   $$\frac{a}{n} < \frac{b}{n} \quad \text{where } n \text{ is a positive number}$$

3. If both sides of a true inequality are multiplied or divided by the same negative number, $-n$, another true inequality in the opposite direction is produced.

   If $a < b$, then

   $$a \cdot (-n) > b \cdot (-n)$$

   and

   $$\frac{a}{(-n)} > \frac{b}{(-n)} \quad \text{where } (-n) \text{ is a negative number}$$

To graph an inequality in two variables:

1. Graph the line that determines the equality. If the inequality includes an equal sign, draw a solid line; otherwise, use a broken line.
2. After trying a few points on either side of the line, shade the portion of the graph that includes the points that make the inequality true.

Review Test 15

Graph the following on a number line. (15.1)

1. $x > -4$
2. $-6 \leq x$
3. $-3 \leq x < 2$
4. $-6 < x \leq -1$

Graph the following on a number line. (15.2)

5. $|x| < 3$
6. $|x| - 3 \geq 4$

Solve the following absolute value equations and graph on a number line. (15.2)

7. $|x| = 3$
8. $|x| - 5 = 2$

Graph the following absolute value equations. (15.2)

9. $y = |x| - 2$

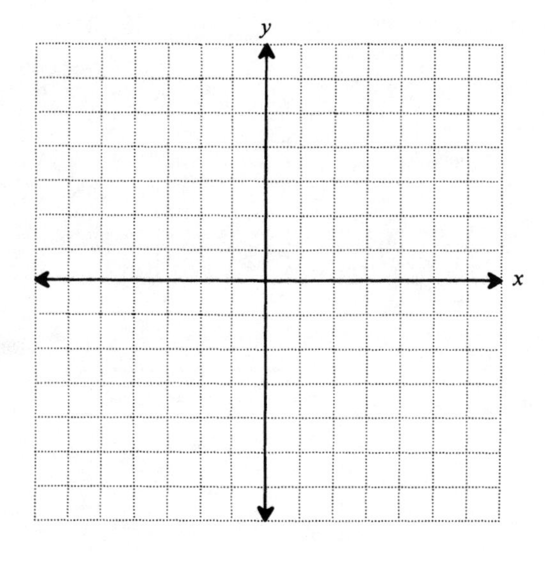

10. $y = |2x| + 3$

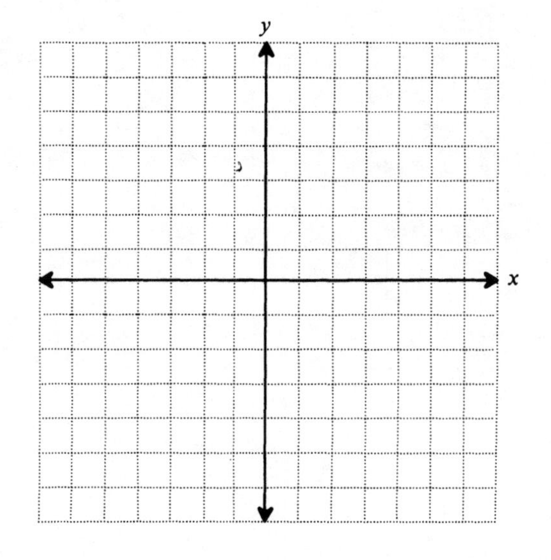

Solve the following inequality equations. (15.3)

11. $x + 4 < 1$
12. $5 - x > 3$
13. $6x - 2 \leq 3x + 4$
14. $\frac{-x}{3} \geq 12$
15. $-6x - 5 < -2x + 3$
16. $5 - 8x > 7 - 3x$

Graph the following linear inequalities. (15.4)

17. $y > 2x + 3$

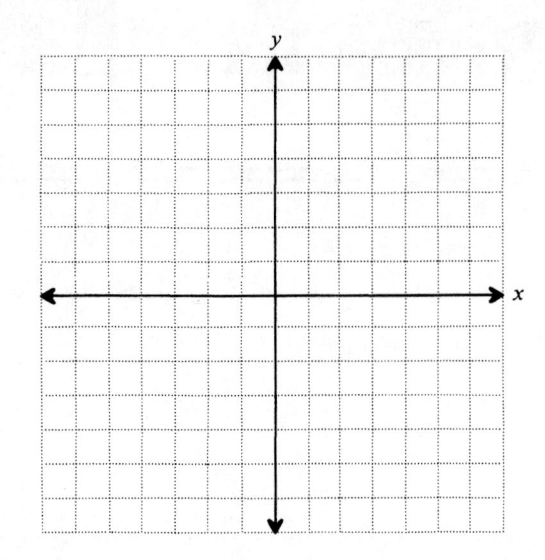

18. $y \leqslant -3x + 5$

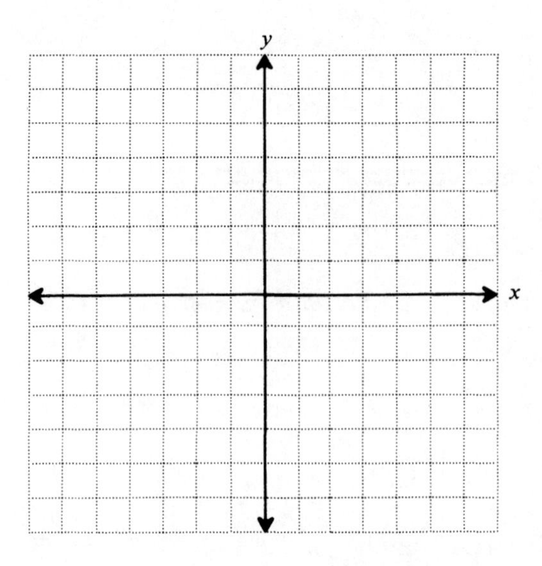

19. $y \leqslant \frac{1}{2}x - 4$

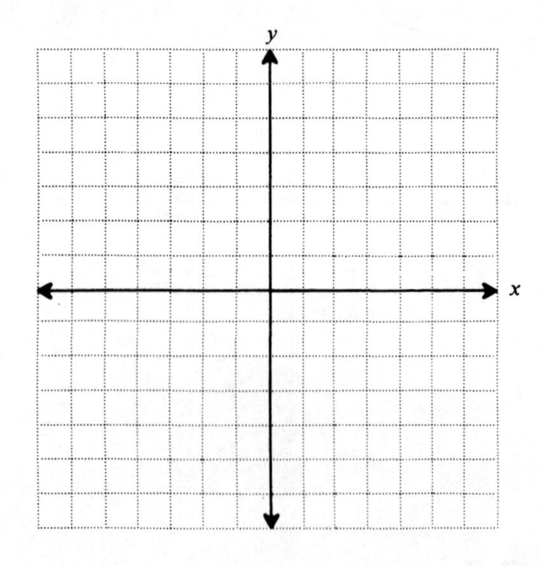

20. $x < -2$

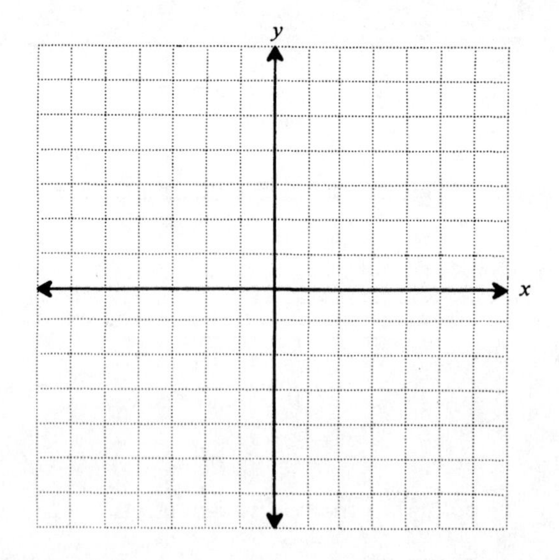

REVIEW YOUR SKILLS

Perform the indicated operations.

21. $\dfrac{4x}{4x-5} - 1$

22. $\dfrac{\dfrac{1}{b} - 1}{1 - \dfrac{1}{b^2}}$

Solve for the variable.

23. $\dfrac{4x-3}{4} - \dfrac{x-5}{2} = 4$

Simplify the following.

24. $\sqrt{75}$

25. $\dfrac{1}{\sqrt{8}}$

26. $\sqrt{\dfrac{18x^3y^5}{2a^3}}$

Simplify the following multiplications and divisions.

27. $(\sqrt{6} - \sqrt{3})(2\sqrt{6} + \sqrt{3})$

28. $\dfrac{2 - \sqrt{x}}{3 + \sqrt{x}}$

Cumulative Review 1–15

Perform any indicated operations and simplify.

1. $-4 + (3 - 5) - 2$ ***(1.4)***
2. $7 - (-2) \cdot (-5)$ ***(1.4)***
3. $(2a^2 + 3a - 6) + (a^2 - 5a - 4)$ ***(2.4)***
4. $(a^2 - 3a + 1) - (2a^2 + a - 4)$ ***(2.4)***
5. $\dfrac{-y}{6} - \dfrac{3y}{10}$ ***(3.3)***
6. $\dfrac{21x^2}{20} \div \dfrac{7x}{30}$ ***(3.2)***
7. $(-3a^{-2}b^3)(2a^2b^3)$ ***(8.1)***
8. $(2st^2)^3$ ***(8.2)***
9. $\dfrac{2x^2y}{x^3y^4}$ ***(8.1)***
10. $\dfrac{10x^4 - 6x^3 + 8x}{2x}$ ***(9.3)***
11. $3x(x^2 + 2x - 4)$ ***(9.1)***
12. $(2x - 3)(3x - 1)$ ***(9.2A)***
13. $(x - 2)(3x^2 + x - 4)$ ***(9.1)***
14. $(2a + 3b)^2$ ***(9.2B)***
15. $\dfrac{2x - 8}{x} \cdot \dfrac{2x^3}{x^2 - 4x}$ ***(11.2)***
16. $\dfrac{a + b}{a^2 + ab - 2b^2} \div \dfrac{a + b}{a^2b + 2ab^2}$ ***(11.3)***
17. $\dfrac{y^2 - 4}{y^2 + y - 6} \cdot \dfrac{y^2 - y - 12}{y^2 - 3y - 4}$ ***(11.2)***
18. $\dfrac{2}{x^2 + 4x + 3} - \dfrac{1}{x^2 + 5x + 6}$ ***(12.2)***
19. $\dfrac{1}{2x + 1} + 2$ ***(12.2)***
20. $\dfrac{1 - \dfrac{1}{a}}{a - \dfrac{1}{a}}$ ***(12.3)***
21. $\sqrt{\dfrac{18x^3}{y}}$ ***(13.2)***
22. $\dfrac{4}{2 - \sqrt{2}}$ ***(13.4B)***
23. $3\sqrt{3} - \sqrt{12}$ ***(13.3)***
24. $(2\sqrt{5} + \sqrt{2})(\sqrt{5} - 3\sqrt{2})$ ***(13.4A)***

Factor the following.

25. $x^2 + x - 6$ ***(10.2)***
26. $6x^2 - 5x - 4$ ***(10.3)***
27. $3y^2 - 27$ ***(10.4)***
28. $2x^4 + 7x^3 + 6x^2$ ***(10.4)***

Solve for the variable.

29. $3x + 1 = 6x - 11$ ***(4.2)***
30. $5(x + 3) = 2 - (x - 1)$ ***(4.2)***
31. $y = mx + b$ for x ***(4.3)***
32. $\dfrac{2}{x} - \dfrac{5}{3x} = \dfrac{1}{3}$ ***(12.4)***
33. $x^2 - 2x - 15 = 0$ ***(14.1)***
34. $6x^2 - x - 12 = 0$ ***(14.1)***
35. $x^2 - 3x + 1 = 0$ ***(14.2 or 14.3)***
36. $1 - 2x \geqslant 10 + x$ ***(15.3)***

37. $P = 2l + 2w$ is the formula for the perimeter of a rectangle. Find the width (w) of a rectangle with a perimeter (P) of 50 centimeters and a length (l) of 12 centimeters. *(4.5)*

38. The sum of a number and twice the number is -39. Find the number. *(5.1)*

39. Ace and Irving take off from the Blythe Airport at the same time but in opposite directions. After 3 hours they are 1365 miles apart. If Ace's plane travels 45 mph slower than Irving's, what is the speed of each plane? *(5.5)*

40. If $f(x) = 2x^2 - 1$, then $f(-3) =$ ______. *(6.1)*

41. Find the equation of the line with slope $\frac{-2}{5}$ and through the point $(5, -3)$. *(6.4B)*

42. Find the equation of the line through the points $(3, -2)$ and $(5, -6)$. *(6.4C)*

43. Sketch the graph. *(6.5)*

$$y = \frac{-2}{5}x - 1$$

44. Solve by graphing. *(7.1)*

$$2x + 3y = 12$$
$$2x - y = 4$$

Solve the following systems.

45. $x + 5y = -5$ *(7.3)*
    $y = 2x - 12$

46. $2x + 3y = -3$ *(7.2)*
    $3x + 5y = -4$

47. A woman has 24 coins in her purse consisting of dimes and quarters. If the total value of the coins is \$4.50, how many of each coin does she have? *(7.4)*

48. Graph on a number line. $|x| - 4 < -2$ ⟵⟶ *(15.2)*

49. Graph $y < \frac{1}{3}x - 2$. *(15.4)*

50. Graph $y = x^2 + 2x + 1$. *(14.5)*

Appendix

An Alternate Approach to Factoring Trinomials

This appendix provides an alternate approach to factoring trinomials. The numbering of the sections and the examples corresponds to the numbering used in Unit 10. Answers to the problem sets are located after the answers to Cumulative Review Units 1–15 in the answers section.

A10.2 FACTORING TRINOMIALS

To factor a trinomial, we do the process of multiplication backward. In our earlier work, we multiplied two binomials. Let's look carefully at how we got the result.

EXAMPLE 1

$$(x + 2)(x + 3) = x^2 + 5x + 6$$
$$(x + 4)(x + 5) = x^2 + 9x + 20$$
$$(x - 4)(x - 5) = x^2 - 9x + 20$$

Notice the form of the binomials. Each is (x + number). We could say the products look like this:

$$(x + a)(x + b) = x^2 + ax + bx + ab$$

ax

bx

$ax + bx$

Let's examine the product of two binomials term by term, paying close attention to how the FOIL method of multiplying works.

First

The first term is always x^2. It results from the product of the first two terms in each binomial factor.

$$(x + \quad)(x + \quad) = x^2 + \underline{\quad\quad} + \underline{\quad\quad}$$

If both trinomials have a first term of x, the resulting product will start with x^2.

Last

The last term is the product of the last terms in each binomial.

$$(\quad + 2)(\quad + 3) = \underline{\qquad} + \underline{\qquad} + 6$$

In general,

$$(x + a)(x + b) = \underline{\qquad} + \underline{\qquad} + ab$$

Outside
+
Inside

The middle term is the sum of the product of the inside terms plus the product of the outside terms.

$$(x + 2)(x + 3) = \underline{\qquad} + 5x + \underline{\qquad}$$

2x
3x
5x

In general,

$$(x + a)(x + b) = x^2 + \underbrace{ax + bx} + ab$$

ax
bx

The process of determining which factors were multiplied to give a trinomial is called factoring.

EXAMPLE 2 Factor $x^2 + 7x + 10$.

First, write the trinomial followed by two empty parentheses.

$$x^2 + 7x + 10 = (\quad)(\quad)$$

The only way to get a first term of x^2 in the trinomial is if both binomials have first terms of x. Therefore, write an x in both parentheses.

$$x^2 + 7x + 10 = (x \quad)(x \quad)$$

Now, look at the last term 10. It is the product of the final terms in each binomial. Either $2 \times 5 = 10$ or $1 \times 10 = 10$.

To determine which pair of factors to use, we need to look at the middle term. It is the sum of the inner product plus the sum of the outer product.

We are looking for the factors of 10 such that their sum is 7 and their product is 10. They are 5 and 2. Write these in the parentheses.

$$x^2 + 7x + 10 = (x + 5)(x + 2)$$

In the general case,

$$(x + a)(x + b) = x^2 + ax + bx + ab$$
$$= x^2 + (a + b)x + ab$$

The last term of the trinomial is the product of the constants in the binomials.

The coefficient of the middle term is the numerical sum or difference of the constants.

EXAMPLE 3 Factor $x^2 - 7x + 10$.

Notice that this trinomial is almost identical to Example 2. The only difference is the sign of the middle term.

Write the trinomial followed by two empty parentheses.

$$x^2 - 7x + 10 = (\quad)(\quad)$$

There is only one way to get a first term of x^2, so we can fill in these factors.

$$x^2 - 7x + 10 = (x \quad)(x \quad)$$

Examine the last term. Look for factors whose product is $+10$. Because the middle term is "$-7x$," we know the sum of the factors is -7.

Since their product is positive, both factors must have the same sign.
$(-)(-) = +$
$(+)(+) = +$

If both factors have the same sign and their sum is negative, they must be negative.

The factors we are seeking are -5 and -2.

$$(-5)(-2) = +10 \qquad (-5) + (-2) = -7$$

Write the trial result and multiply to check your answer.

$$x^2 - 7x + 10 = (x - 5)(x - 2)$$

Check.

$$(x - 5)(x - 2) = x^2 - 7x + 10$$

$-5x$
$-2x$
$-7x$

Factoring requires the ability to select combinations of terms to yield a desired sum.

EXAMPLE 4 Find the proper combination of signs so that the desired sum is produced.

− − +
− + −

1.	2.	3.
$\square 3x$	$\square 3x$	$\square 3x$
$\square 2x$	$\square 2x$	$\square 2x$
$-5x$	$-x$	$+x$

\+ − +
\+ + −

4.	5.	6.
$\square 6x$	$\square 6x$	$\square 7x$
$\square 3x$	$\square 3x$	$\square 5x$
$+9x$	$-3x$	$+2x$

− + +
− − −

7.	8.	9.
$\square 10x$	$\square 12x$	$\square 12x$
$\square\ 6x$	$\square\ 7x$	$\square 15x$
$-16x$	$+5x$	$-3x$

EXAMPLE 5 Find the missing sign so that the desired product is produced.

− + −

1. $(-3)(\square 4) = 12$ 2. $(+3)(\square 4) = 12$ 3. $(+3)(\square 4) = -12$

\+ − −

4. $(+4)(\square 2) = 8$ 5. $(\square 4)(-2) = 8$ 6. $(\square 4)(+3) = -12$

EXAMPLE 6 Find two numbers whose product and sum are as follows.

−3, 5

Product $12 = \underline{\ 3\ } \cdot \underline{\ 4\ }$ Product $-15 = \underline{\quad} \cdot \underline{\quad}$

−3, 5

Sum $7 = \underline{\ 3\ } + \underline{\ 4\ }$ Sum $2 = \underline{\quad} + \underline{\quad}$

−4, 2 (−5), (−4)

Product $-8 = \underline{\quad} \cdot \underline{\quad}$ Product $20 = \underline{\quad} \cdot \underline{\quad}$

−4, 2 −5, (−4)

Sum $-2 = \underline{\quad} + \underline{\quad}$ Sum $-9 = \underline{\quad} + \underline{\quad}$

ALTERNATE PROBLEM SET 10.2A

Note: Answers to alternate problem sets are after the answers to Cumulative Review 1–15.

Find two numbers whose product and sum are as follows.

1. Product 6, Sum 5
2. Product 6, Sum −5
3. Product 2, Sum −3
4. Product 4, Sum −5
5. Product −6, Sum 1
6. Product −6, Sum −1
7. Product 15, Sum −8
8. Product 10, Sum 7
9. Product −12, Sum 1
10. Product −15, Sum −2
11. Product −18, Sum −3
12. Product −36, Sum 5
13. Product −30, Sum −1
14. Product −27, Sum 6
15. Product −33, Sum −8
16. Product −26, Sum 11
17. Product 20, Sum −9
18. Product 18, Sum −9
19. Product −48, Sum 2
20. Product −48, Sum −8
21. Product −45, Sum −4
22. Product −28, Sum 3

EXAMPLE 7 Factor $x^2 + x - 12$.

Rewrite with empty parentheses.

$$x^2 + x - 12 = (\quad)(\quad)$$

Write in the first factor.

$$x^2 + x - 12 = (x\quad)(x\quad)$$

The last term -12 is negative; therefore, one binomial has a $+$ sign and the other a $-$ sign.

Since the middle term is positive, we know the larger factor of -12 must be $+$.

Find two factors of -12 whose sum is $+1$.

$$(-3)(+4) = -12 \quad \text{and} \quad (+3)(-4) = -12$$

but only

$$(-3) + (+4) = +1$$

Write the result and check.

$$\begin{aligned} x^2 + x - 12 &= (x + 4)(x - 3) \\ &= x^2 + x - 12 \end{aligned}$$

EXAMPLE 8 Factor $x^2 - 8x + 15$.

Rewrite with empty parentheses.

$$x^2 - 8x + 15 = (\quad)(\quad)$$

Write the factors of x^2.

$$x^2 - 8x + 15 = (x\quad)(x\quad)$$

Next find two factors of $+15$ whose sum is -8. Since $+15$ is positive, we know that the signs of the factors are either both positive or both negative. We are looking for a total of -8; therefore, we want two negative factors.

$-5, -3$

The factors whose product is $+15$ with a sum of -8 are ______ and ______.

Write the result and check.

$$\begin{aligned} x^2 - 8x + 15 &= (x - 3)(x - 5) \\ &= x^2 - 8x + 15 \end{aligned}$$

EXAMPLE 9 Factor $x^2 + 8x + 15$.

Write with the factors of x^2 filled in.

$$(x^2 + 8x + 15) = (x \quad)(x \quad)$$

+5, +3

The factors of 15 with a sum of +8 are _____ and _____.

Write the result and check.

$$x^2 + 8x + 15 = (x + 3)(x + 5)$$
$$= x^2 + 8x + 15$$

EXAMPLE 10 Factor $x^2 + 2x - 15$.

Write with the factors of the first term of the trinomial filled in.

$$x^2 + 2x - 15 = (x \quad)(x \quad)$$

The constant term is −15. This means the signs of the factors will be different. We need two factors of 15 whose difference is 2. The factors are 3 and 5. Since the middle term is positive, we know the larger factor, 5, should be positive.

+5, −3

Factors of −15 whose sum is +2 are _____ and _____.

$$x^2 + 2x - 15 = (x + 5)(x - 3)$$

EXAMPLE 11 Factor $x^2 - 2x - 15$.

Fill in the first term of each binomial.

$$x^2 - 2x - 15 = (x \quad)(x \quad)$$

Notice that the middle term is negative. This means that the larger factor of the constant term −15 must be negative.

−5, 3

The two factors of −15 whose sum is −2 are _____ and _____.

Write the result.

$(x - 5)(x + 3)$

$$x^2 - 2x - 15 = (\quad)(\quad)$$

ALTERNATE PROBLEM SET 10.2B

Note: Answers to alternate problem sets are after the answers to Cumulative Review 1–15.

Factor the following trinomials. Check your answers by multiplication.

1. $x^2 + 3x + 2$
2. $x^2 - 3x + 2$
3. $x^2 - 4x + 3$
4. $x^2 + 4x + 3$
5. $x^2 + 7x + 6$
6. $x^2 - 7x + 6$
7. $x^2 + 6x + 5$
8. $x^2 + 6x + 8$
9. $x^2 - 6x + 8$
10. $x^2 + 9x + 8$
11. $x^2 - 9x + 8$
12. $x^2 + 11x + 10$
13. $x^2 - 11x + 10$
14. $x^2 - 7x + 10$
15. $x^2 + 7x + 10$
16. $x^2 + 3x - 10$
17. $x^2 - 3x - 10$
18. $x^2 + 9x - 10$
19. $x^2 - 9x - 10$
20. $x^2 - 3x - 4$
21. $a^2 - 10a - 11$
22. $a^2 + 12a - 13$
23. $x^2 - x - 2$
24. $a^2 - a - 6$
25. $x^2 + 3x - 4$
26. $y^2 + 2y - 3$
27. $x^2 + x - 20$
28. $x^2 - 11x + 30$
29. $x^2 + 8x + 7$
30. $x^2 + 8x + 15$
31. $a^2 - 7a - 18$
32. $x^2 - 12x + 35$
33. $x^2 + 2x - 8$
34. $y^2 + 3y - 18$
35. $x^2 + 11x + 28$
36. $x^2 - 11x + 24$

37. $x^2 + 2x - 24$

38. $b^2 - 14b + 24$

39. $y^2 - 13y + 30$

40. $x^2 - 12x + 32$

41. $x^2 - 8x + 12$

42. $x^2 + 13x + 40$

43. $r^2 - 3r - 18$

44. $a^2 - 7a + 12$

45. $x^2 - 13x - 30$

46. $x^2 + 8x - 48$

47. $x^2 + 12x + 20$

48. $m^2 + 12m + 27$

49. $x^2 - 2x - 24$

50. $a^2 - 15a + 54$

51. $a^2 + 14a - 51$

52. $t^2 - t - 42$

53. $x^2 + x - 72$

54. $x^2 + x - 56$

55. $x^2 - 11x - 42$

56. $x^2 + 10x - 56$

57. $b^2 - 8b - 65$

58. $x^2 - 6x - 72$

59. $z^2 + 4z - 77$

60. $x^2 - 18x + 65$

61. $y^2 - 17y + 42$

62. $x^2 - 17x + 66$

63. $x^2 - 6x - 55$

64. $x^2 + 15x + 44$

65. $a^2 - 15a + 26$

66. $x^2 - 18x + 45$

A10.3 FACTORING MORE DIFFICULT TRINOMIALS

In the previous section, the trinomials you were asked to factor had an x^2 term whose coefficient was 1. This is not always the case.

EXAMPLE 12 Factor $3x^2 + 7x - 20$.

Examine the coefficients of the first and the third terms of the trinomial. If either of these terms has only one set of factors, place these factors in parentheses.

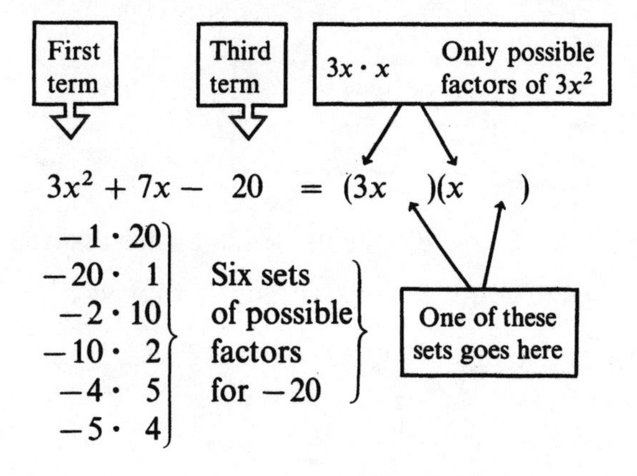

If we try all six sets of possible factors for the third term, one of them may work.

We can usually avoid having to test all possible sets of factors if we make a few observations.

Consider the trinomial $ax^2 + bx + c$.

Observation: To get a positive product, both factors must be positive or both factors must be negative.

Conclusion: If the sign of the third term in the trinomial is positive, the signs in both binomials must be alike. That is, they are both positive or both negative.

$$(+)(+) \Rightarrow \underline{A}\,x^2 \quad \underline{B}\;x + \underline{C}$$

$$\text{or} \quad (-)(-) \Rightarrow \underline{A}\,x^2 \quad \underline{B}\;x + \underline{C}$$

Observation: When you use the FOIL method to multiply binomials, the middle term of the trinomial is the sum of the inner and outer products.

Conclusion: If the signs in both binomials are alike, the sign of the middle term in the trinomial will be the same as the signs of the binomials.

$$(+)(+) \Rightarrow \underline{A}\,x^2 + \underline{B}\,x + \underline{C}$$

$$\text{and} \quad (-)(-) \Rightarrow \underline{A}\,x^2 - \underline{B}\,x + \underline{C}$$

Observation: The only way to get a negative product is for one factor to be positive and the other to be negative.

Conclusion: If the sign of the third term of the trinomial is negative, the sign in one binomial must be positive and the sign in the other binomial must be negative.

$$(+)(-) \Rightarrow \underline{A}\,x^2 \quad \underline{B}\,x - \underline{C}$$

$$(-)(+) \Rightarrow \underline{A}\,x^2 \quad \underline{B}\,x - \underline{C}$$

We can't tell this sign from the information we have.

In the case of $3x^2 + 7x - 20$, -20 tells us the signs are different in the factored form. Experiment with possible combinations of factors that yield -20. Try $+1$ and -20.

$$3x^2 + 7x - 20 \stackrel{?}{=} (3x + 1)(x - 20)$$

$$\begin{array}{r} 1x \\ -60x \\ \hline -59x \end{array}$$

The numerical difference is not the middle term.

We need a middle term of $+7x$. Therefore, try reversing the signs in the factored form.

Try -1 and $+20$.

$$3x^2 + 7x - 20 \stackrel{?}{=} (3x - 1)(x + 20)$$

$$\begin{array}{r} -1x \\ +60x \\ \hline +59x \end{array}$$

Middle term has the correct sign, but it is too large. Try smaller factors of -20 such as -4 and $+5$.

Try -4 and $+5$.

$$3x^2 + 7x - 20 \stackrel{?}{=} (3x - 4)(x + 5)$$

$$\begin{array}{r} -4x \\ +15x \\ \hline +11x \end{array}$$

The middle term is still too large. Reverse the factors.

Try $+5$ and -4.

$$3x^2 + 7x - 20 \stackrel{?}{=} (3x + 5)(x - 4)$$

$$\begin{array}{r} 5x \\ -12x \\ \hline -7x \end{array}$$

Because the last term of the trinomial is negative, we know the middle term is the numerical *difference* of the inner and outer products.

If the last term of the trinomial is positive, the middle term will be the numerical *sum* of the inner and outer products.

The bubbles imply that you may test possible arrangements of factors without signs until you find the correct sum or difference to produce the middle term. Then you can arrange the signs in the binomials to get the correct sign for the middle term.

The middle term is the right magnitude but has the wrong sign. Reverse signs.

$$3x^2 + 7x - 20 \stackrel{?}{=} (3x - 5)(x + 4)$$

$$\begin{array}{r} -5x \\ 12x \\ \hline +7x \end{array}$$

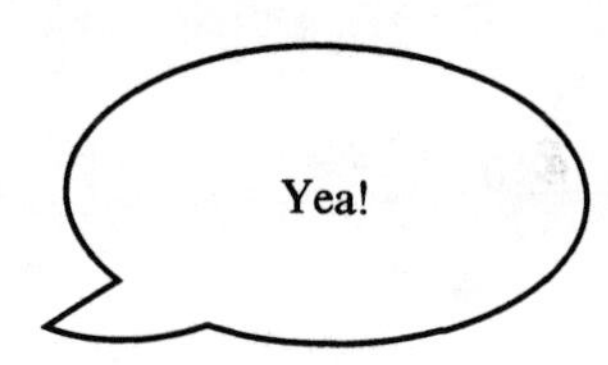

Multiply to check the result.

$$(3x - 5)(x + 4) = 3x^2 + 7x - 20$$

A POINTER FOR EASIER FACTORING

Consider

$$(3x - 5)(x + 4) = 3x^2 + 7x - 20$$

Look at the inner and outer products of the binomials above.

They are

$-5x$	inner product
$+12x$	outer product
$+7x$	middle term of the trinomial

To get a positive middle term in the trinomial, the product with the larger magnitude must be positive.

$+12x$ has the larger magnitude. It must be positive to produce a positive middle term.

Notice that if you reverse the signs in the binomial, you get a similar trinomial product. However, the sign of the middle term of the trinomial is reversed.

$$(3x - 5)(x + 4) = 3x^2 + 7x - 20$$

Changing these signs — Reverses this sign

$$(3x + 5)(x - 4) = 3x^2 - 7x - 20$$

In each case, the sign of the second term in the trinomial depends on the magnitude of the inner and outer products. Since the second term is the sum of the inner and outer products, the sign of the second term is the same as the sign of the product with the larger magnitude.

EXAMPLE 13 Factor $7x^2 - 71x + 10$.

Notice that the only way to get a product of $7x^2$ with positive factors is $7x \cdot x$. However, there are several different combinations whose product is $+10$. Therefore, fill in the x terms.

$$7x^2 - 71x + 10 = (7x \quad)(x \quad)$$

Examine the signs.

+sign in 3rd term ⇒ Signs are *alike*

−sign in 2nd term ⇒ Both signs are −

$$7x^2 - 71x + 10 = (7x - \quad)(x - \quad)$$

Experiment with possible combinations of factors that yield $+10$. Since the middle term of the trinomial has a large magnitude, try combinations with one large factor first.

$$10 \cdot 1 = 10$$

$$1 \cdot 10 = 10$$

$$7x^2 - 71x + 10 \stackrel{?}{=} (7x - 10)(x - 1)$$

$$\begin{array}{r} -10x \\ -7x \\ \hline -17x \end{array}$$

Reverse the factors.

$$7x^2 - 71x + 10 \stackrel{?}{=} (7x - 1)(x - 10)$$

$$\begin{array}{r} -1x \\ -70x \\ \hline -71x \end{array}$$

Check using the FOIL method.

$$(7x - 1)(x - 10) = 7x^2 - 71x + 10$$

ALTERNATE PROBLEM SET 10.3A

Factor the following.

1. $2x^2 + x - 1$
2. $2x^2 - x - 1$
3. $3x^2 - 5x - 2$
4. $3x^2 + 5x - 2$
5. $2x^2 - 5x - 3$
6. $2x^2 - 7x + 3$
7. $7y^2 - 19y + 10$
8. $7y^2 + 3y - 10$
9. $5x^2 + 41x + 8$
10. $5x^2 - 39x - 8$
11. $3x^2 - 7x + 2$
12. $3x^2 + 7x + 2$
13. $3a^2 + 23a + 14$
14. $3a^2 - 23a + 14$
15. $5x^2 - 17x - 12$
16. $3x^2 - 20x + 12$
17. $3b^2 - 19b - 14$
18. $3b^2 + 19b - 14$
19. $2x^2 + 13x + 20$
20. $2x^2 + 3x - 2$
21. $3x^2 + 4x + 1$
22. $3x^2 - 17x + 10$
23. $5x^2 + 47x - 30$
24. $5x^2 - 31x + 30$
25. $7a^2 + 12a - 4$

It isn't always possible to establish the factors of the x^2 term immediately. In that case, check to see if the constant term has only one pair of factors.

EXAMPLE 14 Factor $15x^2 + 32x - 7$.

Disregarding the sign for the moment, the only way to get a product of 7 is $7 \cdot 1$. Hence, we can write the factors immediately.

$$15x^2 + 32x - 7 = (\quad 7)(\quad 1)$$

Examine the signs.

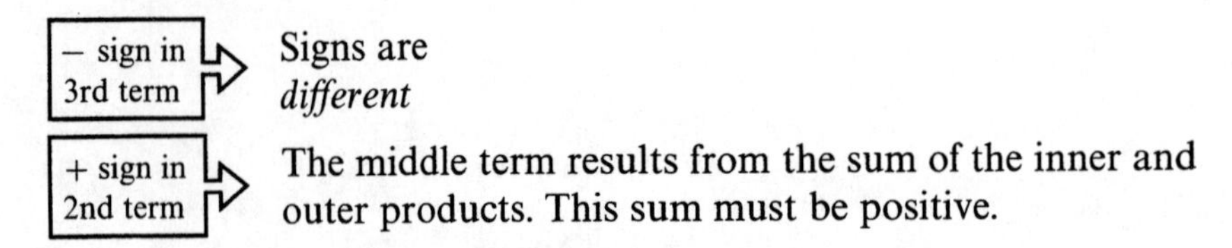

We still don't know whether the negative sign goes with the 7 or the 1. Let's try

$$15x^2 + 32x - 7 \stackrel{?}{=} (\quad -7)(\quad +1)$$

Experiment with possible combinations of factors that yield $15x^2$.

1st factor	·	2nd factor
x	·	$15x$
$15x$	·	x
$3x$	·	$5x$
$5x$	·	$3x$

Just jump in and try the first pair.

$$15x^2 + 32x - 7 \stackrel{?}{=} (x - 7)(15x + 1)$$

$-105x$

$+x$

Stop! This middle term is nowhere near $+32x$.

OK. Then try a combination of more equal factors, $5x \cdot 3x$.

$$15x^2 + 32x - 7 \stackrel{?}{=} (5x - 7)(3x + 1)$$

$-21x$

$+5x$

That can't be done this way; reverse the factors.

Since the signs are different, we need the difference of these products to equal the middle term, $+32x$.

Reversing the first factors,

$$15x^2 + 32x - 7 \stackrel{?}{=} (3x - 7)(5x + 1)$$

$$\begin{array}{r} -35x \\ +3x \\ \hline -32x \end{array}$$

That difference is $-32x$. I want $+32x$.

Then just reverse the signs

Reverse signs and check your answer.

$$(3x + 7)(5x - 1) = 15x^2 + 32x - 7$$

ALTERNATE PROBLEM SET 10.3B

Factor the following.

1. $4x^2 + 4x - 3$
2. $4x^2 - 8x + 3$
3. $6x^2 + x - 2$
4. $6x^2 - x - 2$
5. $6x^2 - 7x + 2$
6. $6x^2 + 7x + 2$
7. $6x^2 - 7x - 3$
8. $6x^2 - 17x - 3$
9. $10x^2 - 27x + 5$
10. $10x^2 + 49x - 5$
11. $8y^2 - 11y + 3$
12. $8y^2 + 10y + 3$
13. $9x^2 - 12x - 5$
14. $9x^2 + 26x - 3$
15. $7x^2 - 12x + 5$
16. $11x^2 - 8x - 3$
17. $7x^2 + 12x + 5$
18. $11x^2 - 14x + 3$
19. $12a^2 + 17a - 5$
20. $12a^2 - 4a - 5$
21. $12a^2 + 17a + 5$
22. $12a^2 - 19a + 5$
23. $12a^2 - 32a + 5$
24. $15y^2 + 17y + 4$
25. $15y^2 - 7y - 4$
26. $15a^2 + 17a - 4$
27. $15a^2 - 23a + 4$

Occasionally there are several possible combinations of factors for both the first and the third term of the trinomial. In this case, we simply have to be willing to try a few more guesses. However, we are not completely in the dark and, as you gain experience, you'll find it takes fewer tries until you have the trinomial factored or you know it's impossible to factor.

EXAMPLE 15 Factor $20x^2 + 37x - 18$.

Since there are several combinations of factors that will produce 18 and several combinations whose product is $20x^2$, we start with a guess at the first terms.

$$20x^2 + 37x - 18 = (5x \quad)(4x \quad)$$

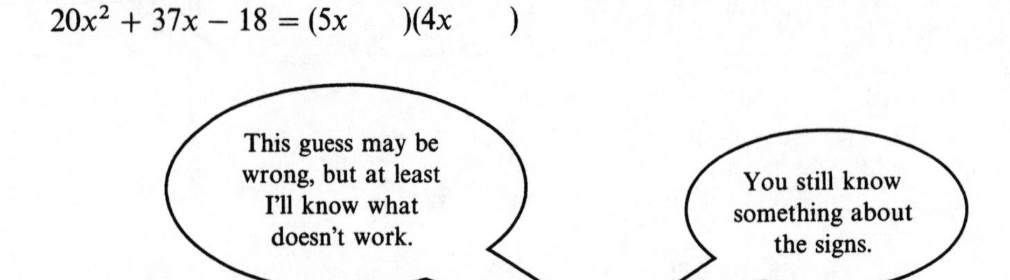

Examine the signs.

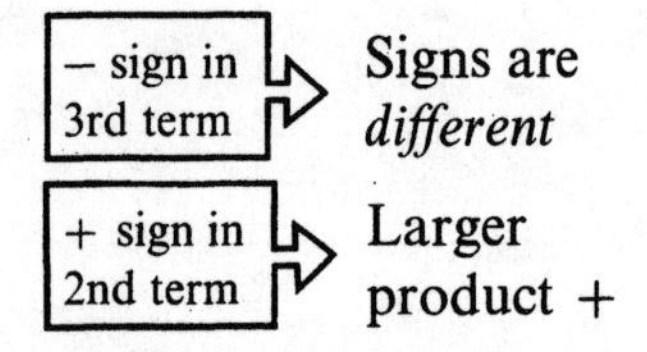

Let's try

$$20x^2 + 37x - 18 \stackrel{?}{=} (5x + \quad)(4x - \quad)$$

Experiment with combinations of factors that produce 18.

$$\left.\begin{matrix} 1 \cdot 18 \\ 2 \cdot 9 \\ 3 \cdot 6 \end{matrix}\right\} \text{ all yield 18}$$

Try $3 \cdot 6$

$$20x^2 + 37x - 18 \stackrel{?}{=} (5x + 3)(4x - 6)$$

$$\begin{array}{r} 12x \\ -30x \\ \hline -18x \end{array}$$

Difference → $-18x$ is way too small

Therefore, try factors that produce larger products.

$$20x^2 + 37x - 18 \stackrel{?}{=} (5x + 2)(4x - 9)$$

$$\begin{array}{r} 8x \\ -45x \\ \hline -37x \end{array}$$

Difference ⇨ $-37x$

Got $37x$, but wrong sign.

Good, now reverse the signs.

$$20x^2 + 37x - 18 \stackrel{?}{=} (5x - 2)(4x + 9)$$

Check.

$$(5x - 2)(4x + 9) = 20x^2 + 37x - 18$$

ALTERNATE PROBLEM SET 10.3C

Factor the following.

1. $4x^2 + 11x + 6$
2. $4x^2 + 5x - 6$
3. $6x^2 + 19x + 10$
4. $6x^2 - 19x + 10$
5. $6x^2 + 11x - 10$
6. $6x^2 - 11x - 10$
7. $6x^2 + 13x - 28$
8. $12a^2 - 8a - 15$
9. $4x^2 + 17x - 15$
10. $20y^2 + 13y - 15$
11. $4t^2 + 4t - 15$
12. $18x^2 - 3x - 10$
13. $18x^2 - 9x - 14$
14. $20x^2 - 9x - 18$
15. $24x^2 + 47x - 21$
16. $24x^2 + 10x - 21$
17. $4a^2 - 21a - 18$
18. $4a^2 - 9a - 9$
19. $8x^2 + 17x - 21$
20. $12x^2 - 7x - 49$
21. $24x^2 + 14x - 3$
22. $16x^2 + 32x + 15$
23. $20x^2 - 121x + 6$
24. $9b^2 + 29b + 20$
25. $8a^2 - 2a - 15$
26. $24x^2 - 26x - 15$
27. $24x^2 + 22x - 21$

A POINTER FOR EASIER FACTORING

Factor $8 + 14x - 15x^2$.

It's easier if you write expressions to be factored in descending order. Rewriting,

$$8 + 14x - 15x^2 = -15x^2 + 14x + 8$$

This way it will look more familiar and you'll spot things quicker.

I just spotted a negative sign in front of $15x^2$. What do I do with $-15x^2$?

Avoid $-15x^2$ by factoring -1 from the trinomial.

$$-15x^2 + 14x + 8 = -1[15x^2 - 14x - 8]$$

Now factor the trinomial in the square brackets.

$$= -1[(5x + 2)(3x - 4)]$$

Notice -1 is still multiplying the quantity inside the square brackets.

It's reasonable to call $-1[(5x + 2)(3x - 4)]$ the factored form. You can also write the factored form as $(-1)(5x + 2)(3x - 4)$ or if you want to avoid writing the negative one explicitly, you might say

$$(-1)(5x + 2)(3x - 4) = (-5x - 2)(3x - 4)$$

or you could have written

$$(-1)(5x + 2)(3x - 4) = (5x + 2)(-3x + 4)$$

That's three different answers. Which one do you want?

That's one answer in three equivalent forms. Any one is okay. In Chapter 11, you'll notice times where a particular form is most useful.

ALTERNATE PROBLEM SET 10.3D

Here is some practice with all types. Factor the following.

1. $5x^2 + 2x - 3$
2. $6x^2 + 11x - 10$
3. $7x^2 - 5x - 2$
4. $3x^2 - 2x - 8$
5. $x^2 - 2x + 1$
6. $2x^2 + 5x - 3$
7. $2x^2 - 5x - 3$
8. $2x^2 + 3x + 1$
9. $2x^2 - 3x + 1$
10. $y^2 + 5y + 6$
11. $y^2 + 14y + 49$
12. $12 + 16x - 3x^2$
13. $10 + 11x - 6x^2$
14. $10x^2 + 27x + 5$
15. $5 + 49x - 10x^2$
16. $6x^2 + 29x + 28$
17. $20x^2 + 56x + 15$
18. $7a^2 + 2a - 5$
19. $6x^2 + 7x - 3$
20. $5 + 28b - 12b^2$
21. $x^2 + 6x + 9$
22. $8x^2 + 11x + 3$
23. $9 + 12x - 5x^2$
24. $6 - x - 15x^2$
25. $8x^2 - 10x + 3$
26. $15 - 2x - 8x^2$
27. $9y^2 - 8y - 20$
28. $x^2 - 6x + 9$
29. $5 - 12y - 9y^2$
30. $24a^2 - 26a - 15$

What do I do now?

You have just completed Alternate Sections 10.2 and 10.3. Now continue with Section 10.4 on page 297.

Answers to Selected Problems

Problem Set 1.1A

1. 25 3. 64 5. 81 7. 9 9. 100
11. 10,000 13. 32 15. 5 17. $x = 4^2$

Problem Set 1.1B

1. 10 3. 13 5. 14 7. 14 9. 20 11. 70
13. 64 15. 38 17. 100 19. 7 21. 7 23. 3
25. 8 27. 2 29. 3 31. 11 33. 3 35. 9
37. 1 39. 8 41. 22 43. 482
45. $x = 2 + 3 \cdot 4$ 47. $z = (8 + 18) \cdot 5$ 49. $z = (16 - 4)^5$ 51. $x = \dfrac{18}{2 \cdot 3}$

Problem Set 1.1C

1. $3 \cdot 4 + 3 \cdot 5$ 3. $7 \cdot 2 + 7 \cdot 8$ 5. $4 \cdot x + 4 \cdot y$ 7. $3 \cdot x + 3 \cdot 2$

Problem Set 1.1D

1. 0 3. 2 5. 3 7. 0 9. $-x$
11. -3 13. 9 15. 7 17. 1 19. -1
21. 10 23. $-x$ 25. -13 27. -13

Problem Set 1.2A

1. 5 3. 1 5. -4 7. -6 9. -11
11. 7 13. 3 15. -16 17. -2 19. -2
21. -12 23. 0 25. -4 27. -60 29. 25.4
31. -5.07 33. 1.48 35. -2.00 37. -69.26

Problem Set 1.2B

1. -1 3. 7 5. 11 7. -8 9. -7 11. -3
13. -7 15. -4 17. 12 19. -10 21. 7 23. 14
25. 5 27. 92 29. -12 31. 5 33. -5 35. 3.06
37. 11.35 39. -1.67 41. 12.53 43. 16.10

Problem Set 1.3A

1. -8 3. 0 5. 18 7. -21 9. 36
11. 0 13. -12 15. -48 17. 48 19. -7
21. -9 23. -150 25. -60 27. 828 29. -6.0
31. 28.8 33. 30.1 35. -10.12 37. 18.88

Problem Set 1.3B

1. -3 3. -6 5. 4 7. 9 9. 8 11. 4 13. -4
15. 8 17. -8 19. 8 21. -14 23. -17 25. 12 27. -5
29. -0.9 31. -1.7 33. 9 35. -2.3 37. 1.4

Problem Set 1.3C

1. -2 3. -2 5. -32 7. -32 9. -4 11. 5
13. -5 15. 15 17. -5 19. -15 21. -5 23. -5
25. -15 27. 36 29. -20 31. 4 33. -30 35. 32
37. -18 39. -2 41. 0 43. -20 45. $x = 4 + (-5)$
47. $y = 12 + (-5)$ 49. $x = 15 \cdot 0$ 51. $x = 30 \div 6$ 53. $x = 30 \div 6$

Problem Set 1.4A

1. -1 3. -3 5. 4 7. -6 9. -11 11. -8
13. -16 15. -4 17. 16 19. -14 21. 6 23. 4
25. -25 27. 2 29. -9.6 31. 2.4 33. 4.3 35. 4.6

Problem Set 1.4B

1. 13 3. 1 5. 1 7. 1 9. 1 11. -125 13. -26
15. 15 17. 5 19. -19 21. -1 23. 0 25. 12 27. -54

Problem Set 1.4C

1. $A = 48$ sq in. 3. $C = 37.68$ ft 5. $A = 314$ sq ft 7. $C = 37$ 9. $d = 256$ ft
11. $C = 69.08$ ft 13. $A = 50.24$ sq ft

Unit 1 Review Test

1. 29 2. -21 3. -9 4. -24 5. -34
6. 6 7. 5 8. 12 9. -60 10. 99
11. 12 12. -22 13. -104 14. 0 15. -15
16. 216 17. -8 18. -4 19. -1 20. -1
21. 13 22. 30 23. -20 24. 20° Celsius 25. 113.04 cm^2

Problem Set 2.1A

1. 3^2 3. $(-2)^2$ 5. $(-4)^4$ 7. $2^3 \cdot 3^2$ 9. x^3 11. x^2y^2
13. xy^3 15. 2^2x^3 17. $-2a^2b$ 19. $(-5)^2a^2b$ 21. $2^2a^2bc^3$

Problem Set 2.1B

1. 10 3. -4 5. -6 7. -14 9. 6 11. 0
13. -5 15. 17 17. 20 19. 8 21. -1 23. 1
25. 5 27. 51 29. 72 31. -44 33. -46 35. 40

Problem Set 2.1C

1.

x	$2x + 6$
-4	-2
-2	2
1	8
2	10
3	12

3.

x	$x^3 + 1$
-2	-7
-1	0
0	1
1	2
2	9

5.

x	$x^3 - 1$
-2	-9
-1	-2
0	-1
1	0
2	7

7.

x	$x^2 + 2x + 3$
-2	3
0	3
1	6
2	11
3	18

9.

x	$2x^3 + x - 3$
-3	-60
-1	-6
0	-3
1	0
2	15

11.

x	$2x^3 + 4x^2 + 4$
-3	-14
-1	6
0	4
1	10
2	36

Problem Set 2.2A

1. $2xy$ 3. $6xy$ 5. $-8xy$ 7. $14ax$ 9. $24z$ 11. $9y$
13. $20t^2$ 15. $-4x^2$ 17. $12y^2$ 19. $7s^2t^4$ 21. $16mn^5$ 23. a^2m^2
25. $6x^4y^3$ 27. a^2b^3 29. -4 31. -36

Problem Set 2.2B

1. x^2 3. x^3 5. b^5 7. $-2x^6$
9. $6a^5$ 11. $-12x^5$ 13. $30x^7$ 15. $-72x^2y^3$
17. $6a^4b^4$ 19. $12x^5y^2$ 21. $-3a^6b^2$ 23. $42ax^2y$
25. $54a^3b$ 27. $-7xa^2$ 29. $6ax^3$ 31. $x = -3b^3 \cdot c^2$

Problem Set 2.3

1. $2a + 8$ 3. $6x - 48$ 5. $-a + 9$ 7. $12a + 8b$ 9. $-24a + 16y$
11. $3a - 4x$ 13. $2a + 3b$ 15. $4x - y$ 17. $x^6 + x^3$ 19. $14x^2y^3 + 6x^3y^2$
21. $-12x^2y^3 + 24x^3y^3$ 23. $-15a^2b^3x + 18a^2b^2y^3$
25. $12xb^2a^3 + 18ax^2b^2$ 27. $6x^3 - 8x^2 + 2x$
29. $-28x^6 + 32x^4 - 4x^3$ 31. $81a^2x^6 - 72a^3x^3 + 54a^2x^3$
33. $-9a^4b^5c^2 + 15a^4b^6c^2 - 12a^7b^5c$ 35. $12yx^3a^3 - 18ay^3x^5 + 30x^6y^3a^2$

Problem Set 2.4A

1. $8a$ 3. $-2y$ 5. $-x$ 7. $-a^2$ 9. $7abc$

Problem Set 2.4B

1. $3x$ 3. $9x$ 5. $-6b$ 7. $-9a^2$
9. $5x - 5y$ 11. $-10x + 7a$ 13. $4a - 6b + 1$ 15. $-2x^2 + 2x$
17. $10x^3 + 2x^2 - 5x - 2$ 19. $5a^2b + 9ab^2 - 7ab$ 21. $-a^2b + 5ab^3 - 6ab$
23. $-3x^2 - 3x$ 25. $9x^2y^2 - 6x^2y - 7xy + 3x$ 27. $-x^2y + 3xy^2 - x + y$
29. $3a - b - 8ab - 2a^2$

Problem Set 2.4C

1. $-3x + 9$ 3. $-10x^2 - 3$ 5. $4x^2 - 2x - 7$ 7. $10x^2 - 2x - 12$
9. $-2xy - 7x^2$ 11. $4x - 4$ 13. $4x^2 + 3x - 12$ 15. $-5x - 17$
17. $-4xy + 1$ 19. $x^2 + x + 7$ 21. $-8xy + 2y^2$ 23. $6a^2 - 10a - 7$
25. $6y - 4y^2 - 4x^2$ 27. $20x - 31$ 29. $-8x - 20y$ 31. $-5x^2 + 2y^2$
33. $x = (3b + 4) + (-6b + 5)$

Unit 2 Review Test

1. 7^4 2. $(-6)^3$ 3. 3^2ab^2 4. $(-2)^2a^3b$
5. -13 6. 4 7. 0 8. 544

9.

x	$x^2 - 2$
-4	14
-2	2
0	-2
1	-1
3	7

10.

x	$2x^3 - x + 4$
-2	-10
-1	3
0	4
1	5
2	18

11. $-12x^3y^2$
12. $-5x^2a^4$
13. $27x^4y$
14. $6x^3y^2$
15. $15a^3b^2c^6$
16. $-6a^2c^3$
17. $-12x + 18$
18. $a^3b^3 + 2a^2b^3$
19. $x^5 - x^4 + 3x^3$
20. $-2x^3 - 8x^2 - 12x$
21. $-24x^4y^7 + 32x^5y^8$
22. $2a^4b^2c^4 - 10a^2b^4c^4 + 8a^6bc^6$
23. $4x + 3$
24. $-x + 6$
25. $3a^2 - 3a + 7$
26. $5y^2 - 2$
27. $2x^2 + 2x - 3$
28. $7a + 15$
29. $-2x + 11$
30. $x^2 + 3y - 10$
31. $-9x^4 - 2x^2 + 2x - 3$
32. $-x + y - xy - x^2y$

Problem Set 3.1A

1. $\frac{3}{6}$
3. $\frac{14}{21}$
5. $\frac{20}{24}$
7. $\frac{-48}{56}$
9. $\frac{-24a}{32}$
11. $\frac{16x}{28}$
13. $\frac{-16x}{28}$
15. $\frac{18}{24}$
17. $\frac{25}{40}$
19. $\frac{54}{99}$
21. $\frac{-12}{20}$
23. $\frac{-30x}{54}$
25. $\frac{56a}{-63}$
27. $\frac{-48xy}{56}$
29. $\frac{72xa}{81}$
31. $\frac{-80ay}{88}$

Problem Set 3.1B

1. $\frac{2}{3}$
3. $\frac{3}{4}$
5. $\frac{-6}{7}$
7. $\frac{3}{7}$
9. $\frac{-3}{4}$
11. $\frac{3}{4}$
13. $\frac{3}{4}$
15. $\frac{-2}{3}$
17. $\frac{-7x}{6}$
19. $\frac{9}{8}$
21. $\frac{-3}{5}$
23. $\frac{-8xy}{3}$

Problem Set 3.2A

1. $+$
3. $-$
5. $+$
7. $-$
9. $-$

Problem Set 3.2B

1. $\frac{3}{10}$
3. $\frac{5}{14}$
5. $\frac{-1}{3}$
7. 14
9. $3a$
11. $\frac{-24x}{7}$
13. $\frac{-4x}{5}$
15. $\frac{-4y}{5}$
17. $\frac{-2ab}{5}$
19. $28x^2$
21. $-3a^2b$
23. $\frac{-22a^2x^3}{3}$
25. 42
27. 6

Problem Set 3.2C

1. 3
3. 10
5. 1
7. $3a + 10$
9. $-x + 3$
11. $-6 - 3a$
13. $\frac{-2a}{3} - \frac{b}{2}$
15. $-6 + 4a$
17. $9 - 3a$
19. $-12 - 6a$
21. $\frac{7a}{2} - \frac{21}{2}$
23. $\frac{a^2}{4} + \frac{a}{2}$
25. $4a^2 - 6ax$
27. $9x^2 + 2ax$

Problem Set 3.2D

1. $\frac{1}{4}$
3. 3
5. $\frac{-1}{3}$
7. $\frac{4}{3}$
9. $\frac{-9}{8}$
11. $\frac{1}{6}$
13. $\frac{1}{a}$
15. None
17. -3
19. $\frac{-4}{3}$

21. $\frac{1}{32}$ 23. 8 25. $\frac{1}{27}$ 27. 16 29. $\frac{8}{9}$

31. $\frac{-20}{21}$ 33. $\frac{-7a}{9}$ 35. $\frac{10x}{3}$ 37. $\frac{10a}{9}$ 39. $\frac{-14a}{5}$

Problem Set 3.3A

1. $\frac{6}{7}$ 3. $\frac{1}{4}$ 5. $\frac{1}{2}$ 7. $\frac{1}{3}$ 9. $\frac{1}{8}$ 11. $\frac{1}{4}$

13. $\frac{-4}{3}$ 15. $\frac{-29a+5}{21}$ 17. $\frac{x}{2}$ 19. $\frac{7x}{15}$ 21. $\frac{-2ab}{9}$ 23. $-ax$

Problem Set 3.3B

1. $\frac{1}{5}$ 3. $\frac{1}{3}$ 5. $\frac{-1}{3}$ 7. $\frac{4a}{7}$ 9. $\frac{-3x}{8}$

11. $\frac{14}{11}$ 13. $\frac{-2}{3}$ 15. $\frac{-6a-7}{12}$ 17. $\frac{17ax+11}{18}$ 19. $\frac{-19xy+23}{24}$

21. $\frac{2x}{5}$ 23. $\frac{-14ab}{9}$ 25. ax 27. x^2

Problem Set 3.3C

1. $\frac{7}{8}$ 3. $\frac{17}{20}$ 5. $\frac{2}{9}$ 7. $\frac{-13}{24}$ 9. $\frac{-1}{6}$

11. $\frac{25}{24}$ 13. $\frac{20a-21}{30}$ 15. $\frac{14-3a}{18}$ 17. $\frac{-21+20a}{24}$ 19. $\frac{33a-35}{45}$

21. $\frac{-95a-28}{60}$ 23. $\frac{-14xy+33}{36}$ 25. $\frac{23a}{18}$ 27. $\frac{-5x}{24}$ 29. $\frac{-31bx}{28}$

31. $\frac{10ax^2}{21}$ 33. $\frac{-11a}{24}$ 35. $\frac{-x}{24}$ 37. $\frac{11by}{24}$ 39. $\frac{-11a^2b}{72}$

41. 2 43. 4

Problem Set 3.4A

1. 4 3. 1 5. 2 7. -6 9. -6 11. -7

13. -1 15. -8 17. $\frac{2}{3}$ 19. -10 21. 2 23. -98

Problem Set 3.4B

1. $\frac{3x}{y}$ 3. Cannot be reduced 5. Cannot be reduced 7. $\frac{3x}{(x+y)}$

9. $\frac{12x+1}{x}$ 11. $\frac{64x^3}{9y^2}$ 13. $-x$ 15. $x-20$

Unit 3 Review Test

1. $\frac{-3}{2}$ 2. $\frac{1}{5}$ 3. $\frac{-36}{45}$ 4. $\frac{42a}{48}$ 5. $\frac{-2}{3}$

6. $\frac{4}{5}$ 7. $\frac{-8a}{3}$ 8. $\frac{-3xy}{11}$ 9. $\frac{-8}{3}$ 10. $\frac{9ax^2}{2}$

11. $4a^3b$ 12. $5x^3$ 13. $\frac{-1}{18}$ 14. $\frac{15x}{8}$ 15. $\frac{4x}{3}$

16. $\frac{-9x^2}{5}$ 17. $15a - 8$ 18. $18a - 9x$ 19. $\frac{-3a^2}{2} + \frac{7a}{3}$ 20. $4a^3 - \frac{7a^2}{27}x$

21. $\frac{-ab}{4}$ 22. $\frac{12x + 7}{15}$ 23. $\frac{1}{24}$ 24. $\frac{-22b - 39}{24}$ 25. $\frac{-85ax - 44}{60}$

26. $\frac{-41x^2}{45}$ 27. $\frac{-23ax}{10}$ 28. $\frac{-7x}{10}$ 29. $\frac{-17x^2}{24}$ 30. $\frac{-47ax^2}{56}$

31. 3 32. 3 33. -11 34. -2 35. -2
36. 9 37. 35 38. 17

Problem Set 4.1

1. $x = 12$ 3. $x = 12$ 5. $y = 1$ 7. $x = -4$ 9. $t = 9$
11. $x = 1$ 13. $x = -3$ 15. $b = -11$ 17. $x = 9$ 19. $c = -2$
21. $x = 14$ 23. $a = -12$ 25. $x = 13$ 27. $x = -24$ 29. No solution
31. -6 33. 8

Problem Set 4.2A

1. $x = 4$ 3. $x = \frac{3}{2}$ 5. $x = \frac{1}{2}$ 7. $x = -6$ 9. $x = -8$

11. $x = 0$ 13. $x = 8$ 15. $x = \frac{7}{2}$ 17. $x = \frac{9}{2}$ 19. $x = 3$

Problem Set 4.2B

1. $x = 3$ 3. $x = 2$ 5. $y = 3$ 7. $x = -2$ 9. $z = 3$

11. $x = -1$ 13. $b = 2$ 15. $x = -7$ 17. $x = 5$ 19. $x = \frac{2}{3}$

Problem Set 4.2C

1. $r = 3$ 3. $r = 1$ 5. $r = 8$ 7. $x = \frac{9}{2}$ 9. $s = \frac{20}{17}$

11. $x = -2$ 13. $x = 15$ 15. $x = 5$ 17. $4x + 12$ 19. $-2a + 3$

Problem Set 4.2D

1. $x = 5$ 3. $x = 3$ 5. $a = \frac{15}{4}$ 7. $x = 2$ 9. $h = 0$ 11. $x = -2$

13. $x = 2$ 15. $y = 0$ 17. $x = -3$ 19. $3a$ 21. 3

Problem Set 4.3

1. $t = \frac{d}{r}$ 3. $m = \frac{f}{a}$ 5. $K = \frac{y}{x}$ 7. $b = \frac{A}{h}$ 9. $l = \frac{V}{wh}$

11. $r = \frac{C}{2\pi}$ 13. $a = c - b$ 15. $x = \frac{-b}{a}$ 17. $l = \frac{P - 2w}{2}$ 19. $g = \frac{V - a}{t}$

21. $d = \frac{s - a}{n - 1}$ 23. $h = \frac{2A}{B + b}$ 25. $a = \frac{2S - hn}{h}$ or $\frac{2S}{h} - n$

Problem Set 4.4A

1. 7%

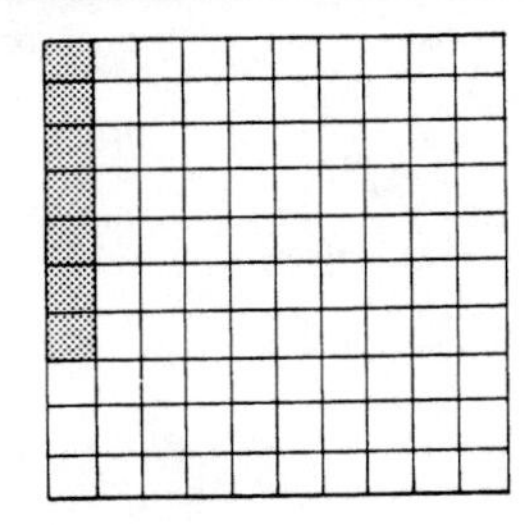

3. 30%

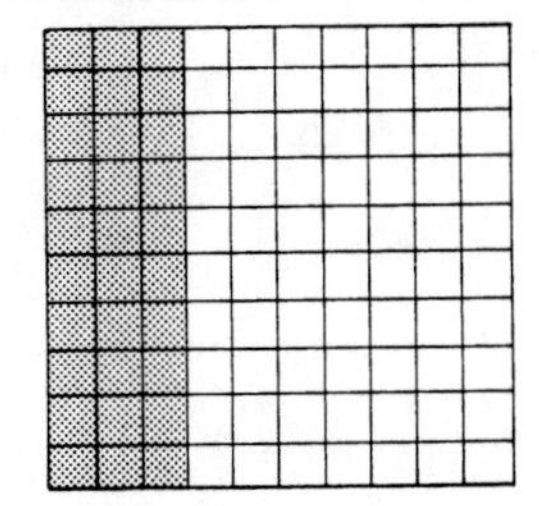

5. 10%

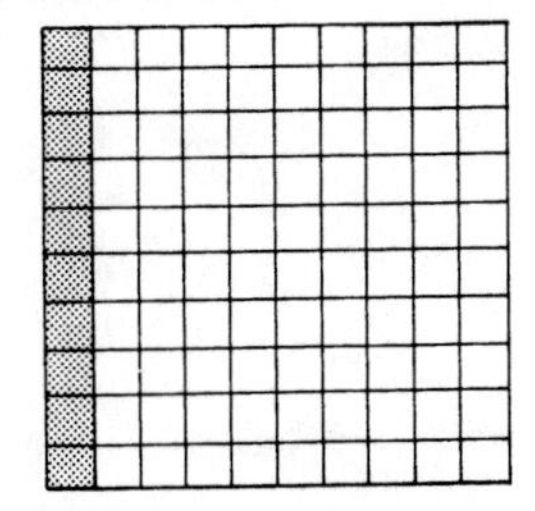

Problem Set 4.4B

1. 20%

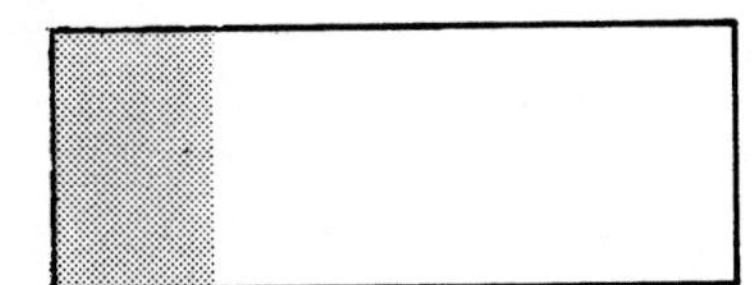

3. 3%

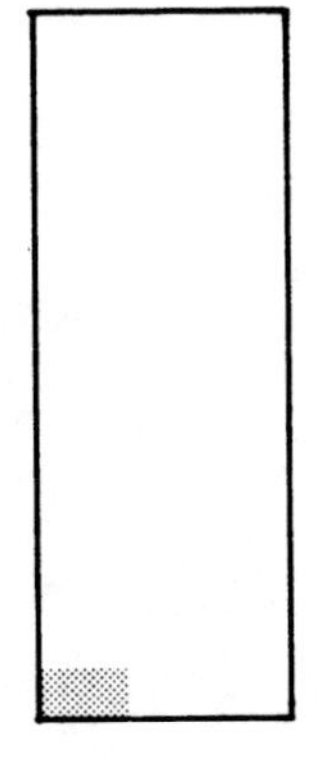

Problem Set 4.4C

1. $\frac{1}{2}$ 3. $\frac{3}{25}$ 5. $\frac{33}{100}$ 7. $\frac{3}{20}$ 9. $\frac{77}{100}$ 11. $\frac{1}{3}$

13. $\frac{1}{6}$ 15. $\frac{3}{8}$ 17. $\frac{1}{500}$ 19. $\frac{3}{400}$ 21. 1 23. $\frac{3}{2}$

Problem Set 4.4D

1. 0.1 3. 0.15 5. 0.07 7. 0.056 9. 0.3025 11. 0.165

13. 0.7875 15. 0.0625 17. 0.375 19. 0.005 21. 0.001 23. 0.0025

Problem Set 4.4E

1. 45% 3. 50% 5. $12\frac{1}{2}\%$ 7. 6% 9. 20% 11. 360%

13. 356% 15. $16\frac{1}{4}\%$ 17. $\frac{1}{2}\%$ 19. 500% 21. $\frac{6}{17}$ 23. $\frac{5}{4}$

Problem Set 4.4F

1. 50% 3. 20% 5. 80% 7. $33\frac{1}{3}\%$ 9. $83\frac{1}{3}\%$ 11. $31\frac{1}{4}\%$

13. 0.7% 15. 87.5%

Problem Set 4.4G

1. 42 3. 25% 5. 140 7. 88 9. 120 11. $137\frac{1}{2}$

Problem Set 4.5A

1. 80% 3. 88 5. 35 7. \$375 9. 6% 11. \$3,600
13. 20 15. 12.5% 17. \$10.20 19. 20% 21. $58\frac{1}{3}\%$ 23. \$1,520
25. \$170 27. \$145.50 29. 2% 31. 5% 33. 0.07 35. 0.125

Problem Set 4.5B

1. \$2,000 3. \$260 5. 3.5 hr 7. 2 cm 9. 14
11. 6.31 m 13. $7\frac{1}{2}$ cm 15. \$950 at 4% 17. \$100 + 5% plan

Unit 4 Review Test

1. -24 2. -8 3. -8 4. 14 5. -5
6. $\frac{4}{3}$ 7. -18 8. 3 9. 3 10. -5
11. 6 12. -4 13. $y = z - x$ 14. $l = \frac{P - 2w}{2}$ 15. $x = ky$
16. $w = \frac{V}{lh}$ 17. $\frac{l - a + d}{d} = n$ 18. $\frac{S}{2r} - r = h$ 19. $\frac{9}{50}$ 20. $\frac{7}{8}$
21. $\frac{1}{250}$ 22. $\frac{7}{5}$ 23. 0.18 24. 0.062 25. 0.0875
26. 0.006 27. 39% 28. 60% 29. 3% 30. $24\frac{3}{4}\%$
31. 80% 32. 0.3% 33. 56.25% 34. 21.875% 35. 52.7
36. $26\frac{2}{3}$ 37. 252 38. 35 39. $8\frac{3}{4}\%$ 40. 24 in.
41. 32 cm 42. \$1.13 43. \$3,260 44. 16% 45. -24
46. 20 47. $3x^4y^4a^4 - 15x^2ya^7 + 12x^5y^3a^4$ 48. $-2x - 13xy - 5x^2y$

Cumulative Review Units 1–4

1. -1 2. -3 3. -13 4. -72 5. 13 6. -8
7. 11 8. 33 9. 21 10. -48 11. 2 12. 1
13. $-10x^4y^3$ 14. $24a^2b^5c$ 15. $\frac{5}{6}$ 16. $\frac{-6ay}{5}$ 17. $-x^3y$ 18. a^2b^2
19. $-2x^4 + 6x^3 - 8x^2$ 20. $4a^2b^4 - 8a^5b^5c + 12a^5b^3c^2$ 21. $10a^2 - 6x$
22. $3x + 6$ 23. $2y^2 - 10y + 4$ 24. $-4a^2 + a - b + 4$
25. $3x^2y - 2xy - x + 4$ 26. $\frac{-xy}{2}$ 27. $\frac{-4ab^2}{45}$
28. $a = -5$ 29. $x = \frac{9}{2}$ 30. $a = -12$ 31. $x = \frac{16}{3}$ 32. $h = \frac{V}{lw}$
33. $d = \frac{S - 2a}{n - 1}$ 34. $\frac{5}{8}$ 35. 0.125 36. 34.75% 37. 162
38. 20% 39. 20° Celsius 40. 32 in.

Problem Set 5.1A

1. $n + 10$ 3. $n + 6$ 5. $15 - n$ 7. $5n$ 9. $\frac{n}{-8}$

11. $2n - 16$ 13. $2n + 7$ 15. $-8 \cdot 2n$ 17. $2n + \frac{1}{2}n$ 19. $\frac{2n+7}{8}$

21. $2n(n + 8)$ 23. $3[n + (-6)]$ 25. $n = \frac{-7}{2}$ 27. $n = 9$

Problem Set 5.1B

1. 7 3. 24 5. 8 7. 12 9. 69 11. 13
13. 18 15. 2 17. 36 19. 20 21. 7 23. 5

Problem Set 5.2

1. 7, 8 3. 17, 19 5. −17, −15 7. 5, 7, 9
9. 18, 20, 22 11. 7, 8, 9 13. 1, 2 15. 3, 5, 7
17. 20, 21 19. 8, 10 21. 12, 13 23. −7, −5, −3

Problem Set 5.3A

1. $20 + 4 = 24$ 3. $10 - 6 = 4$ 5. $25 + x$ 7. $x - 5$ 9. $25 + 3 = 28$
11. $28 - 6 = 22$ 13. $n - 10$ 15. $x = 12$

Problem Set 5.3B

1. 10 3. 12 5. 11 7. 7
9. 30 11. 32 13. 18 15. Maria: 12
Louise: 48

17. Susan: 12
brother: 8 19. Manuel: 15
Jose: 21 21. Tony: 5
brother: 13
father: 39 23. $x = 8$

Problem Set 5.4

1. $8 \cdot 10 = 80$ cents 3. $26 \cdot 5 = 130$ cents 5. $9 \cdot 25 = 225$ cents 7. $5n$
9. \$1.00 11. \$1.92 13. $(5n + 10d + 25q)$ cents
15. a. $12 - x$ b. $(12 - x)5$ c. 8 dimes, 4 nickels 17. 6 dimes, 18 nickels
19. 8 nickels, 11 quarters 21. 21 dimes, 7 nickels 23. 3 quarters, 7 dimes, 4 nickels
25. 16 nickels, 8 dimes, 48 pennies 27. 640 adults 29. 8 children
31. 120 reserved, 640 general

Problem Set 5.5A

1. 182 mi 3. 243 mi 5. 48 mph 7. $3\frac{1}{4}$ hr

Problem Set 5.5B

1. 3 hr 3. $1\frac{1}{5}$ hr 5. Jill: 40 mph
Jack: 50 mph 7. 4 hr

9. 6 mi 11. 5 hr 13. $1\frac{1}{3}$ hr

Unit 5 Review Test

1. $n - 2$ 2. $3n - 25$ 3. $2n + 12$ 4. $\frac{n+8}{7}$ 5. $6 - 3x$
6. $n + n + 1$ 7. $35 - x$ 8. $n + 6$ 9. $x + y$ 10. $42 - x$

11. $3x - x$
12. $10x$
13. $5n$
14. $5y + 25n$
15. $10n + 5(2n)$
16. $20 + x$
17. $2x$
18. 17
19. 21, 23, 25
20. 14 years old
21. Laura is 8
brother is 4
22. 13 quarters
39 nickels
23. 1500 student
720 adult
24. 52 mph
25. Joe: 150 mph
Ted: 180 mph
26. -180
27. $-12a^2x^4y^4 + 18a^4x^6y^2 + 15a^5x^3y^4$
28. $\dfrac{45x^2}{4}$
29. $\dfrac{-x^2}{20}$

Problem Set 6.1

1. 4
3. -2
5. -9
7. -17
9. 18
11. 6
13. -9
15. 0
17. 9
19. 7

21.

x	$f(x) = 4x + 2$	(x, y)
3	$f(3) = 14$	$(3, 14)$
-4	$f(-4) = -14$	$(-4, -14)$
0	$f(0) = 2$	$(0, 2)$
-6	$f(-6) = -22$	$(-6, -22)$
2	$f(2) = 10$	$(2, 10)$
1	$f(1) = 6$	$(1, 6)$

Problem Set 6.2A

1.

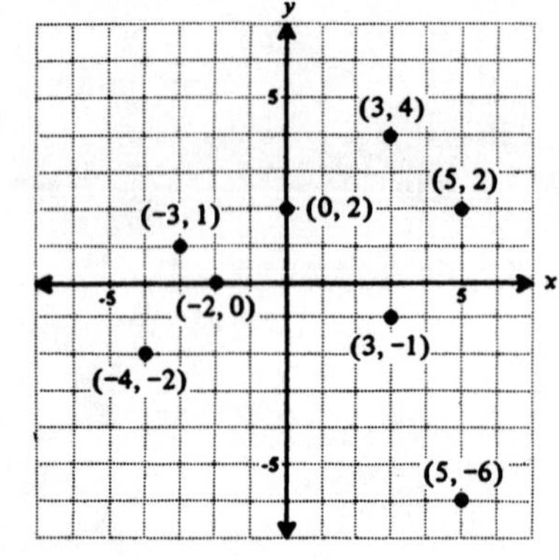

3. $A(5, 4)$ $B(3, 10)$
$C(-3, -4)$ $D(-9, 9)$
$E(7, -8)$ $F(3, -2)$
$G(-9, -9)$ $H(-2, 3)$
$I(0, -6)$ $J(10, 7)$
$K(-8, 0)$ $L(-2, -8)$

Problem Set 6.2B

1.

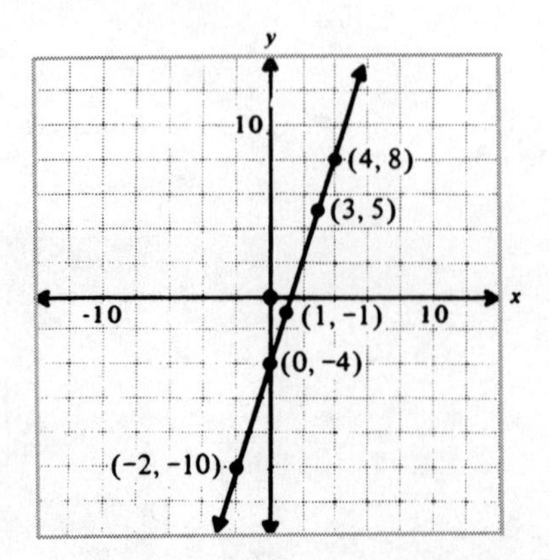

3.

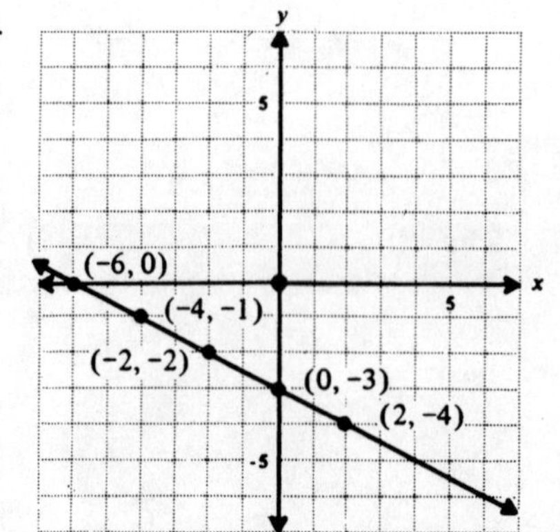

5.

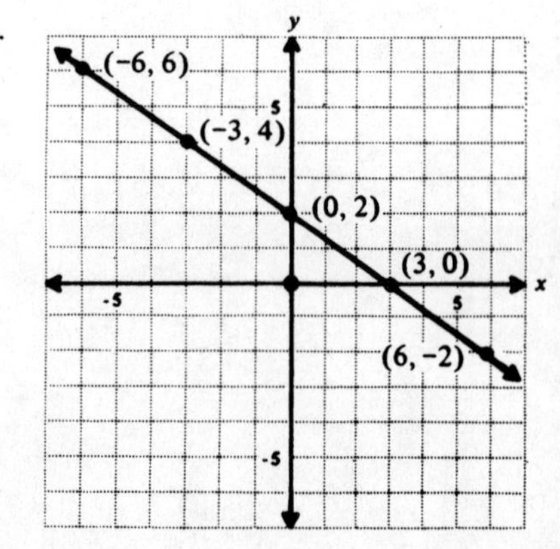

7. $y = x + 2$

x	y
-2	0
0	2
2	4

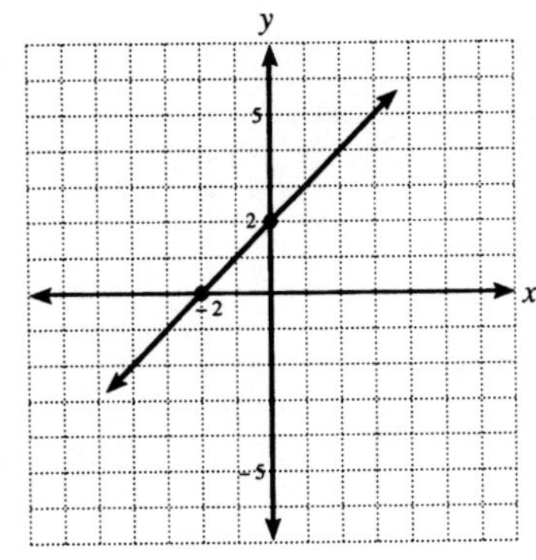

9. $y = 2x - 4$

x	y
-2	-8
0	-4
3	2

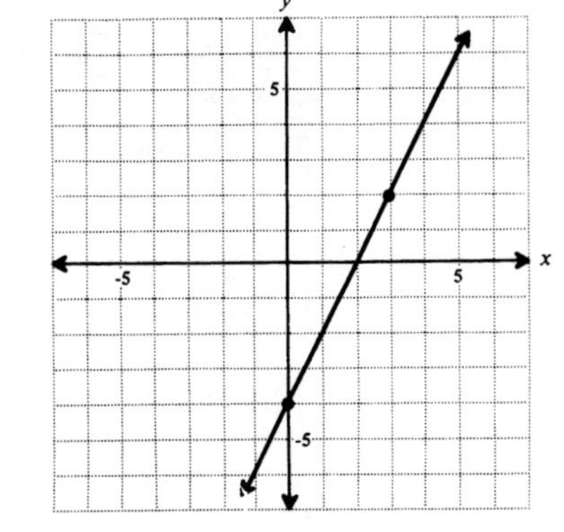

11. $y = \frac{3}{4}x - 2$

x	y
-4	-5
0	-2
4	1

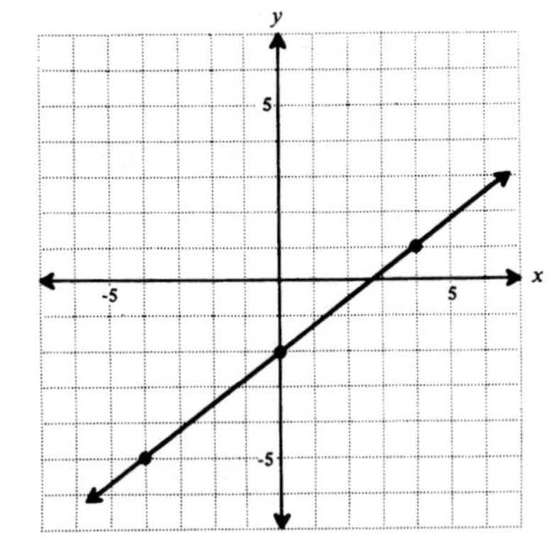

13. $y = -\frac{2}{3}x + 3$

x	y
-3	5
0	3
3	1

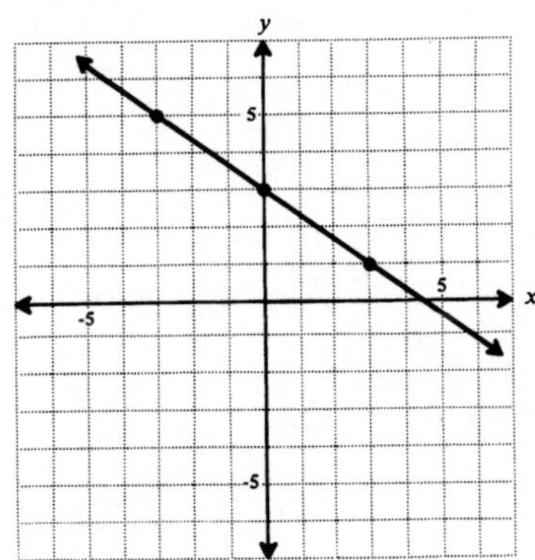

15. $x + y = 4$

x	y
-1	5
0	4
4	0

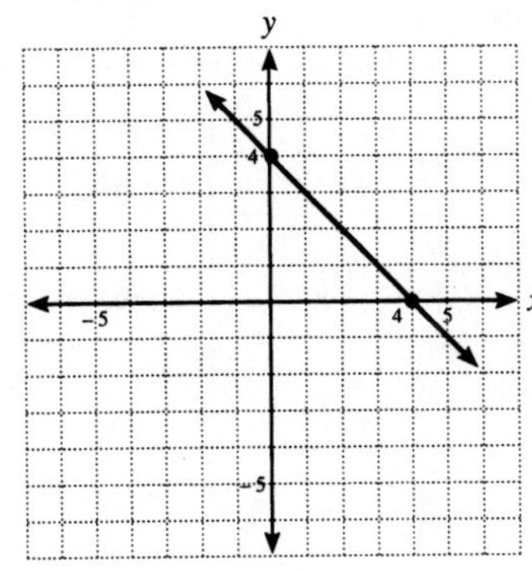

17. $4x - y = 5$

x	y
0	-5
1	-1
2	3

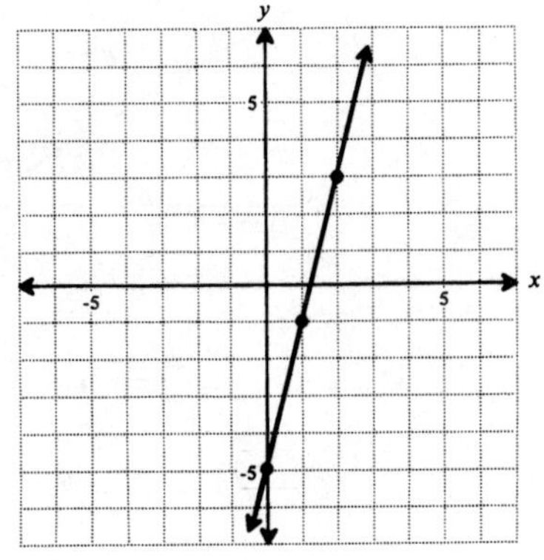

19. $4x + 3y = 9$

x	y
-3	7
0	3
3	-1

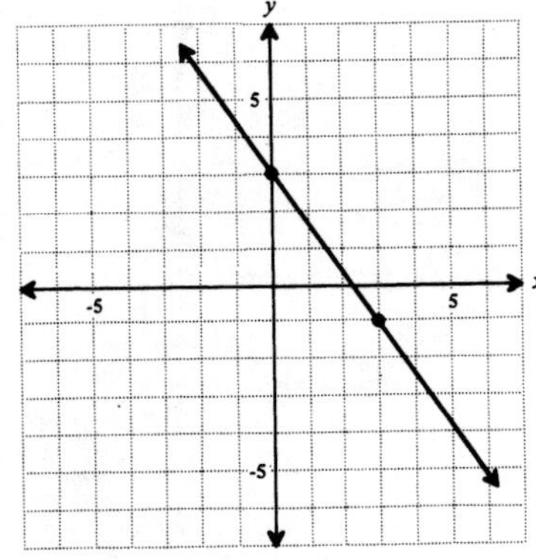

21. $2x - 4y = -12$

x	y
-4	1
0	3
2	4

23. $\frac{2}{3}$

Problem Set 6.3

1. $m = \frac{2}{5}$

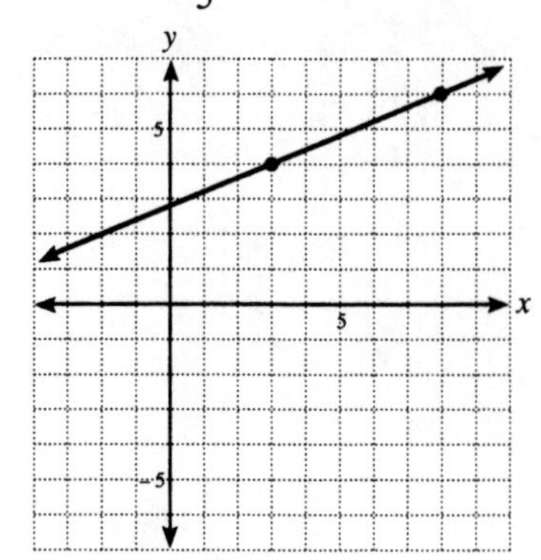

3. $m = \frac{2}{5}$

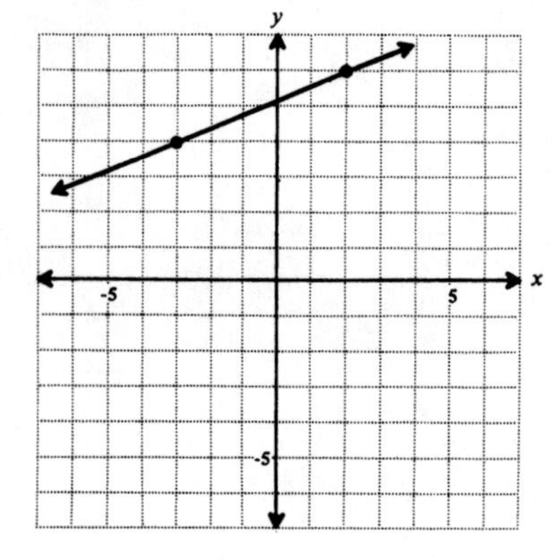

5. $m = \frac{-9}{2}$

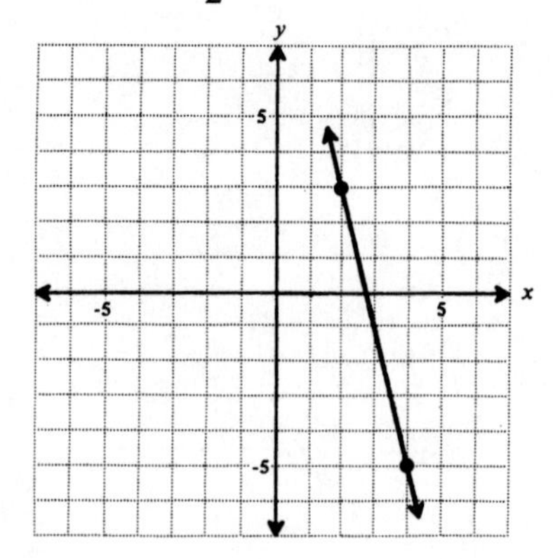

7. $m = \frac{-4}{7}$

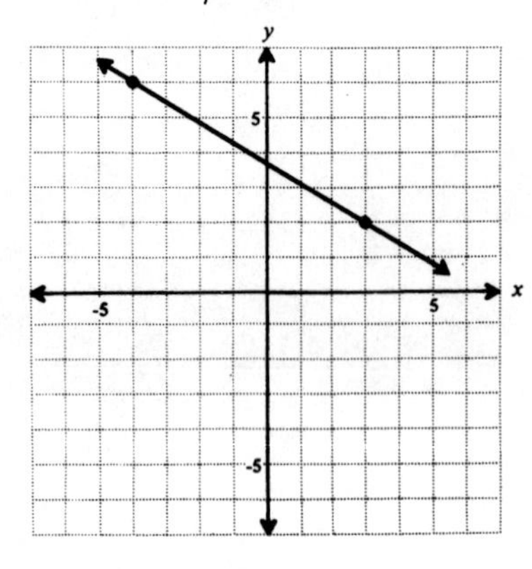

9. $m = \frac{1}{2}$

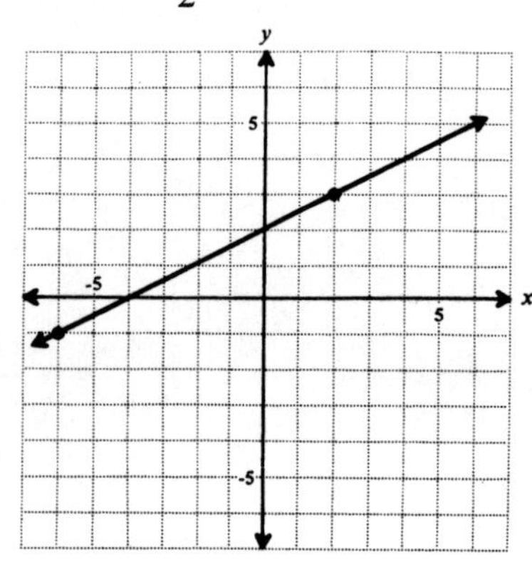

11. $m = \frac{-2}{7}$

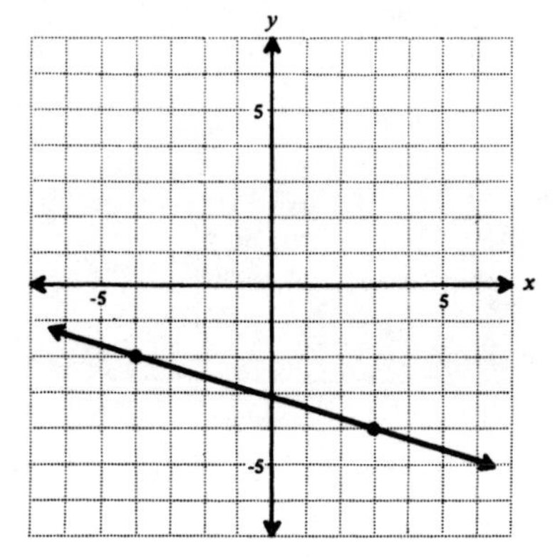

13. $m = \frac{-8}{5}$

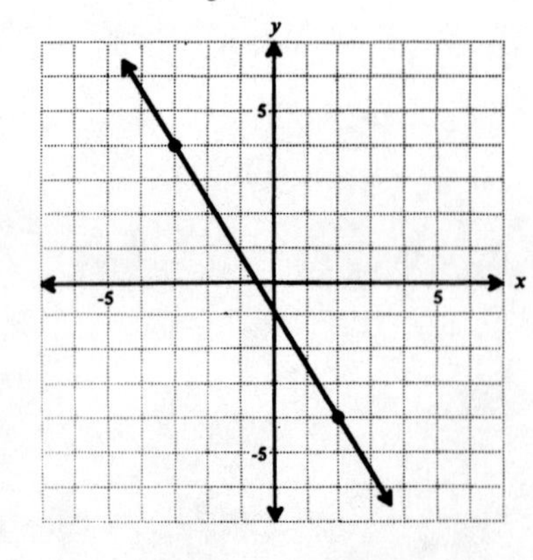

15. $m = \frac{-1}{4}$

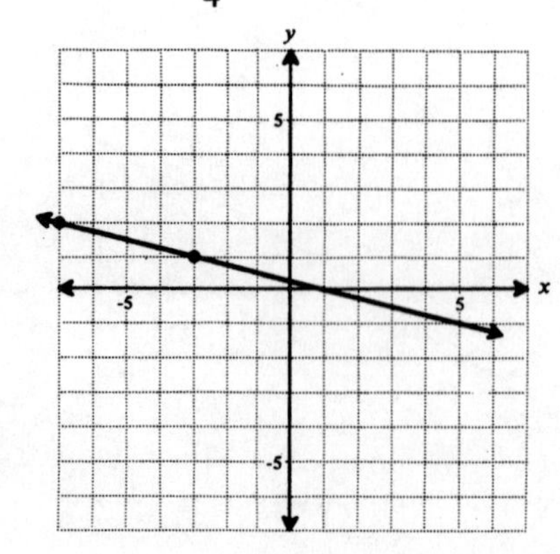

17. $m = -4$

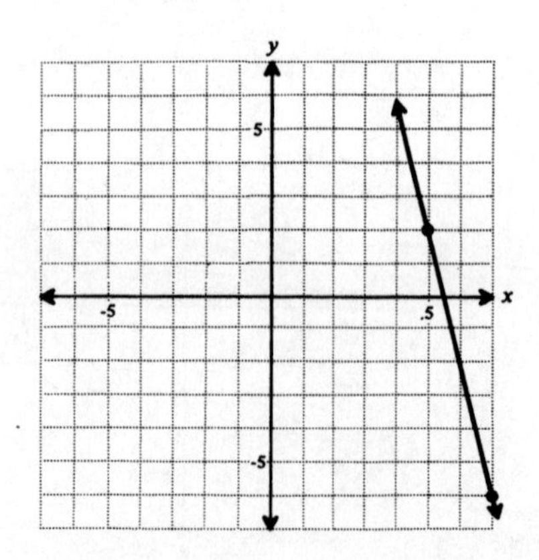

19. $m = \frac{11}{4}$

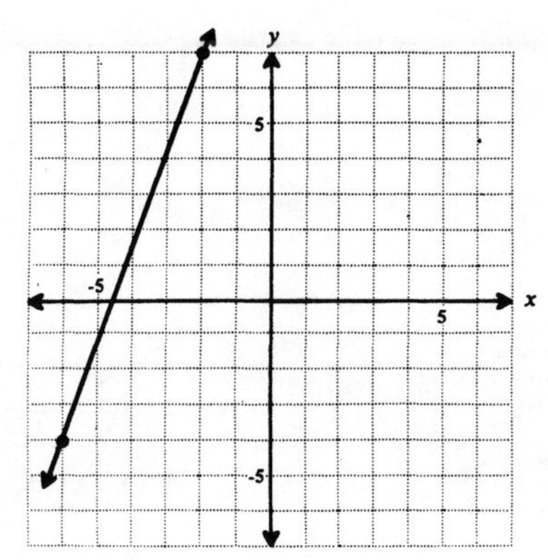

Problem Set 6.4A

1. $y = 2x + 4$
3. $y = -5x + 6$
5. $y = -3x - 4$
7. $y = \frac{3}{4}x + 4$
9. $y = \frac{-4}{7}x - 5$
11. $y = \frac{-3}{2}x + 6$
13. $y = 4x$
15. $y = 5$
17. $y = -2x + 3$
19. $y = \frac{-2}{3}x$
21. 0

Problem Set 6.4B

1. $y = 2x + 1$
3. $y = -2x + 2$
5. $y = \frac{2}{3}x + 5$
7. $y = \frac{-1}{2}x + 5$
9. $y = \frac{-4}{7}x - 2$
11. $y = \frac{-4}{5}x$
13. $y = \frac{2}{5}x + \frac{11}{5}$
15. $y = \frac{5}{4}x + \frac{19}{2}$
17. $m = \frac{4}{9}$

Problem Set 6.4C

1. $y = \frac{-3}{4}x + 3$
3. $y = \frac{-1}{5}x + 4$
5. $y = \frac{-9}{4}x - 3$
7. $y = \frac{12}{5}x + 6$
9. $y = \frac{1}{2}x + \frac{11}{2}$
11. $y = \frac{-9}{4}x - \frac{33}{4}$
13. $y = \frac{-4}{5}x - \frac{9}{5}$
15. $y = x - 1$
17. $y = -x + 5$
19. $y = 7x - 11$
21. $y = -x + 5$
23. $\frac{-3}{4}x + \frac{3}{2}$

Problem Set 6.4D

1. $y = 2x - 4$
3. $y = -2x + 13$
5. $y = \frac{2}{3}x$
7. $y = \frac{3}{5}x + \frac{26}{5}$
9. $y = \frac{-3}{2}x + 4$
11. $y = \frac{-3}{7}x + \frac{32}{7}$
13. $y = -7$

Problem Set 6.4E

1. $y = 2x - 3$
3. $y = 3x + 19$
5. $y = -4x + 26$
7. $y = \frac{3}{5}x - \frac{28}{5}$
9. $y = \frac{-2}{7}x + \frac{34}{7}$
11. $y = \frac{7}{3}x + \frac{43}{3}$
13. $y = \frac{-4}{3}x + \frac{10}{3}$

Problem Set 6.5A

1. y-intercept 2, slope 1

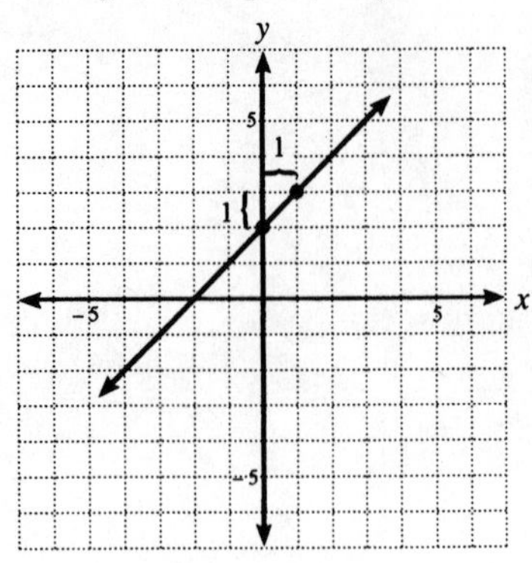

3. y-intercept 1, slope 2

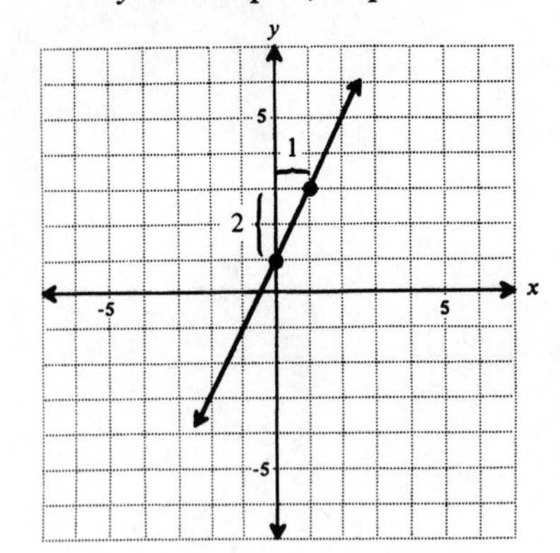

5. y-intercept -1, slope -2

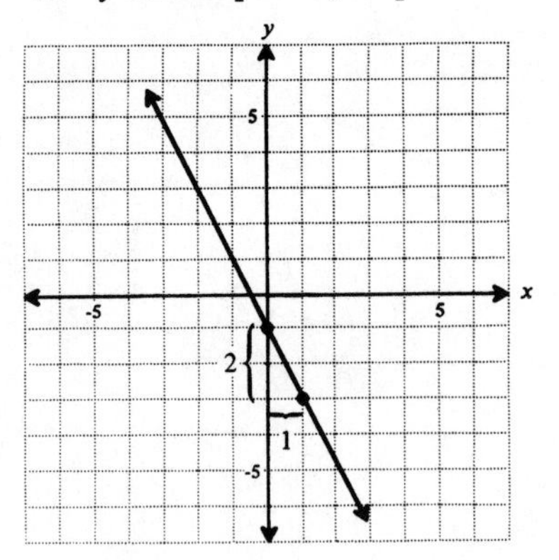

7. y-intercept -2, slope $\dfrac{2}{3}$

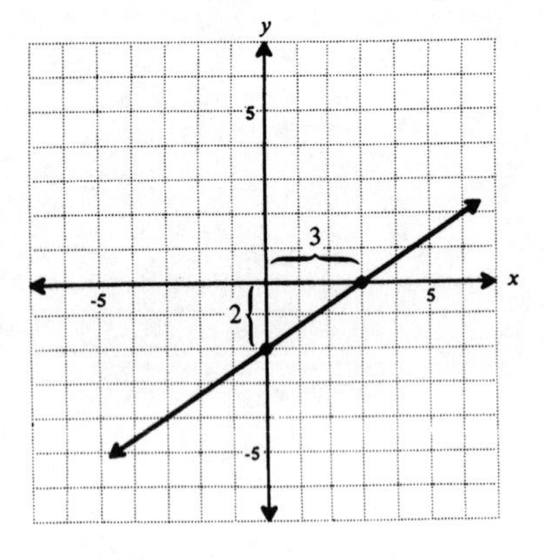

9. y-intercept 4, slope $\dfrac{-3}{5}$

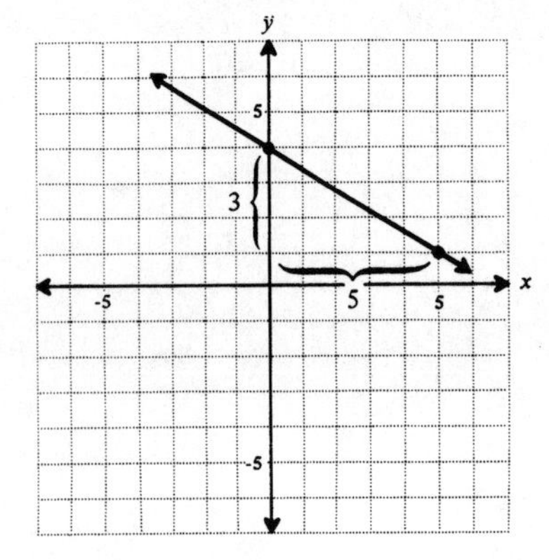

11. y-intercept -4, slope $\dfrac{1}{2}$

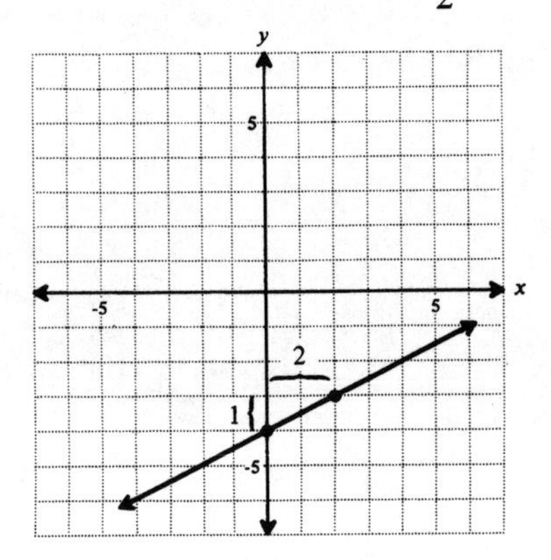

13. y-intercept 1, slope 3

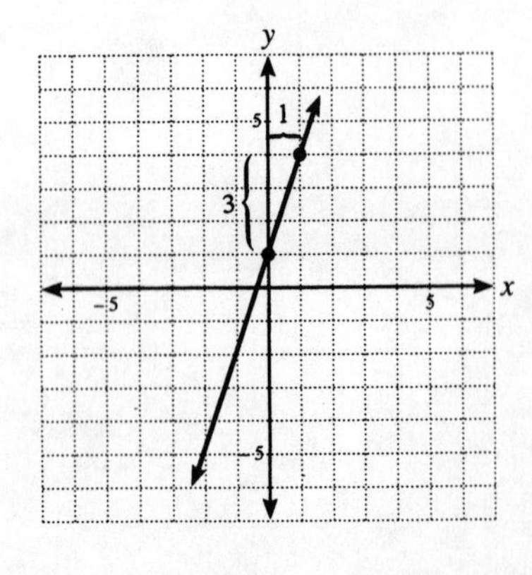

15. y-intercept 3, slope $\dfrac{3}{2}$

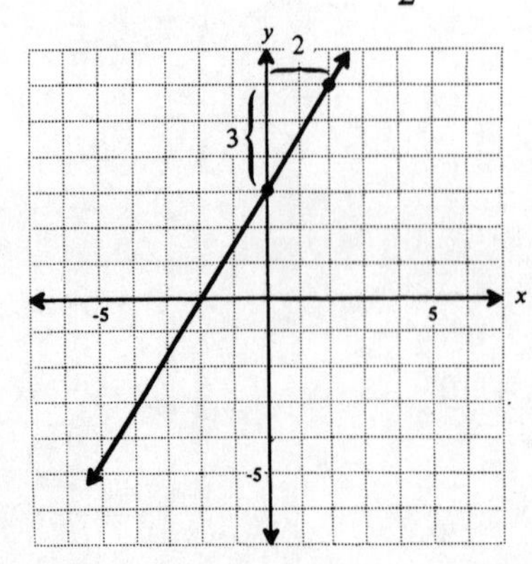

17. y-intercept 4, slope $\dfrac{-4}{3}$

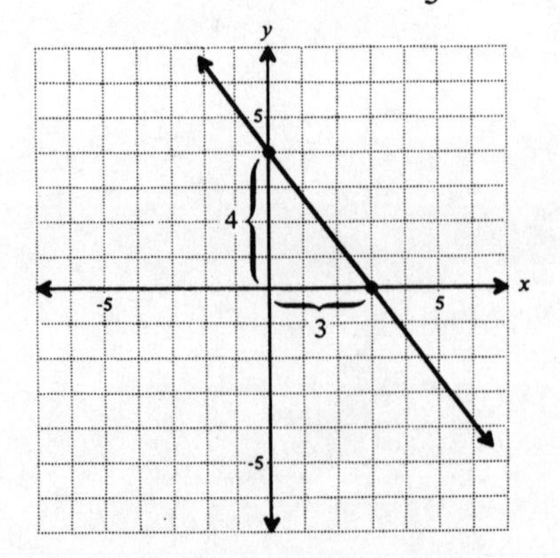

19. y-intercept 4, slope $\frac{1}{2}$

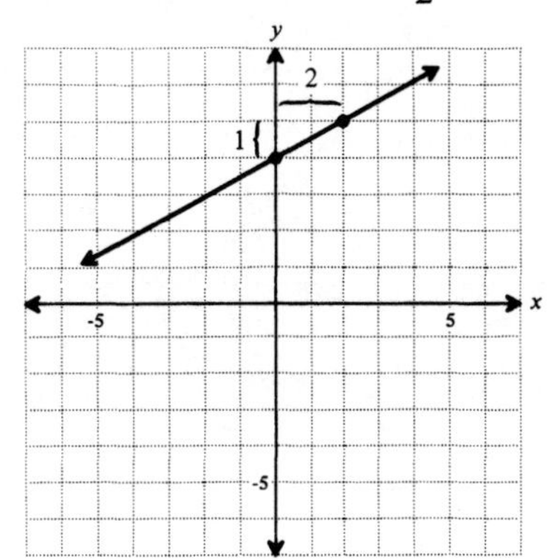

21. y-intercept 0, slope -3

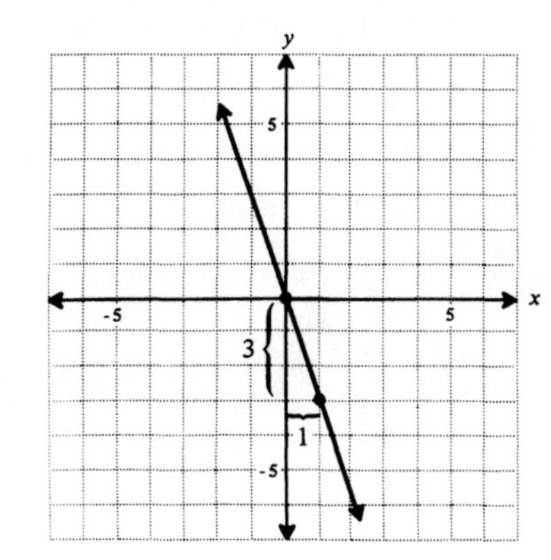

23. y-intercept 2, slope $\frac{2}{3}$

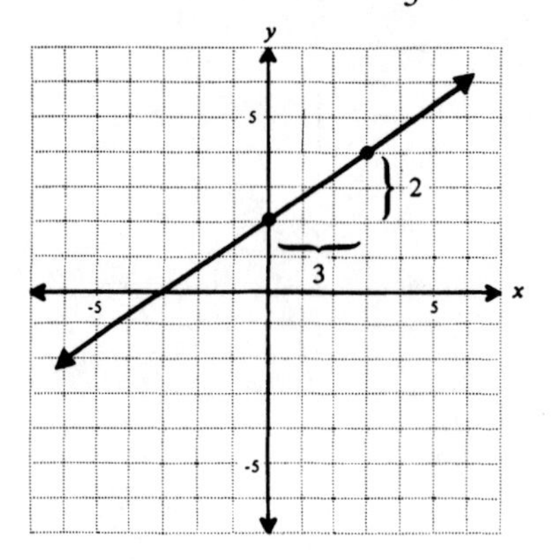

25. -7

27. 9

Problem Set 6.5B

1. Intercepts (0, 4), (4, 0)

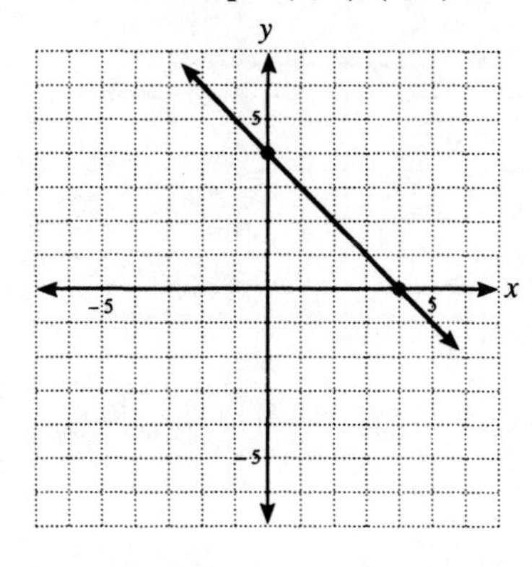

3. Intercepts (0, 2), (3, 0)

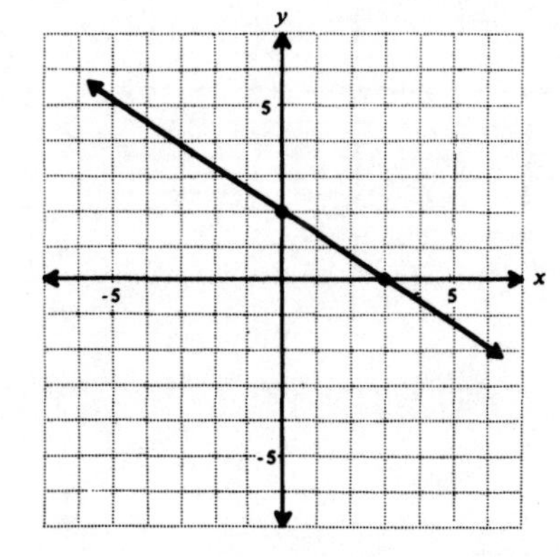

5. Intercepts (0, -3), (2, 0)

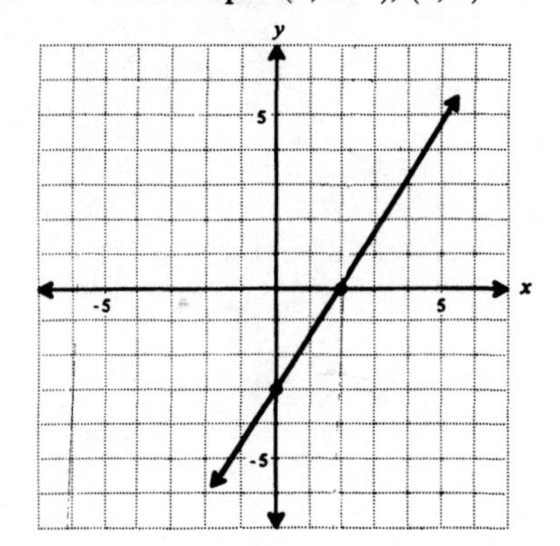

7. Intercepts (0, -2), (1, 0)

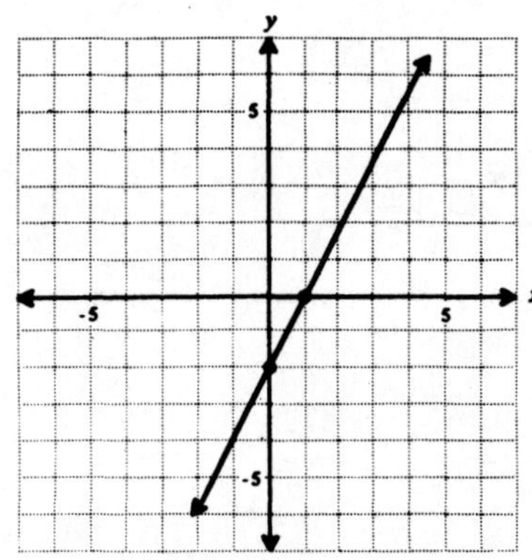

9. Intercepts (0, 3), (-4, 0)

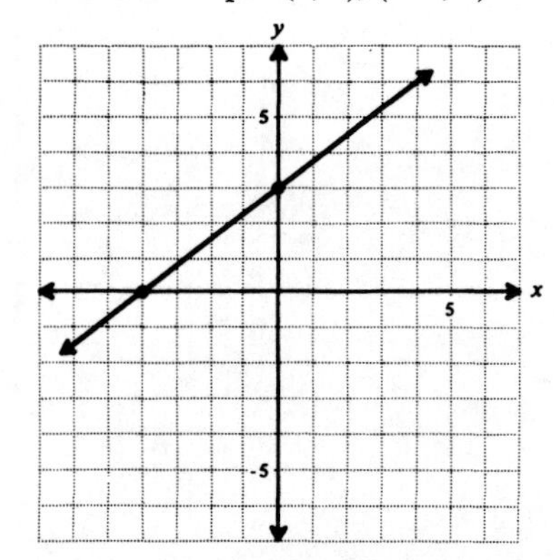

11. Intercepts (0, -4), (2, 0)

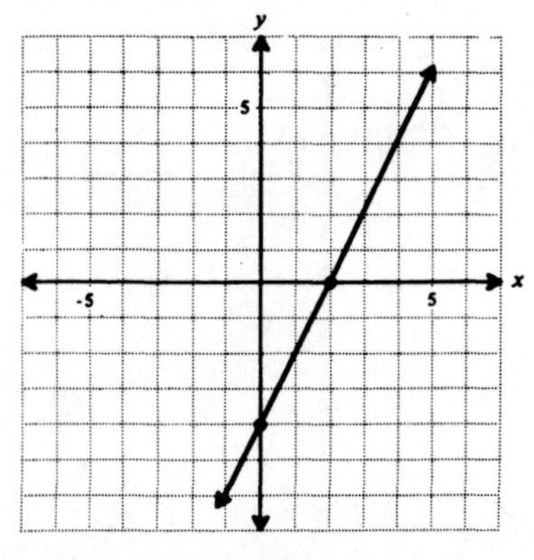

13. Intercepts $(0, -9)$, $(3, 0)$

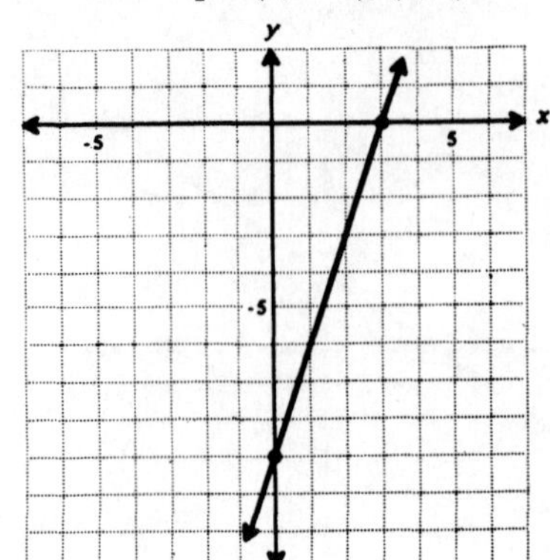

15. Intercepts $(0, -8)$, $(4, 0)$

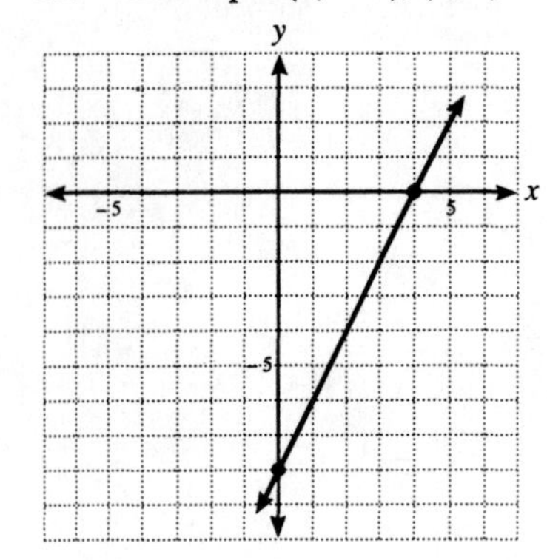

17. Intercepts $(0, 7)$, $(7, 0)$

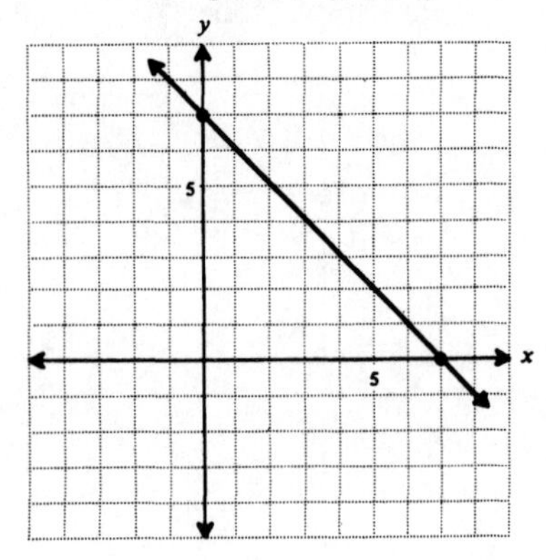

19. Intercepts $(0, -7)$, $(7, 0)$

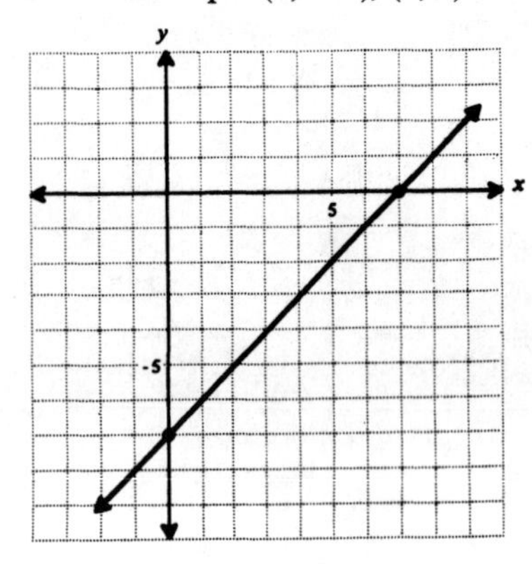

21. Intercepts $(-5, 0)$, no y-intercept

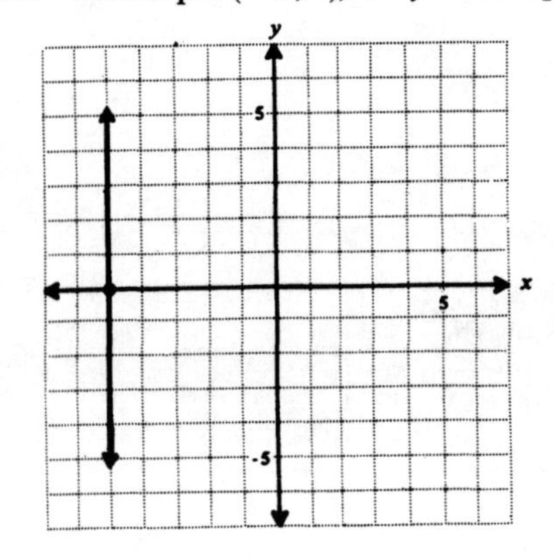

23. Intercepts $(0, 1)$, no x-intercept

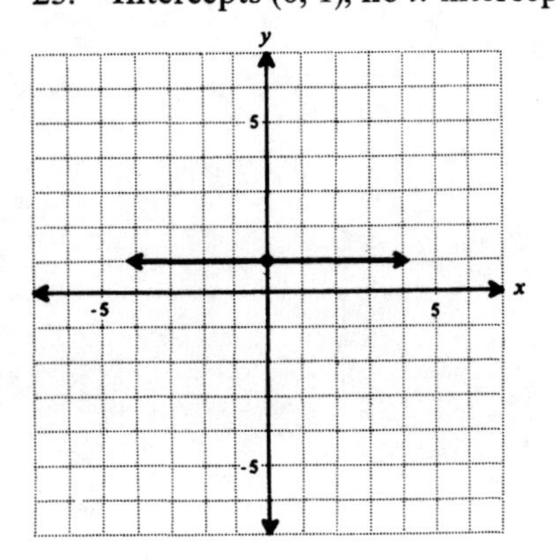

25. Intercepts $(0, 0)$, no y-intercept

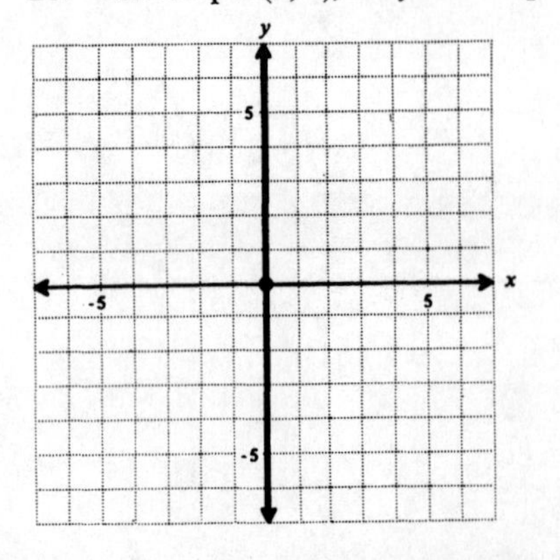

27. Intercepts $(3, 0)$, $(0, -6)$

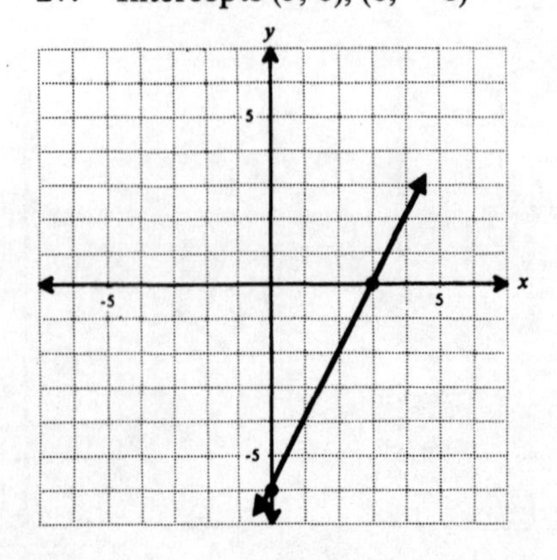

Problem Set 6.5C

1.

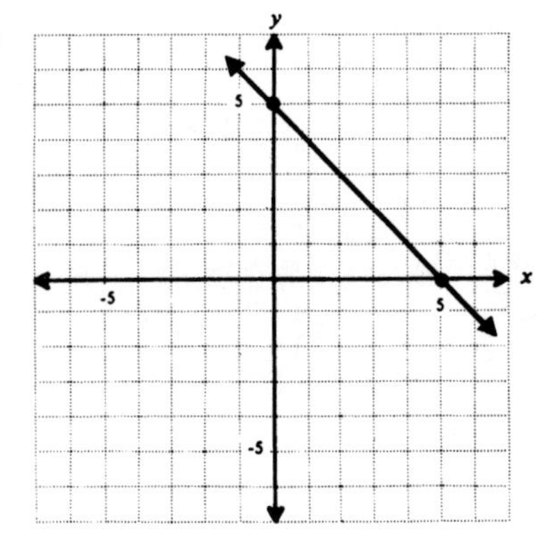

3.

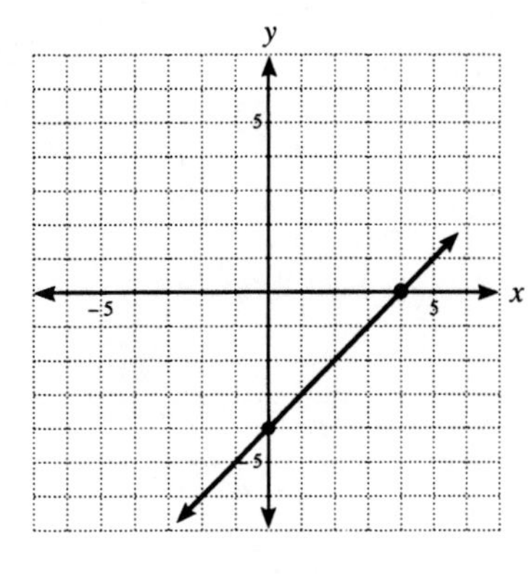

5.

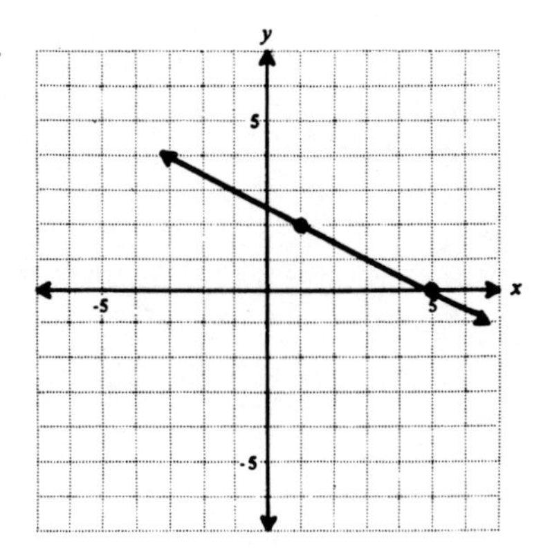

7.

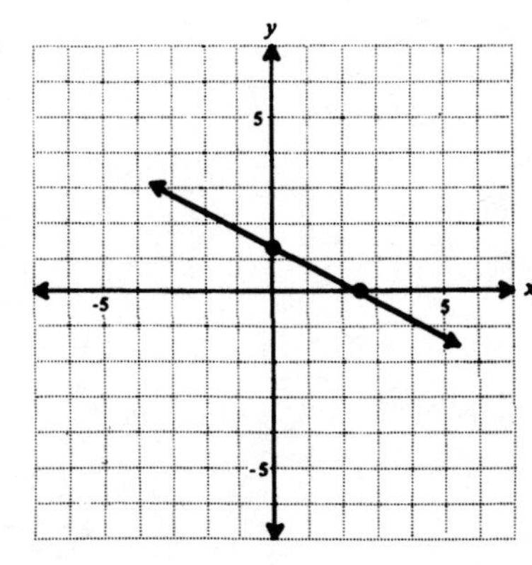

9.

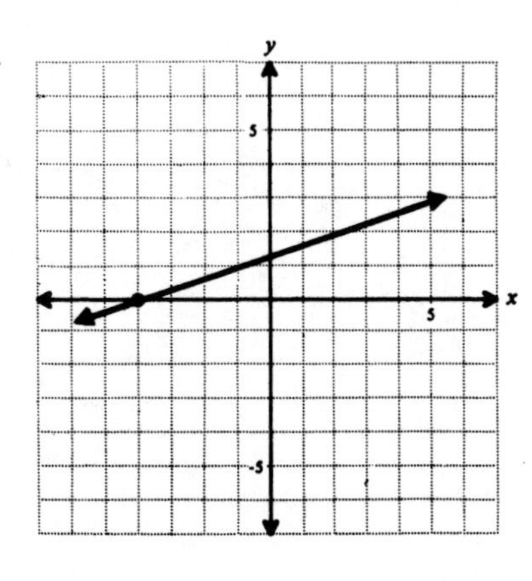

11.

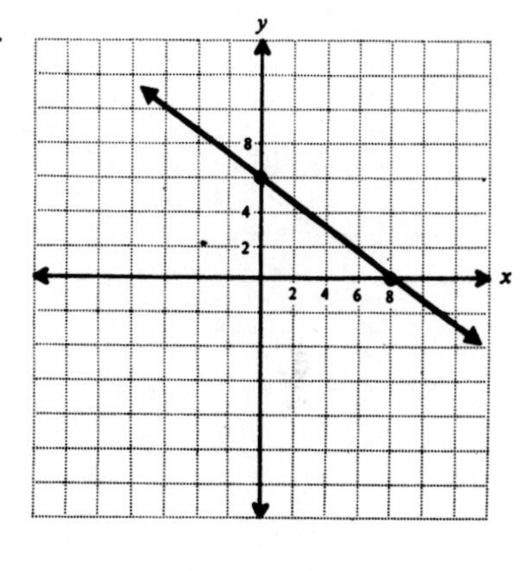

13.

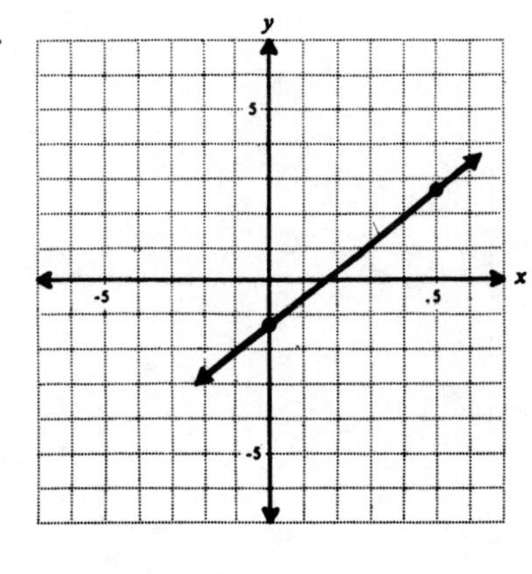

15.

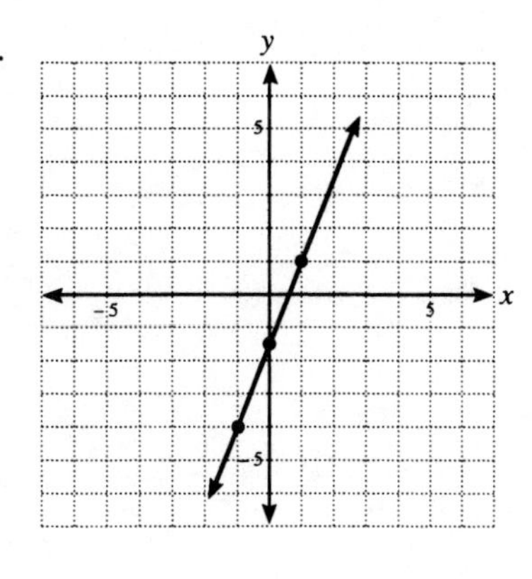

17.

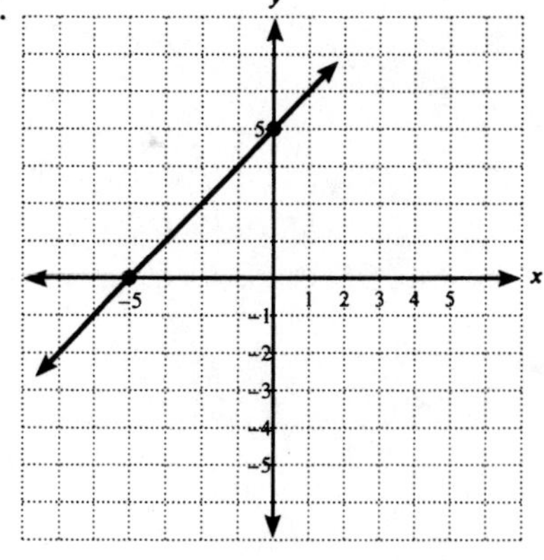

19.

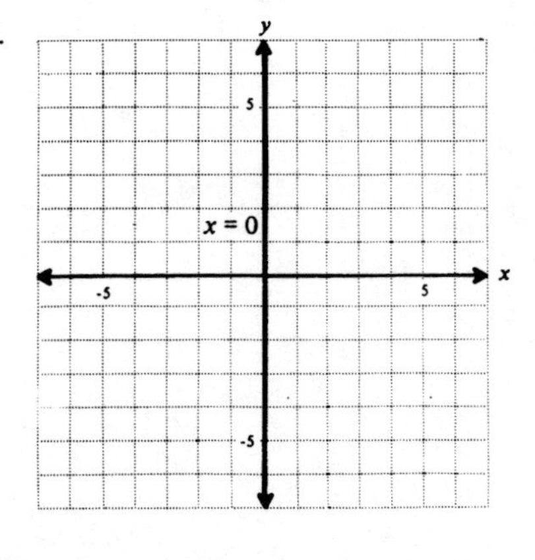

21.

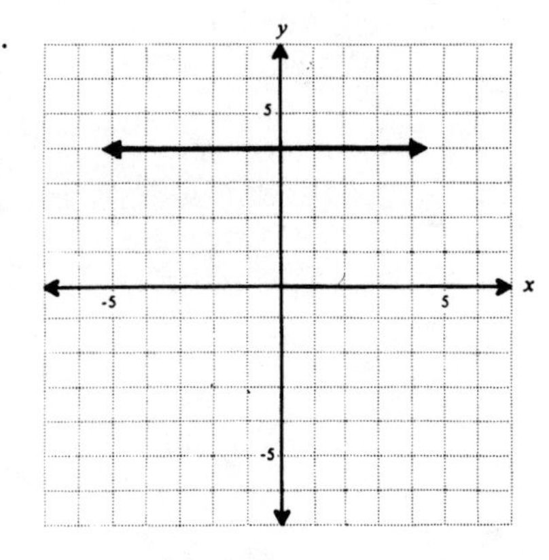

23.

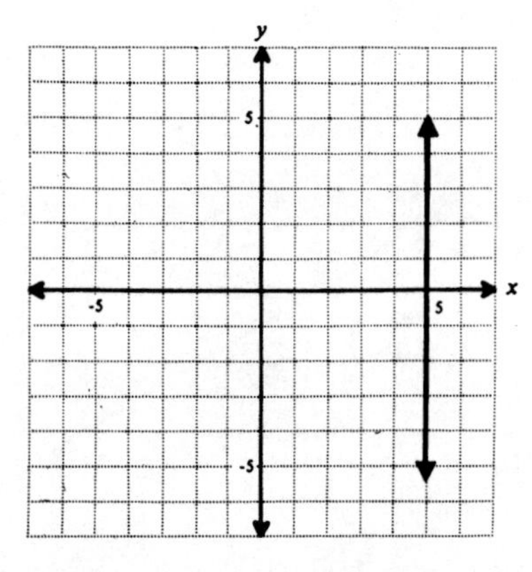

Problem Set 6.6A

1.

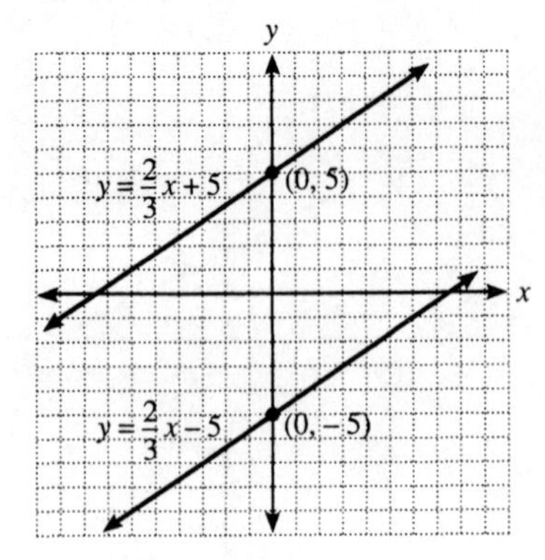

3.

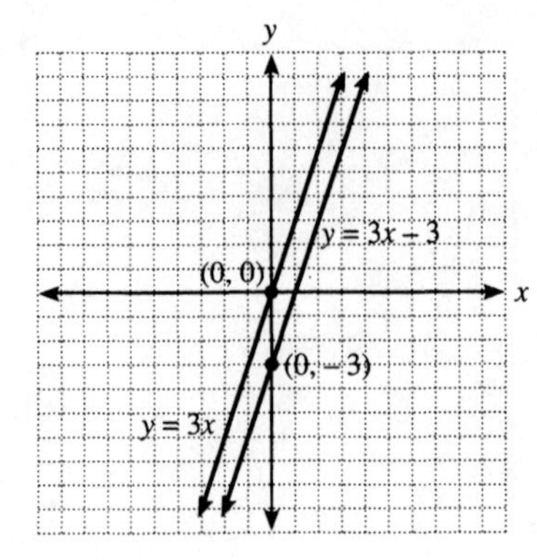

5.

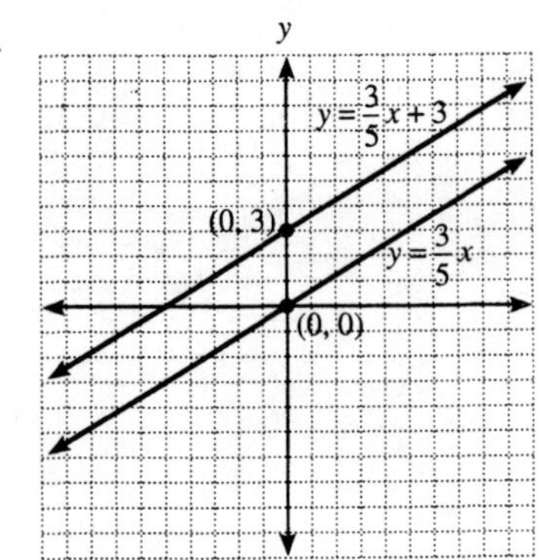

7.

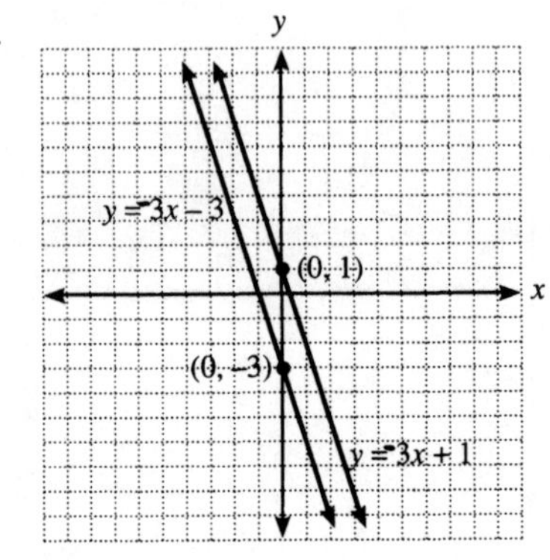

Problem Set 6.6B

1.

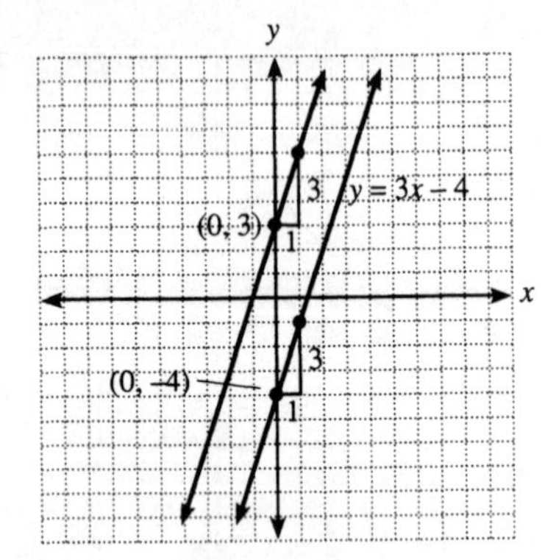

3.

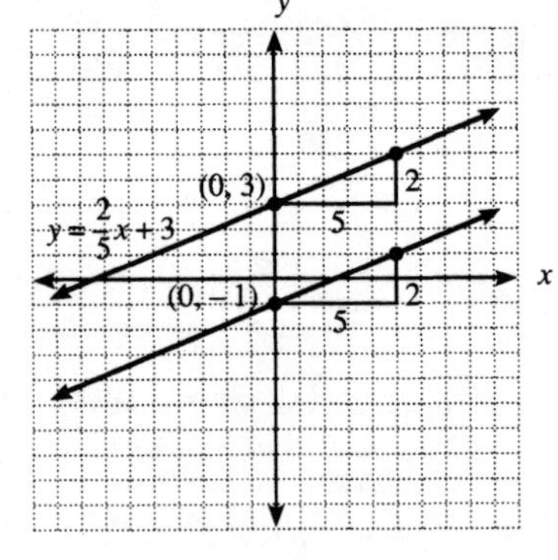

5.

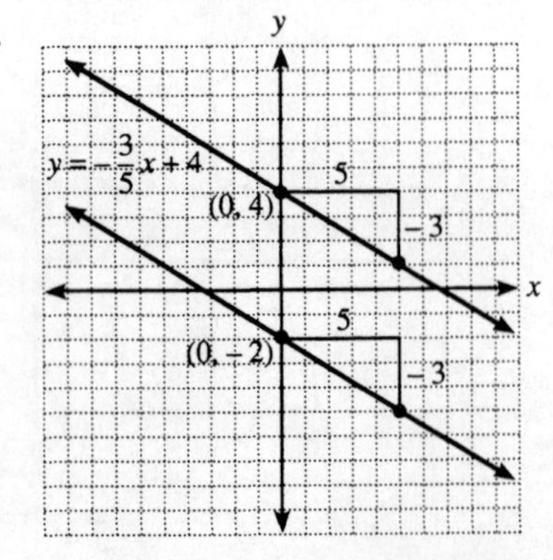

7.

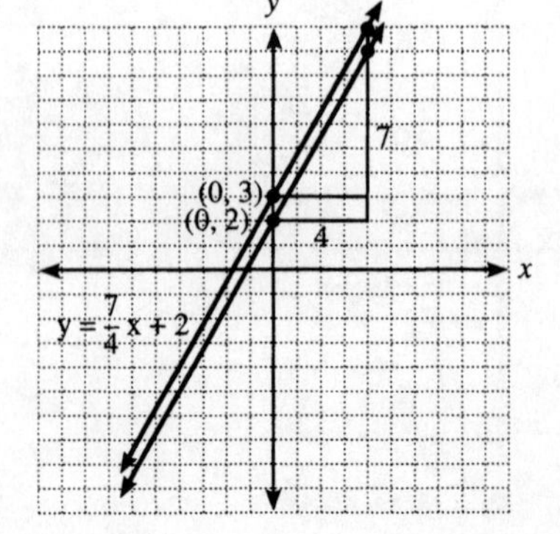

Problem Set 6.6C

1.

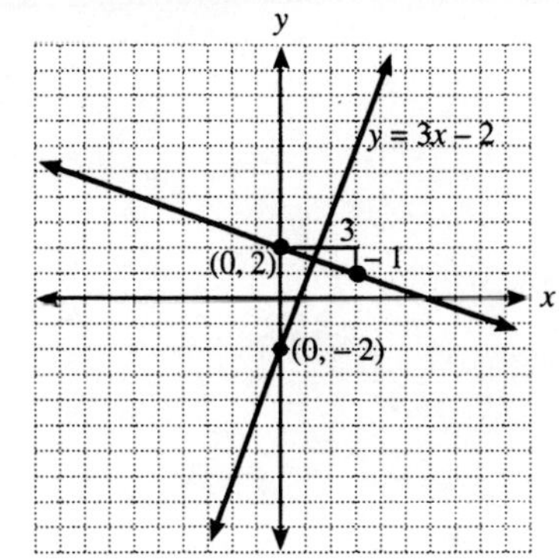

3.

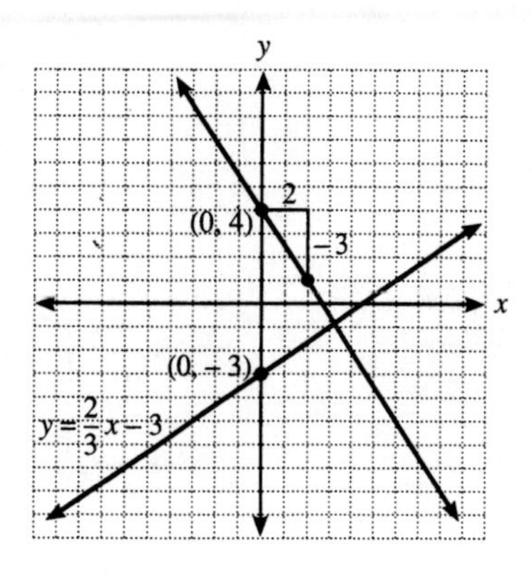

5.

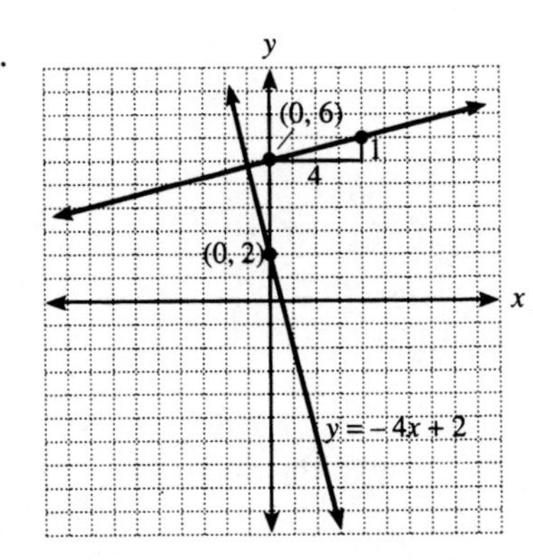

7.

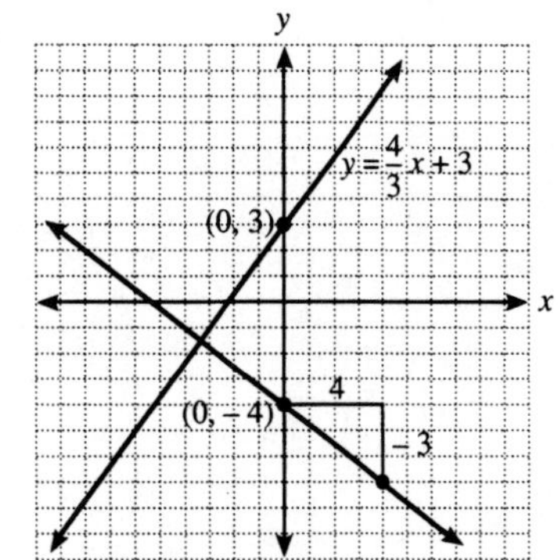

Problem Set 6.6D

1. a) $y = \frac{2}{3}x + 8$
   b) $y = \frac{-3}{2}x - 5$

3. a) $y = \frac{1}{2}x - 4$
   b) $y = -2x + 1$

5. a) $y = \frac{-2}{3}x - \frac{10}{3}$
   b) $y = \frac{3}{2}x - 12$

7. a) $y = \frac{2}{3}x + \frac{10}{3}$
   b) $y = \frac{-3}{2}x + \frac{11}{2}$

9. $y = \frac{2}{3}x + \frac{4}{3}$

11. $y = \frac{4}{3}x + 3$

13. $y = \frac{3}{4}x + \frac{3}{2}$

15. $y = \frac{2}{5}x + \frac{6}{5}$

17. $x = -3$

19. $x = -5$

21. $y = -2$

23. $y = 4$

Unit 6 Review Test

1.

x	$f(x)$
-2	10
0	4
3	-5
-4	16
2	-2

2. $f(0) = -3$
   $f(1) = -1$
   $f(-6) = -15$
   $f(-4) = -11$
   $f(3) = 3$

3.

4.

x	$f(x)$
0	-3
2	-2
-4	-5

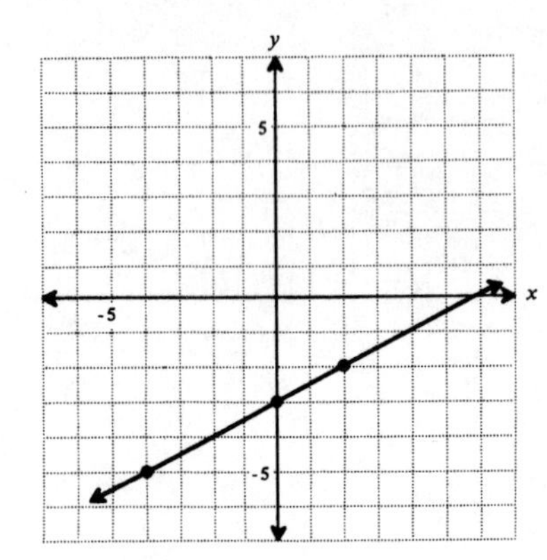

5.

x	$f(x)$
0	3
3	-3
-1	5

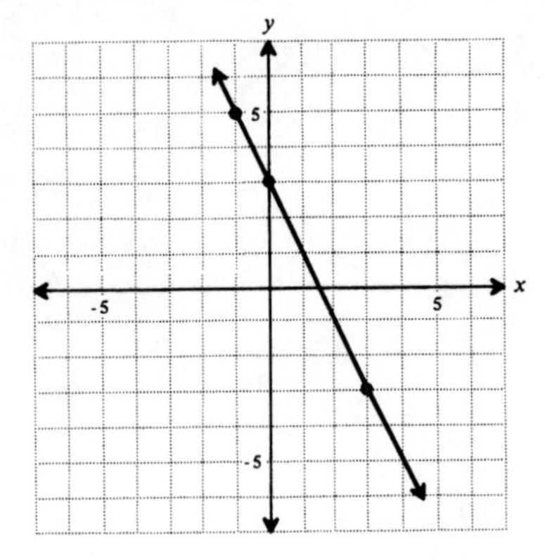

6. 2

7. $\dfrac{-7}{4}$

8. $y = -4x - 3$

9. $y = \dfrac{2}{3}x + 4$

10. $y = \dfrac{2}{3}x + 3$

11. $y = \dfrac{-3}{4}x - \dfrac{3}{2}$

12. $y = \dfrac{2}{3}x + \dfrac{13}{3}$

13. $y = \dfrac{-1}{7}x - \dfrac{11}{7}$

14. y-intercept -1, slope $\dfrac{3}{4}$

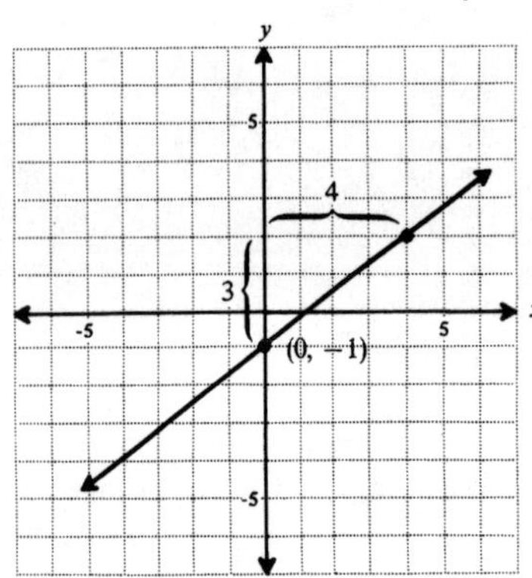

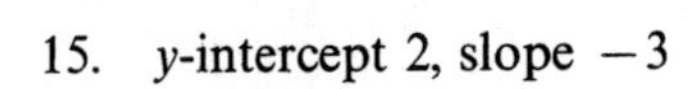
15. y-intercept 2, slope -3

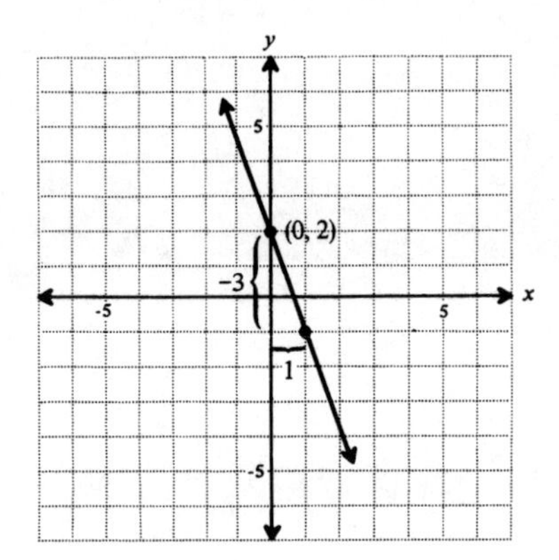

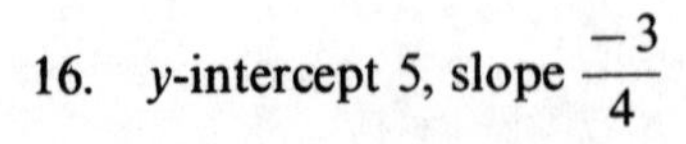
16. y-intercept 5, slope $\dfrac{-3}{4}$

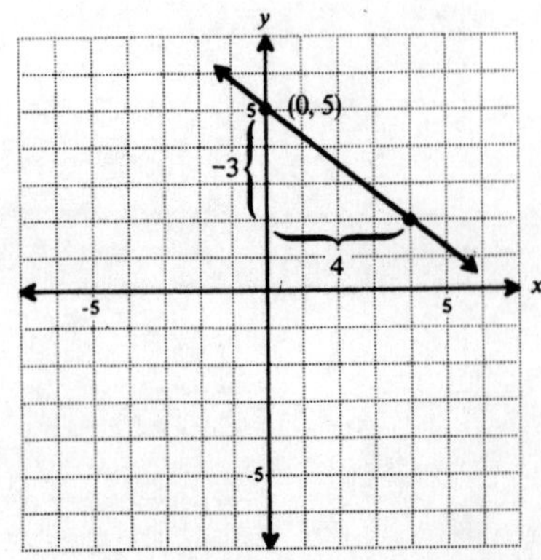

17. y-intercept -3, slope $\dfrac{7}{2}$

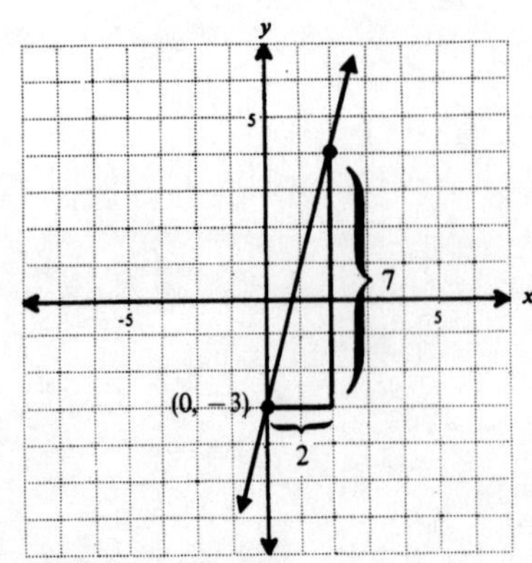

18. Intercepts (0, 6), (4, 0)

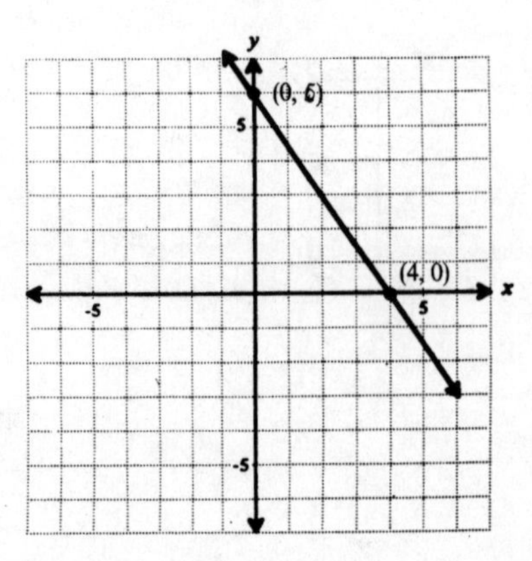

19. Intercepts $(0, -2)$, $(5, 0)$

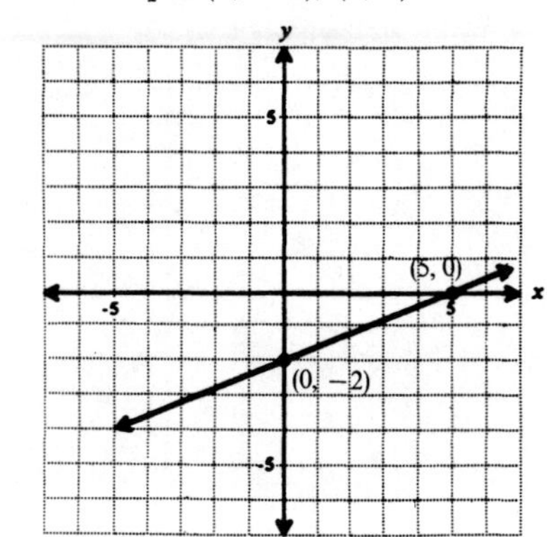

20. Intercepts $(-2, 0)$, no y-intercept

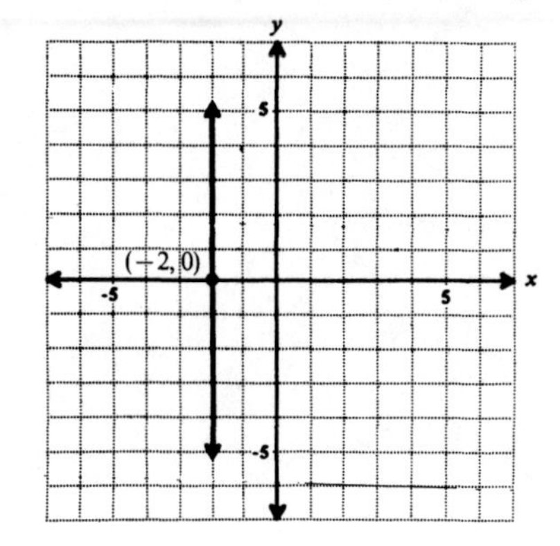

21. Intercepts $(0, 4)$, no x-intercept

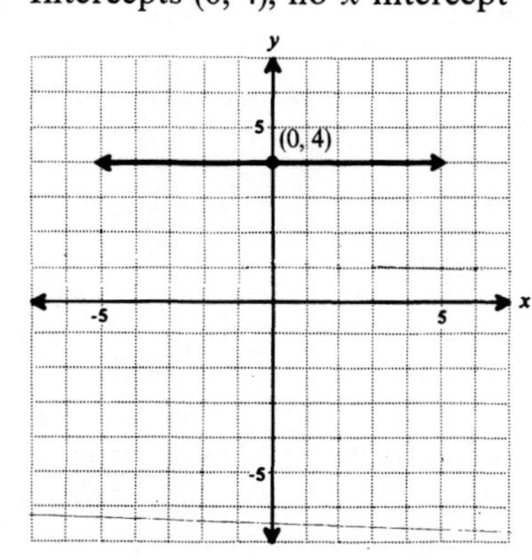

22.

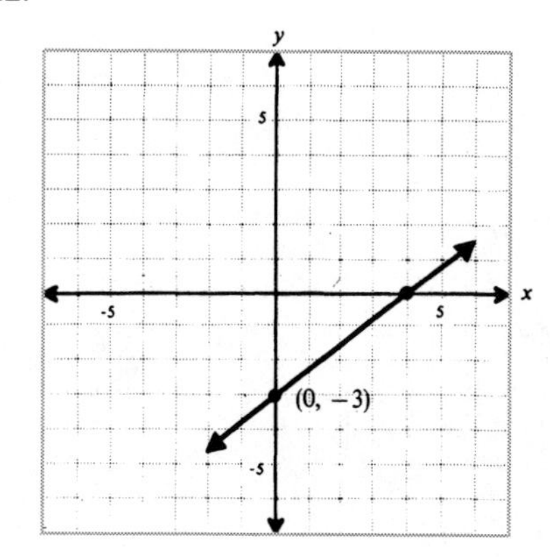

23.

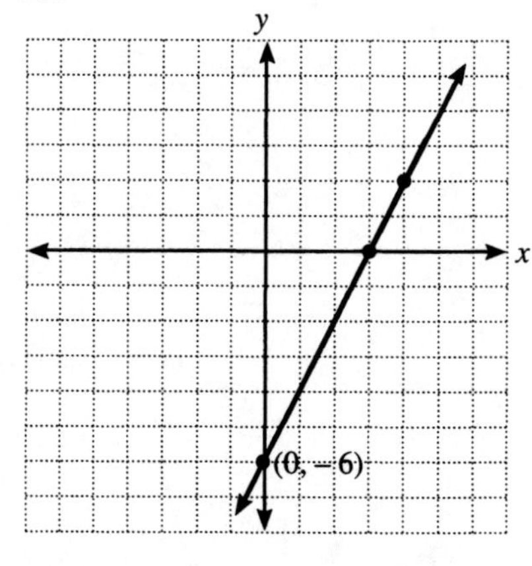

24.

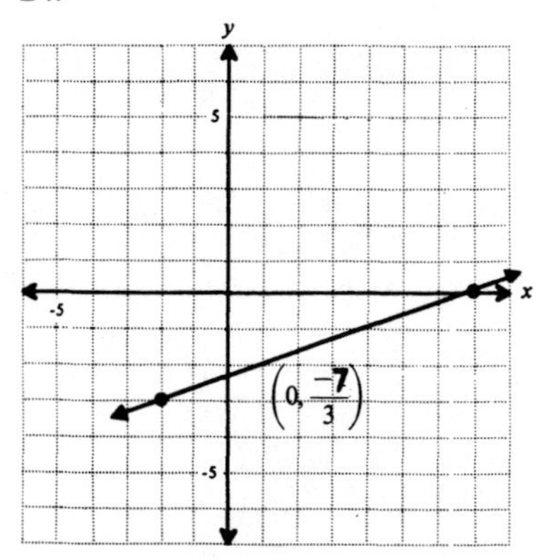

25.

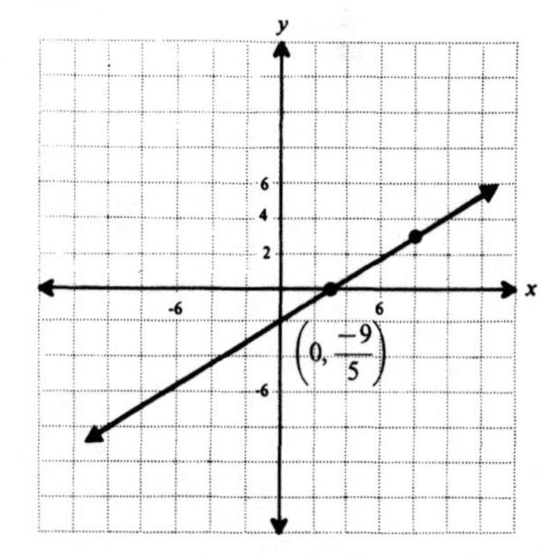

26. $l = \dfrac{2S}{n} - n$

27. $d = \dfrac{l - a}{n - 1}$

28. $\dfrac{2}{3}$

29. $\dfrac{1}{8}$

30. 16

31. $y = \dfrac{2}{5}x + \dfrac{12}{5}$

32. $y = \frac{-3}{5}x - \frac{29}{5}$

33.

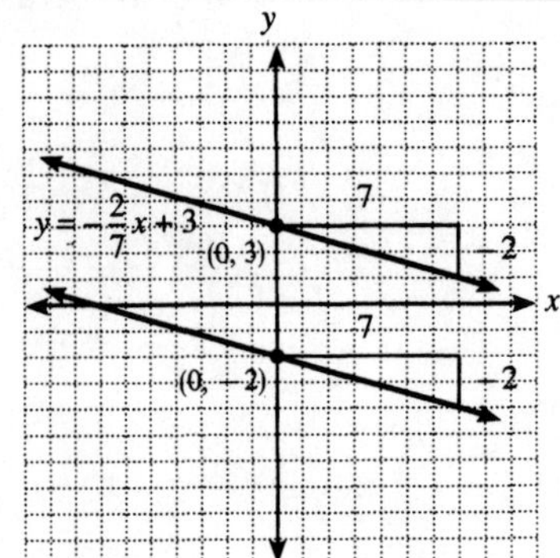

34.

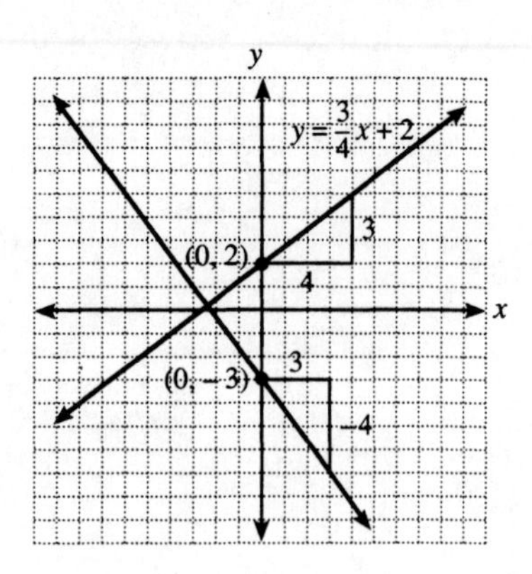

Problem Set 7.1

1. Yes

3. No

5. No

7. Solution (3,5)

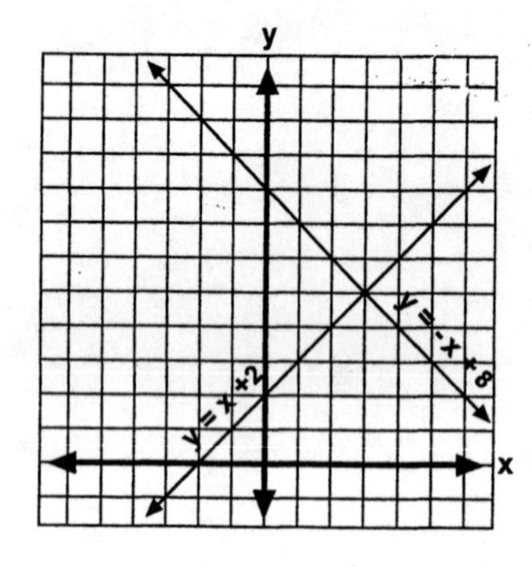

9. Solution $(-2, 4)$

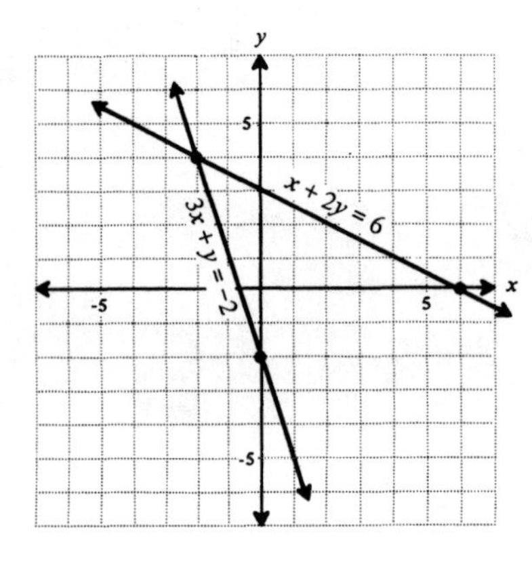

11. Solution $(4, -2)$

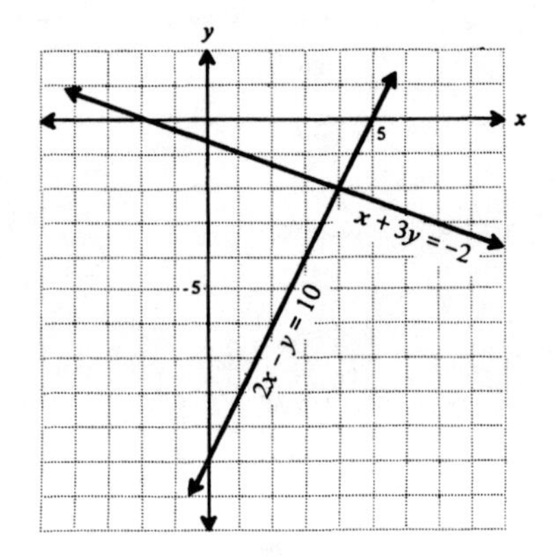

13. Parallel lines

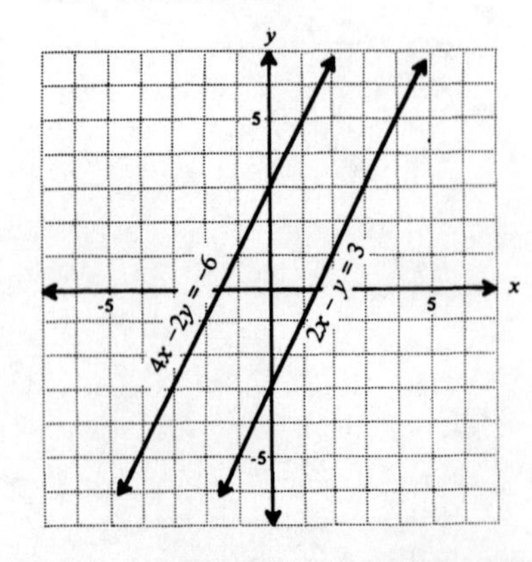

15. Solution $(0, -1)$

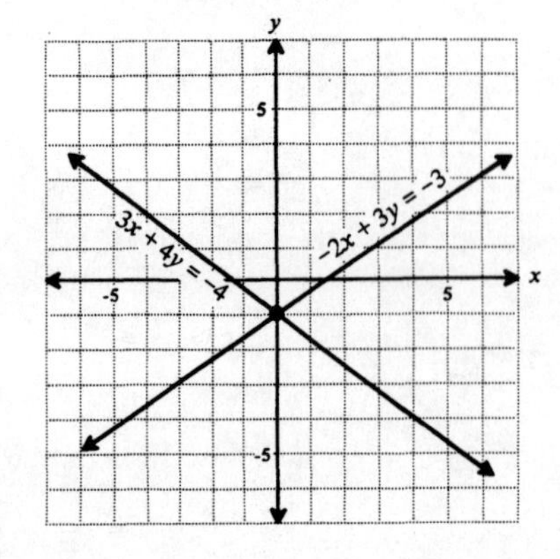

17. Same line

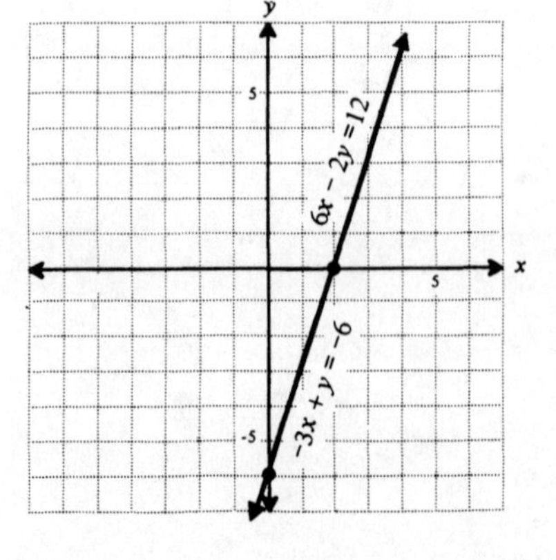

19. Solution (2, −6)

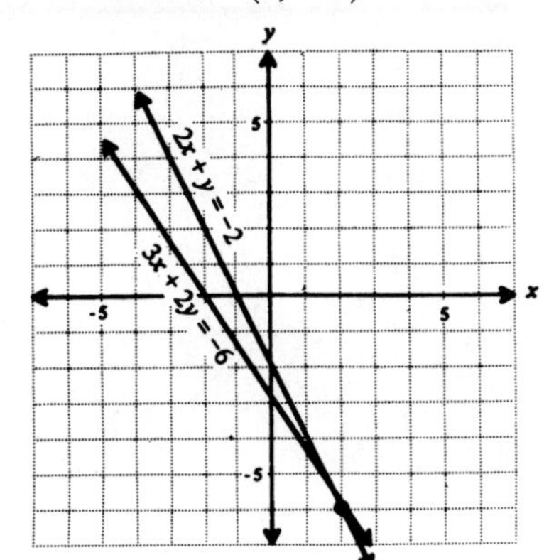

21. Same line

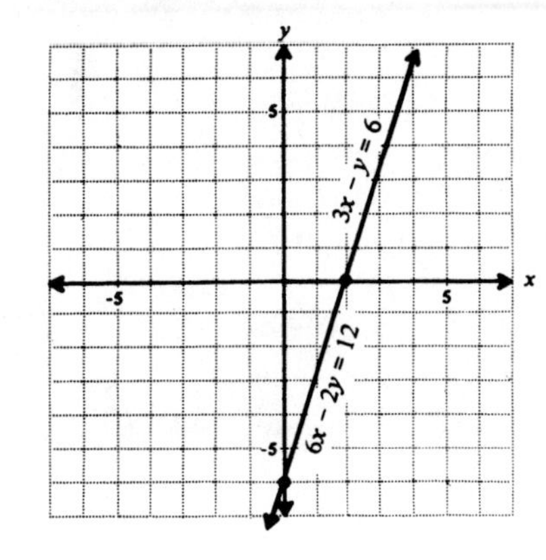

23. Solution (−1, 2)

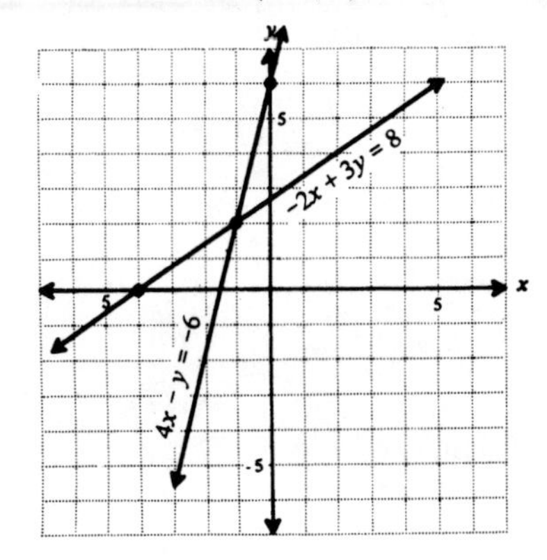

25. Same line

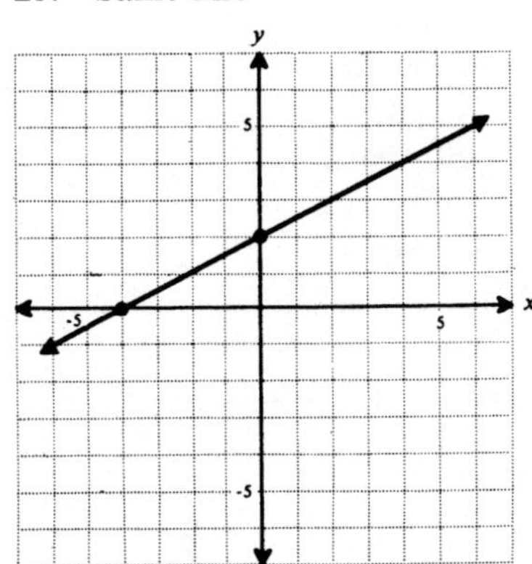

27. Solution (1, 4)

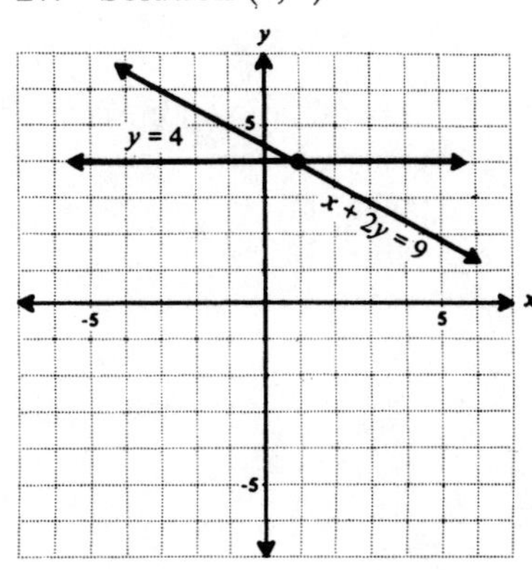

29. Solution (3, −2)

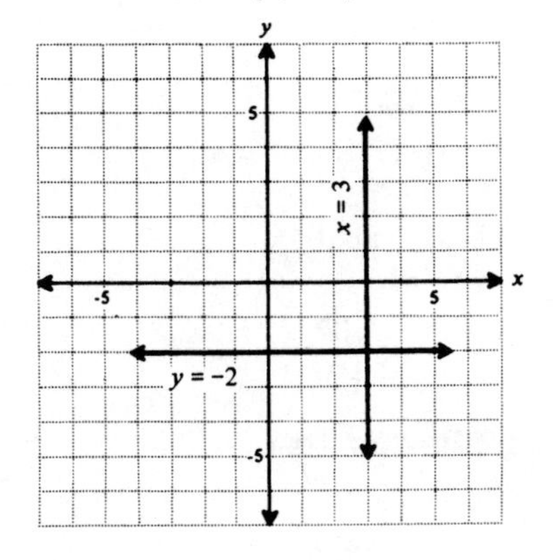

31. Parallel lines

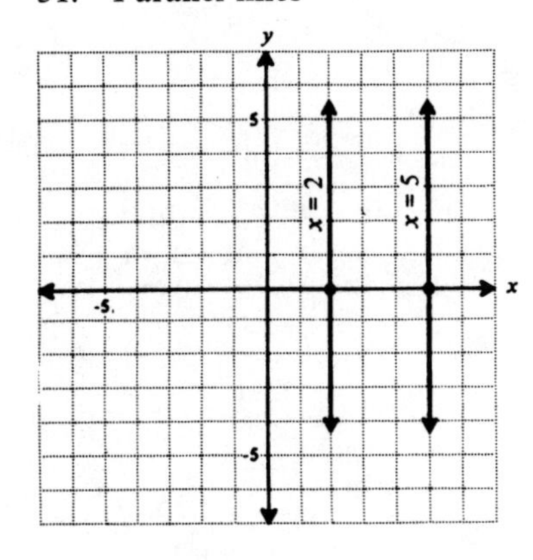

33. Solution (2, 0)

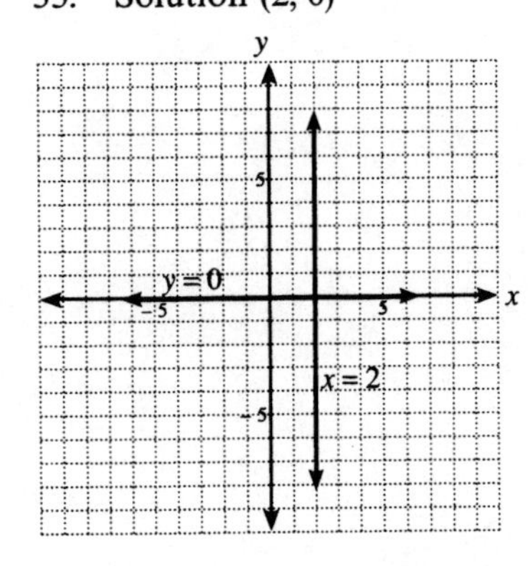

Problem Set 7.2

1. (5, 1)
3. (−4, 0)
5. Parallel lines
7. (−2, 3)
9. (0, 5)
11. Same equation
13. (−2, 1)
15. (−3, 4)
17. (−4, 3)
19. (−2, 4)
21. (8, −1)
23. Parallel lines
25. (−1, 4)
27. Parallel lines
29. (0, 4)
31. (1, 0)
33. $\left(\frac{-1}{2}, \frac{3}{5}\right)$
35. $\left(\frac{2}{3}, \frac{-3}{4}\right)$
37. Same equation
39. (5, 0)
41. 3

Problem Set 7.3

1. (4, 2)
3. (−2, 5)
5. (0, 2)
7. Same equation
9. $\left(\frac{1}{2}, 3\right)$
11. (−4, 0)
13. (2, 1)
15. (6, −2)
17. Parallel lines
19. $\left(\frac{2}{3}, -3\right)$
21. (−4, 1)
23. (−7, 6)
25. Parallel lines
27. (−3, −2)
29. $\left(2, \frac{1}{3}\right)$
31. Same equation
33. $\left(\frac{3}{4}, -1\right)$
35. (−2, −3)

Problem Set 7.4

1. 7, 11
3. 16, 32
5. 17, 51
7. 32, 40
9. 1, −4
11. 13 dimes
20 nickels
13. 48 dimes
16 quarters
15. 400 general admission seats
240 reserved seats
17. 440 adult
510 student
19. \$12,000 at 12%
\$6,000 at 10%
21. \$25,000 at 16%
\$20,000 at 12%
23. 4 lb walnuts
3 lb pecans
25. $2^4a^2b^3$
27. 26

Unit 7 Review Test

1. (−2, 3)
2. (3, −1)
3. (5, 0)
4. (−1, −4)
5. $\left(\frac{3}{2}, \frac{-1}{3}\right)$
6. Same equation
7. (−3, 2)
8. (0, −3)
9. Parallel lines
10. $\left(\frac{-4}{9}, \frac{11}{18}\right)$
11. (4, −3)
12. Parallel lines
13. Same line
14. $\frac{-1}{3}, \frac{8}{3}$
15. 18 nickels, 15 dimes
16. 45 ft, 27 ft
17. 12 lb @ \$4.25/lb
8 lb @ \$3.60/lb
18. 17
19. $y = \frac{17}{3}$
20. Jane: 140 mph
Jim: 160 mph

Problem Set 8.1A

1. $\frac{1}{a^3}$
3. $\frac{1}{x^4}$
5. y^5
7. $\frac{y^4}{x^3}$
9. $\frac{1}{x^3y^4}$
11. x^3y^4
13. $\frac{x}{y^2z}$
15. $\frac{x^2z^4}{y^3}$

Problem Set 8.1B

1. 4
3. 8
5. 9
7. $\frac{-1}{27}$
9. 1
11. $\frac{-2}{3}$
13. $\frac{9}{2}$
15. $\frac{-1}{9}$
17. $\frac{4}{3}$
19. $\frac{-9}{8}$
21. $\frac{-2}{3}$
23. $\frac{-1}{9}$
25. 27
27. 72
29. $\frac{-27}{4}$
31. −1
33. −4
35. 1

Problem Set 8.1C

1. x^3 3. x^5 5. $\frac{1}{x}$ 7. x 9. $\frac{1}{x^3}$ 11. $\frac{1}{a^4}$

13. 1 15. $\frac{y}{x^2}$ 17. x^2y^3 19. $\frac{r}{t^4}$ 21. a^3b 23. $\frac{1}{x^2y^2}$

25. $2xy^3$ 27. $\frac{-7}{y}$ 29. $\frac{3}{5b^3}$ 31. $\frac{2}{3r^2}$ 33. $\frac{x}{y}$ 35. x^5y

37. $\frac{b^7}{a^4}$ 39. $2a^5b$ 41. $-6r^{10}$ 43. $\frac{4b}{a}$ 45. $\frac{6y^2}{x^5}$

Problem Set 8.2A

1. a^6 3. x^{20} 5. b^{12} 7. x^6y^3 9. x^6y^4 11. $a^{15}b^3c^6$
13. $a^8b^4c^2$ 15. $8x^3$ 17. $4x^8$ 19. $-x^6$ 21. $-8x^3$ 23. $4x^2$
25. $16x^4y^2$ 27. $-64n^{12}p^6$ 29. $36m^{12}n^6$ 31. $125s^{15}t^3$ 33. $-27x^6y^9$ 35. 324
37. 36 39. 64 41. 144 43. -64 45. 512 47. 1
49. 0 51. -1 53. -125 55. 0 57. 64 59. 9
61. -43 63. 81

Problem Set 8.2B

1. $\frac{a^2}{b^2}$ 3. $\frac{a^3}{b^3}$ 5. $\frac{a^2}{b^6}$ 7. $\frac{x^2y^2}{z^2}$ 9. $\frac{x^4y^{12}}{z^8}$ 11. $\frac{a^{18}}{b^6c^{12}}$

13. $\frac{a^{10}b^{15}}{c^{25}d^5}$ 15. $\frac{4x^2}{y^2}$ 17. 1 19. $\frac{-27a^{12}}{b^6}$ 21. $\frac{-8x^9}{y^6}$ 23. $\frac{4a^2}{9b^2}$

25. 1 27. $\frac{-27x^6}{8y^9}$ 29. 1 31. $\frac{4}{9}$ 33. $\frac{-8}{27}$ 35. $\frac{-1}{8}$

37. 1 39. $\frac{1}{8}$ 41. 9 43. 1 45. $\frac{64}{81}$

Problem Set 8.3A

1. 40 3. 365.4 5. 8,620 7. 82.65 9. 38,600 11. 56,000
13. 400 15. 9,230 17. 10^2 19. 10^4 21. 10^3 23. 10^2
25. 10^3 27. 10^2 29. 10^1

Problem Set 8.3B

1. 3,200 3. 76,700 5. 560 7. 0.00081 9. 36.78
11. -0.00004 13. 12,000 15. 0.002679 17. 6.54×10^{-2} 19. 9.46×10^2
21. 3.46×10 23. 7.6×10^{-1} 25. 9.58×10^5 27. 5.6×10^{-5} 29. -2.18×10^2
31. -8.6×10^{-4} 33. -2×10^4 35. 1×10^{-6} 37. 0.00000000392
39. 409,400,000,000 41. 0.000000000045 43. 18,000 45. 0.012
47. 180 49. 0.0000061143 51. 28 yrs 53. 330,000
55. $ax + 2a$ 57. $6a^2 - 2ab$

Unit 8 Review Test

1. $\frac{1}{x^2}$ 2. $4x^2$ 3. x^{12} 4. $\frac{1}{x^4}$ 5. $\frac{x^{20}}{y^8}$

6. $\frac{m^4}{n^2}$ 7. a^2b^6 8. y^3 9. $\frac{4b^2}{a}$ 10. $\frac{-3b}{4a^3}$

11. $\frac{x^{15}y^5}{z^{10}}$ 12. $\frac{-8a^6}{27b^9}$ 13. $16a^8b^4$ 14. $\frac{-6}{x^4y^4}$ 15. $\frac{-2}{9}$

16. -27 17. 16 18. $\frac{-9}{8}$ 19. 4 20. $\frac{9}{4}$

21. 0.00567 22. 81,460 23. 2.678×10^6 24. 7.43×10^{-3}

25. 0.1708 26. $m = \frac{-9}{8}$ 27. $y = \frac{-3}{4}x - 5$ 28. $y = \frac{-5}{4}x$

29. $y = -5x - 21$ 30. 36 years old

Cumulative Review Units 1–8

1. -1 2. -9 3. 15 4. -8
5. 7 6. 15 7. -5 8. 8
9. $x^5 + 6x^4$ 10. $-2x^3 + 8x^2 - 10x$ 11. $3x^2 + 2x - 11$ 12. $y^2 - y - 8$
13. $2x + 11$ 14. $\frac{4b}{3}$ 15. $\frac{9}{2}$ 16. $\frac{9a + 2}{12}$
17. $\frac{-x}{2}$ 18. $x = -4$ 19. $y = 2$ 20. $x = 7$
21. $a = 2$ 22. $t = -18$ 23. $y = \frac{x + 2}{4}$ 24. $h = 3$
25. $2n + 4$ 26. $x - 10$ 27. $10d + 25q$ 28. $-42, -43$
29. 6, 15 30. 40 mph, 50 mph 31. -14

32.

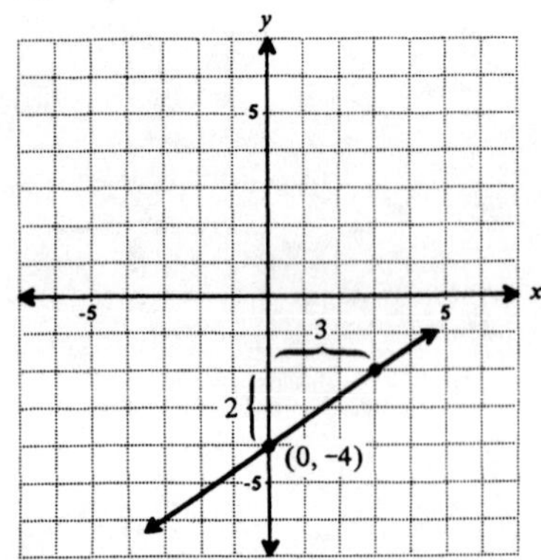

33.

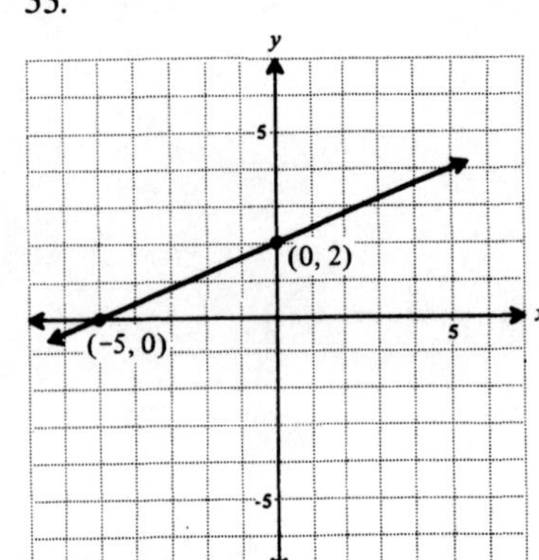

34.

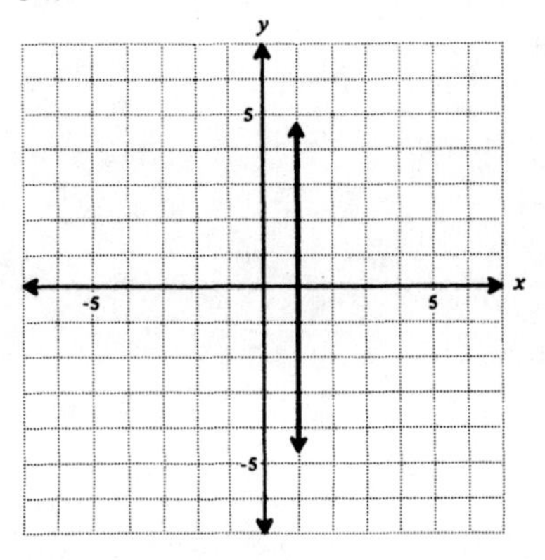

35. $m = \frac{1}{3}$ 36. $y = 3x - 4$ 37. $y = \frac{-2}{3}x + 5$ 38. $y = \frac{1}{4}x - 2$
39. (2, 3) 40. $(-3, -4)$ 41. (2, 1) 42. $(2, -2)$
43. $(-2, 1)$ 44. $(-3, 1)$ 45. $d = 12, n = 36$ 46. 18 lbs peanuts
12 lbs cashews
47. $\frac{20}{x}$ 48. $9a^2b^6$ 49. $\frac{4}{x^3}$ 50. $\frac{-8m^{12}}{27n^9}$
51. 4.21×10^{-4} 52. 161.2

Problem Set 9.1

1. $12a^2b$ 3. $42x^3y^3$ 5. $-56a^2x^2y$ 7. $-3x^5y^3$
9. $3x^2 + 4x$ 11. $10x^2 + 15x$ 13. $-6y^3 - 8y^2$ 15. $21x^3 - 14x^2$
17. $2a^4 + 16a^3$ 19. $-12x^3y + 8xy^3$ 21. $6x^2 + 13x + 6$ 23. $x^2 + xy - 6y^2$

25. $20x^2 - 11xy - 3y^2$
27. $18a^2 - 18ab + 4b^2$
29. $-6a^2 + 23ax - 20x^2$
31. $x^2 - x - 12$
33. $12x^2 + 5x - 28$
35. $2x^2 - 3xy - 5y^2$
37. $12x^2 + 19xy - 21y^2$
39. $x^3 - 5x^2 + 7x - 3$
41. $2x^3 + 9x^2 - 11x + 3$
43. $15x^3 + 28x^2 - 18x - 36$
45. $10x^3 - x^2 - 39x + 27$
47. $x^4 - x^2 + 4x - 4$
49. $x^4 + 6x^3 - 22x + 15$
51. $6a^4 - 4a^3 - 47a^2 + 63a - 21$

Problem Set 9.2A

1. $x^2 - x - 6$
3. $a^2 - 5a + 6$
5. $y^2 + 6y + 8$
7. $x^2 - x - 12$
9. $x^2 - 7x + 12$
11. $3b^2 - 11b - 4$
13. $6y^2 + 11y + 3$
15. $16x^2 - 6x - 7$
17. $18x^2 + 9x - 20$
19. $28a^2 - 23a - 15$
21. $40t^2 + 77t + 36$
23. $2x^2 - xy - y^2$
25. $8x^2 + 6xy + y^2$
27. $6x^2 - 25xy + 24y^2$
29. $15x^2 + 2xy - 24y^2$
31. $6a^2 + 17ab + 12b^2$
33. $6a^2 + ab - 12b^2$
35. $16s^2 - 24st + 9t^2$

Problem Set 9.2B

1. $a^2 + 2a + 1$
3. $x^2 + 6x + 9$
5. $x^2 - 9$
7. $4a^2 - 4a + 1$
9. $9x^2 - 4$
11. $25y^2 - 16$
13. $36a^2 - 25$
15. $x^2 - 2xy + y^2$
17. $a^2 - 4ab + 4b^2$
19. $9r^2 - s^2$
21. $9x^2 + 12xy + 4y^2$
23. $x^2 - 14xy + 49y^2$
25. $a^2 - 4b^2$
27. $16x^2 - 25y^2$
29. $x^2 - 9y^2$
31. $16x^2 + 8xy + y^2$
33. $x^4 - 6x^2 + 9$
35. $9x^4 - 4$
37. $16x^4 - 9y^2$
39. $25x^2 + 30xy^3 + 9y^6$
41. $x^2 + 2x - 3$

Problem Set 9.3

1. $x + 1$
3. $3a + 4b$
5. $5a + 10$
7. $2x^2 - 3x$
9. $4x + 12$
11. $x^2 + 3x$
13. $8x - 5 + \frac{2}{x}$
15. $5x - 1$
17. $5x - 1$
19. $3a - \frac{2a}{b}$
21. $6ab^3 - 9b$
23. $3a^2 - 4ab + 2b^2$
25. $4 - 3a + 6a^2$
27. $3a - \frac{4b}{a} + 5$
29. $-ab^2 + 3 - 6a^2b$
31. $x^2 - \frac{7}{y} + 4xy^2$
33. $-y + 3x - \frac{4y^2}{x}$
35. $\frac{-x^3}{y^3} + 2x^4 - 3y$
37. $10 + \frac{5b^2}{a} - \frac{4a^3}{b}$
39. $\frac{2}{b} - a^2 + \frac{7ab}{3}$

Problem Set 9.4

1. $x - 6$
3. $x + 5$
5. $x - 2$
7. $2x + 3$
9. $2x + 5$
11. $3x - 6$
13. $3x + 4 + \frac{5}{2x + 5}$
15. $6x + 5 + \frac{-4}{7x - 3}$
17. $x^2 - 2x + 3$
19. $x^2 + 4x - 5$
21. $x^2 + 3x - 5$
23. $x^2 - 2x + 8$
25. $4x^2 - 6x + 5 + \frac{-3}{2x + 3}$
27. $2x^2 + 1 + \frac{2}{5x - 2}$
29. $2x^3 + 3x + 2$
31. $4x + 8$
33. $x^2 - 2x - 8$

Unit 9 Review Test

1. $28a^2 - 12a$
2. $2x^2 + 9x - 5$
3. $15x^2 + 29x + 12$
4. $a^2 - 36$
5. $9x^2 - 12x + 4$
6. $3x^4 + 12x^3 + 21x^2$
7. $-6x^4 + 8x^3 + 10x^2$
8. $9x^2 - 4y^2$
9. $x^2 + 8xy + 15y^2$
10. $3a^2 - 5ab - 2b^2$
11. $25a^2 - 16b^2$
12. $25a^2 + 60ab + 36b^2$
13. $6x^2 - 23xy + 20y^2$
14. $12a^3b^2 - 6a^2b^3 + 6ab^4$
15. $-15x^4 + 9x^3 - 12x^2$
16. $4x^3 - 10x^2 + 24x - 10$
17. $6x^3 - x^2 - 27x + 20$
18. $3a^4 - 4a^3 - 7a^2 + 46a - 35$
19. $5x^2 - 3x + 1$
20. $2ab - 3b^5$
21. $a^2 - \frac{2b}{a} - 3a^3b$

22. $2x + 1$

23. $2x^2 + 4x - 5 + \frac{-2}{3x - 1}$

24. $4x^2 + 2x + 1$

25. $3x^2 + 2x - 4 + \frac{-12}{2x - 3}$

26. $y = \frac{-4}{3}x - 3$

27. $3x + 2y = 8$

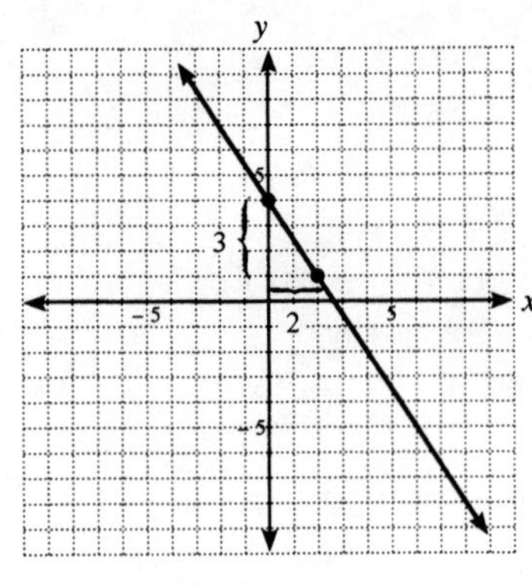

28. Same line

29. $\left(3, \frac{1}{2}\right)$

30. \$36,000 at 10%
\$12,000 at 8%

Problem Set 10.1

1. $2(x + 2)$
3. $2(2x - 5)$
5. $2(3a - 5)$
7. $x(5x + 6)$
9. $4x(x - 1)$
11. $a^2(a - 4)$
13. $y(y^2 - 2y - 5)$
15. $4x(2x^2 - 3)$
17. $2x(2x^2 - x + 5)$
19. $7y(y^3 - 2y - 3)$
21. $b(x^2 + 3x - 5)$
23. $2ay^2(y^2 + 3y - 5)$
25. $4a^2y^2(3ay^2 + 4y - 5)$
27. $5a^3(4a^2b^6 + 5ab^2 - 6)$
29. -1
31. -8
33. 24
35. 84

Problem Set 10.2A

1. 2, 3
3. −2, 3
5. −3, −5
7. 4, −3
9. −6, 3
11. −6, 5
13. −4, −5
15. 8, −6
17. −9, 5

Problem Set 10.2B

1. $(x + 1)(x + 2)$
3. $(x + 1)(x + 6)$
5. $(x + 1)(x + 5)$
7. $(x - 4)(x - 2)$
9. $(x - 1)(x - 8)$
11. $(x - 1)(x - 10)$
13. $(x + 2)(x + 5)$
15. $(x + 2)(x - 5)$
17. $(x + 1)(x - 10)$
19. $(x - 2)(x + 1)$
21. $(x - 1)(x + 4)$
23. $(x - 4)(x + 5)$
25. $(x + 1)(x + 7)$
27. $(a - 9)(a + 2)$
29. $(x - 2)(x + 4)$
31. $(x + 7)(x + 4)$
33. $(x - 4)(x + 6)$
35. $(y - 3)(y - 10)$
37. $(x - 6)(x - 2)$
39. $(r + 3)(r - 6)$
41. $(x - 15)(x + 2)$
43. $(x + 2)(x + 10)$
45. $(x + 4)(x - 6)$
47. $(x + 9)(x - 8)$
49. $(x + 3)(x - 14)$
51. $(b + 5)(b - 13)$
53. $(z - 7)(z + 11)$
55. $(y - 3)(y - 14)$
57. $(x + 5)(x - 11)$
59. $(a - 2)(a - 13)$

Problem Set 10.3

1. $(x + 1)(2x - 1)$
3. $(x + 1)(2x + 1)$
5. $(3x + 1)(x + 2)$
7. $(3x + 2)(x + 1)$
9. $(x + 1)(3x - 2)$
11. $(3x + 1)(x - 2)$
13. $(4x + 1)(2x + 3)$
15. $(4x + 1)(2x - 3)$
17. $(2x + 1)(4x + 3)$
19. $(2x - 1)(4x + 3)$
21. $(2x - 1)(3x - 2)$
23. $(2x - 1)(3x + 2)$
25. $(7y - 10)(y - 1)$
27. $(7y - 10)(y + 1)$
29. $(5y + 2)(y + 2)$
31. $(5y - 2)(y + 2)$
33. $(7y - 5)(y - 2)$
35. $(x - 8)(3x - 1)$
37. $(b - 7)(3b + 2)$
39. $(x - 6)(3x + 1)$
41. $(3b - 2)(b + 7)$
43. $(3x - 4)(2x + 5)$
45. $(11x + 3)(x - 1)$
47. $(4a - 1)(3a + 5)$
49. $(8x - 3)(3x + 7)$
51. $(6x - 1)(4x + 3)$
53. $(20x - 1)(x - 6)$
55. $(1 + 10x)(5 - x)$
57. $(10x + 3)(2x + 5)$
59. $(5 + 3y)(1 - 3y)$

Problem Set 10.4A

1. $(x + 2)(x - 2)$
3. $(x + 6)(x - 6)$
5. $(y - 8)(y + 8)$
7. $(3x + 2)(3x - 2)$
9. $(x + y)(x - y)$
11. $(x - 4y)(x + 4y)$
13. $(a - 6b)(a + 6b)$
15. $(5x - 2y)(5x + 2y)$

Problem Set 10.4B

1. $2(x + 2)(x - 2)$
3. $2(x - 1)(x - 1)$
5. $2(b + 3a)(b - 3a)$
7. $x(2y + 3)(y + 3)$
9. $2(x + 3)(x + 3)$
11. $5(2 + x)(3 - x)$
13. $3x(3x - 1)(x - 2)$
15. $2a(2x - 1)(x + 4)$
17. $4x(2y + 3)(2y + 3)$
19. $8b(3x + 2b)(3x - 2b)$
21. $ab(3 + b)(6 - b)$
23. $8x^2(4x - 3)(4x + 3)$
25. $6xy^2(2y - 3)(2y - 3)$
27. $3x^3(7x - 1)(-3x - 2)$
29. $3a^2b^2(4a + 1)(a - 3)$
31. $8ab^2(4b - a)(4b - a)$
33. $7ax^2(4x - 3)(2x + 3)$
35. $6b^2y(2b - 5)(2b + 5)$
37. $4a^3x(3x + 5)(3x - 5)$
39. $x^3a(a - 5x)(3a + 4x)$
41. $3a^3b(a^2 - 2b)(a^2 - 2b)$
43. $2a^4b^3(a^2 - a + 1)$
45. $3ab^2(b - 3)(b - 3)$
47. $5xy(x - y)(x - 9y)$
49. $9y^2(3 + y)(3 - y)$
51. $4a^2y^2(3a + 7)(a - 2)$
53. $5ax^3(5 + 2a)(5 - 2a)$
55. $3a^2(2a + 3b)(3a - 2b)$
57. $7a^2b^3(3a - 2b^2)(3a + 2b^2)$
59. $2a$
61. $\dfrac{9a^3x^3}{4}$

Unit 10 Review Test

1. $3x(x^3 - 5x + 1)$
2. $2ay^2(y^2 - 3y + 5)$
3. $(x - 6)(x + 4)$
4. $(x - 2)(x - 3)$
5. $(x + 7)(x - 3)$
6. $(3x + 4)(x + 2)$
7. $(5x - 4)(x - 5)$
8. $a(6a - 1)(a - 3)$
9. $2x(x + 4)(x - 3)$
10. $(2x + 1)(3x - 2)$
11. $x^2(4x - 3)(2x + 1)$
12. $2ay(6y - 5)(2y + 1)$
13. $3(2x - 1)(x - 4)$
14. $(2y - 3)(2y - 3)$
15. $(x - 4)(2x + 3)$
16. $(3a + 1)(2a + 3)$
17. $(5x + 4)(5x - 4)$
18. $4xy(y + 3)(y - 3)$
19. $6y(2y - 3a)(2y + 3a)$
20. $4bx^3(3x + 2)(3x + 2)$
21. $\left(\dfrac{-2}{3}, 3\right)$
22. $\left(\dfrac{-22}{7}, \dfrac{9}{28}\right)$
23. 16, 48
24. $-9x^2$
25. $\dfrac{x^4}{y^6}$
26. $\dfrac{1}{x^2y}$
27. $\dfrac{-2b^2}{3a}$
28. -21
29. 0

Problem Set 11.1A

1. $\dfrac{3}{8}$
3. $\dfrac{-16}{3}$
5. $\dfrac{18}{7}$
7. $\dfrac{2x}{3y}$
9. $\dfrac{4x}{y}$
11. $\dfrac{x^3y}{z}$
13. $\dfrac{x^3y}{z}$
15. $\dfrac{-r^2}{2s}$

Problem Set 11.1B

1. $\dfrac{1}{4}$
3. $\dfrac{2}{3}$
5. $\dfrac{-3}{5}$
7. $\dfrac{a}{b}$
9. $\dfrac{2}{x}$
11. $\dfrac{2x^2}{3}$
13. $\dfrac{3}{2a^2}$
15. $3a$
17. $\dfrac{a}{2b^2}$
19. $\dfrac{3y}{8z^2}$
21. $2(x - 3)$
23. $3a(3x - y)(3x + y)$
25. $3(x - 2)(x - 3)$

Problem Set 11.1C

1. $\dfrac{x + 2}{3}$
3. $\dfrac{3}{2x - 1}$
5. $\dfrac{x + y}{x - y}$
7. -1
9. $\dfrac{-3}{4}$
11. $x + 2$
13. $\dfrac{y - 4}{y}$
15. $\dfrac{x + y}{x - y}$

17. $\frac{x+y}{x-y}$ 19. $\frac{-a}{b}$ 21. $\frac{x}{y}$ 23. $\frac{-x}{y}$ 25. $\frac{1}{2x}$ 27. $\frac{-1}{x}$

29. $\frac{a(a-b)}{a+b}$ 31. $\frac{1}{x+2}$ 33. $\frac{-1}{x+1}$ 35. $\frac{x+2}{x-4}$ 37. $\frac{b-a}{a+b}$ 39. $\frac{x-3}{x+2}$

41. $\frac{x+5}{x-6}$ 43. $\frac{a+3}{a-4}$ 45. $\frac{2x-3y}{3x+y}$ 47. $\frac{-(x+y)}{y}$ 49. $\frac{2a}{3}$ 51. $\frac{36a^2}{5}$

Problem Set 11.2

1. $\frac{3}{4}$ 3. $\frac{4}{3}$ 5. 1 7. $\frac{ac}{bd}$ 9. $\frac{1}{b^2}$

11. 1 13. $\frac{1}{2s}$ 15. $\frac{2}{x}$ 17. $\frac{2s^2}{3r^3t}$ 19. $\frac{x+1}{3}$

21. $\frac{x}{x+y}$ 23. $\frac{2}{3}$ 25. -1 27. $\frac{1}{2}$ 29. $\frac{3(x-4)}{4(x+3)}$

31. $\frac{d+3}{9}$ 33. $\frac{2x(x-3)}{3(x+3)(4-x)}$ 35. $\frac{-6}{x-2}$ or $\frac{6}{2-x}$ 37. ay

39. $\frac{y-1}{y}$ 41. $\frac{x+9}{x+3}$ 43. $\frac{x+4}{x-8}$ 45. $\frac{x-3}{4(x-12)}$

47. $\frac{x(x+y)}{x^2+y^2}$ 49. $\frac{4}{x-4}$ 51. $\frac{-1}{3x^2}$ 53. $\frac{-m}{2}$

55. $\frac{2x-1}{x+3}$

Problem Set 11.3A

1. $\frac{1}{3}$ 3. $\frac{8}{3}$ 5. $\frac{-3}{4}$ 7. b 9. $\frac{3y}{4x}$ 11. $\frac{7yz}{6x^2}$

13. $\frac{1}{3-a}$ 15. $b-3$ 17. $\frac{1}{x+y}$

19. $\frac{3x-y}{x+2y}$ 21. None, undefined $\left(\frac{1}{0}\right)$ 23. -1

25. $\frac{1}{5}$ 27. $6b$ 29. $\frac{5xy^2}{2}$

Problem Set 11.3B

1. $\frac{2}{3}$ 3. $\frac{1}{9}$ 5. 6 7. $\frac{ad}{bc}$ 9. a^2 11. 1 13. $\frac{a}{y}$ 15. $\frac{z}{x}$ 17. $\frac{2s}{t}$

Problem Set 11.3C

1. $\frac{x+y}{y}$ 3. $\frac{b}{a}$ 5. 1 7. $\frac{-1}{a}$ 9. $\frac{x}{2}$ 11. $\frac{3x+1}{3(x+1)}$

13. $\frac{x^2}{x+1}$ 15. $\frac{-y}{x}$ 17. $\frac{x+3}{x^2-3}$ 19. $\frac{y+2}{2}$ 21. $\frac{a(a+3)}{a^2+3}$ 23. $\frac{2y-x}{2x-y}$

25. $\frac{3y^2(y-2)}{(y+1)(y+2)(y+3)}$ 27. $\frac{x+1}{x-2}$ 29. $\frac{x^2+1}{x^3}$

31. $\frac{(x+1)^2}{x(x+2)}$ 33. $\frac{1}{4}$ 35. $\frac{1}{3}$

Unit 11 Review Test

1. $\frac{a+3}{3}$ 2. $\frac{a}{b}$ 3. $2-x$ 4. $\frac{x+4}{x-6}$

5. $\frac{1}{3}$ 6. $\frac{2ab(a-b)}{a^2+b^2}$ 7. $\frac{y-2}{y}$ 8. $\frac{-1}{a}$

9. ab 10. $\frac{a+2}{2}$ 11. $\frac{-1}{x}$ 12. $\frac{-x}{2}$

13. $\frac{2x}{x+5}$ 14. $\frac{1}{x^8y^2}$ 15. $\frac{-27x^6}{8y^3w^3}$ 16. -6

17. $30a^2-24a$ 18. b^2-16 19. $4x^2-9y^2$

20. $6x^3+x^2-30x+24$ 21. $\frac{3}{2b}-\frac{5ab}{2}-\frac{3b^2}{a}$ 22. $2x^2-x-2+\frac{-9}{3x-2}$

Problem Set 12.1

1. $\frac{4}{5}$ 3. $\frac{1}{4}$ 5. 1 7. $\frac{x+y}{9}$ 9. $\frac{x-y}{9}$

11. 0 13. $\frac{x^2+x}{9}$ 15. $\frac{6}{x+2}$ 17. $\frac{4}{a+5}$ 19. $\frac{-2}{y+1}$

21. $\frac{-3}{y-3}$ 23. $\frac{7}{x-1}$ 25. 2 27. $\frac{2y-2}{y-2}$ 29. $\frac{2}{m-1}$

31. $\frac{3}{x-3}$ 33. $\frac{2x+1}{x+2}$ 35. $\frac{1}{x+1}$ 37. $\frac{-27+28a}{24}$ 39. $\frac{10ay^2}{21}$

Problem Set 12.2

1. $\frac{7}{6}$ 3. $\frac{1}{2}$ 5. $\frac{x}{18}$ 7. $\frac{y+x}{xy}$ 9. $\frac{7}{6x}$

11. $\frac{x+1}{2x}$ 13. $\frac{3y+x}{2xy}$ 15. $\frac{1+x}{x}$ 17. $\frac{1+2x}{x^2}$ 19. $\frac{8y-7x}{x^2y}$

21. $\frac{6y-5x}{x^2y}$ 23. $\frac{x+4}{2(x+2)}$ 25. $\frac{5x+2}{2x(x-2)}$ 27. $\frac{-32}{(x+4)(x-4)}$ 29. $\frac{4}{x-4}$

31. $\frac{9}{x+3}$ 33. $\frac{42}{x+6}$ 35. $\frac{3}{2(c+1)}$ 37. $\frac{1}{x}$ 39. $\frac{19}{4(y-4)}$

41. $\frac{3}{x+1}$ 43. $\frac{8}{x-4}$ 45. $\frac{5}{2(x-1)}$ 47. $\frac{1}{x+2}$

49. $\frac{2(x-6)}{(x+3)^2(x-3)}$ 51. $\frac{4}{x(x-y)}$ 53. $\frac{2}{(x+2)(x-2)}$ 55. $\frac{3}{(x+2)(x+5)}$

57. $\frac{1}{x(x-3)}$ 59. $\frac{2}{x+2}$ 61. $\frac{-12}{35}$ 63. $\frac{-5a}{4}$

Problem Set 12.3

1. $\frac{2}{3}$ 3. $\frac{4}{5}$ 5. $\frac{9}{16}$ 7. 6 9. $\frac{2}{7}$ 11. $\frac{5}{21}$

13. $\frac{5}{2}$ 15. 18 17. $\frac{3x}{2}$ 19. $\frac{1}{2}$ 21. $\frac{4}{x}$ 23. $\frac{3}{5}$

25. $x+1$ 27. $\frac{a-1}{a^2}$ 29. $\frac{x^2}{x-3}$ 31. $\frac{2-3a}{1+2a}$ 33. $\frac{1+2x}{x}$ 35. $\frac{2x-1}{2}$

37. $\frac{1}{x+1}$ 39. $y-1$ 41. $\frac{3(3x-1)}{x+3}$ 43. $\frac{x}{y}$ 45. $\frac{b+a}{b-a}$ 47. -1

49. $x=4$ 51. $x=\frac{3}{2}$

Problem Set 12.4

1. $12=x$ 3. $x=2$ 5. $x=21$ 7. $x=-15$ 9. $x=2$

11. $\frac{-1}{2}=x$ 13. $t=4$ 15. $x=2$ 17. $x=3$ 19. $x=3$

21. $t=\frac{1}{2}$ 23. $x=19$ 25. $t=24$ 27. $x=4$ 29. $x=-1$

31. $x=-6$ 33. $x=3$ 35. No solution 37. $x=4$ 39. $x=2$

41. $x=3$ 43. $x=-5$ 45. $x=2$ 47. $x=10$ 49. $x=\frac{22}{3}$ or $7\frac{1}{3}$

51. $y=\frac{-2}{3}$ 53. $x=2$ 55. $y=\frac{13}{2}$ or $6\frac{1}{2}$ 57. $x=\frac{5}{4}$ 59. $x=2$

Problem Set 12.5

1. 60 3. $\frac{8}{12}$ 5. 7 7. $\frac{16}{32}$ 9. 24 students 11. 6 hours

13. $2\frac{2}{5}$ hours 15. $285\frac{5}{7}$ hours 17. 12 seconds 19. $39\frac{3}{23}$ minutes 21. $74.25

Unit 12 Review Test

1. $\frac{-1}{3x+1}$ 2. $\frac{2}{x-2}$ 3. $\frac{-3}{x+3}$ 4. $\frac{-3}{x(x+3)}$ 5. $\frac{4x+1}{2}$

6. $\frac{3x+1}{3-x}$ 7. $\frac{-(1+a)}{a}$ 8. $x=2$ 9. $x=2$ 10. $x=-3$

11. $\frac{11}{17}$ 12. $8\frac{4}{7}$ seconds 13. $-32x^3y^6+16x^2y^7-24x^4y^9$

14. $10x^3-31x^2+45x-18$ 15. $\frac{4ab^2}{3}-\frac{2a^2}{b}-3b^3$ 16. $2x^2+4x+3+\frac{18}{4x-3}$

17. $2ax^2(2x-5)(x+3)$ 18. $2ax^2(2ax^2-3x+4a^2)$ 19. $(4x-5y)^2$

20. $2a^2x(3x+4)(3x-4)$

1. -20
2. -1
3. -8
4. 60 square meters
5. $12b^7$
6. $48a^3x^7 - 54a^2x^6 + 36ax^3$
7. $-5xy^2 - 4x^2y + 10xy - 2x - y$
8. $-16x^2 + 15xy + 7y^2$
9. $\dfrac{28a^3x^3}{3}$
10. $-3by$
11. $-9a^2 + \dfrac{9}{14}ab$
12. $\dfrac{-ax^2}{21}$
13. -2
14. 5
15. $B = \dfrac{2A - hb}{h}$
16. \$12,500
17. 15
18. Jamie: 8 years old
Janiece: 5 years old
19. 6 nickels
11 dimes
7 quarters
20. 110 miles
21. (a) 7, (b) 4, (c) $\dfrac{1}{4}$
22. $\dfrac{-3}{4}$
23. $y = 3x - 2$
24.

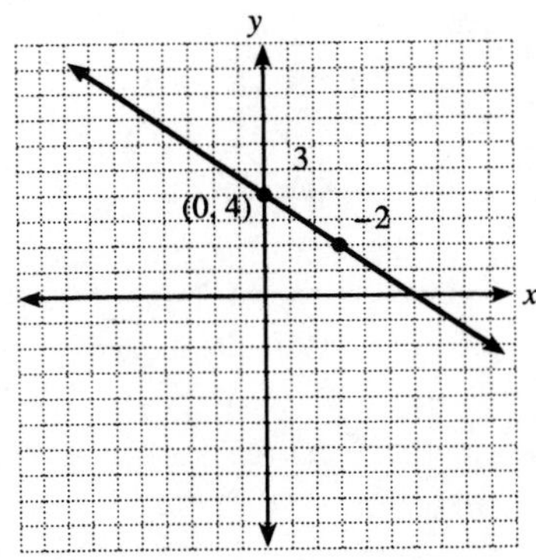

25.

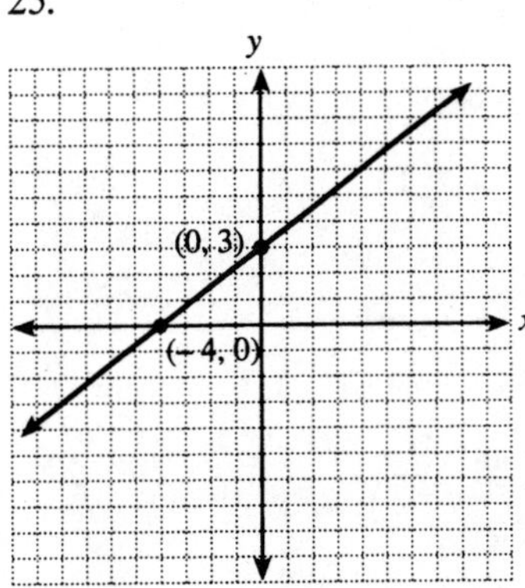

26. $\left(\dfrac{-10}{7}, \dfrac{-1}{21}\right)$
27. $(3, -1)$
28. $\left(\dfrac{1}{2}, -4\right)$
29. \$17,000 at 14%
\$8,000 at 8%
30. $\dfrac{-3}{x}$
31. $-125a^9c^{15}$
32. $\dfrac{5x^3}{6y^3}$
33. $\dfrac{9m^{10}}{16n^6}$
34. 25,000
35. -28
36. 46
37. 31
38. $\dfrac{-9}{8}$
39. -2
40. $\dfrac{1}{4}$
41. $9x^2 - 30xy^2 + 25y^4$
42. $16a^2 - 4ab - 6b^2$
43. $12x^3 - 7x^2 - 18x + 8$
44. $\dfrac{a}{b} - \dfrac{3}{2b^2} + \dfrac{5a^2b}{2}$
45. $2x^2 - x + 3$
46. $-4ax^3(3a^2x^2 - 5a^3 + 8x)$
47. $(2x + 3)(x - 4)$
48. $(6a - 5)(a + 3)$
49. $(8x^2 - 3b)(8x^2 + 3b)$
50. $-6a^3b(a^2 - 2b)^2$
51. $\dfrac{y(x - y)}{x + y}$
52. -1
53. $\dfrac{2a(a + 4)}{a + 3}$
54. $\dfrac{a + 4}{a^2 - 4}$
55. $\dfrac{2x - 1}{x}$
56. $\dfrac{5}{x + 3}$
57. $\dfrac{1}{x - 5}$
58. $\dfrac{1 - a}{a}$
59. -3
60. 12

Problem Set 13.1

1. 5 3. -2 5. 7 7. 9 9. -10 11. $\frac{3}{2}$ 13. $\frac{-5}{4}$ 15. 4 17. 1

19. 6 21. $\frac{5}{6}$ 23. $\frac{4}{5}$ 25. $\frac{27}{28}$ 27. $\frac{3}{8}$ 29. $\frac{21}{10}$ 31. $\frac{1}{3}$ 33. 2 35. 1

Problem Set 13.2A

1. $2\sqrt{2}$ 3. $2\sqrt{6}$ 5. $2\sqrt{5}$ 7. $5\sqrt{5}$ 9. $4\sqrt{2}$ 11. $6\sqrt{2}$

13. $7\sqrt{3}$ 15. $4\sqrt{3}$ 17. $9\sqrt{3}$ 19. $\frac{\sqrt{6}}{6}$ 21. $\frac{\sqrt{5}}{5}$ 23. $\frac{\sqrt{2}}{4}$

25. $\frac{\sqrt{2}}{2}$ 27. $\frac{2\sqrt{5}}{5}$ 29. $\frac{\sqrt{3}}{2}$ 31. $\frac{\sqrt{14}}{6}$ 33. $\frac{3\sqrt{2}}{2}$ 35. $\frac{3\sqrt{2}}{5}$

37. $\frac{5\sqrt{2}}{14}$ 39. $\frac{3\sqrt{2}}{4}$ 41. $\frac{\sqrt{3}}{3}$ 43. $\frac{\sqrt{5}}{3}$ 45. $\frac{2\sqrt{5}}{5}$ 47. $\frac{\sqrt{10}}{5}$

49. $\frac{3\sqrt{10}}{10}$ 51. $\frac{3\sqrt{10}}{20}$ 53. 1 55. 1 57. $\frac{1}{2}$

Problem Set 13.2B

1. x^2 3. a^4 5. xy^2 7. $a\sqrt{a}$ 9. $xy\sqrt{y}$

11. $5\sqrt{x}$ 13. $3a\sqrt{a}$ 15. $2x\sqrt{3}$ 17. $5x\sqrt{2x}$ 19. $2a^2b\sqrt{2a}$

21. $-3x^2a^2\sqrt{5x}$ 23. $-5xa^2\sqrt{2xa}$ 25. $\frac{x\sqrt{xy}}{y}$ 27. $\frac{x^2\sqrt{y}}{y^2}$ 29. $\frac{y^2\sqrt{xa}}{a^3}$

31. $\frac{2a\sqrt{2ab}}{b^3}$ 33. $\frac{3a^2x^2\sqrt{3xy}}{y^4}$ 35. $\frac{3b\sqrt{5aby}}{xy^2}$ 37. $\frac{7x^2\sqrt{10axb}}{5b^2y^2}$ 39. $\frac{y\sqrt{5axy}}{2a^2b}$

41. $\frac{b\sqrt{2aby}}{3xy^2}$ 43. $\frac{5x^3\sqrt{xay}}{6by^2}$ 45. $\frac{xb\sqrt{3by}}{2a^2y^4}$ 47. $\frac{3a^2y\sqrt{5ayx}}{7b^2x^3}$

Problem Set 13.3

1. $7\sqrt{2}$ 3. $-5\sqrt{5}$ 5. $14\sqrt{5}$ 7. $5\sqrt{2}$

9. $7\sqrt{2}$ 11. $3\sqrt{5}+3\sqrt{3}$ 13. $5\sqrt{2}-3\sqrt{5}$ 15. $8\sqrt{5}-4\sqrt{2a}$

17. $-2\sqrt{2}$ 19. $9x\sqrt{3x}$ 21. $2x\sqrt{3x}$ 23. $-x\sqrt{2x}$

25. $3x^2\sqrt{x}$ 27. 0 29. $3\sqrt{7a}-2a\sqrt{7a}$ 31. $6a^2-12$

33. $5a^4-15a^2$

Problem Set 13.4A

1. $3\sqrt{2}-3$ 3. $6-3\sqrt{5}$ 5. $2\sqrt{2}+2$ 7. $3\sqrt{2}-\sqrt{6}$ 9. $10-2\sqrt{5}$

11. $2\sqrt{3}+3\sqrt{2}$ 13. $3\sqrt{2x}-2x$ 15. $3a-a\sqrt{6}$ 17. 9 19. $3a-3a\sqrt{2}$

21. $-3a$ 23. $2-3\sqrt{2}$ 25. $3+2\sqrt{6}$ 27. $5+2\sqrt{10}$ 29. $16-y$

31. $11+6\sqrt{2}$ 33. $\sqrt{6}-1$ 35. 3

Problem Set 13.4B

1. $\frac{3-\sqrt{2}}{7}$ 3. $2(\sqrt{5}+1)$ 5. $\frac{3\sqrt{3}+3}{2}$

7. $\frac{2\sqrt{x}-2}{x-1}$ 9. $2\sqrt{2}+2$ 11. $4\sqrt{6}+6\sqrt{2}$

13. $6\sqrt{3} - 6\sqrt{2}$

15. $12\sqrt{2} + 8\sqrt{3}$

17. $\dfrac{15\sqrt{3} - 9\sqrt{5}}{2}$

19. $9 - 4\sqrt{5}$

21. $\dfrac{3\sqrt{3} + 6 - \sqrt{6} - 2\sqrt{2}}{7}$

23. $\dfrac{5\sqrt{3} - 3}{22}$

25. $\dfrac{17 + 4\sqrt{10}}{3}$

27. $8 - 3\sqrt{6}$

29. $\dfrac{6 + 3\sqrt{2} - \sqrt{6} - 2\sqrt{3}}{4}$

31. $\dfrac{24 - 6\sqrt{x}}{16 - x}$

33. $\dfrac{x + 5\sqrt{x} + 6}{x - 9}$

Problem Set 13.5

1. 5
3. 5
5. 13
7. $\sqrt{58}$
9. $\sqrt{13}$
11. 10
13. $4\sqrt{5}$
15. 12
17. $8 + 4\sqrt{2}$
19. $9 + \sqrt{41}$
21. $\sqrt{34}$
23. $2\sqrt{6}$
25. 15 feet
27. 8 feet
29. $3\sqrt{5}$ feet
31. $(4x - 3)(x + 2)$
33. $(2x - 3)(2x + 3)$
35. $(x - 4)^2$

Unit 13 Review Test

1. 2
2. 4
3. $2\sqrt{6}$
4. $\dfrac{\sqrt{7}}{7}$
5. $\dfrac{7\sqrt{2}}{6}$
6. $\dfrac{-2\sqrt{6}}{5}$
7. $2a^2b\sqrt{5b}$
8. $\dfrac{2ax^2\sqrt{axy}}{y^3}$
9. $-3\sqrt{2}$
10. $6\sqrt{a} + 3\sqrt{2a}$
11. $7\sqrt{3}$
12. $\sqrt{3a}$
13. $\sqrt{6} - 3\sqrt{3}$
14. $3a + 5\sqrt{3a} + 4$
15. $5\sqrt{10} - 7$
16. -2
17. $6 - 3\sqrt{3}$
18. $-(6\sqrt{3} + 9\sqrt{2})$
19. $\dfrac{3\sqrt{x} + 15}{x - 25}$
20. $\dfrac{x - 9\sqrt{x} + 18}{x - 9}$
21. $2\sqrt{34}$
22. $4\sqrt{5}$
23. $-(2x + 1)$
24. $\dfrac{-y}{x^2}$
25. $\dfrac{a}{b - a}$
26. $\dfrac{-2}{2x - 3}$
27. $\dfrac{-(x + 1)}{x}$
28. $x = -3$
29. $x = 5$

Problem Set 14.1

1. $x = 3$ or $x = -2$
3. $x = -2$ or $x = 9$
5. $x = -3$ or $x = 4$
7. $x = 6$ or $x = -1$
9. $x = -8$ or $x = 1$
11. $x = 2$ or $x = -6$
13. $x = \dfrac{1}{2}$ or $x = -3$
15. $x = \dfrac{1}{3}$ or $x = -2$
17. $x = \dfrac{2}{3}$ or $x = 1$
19. $x = -2$ or $x = \dfrac{7}{3}$
21. $x = \dfrac{3}{4}$ or $x = -1$
23. $x = \dfrac{4}{3}$ or $x = 2$
25. $x = 3$ or $x = -3$
27. $x = \dfrac{1}{2}$ or $x = \dfrac{-1}{2}$
29. $x = 0$ or $x = 4$
31. $x = \dfrac{3}{4}$ or $x = \dfrac{-3}{4}$
33. $x = 10$ or $x = \dfrac{3}{2}$
35. $x = 0$ or $x = 5$
37. $x = \dfrac{6}{5}$ or $x = \dfrac{-6}{5}$
39. $x = \dfrac{-1}{2}$ or $x = \dfrac{4}{3}$
41. $x = \dfrac{3}{4}$ or $x = \dfrac{-1}{2}$
43. $x = 0$ or $x = \dfrac{4}{3}$
45. $x = \dfrac{3}{8}$ or $x = -3$
47. $x = \dfrac{-5}{6}$ or $x = \dfrac{3}{2}$

49. $x = \frac{-2}{3}$ or $x = -4$
51. $x = \frac{2}{3}$ or $x = -5$
53. $x = \frac{3}{4}$ or $x = -4$
55. $x = \frac{-7}{4}$ or $x = 3$
57. $x = \frac{1}{9}$ or $x = -1$
59. $x = \frac{5}{3}$ or $x = \frac{-3}{5}$

Problem Set 14.2A

1. ± 6
3. $\pm\sqrt{23}$
5. $\pm\sqrt{17}$
7. $\pm 3\sqrt{2}$
9. $3, -7$
11. $-5, -25$
13. $-3, -11$
15. $-10, 2$
17. $0, 12$
19. $-1, -5$
21. $-13, 3$
23. $5 \pm \sqrt{6}$
25. $-7 \pm 2\sqrt{6}$
27. $-3 \pm \sqrt{10}$

Problem Set 14.2B

1. 16
3. 25
5. 81
7. 64
9. $\frac{25}{4}$
11. $\frac{9}{4}$

Problem Set 14.2C

1. $2, -4$
3. $-2, 8$
5. $3, 5$
7. $-3, -7$
9. $0, 10$
11. $-7, 5$
13. $-1, 11$
15. $\frac{-1}{3}, 5$
17. $\frac{-1}{2}, 4$
19. $\frac{3}{2}, -6$
21. $-4, 4$
23. $0, 4$
25. $\frac{1}{2}, \frac{3}{2}$
27. $\frac{-5 \pm \sqrt{97}}{12}$
29. $\frac{2 \pm \sqrt{14}}{5}$
31. $0, 4$
33. $\frac{-6 \pm 5\sqrt{2}}{2}$
35. $\frac{-1 \pm \sqrt{10}}{3}$

Problem Set 14.3

1. $-5, 2$
3. $5, 6$
5. $\frac{-3}{2}, 5$
7. $\frac{2 \pm \sqrt{14}}{2}$
9. $0, \frac{-9}{4}$
11. $\frac{5}{6}, -1$
13. $\frac{-3}{2}, \frac{2}{3}$
15. $\frac{5}{6}, \frac{-1}{2}$
17. $\frac{2 \pm \sqrt{14}}{10}$
19. $-2\sqrt{2}, 2\sqrt{2}$
21. $0, \frac{9}{4}$
23. $\frac{-5 \pm \sqrt{57}}{8}$
25. $\frac{-7 \pm \sqrt{73}}{4}$
27. $\frac{3 \pm \sqrt{15}}{3}$
29. $-2, 2$
31. $\frac{-9 \pm \sqrt{41}}{10}$
33. $-3, \frac{3}{4}$
35. $\frac{1 \pm \sqrt{10}}{3}$

Problem Set 14.4

1. $4, -5$
3. $6, 8$
5. $6, 7$
7. 6
9. 4
11. 5 meters by 13 meters
13. 12 meters by 6 meters
15. 24 meters by 32 meters
17. 20 meters/second
19. 4 seconds
21. 12 centimeters
23. 32 meters
25. 5
27. 5
29. -1

Problem Set 14.5

1.

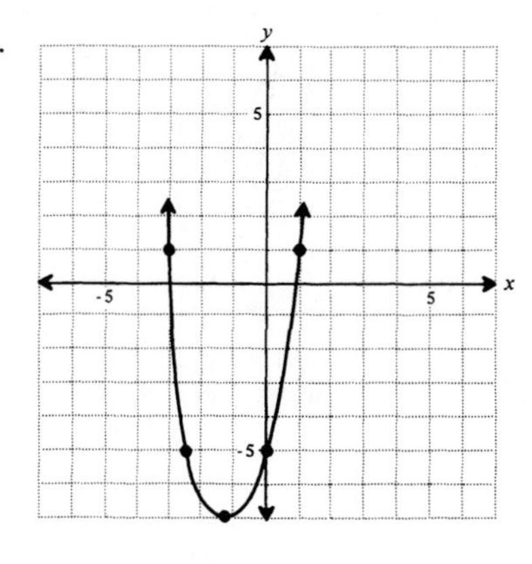

3.

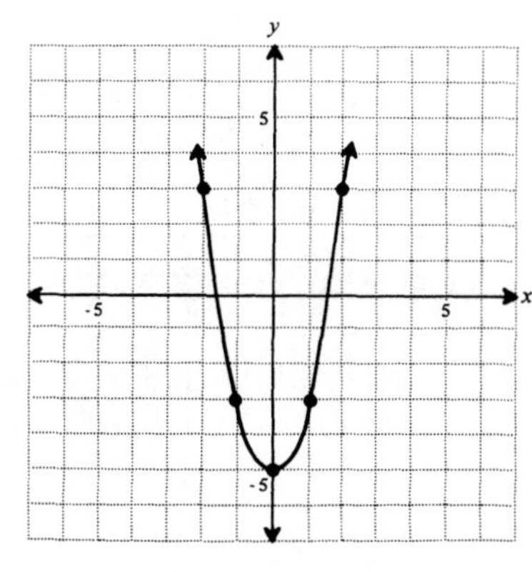

5.

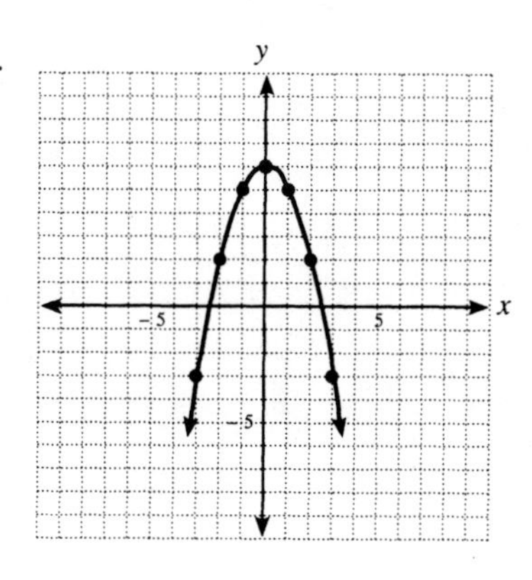

7., 9., 11., 13.

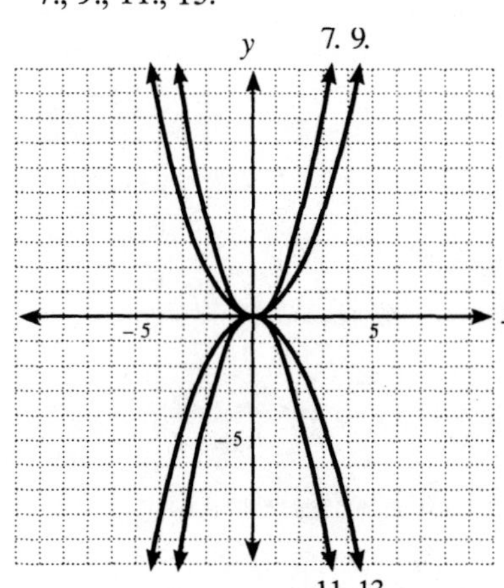

15., 17., 19.

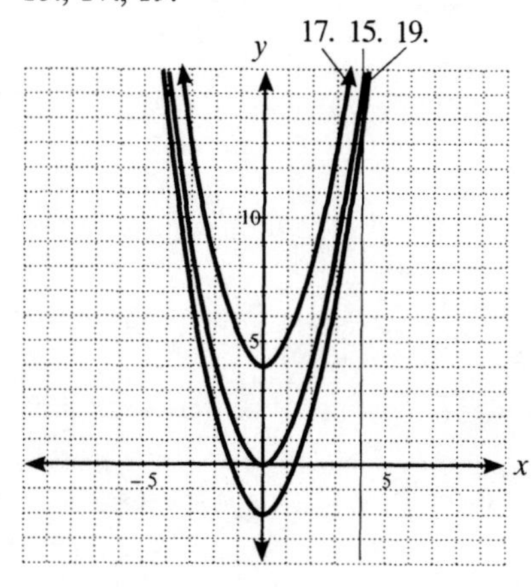

21.

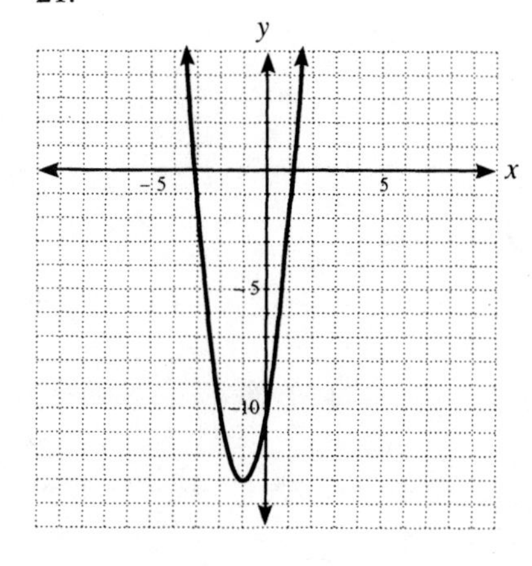

23.

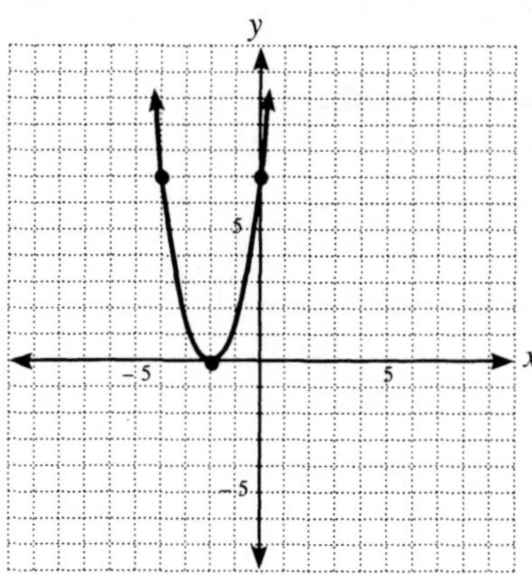

25.

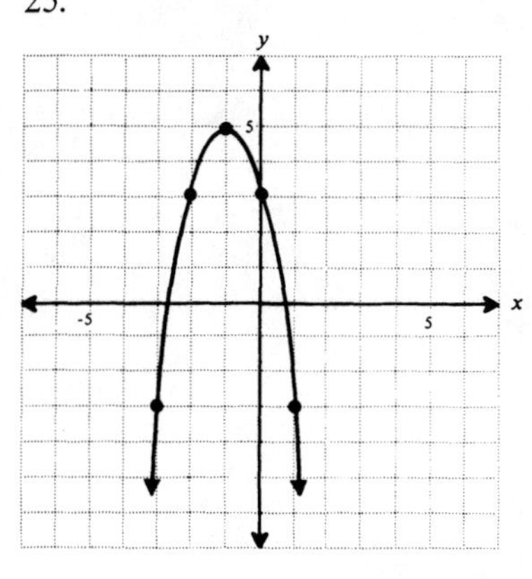

27.

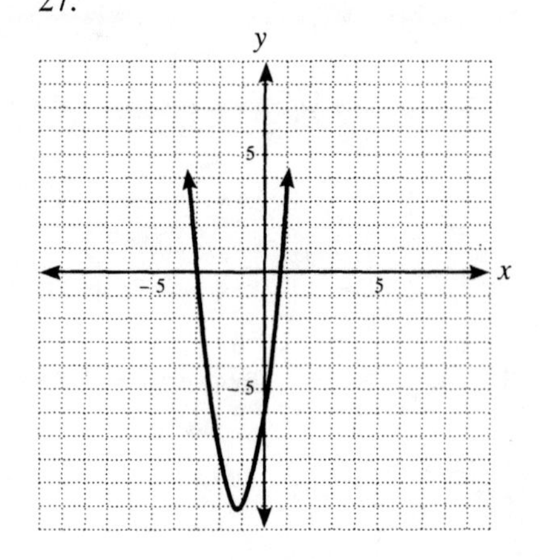

Unit 14 Review Test

1. $1, -4$
2. $-2, -3$
3. $7, -1$
4. 36
5. 100
6. $\frac{49}{4}$
7. $-3 \pm \sqrt{2}$
8. $\frac{2 \pm \sqrt{10}}{3}$
9. $\frac{-b \pm \sqrt{b^2 - 4ac}}{2a}$
10. $\frac{5 \pm \sqrt{13}}{2}$
11. $\frac{-4 \pm \sqrt{6}}{5}$
12. $\frac{-1}{6}, \frac{3}{2}$
13. $-4 \pm \sqrt{11}$
14. $-4, 2$
15. $\pm 5\sqrt{2}$
16. 6

17. 100 meters by 25 meters

18.

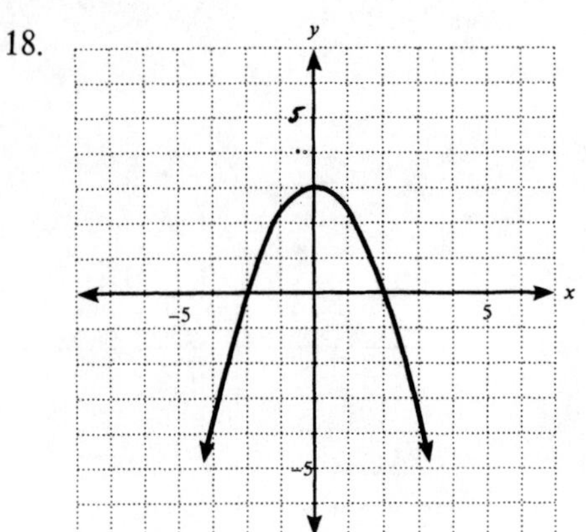

19.

20. $\dfrac{-x}{x+2}$

21. $\dfrac{x(x+1)(x+2)}{x-1}$

22. $\dfrac{3}{x-2}$

23. $\dfrac{2x-1}{2-x}$

24. $x = \dfrac{-5}{2}$

Problem Set 15.1

1.

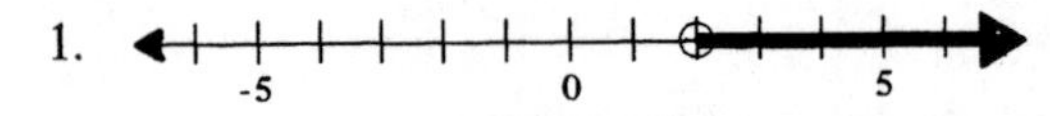

3.

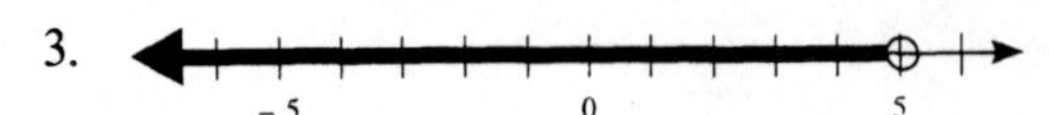

5.

7.

9.

11.

13.

15.

17.

19.

21.

23.

25.

27.

29.

31.

33.

35.

Problem Set 15.2

1. $x = 1$ or $x = -1$

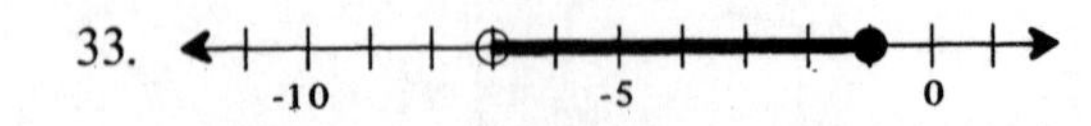

3. $x = 2$ or $x = -2$

5. $x = 1$ or $x = -1$

7. $x = -2$ or $x = 2$

9. $x = 5$ or $x = -5$

11. $x = 1$ or $x = -1$

13. $x = 6$ or $x = -6$

-5 0 5

15. -5 0 5

17. -5 0 5

19. -5 0 5

21. -5 0 5

23. -5 0 5

25. -5 0 5

27. -5 0 5

29. -5 0 5

31. -5 0 5

33. -5 0 5

35. -8 0 8

37.

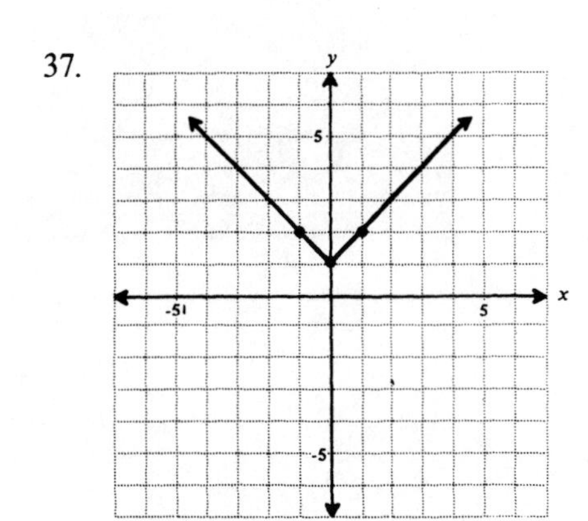

39.

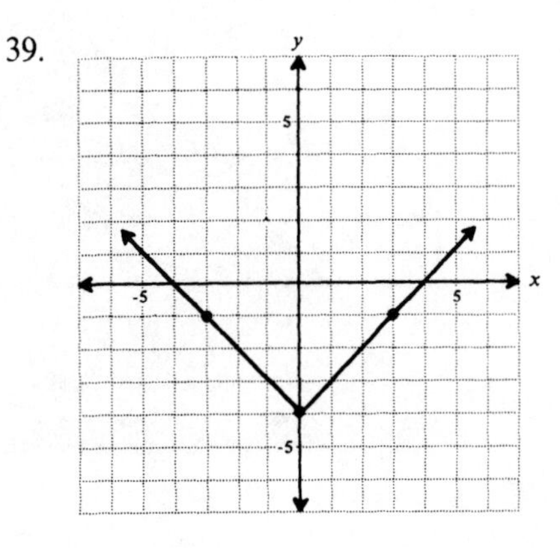

41.

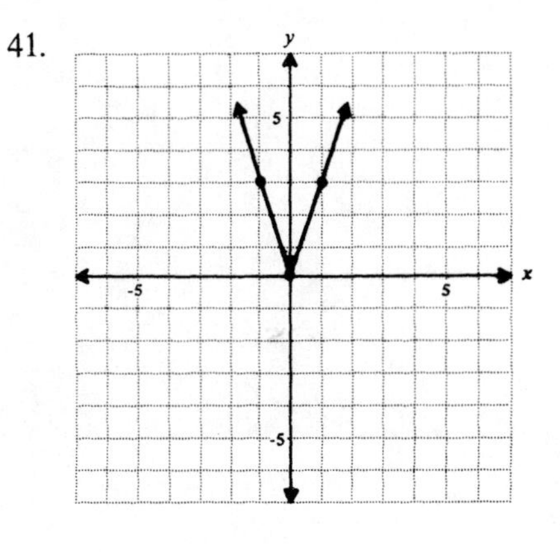

43.

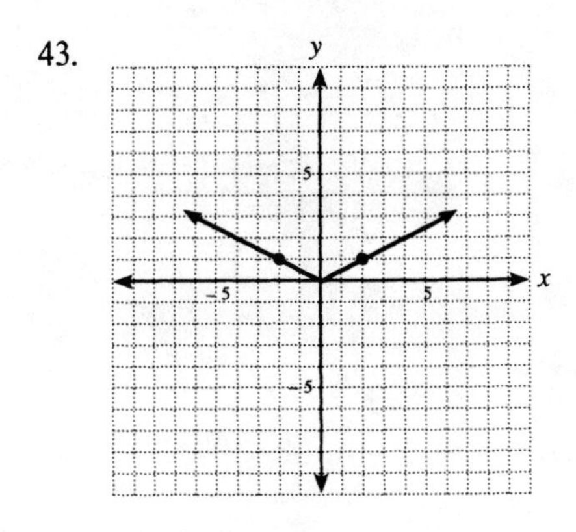

45.

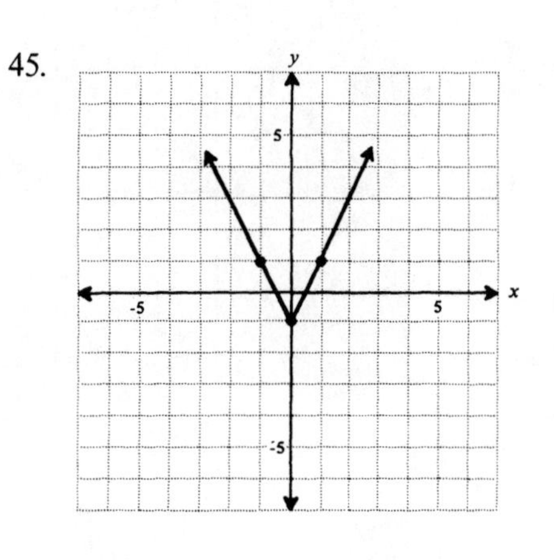

47.

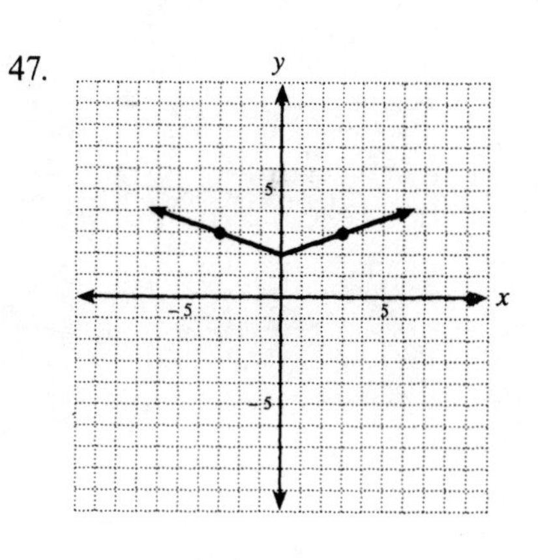

Problem Set 15.3

1. $x < 4$	3. $x > 3$	5. $x < 2$	7. $x > 1$	9. $x \leqslant 12$
11. $x \geqslant 4$	13. $x < 6$	15. $x \geqslant 2$	17. $x < -13$	19. $x \leqslant 11$
21. $x > 8$	23. $x \geqslant 5$	25. $x < 6$	27. $x < -2$	29. $x \geqslant -3$
31. $x > 36$	33. $x \leqslant -22$	35. $x < 16$	37. $x \geqslant -14$	39. $x \geqslant \frac{-2}{5}$
41. $x \leqslant 3$	43. $x > 3$	45. $x > 0$	47. $x \geqslant \frac{10}{7}$	49. $x \geqslant 2$

Problem Set 15.4

1.

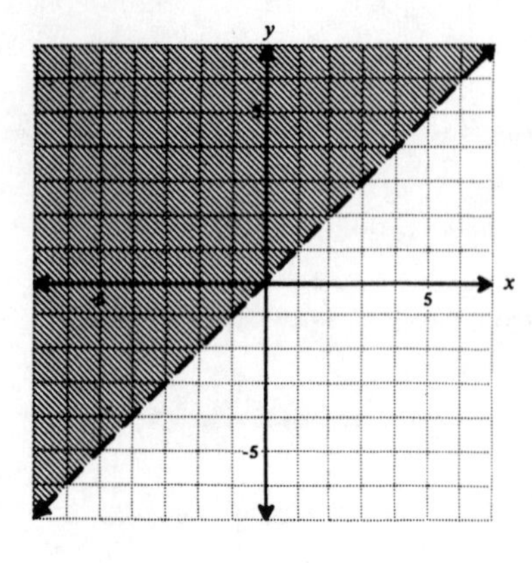

3.

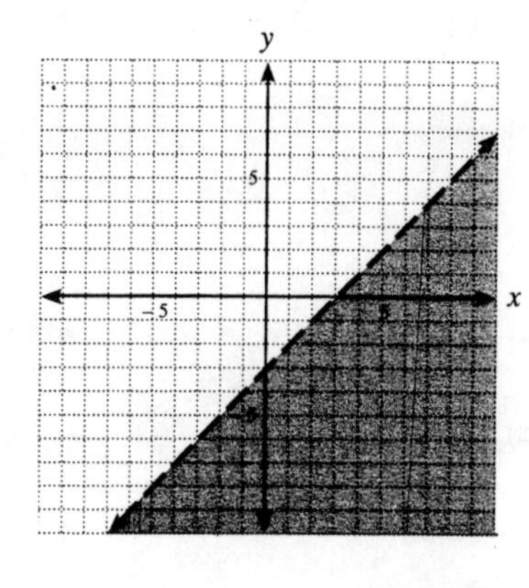

5.

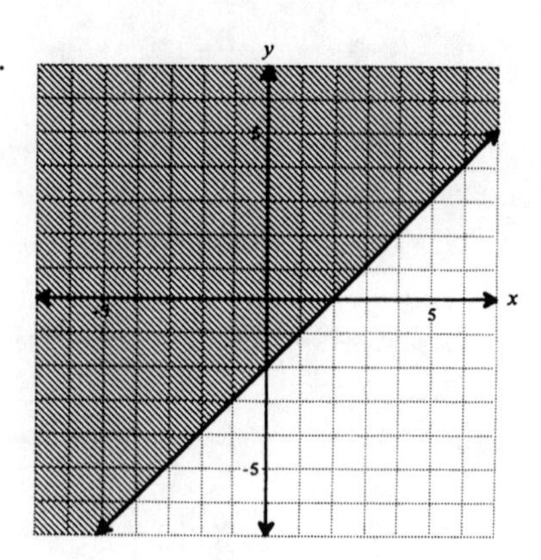

7.

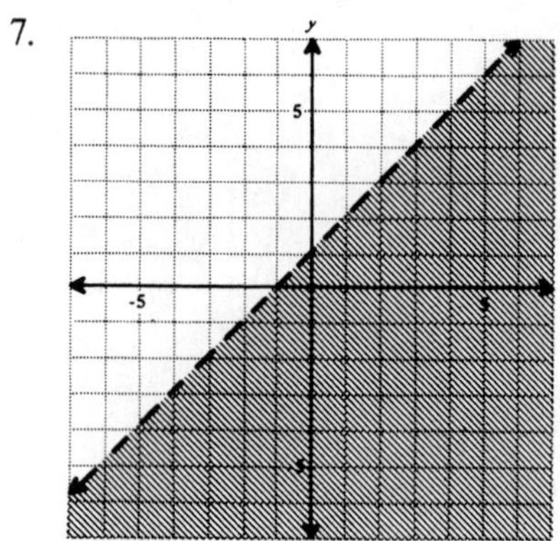

9.

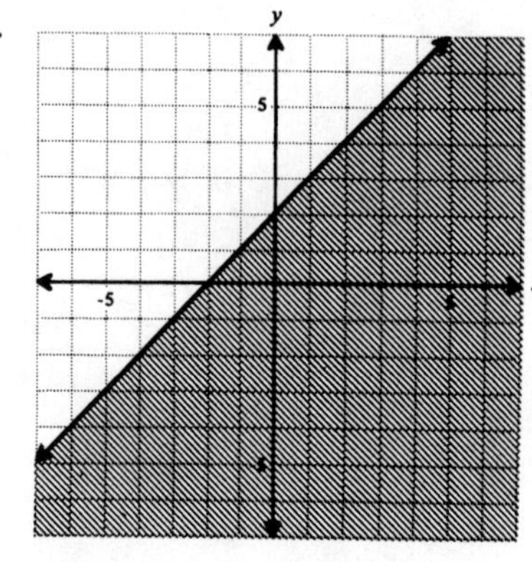

11.

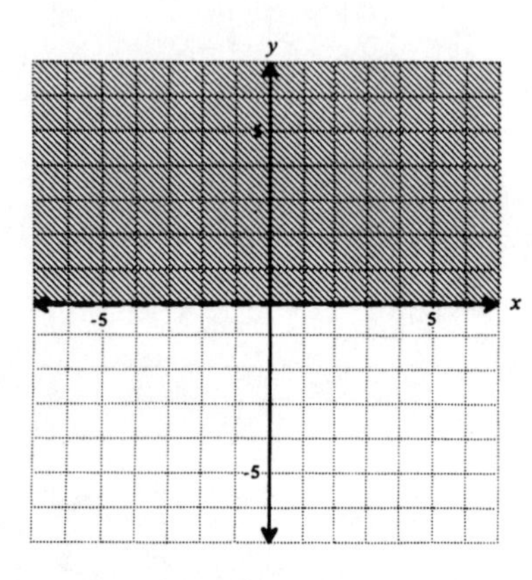

13.

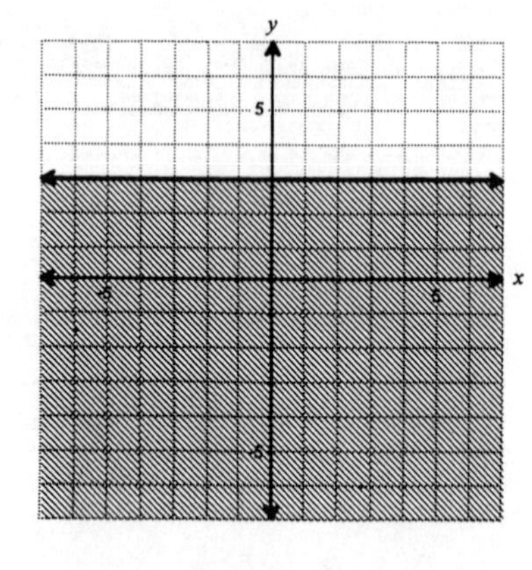

15.

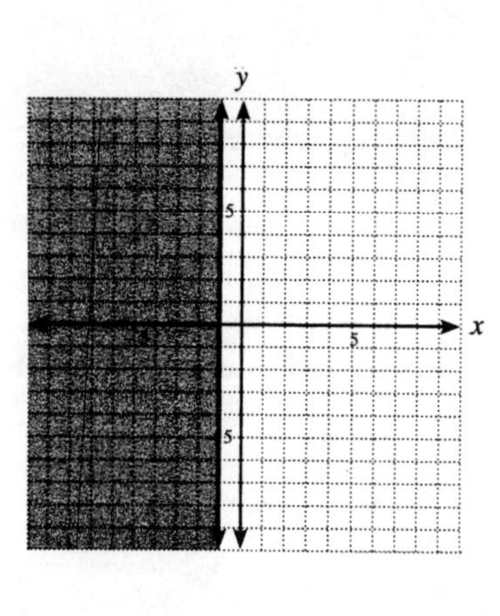

17.

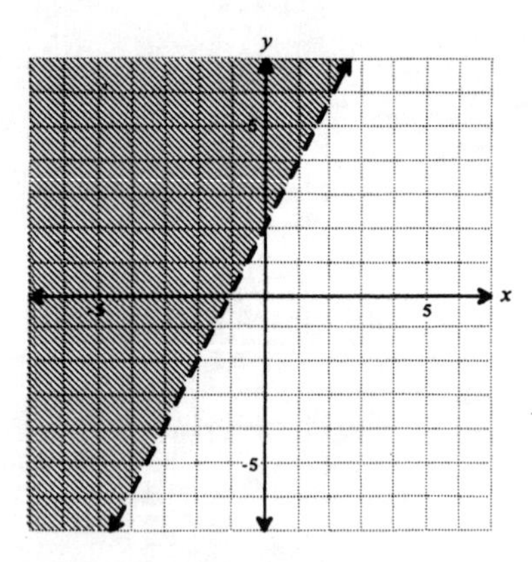

19.

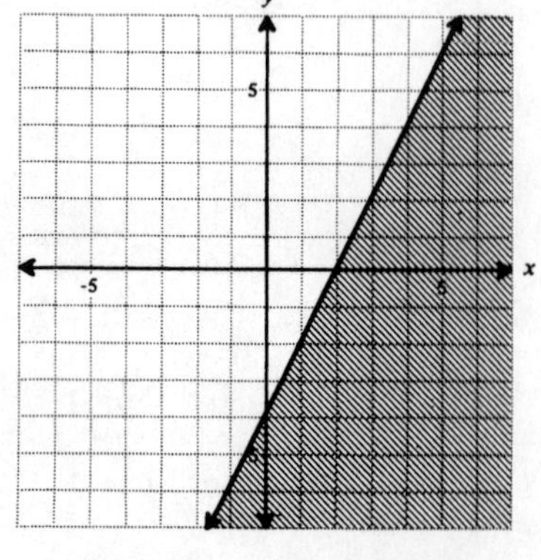

21.

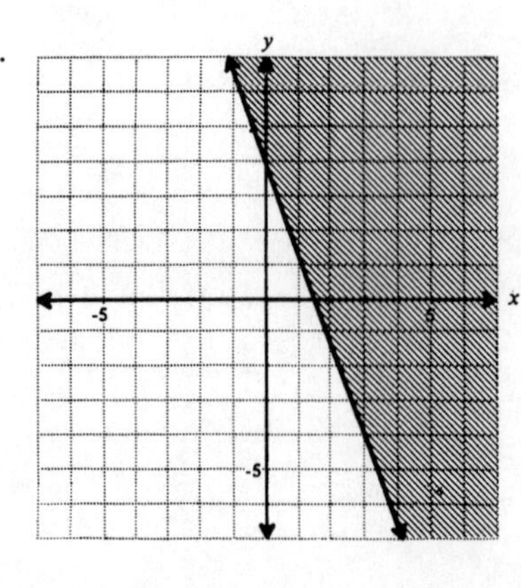

23.

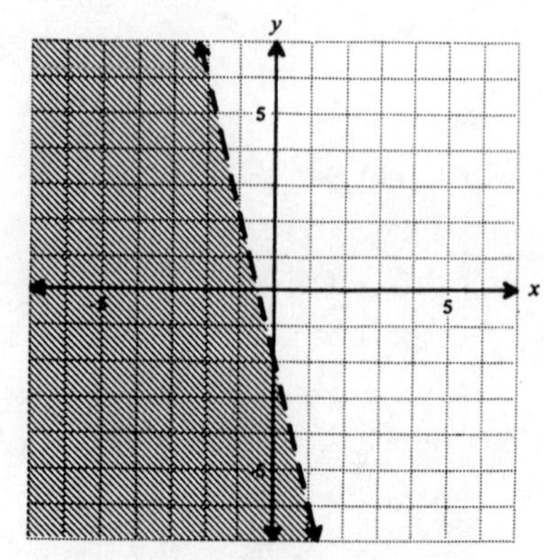

25.

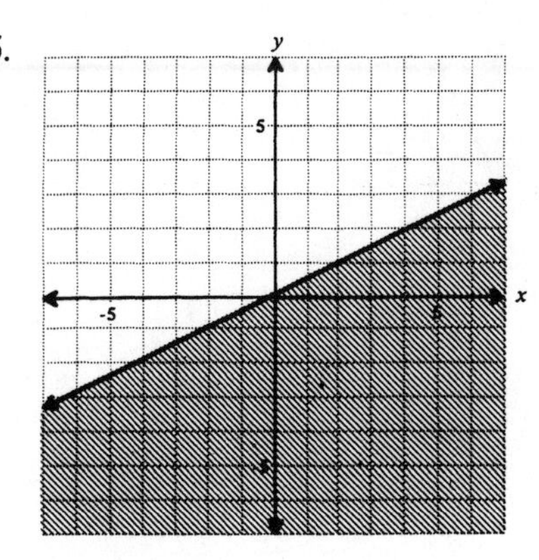

27.

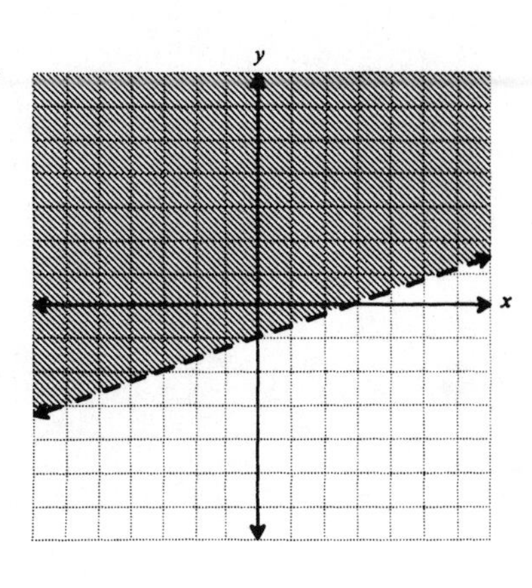

29.

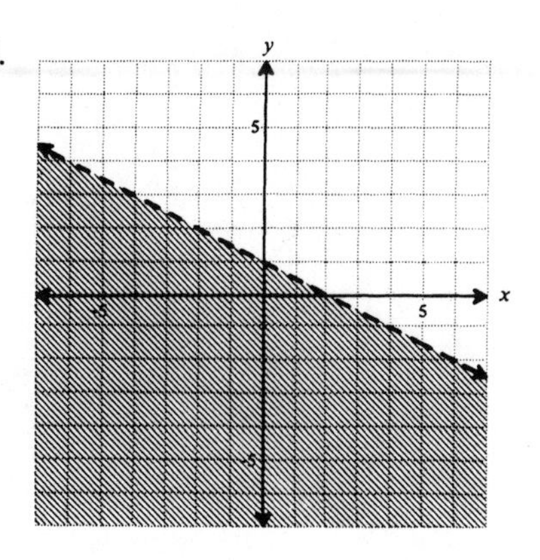

31.

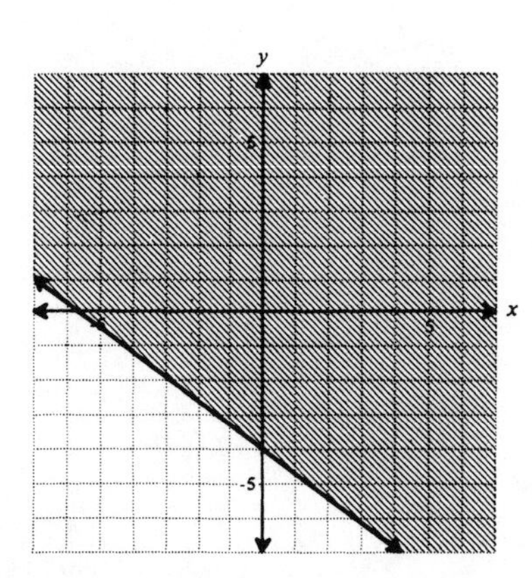

33.

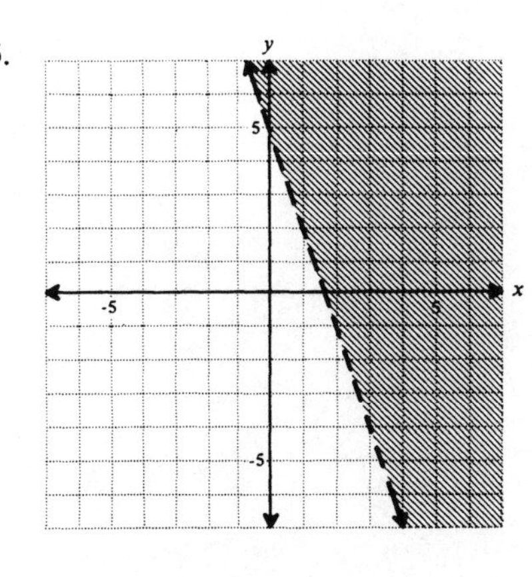

35.

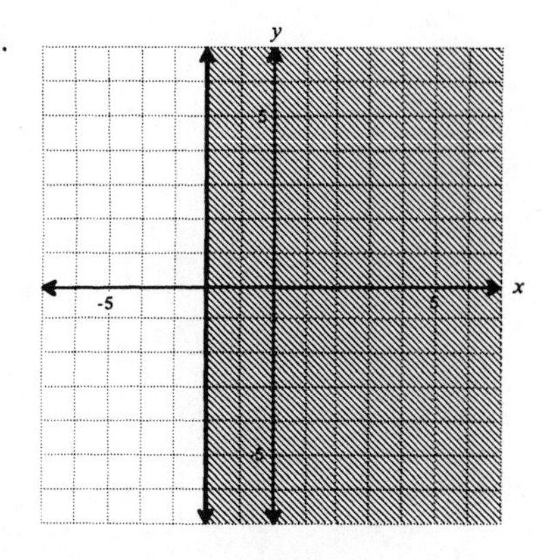

37.

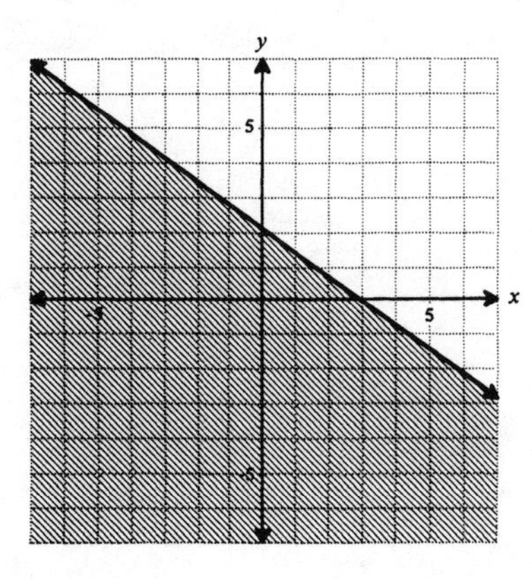

39.

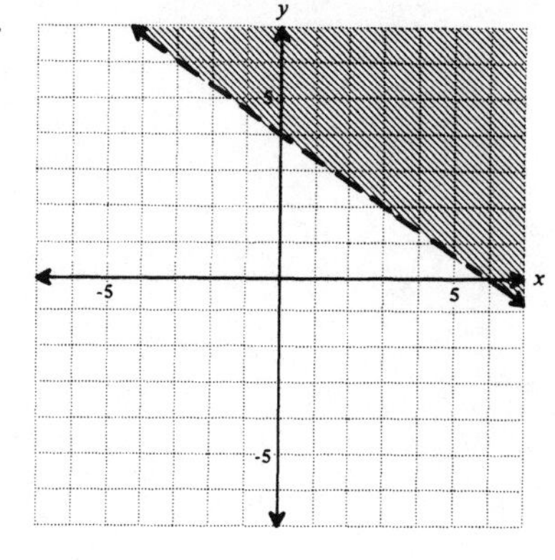

Unit 15 Review Test

1. 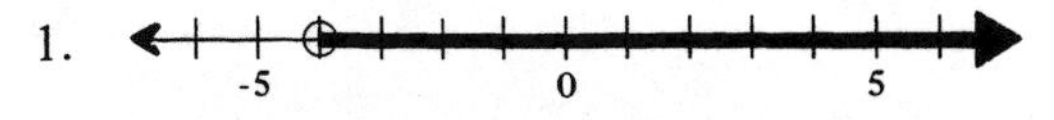

2. 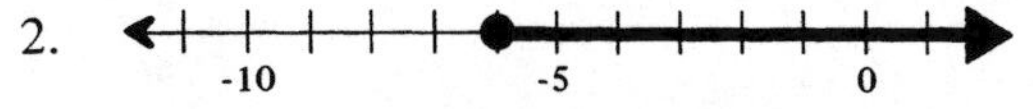

3. 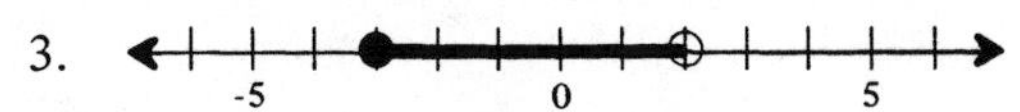

4. 

5.

6.

7. $x = -3$ or $x = 3$

8. $x = -7$ or $x = 7$

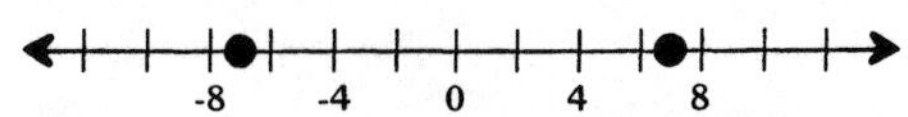

9.
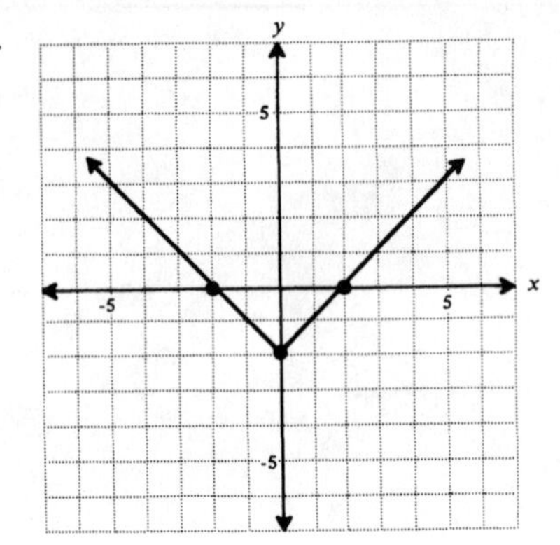

10.
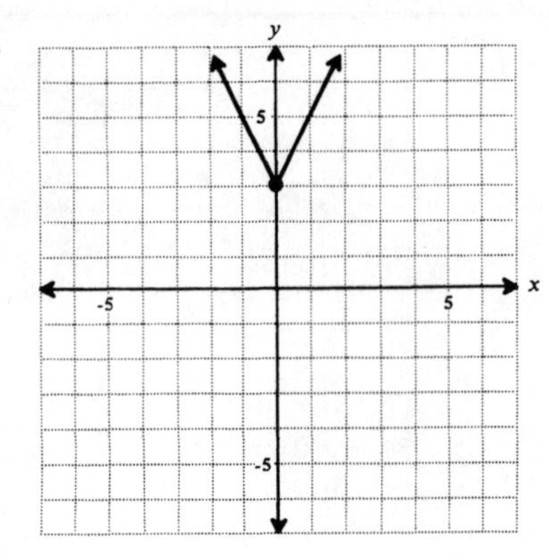

11. $x < -3$

12. $x < 2$

13. $x \leqslant 2$

14. $x \leqslant -36$

15. $x > -2$

16. $x < \dfrac{-2}{5}$

17.
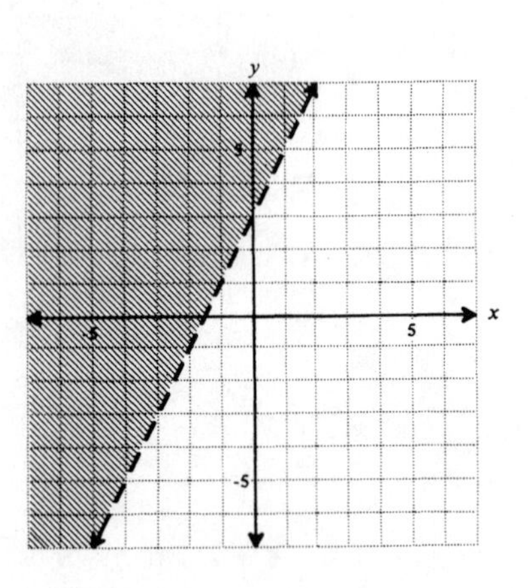

18.
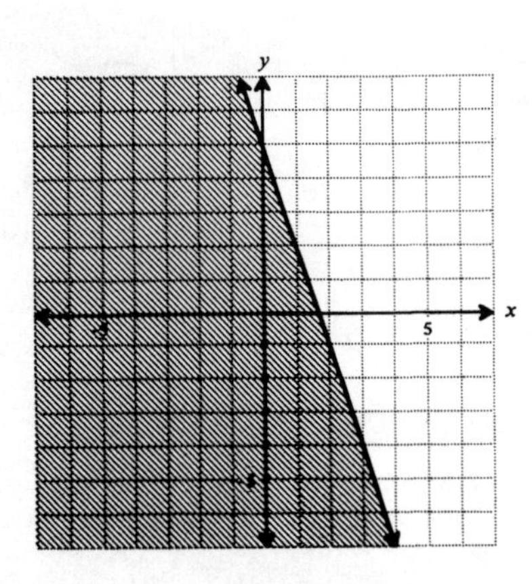

19.
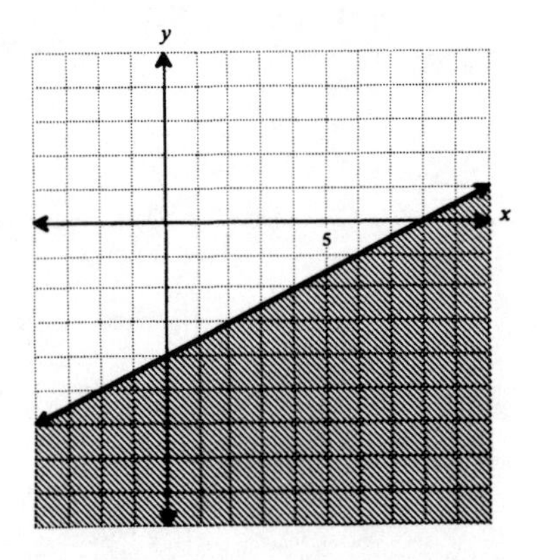

20.
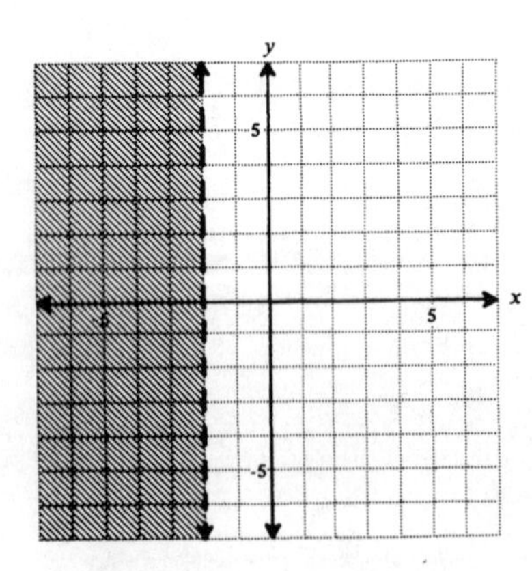

21. $\dfrac{5}{4x - 5}$

22. $\dfrac{-b}{b + 1}$

23. $\dfrac{9}{2}$

24. $5\sqrt{3}$

25. $\dfrac{\sqrt{2}}{4}$

26. $\dfrac{3xy^2\sqrt{axy}}{a^2}$

27. $9 - 3\sqrt{2}$

28. $\dfrac{6 - 5\sqrt{x} + x}{9 - x}$

1. -8
2. -3
3. $3a^2 - 2a - 10$
4. $-a^2 - 4a + 5$
5. $\frac{-7y}{15}$
6. $\frac{9x}{2}$
7. $-6b^6$
8. $8s^3t^6$
9. $\frac{2}{xy^3}$
10. $5x^3 - 3x^2 + 4$
11. $3x^3 + 6x^2 - 12x$
12. $6x^2 - 11x + 3$
13. $3x^3 - 5x^2 - 6x + 8$
14. $4a^2 + 12ab + 9b^2$
15. $4x$
16. $\frac{ab}{a - b}$
17. $\frac{y + 2}{y + 1}$
18. $\frac{1}{(x + 1)(x + 2)}$
19. $\frac{4x + 3}{2x + 1}$
20. $\frac{1}{a + 1}$
21. $\frac{3x\sqrt{2xy}}{y}$
22. $4 + 2\sqrt{2}$
23. $\sqrt{3}$
24. $4 - 5\sqrt{10}$
25. $(x + 3)(x - 2)$
26. $(3x - 4)(2x + 1)$
27. $3(y - 3)(y + 3)$
28. $x^2(2x + 3)(x + 2)$
29. 4
30. -2
31. $x = \frac{y - b}{m}$
32. 1
33. $5, -3$
34. $\frac{-4}{3}, \frac{3}{2}$
35. $\frac{3 \pm \sqrt{5}}{2}$
36. $x \leqslant -3$
37. 13 centimeters
38. -13
39. Ace: 205 mph
    Irv: 250 mph
40. 17
41. $y = \frac{-2}{5}x - 1$
42. $y = -2x + 4$
43.

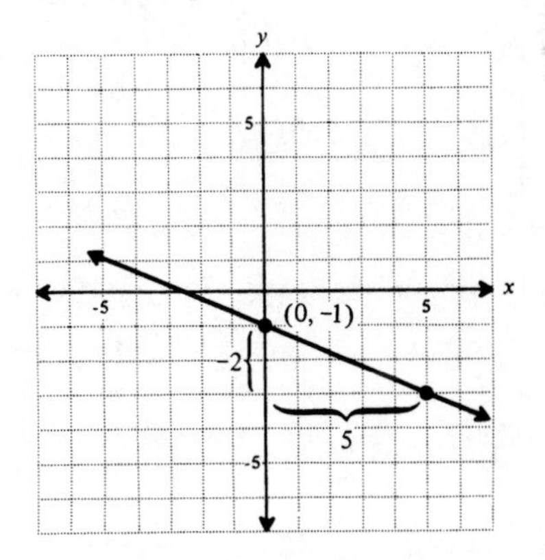

44.

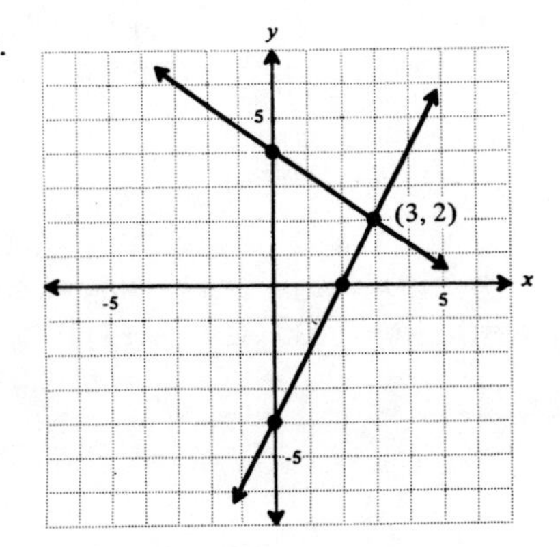

45. $(5, -2)$
46. $(-3, 1)$
47. $d = 10, q = 14$

48. 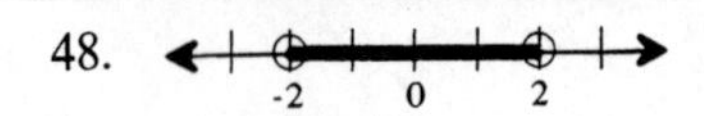

49. 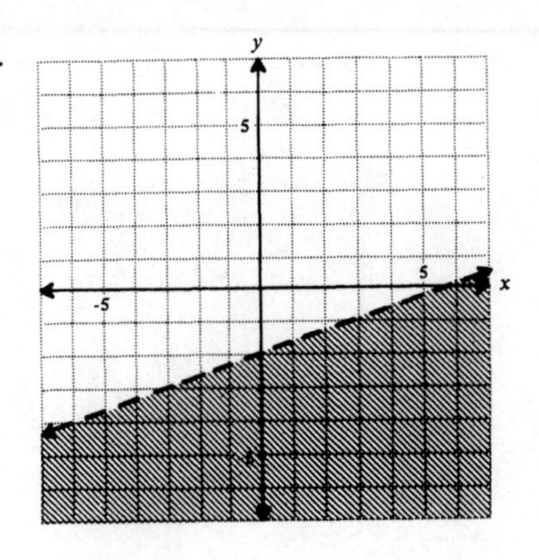

50. 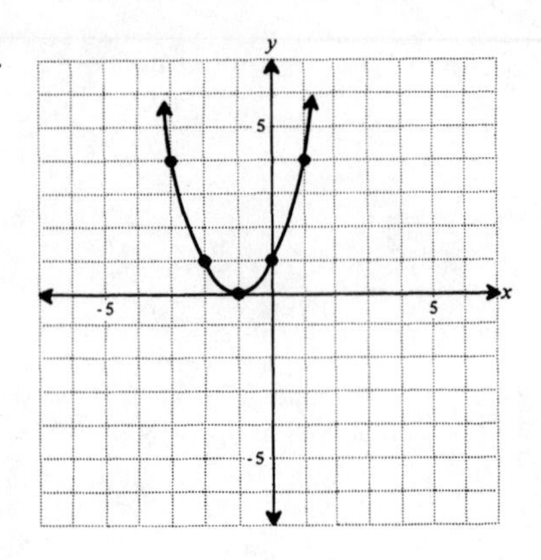

Alternate Problem Set 10.2A

1. 2, 3   3. $-1, -2$ 5. $-2, 3$ 7. $-3, -5$ 9. $4, -3$ 11. $-6, 3$
13. $-6, 5$ 15. $3, -11$ 17. $-4, -5$ 19. $8, -6$ 21. $-9, 5$

Alternate Problem Set 10.2B

1. $(x + 1)(x + 2)$ 3. $(x - 1)(x - 3)$ 5. $(x + 1)(x + 6)$ 7. $(x + 1)(x + 5)$
9. $(x - 4)(x - 2)$ 11. $(x - 1)(x - 8)$ 13. $(x - 1)(x - 10)$ 15. $(x + 2)(x + 5)$
17. $(x + 2)(x - 5)$ 19. $(x + 1)(x - 10)$ 21. $(a + 1)(a - 11)$ 23. $(x - 2)(x + 1)$
25. $(x - 1)(x + 4)$ 27. $(x - 4)(x + 5)$ 29. $(x + 1)(x + 7)$ 31. $(a - 9)(a + 2)$
33. $(x - 2)(x + 4)$ 35. $(x + 7)(x + 4)$ 37. $(x - 4)(x + 6)$ 39. $(y - 3)(y - 10)$
41. $(x - 6)(x - 2)$ 43. $(r + 3)(r - 6)$ 45. $(x - 15)(x + 2)$ 47. $(x + 2)(x + 10)$
49. $(x + 4)(x - 6)$ 51. $(a - 3)(a + 17)$ 53. $(x + 9)(x - 8)$ 55. $(x + 3)(x - 14)$
57. $(b + 5)(b - 13)$ 59. $(z - 7)(z + 11)$ 61. $(y - 3)(y - 14)$ 63. $(x + 5)(x - 11)$
65. $(a - 2)(a - 13)$

Alternate Problem Set 10.3A

1. $(x + 1)(2x - 1)$ 3. $(x - 2)(3x + 1)$ 5. $(2x + 1)(x - 3)$ 7. $(7y - 5)(y - 2)$
9. $(x + 8)(5x + 1)$ 11. $(x - 2)(3x - 1)$ 13. $(a + 7)(3a + 2)$ 15. $(5x + 3)(x - 4)$
17. $(b - 7)(3b + 2)$ 19. $(x + 4)(2x + 5)$ 21. $(x + 1)(3x + 1)$ 23. $(5x - 3)(x + 10)$
25. $(7a - 2)(a + 2)$

Alternate Problem Set 10.3B

1. $(2x - 1)(2x + 3)$ 3. $(2x - 1)(3x + 2)$ 5. $(2x - 1)(3x - 2)$ 7. $(3x + 1)(2x - 3)$
9. $(2x - 5)(5x - 1)$ 11. $(8y - 3)(y - 1)$ 13. $(3x - 5)(3x + 1)$ 15. $(7x - 5)(x - 1)$
17. $(7x + 5)(x + 1)$ 19. $(3a + 5)(4a - 1)$ 21. $(12a + 5)(a + 1)$ 23. $(2a - 5)(6a - 1)$
25. $(3y + 1)(5y - 4)$ 27. $(5a - 1)(3a - 4)$

Alternate Problem Set 10.3C

1. $(4x + 3)(x + 2)$ 3. $(3x + 2)(2x + 5)$ 5. $(3x - 2)(2x + 5)$ 7. $(2x + 7)(3x - 4)$
9. $(4x - 3)(x + 5)$ 11. $(2t - 3)(2t + 5)$ 13. $(6x - 7)(3x + 2)$ 15. $(3x + 7)(8x - 3)$
17. $(a - 6)(4a + 3)$ 19. $(8x - 7)(x + 3)$ 21. $(6x - 1)(4x + 3)$ 23. $(20x - 1)(x - 6)$
25. $(2a - 3)(4a + 5)$ 27. $(12x - 7)(2x + 3)$

Alternate Problem Set 10.3D

1. $(5x - 3)(x + 1)$ 3. $(7x + 2)(x - 1)$ 5. $(x - 1)(x - 1)$ 7. $(x - 3)(2x + 1)$
9. $(x - 1)(2x - 1)$ 11. $(y + 7)(y + 7)$ 13. $(-2x + 5)(3x + 2)$ 15. $(1 + 10x)(5 - x)$
17. $(2x + 5)(10x + 3)$ 19. $(3x - 1)(2x + 3)$ 21. $(x + 3)(x + 3)$ 23. $(-5x - 3)(x - 3)$
25. $(4x - 3)(2x - 1)$ 27. $(9y + 10)(y - 2)$ 29. $(5 + 3y)(1 - 3y)$ 31. $x^2 - 16$
33. $4x^2 - 9$

Index